REPRODUCTION:
A BEHAVIORAL AND
NEUROENDOCRINE PERSPECTIVE

ANNALS OF THE NEW YORK ACADEMY OF SCIENCES
Volume 474

REPRODUCTION: A BEHAVIORAL AND NEUROENDOCRINE PERSPECTIVE

Edited by Barry R. Komisaruk, Harold I. Siegel, Mei-Fang Cheng, and Harvey H. Feder

The New York Academy of Sciences
New York, New York
1986

Copyright © 1986 by the New York Academy of Sciences. All rights reserved. Under the provisions of the United States Copyright Act of 1976, individual readers of the Annals are permitted to make fair use of the material in them for teaching or research. Permission is granted to quote from the Annals provided that the customary acknowledgment is made of the source. Material in the Annals may be republished only by permission of the Academy. Address inquiries to the Executive Editor at the New York Academy of Sciences.

Copying fees: For each copy of an article made beyond the free copying permitted under Section 107 or 108 of the 1976 Copyright Act, a fee should be paid through the Copyright Clearance Center, 21 Congress Street, Salem, Mass. 01970. For articles of more than 3 pages, the copying fee is $1.75.

Cover artist: *Margaret M. McCarthy* (paper cover only)

Library of Congress Cataloging-in-Publication Data

Reproduction: a behavioral and neuroendocrine perspective.

(Annals of the New York Academy of Sciences, ISSN 0077-8923 ; v. 474)
"This volume is the result of a conference entitled Jay S. Rosenblatt-Institute of Animal Behavior 25th Anniversary Alumni Conference held on June 6-8, 1984 by Rutgers—the State University of New Jersey in Newark, New Jersey"—P.
Includes bibliographies and index.
1. Sexual behavior in animals—Congresses.
2. Reproduction—Congresses. 3. Neuroendocrinology—Congresses. I. Komisaruk, Barry R. II. Jay S. Rosenblatt-Institute of Animal Behavior 25th Anniversary Alumni Conference (1984: Rutgers—The State University of New Jersey) III. Series.
Q11.N5 vol. 474   500 s   86-12470
[QL761]   [599'.056]
ISBN 0-89766-343-8
ISBN 0-89766-344-6 (pbk.)

PCP
Printed in the United States of America
ISBN 0-89766-343-8 (cloth)
ISBN 0-89766-344-6 (paper)
ISSN 0077-8923

ANNALS OF THE NEW YORK ACADEMY OF SCIENCES

Volume 474
December 31, 1986

REPRODUCTION: A BEHAVIORAL AND NEUROENDOCRINE PERSPECTIVE[a]

Editors and Conference Organizers

BARRY R. KOMISARUK, HAROLD I. SIEGEL, MEI-FANG CHENG, AND HARVEY H. FEDER

CONTENTS

Preface. By BARRY R. KOMISARUK, HAROLD I. SIEGEL, MEI-FANG CHENG, and HARVEY H. FEDER ix

In Honor of Jay S. Rosenblatt. By BARRY R. KOMISARUK, HAROLD I. SIEGEL, MEI-FANG CHENG, and HARVEY H. FEDER xi

Historical Perspective of Daniel S. Lehrman, Founder of the Institute of Animal Behavior. By COLIN G. BEER, J. WAYNE LAZAR, and CAROL DIAKOW.................................. xiii

Part I. Behavior as a Regulator of Hormonal Mechanisms and Reproduction

Introduction. By BARRY R. KOMISARUK and MEI-FANG CHENG ... 1

Individual Behavioral Response Mediates Endocrine Changes Induced by Social Interaction. By MEI-FANG CHENG 4

Social Induction of the Ovarian Response in the Female Ring Dove. By CARL J. ERICKSON 13

The Effects of Copulatory Behavior on Sperm Transport and Fertility in Rats. By NORMAN T. ADLER and JAMES PATRICK TONER, JR. 21

The Role of Ultrasonic Vocalizations in the Regulation of Reproduction in Rats. By RONALD J. BARFIELD and DAVID A. THOMAS.. 33

Female Vocalizations and Their Role in the Avian Breeding Cycle. By BARBEE L. INMAN 44

[a]This volume is the result of a conference entitled Jay S. Rosenblatt-Institute of Animal Behavior 25th Anniversary Alumni Conference held on June 6-8, 1984 by Rutgers-The State University of New Jersey in Newark, New Jersey.

Role of the Preoptic Area in the Neuroendocrine Regulation of Reproduction: An Analysis of Functional Preoptic Homografts. *By* MARIE J. GIBSON 53

Genital Stimulation as a Trigger for Neuroendocrine and Behavioral Control of Reproduction. *By* BARRY R. KOMISARUK and JUDITH L. STEINMAN... 64

Neuroendocrinology of Coitally and Noncoitally Induced Pseudopregnancy. *By* JOSEPH TERKEL 76

Licking, Touching, and Suckling: Contact Stimulation and Maternal Psychobiology in Rats and Women. *By* JUDITH M. STERN 95

Interaction of Species-Typical Environmental and Hormonal Factors in Sexual Differentiation of Behavior. *By* CELIA L. MOORE 108

Prenatal Stress Disrupts Reproductive Behavior and Physiology in Offspring. *By* LORRAINE ROTH HERRENKOHL................. 120

Hormonal and Metabolic Factors Underlying Intraspecific Variation in Reproductive Social Behavior. *By* DALE F. LOTT 129

Advantages to Female Rodents of Male-Induced Pregnancy Disruptions. *By* ANNE STOREY.................................. 135

Reproductive Behavior as a Phenotypic Correlate of T-Locus Genotype in Wild House Mice: Implications for Evolutionary Models. *By* SARAH LENINGTON 141

The Role of the Vomeronasal Organ in Behavioral Control of Reproduction. *By* MARGARET A. JOHNS........................ 148

Experiential Influences on Hormonally Dependent Ring Dove Parental Care. *By* GEORGE F. MICHEL 158

The Role of Perception in Habitat Selection in Colonial Birds. *By* JOANNA BURGER... 170

Comparative Behavioral Endocrinology. *By* DAVID CREWS.......... 187

Part II. Hormones as Regulators of Reproductive Behavior

A. PARENTAL BEHAVIOR

Introduction. *By* JAY S. ROSENBLATT............................... 199

Hormonal Basis of Maternal Behavior in the Rat. *By* HAROLD I. SIEGEL.. 202

Maternal Responsiveness and Nest Defense during the Prepartum Period in Laboratory Rats. *By* ANNE D. MAYER................ 216

The Role of the Medial Preoptic Area in the Regulation of Maternal Behavior in the Rat. *By* MICHAEL NUMAN 226

Psychobiology of Rat Maternal Behavior: How and Where Hormones Act to Promote Maternal Behavior at Parturition. *By* ALISON FLEMING 234

Role of Prolactin in Avian Incubation Behavior and Care of Young: Is There a Causal Relationship? *By* JOHN D. BUNTIN 252

B. Sexual Behavior

Introduction. *By* BARRY R. KOMISARUK and CARLOS BEYER 268

Elevation in Hypothalamic Cyclic AMP as a Common Factor in the Facilitation of Lordosis in Rodents: A Working Hypothesis. *By* CARLOS BEYER and GABRIELA GONZÁLEZ-MARISCAL 270

A Reevaluation of the Concept of Separable Periods of Organizational and Activational Actions of Estrogens in Development of Brain and Behavior. *By* CHRISTINA L. WILLIAMS 282

Extended Organizational Effects of Estrogen at Puberty. *By* JORGE F. RODRIGUEZ-SIERRA 293

A Two-Step Model for Sexual Differentiation. *By* JACQUES BALTHAZART and MICHAEL SCHUMACHER 308

Reciprocity in Neuroendocrine and Evolutionary Approaches to the Study of Reproductive Behavior. *By* CAROL DIAKOW 325

A Concept of Physiological Time: Rhythms in Behavior and Reproductive Physiology. *By* L. P. MORIN 331

Part III. Biochemical Bases of Hormonal Action on Reproductive Behavior

Introduction. *By* HARVEY H. FEDER 352

Hormonal Control of Behavior—A Cautionary Note. *By* RICHARD E. WHALEN 354

Heterotypical Sexual Behavior: Implications from Variations. *By* JANICE E. THORNTON 362

The Role of Androgen Metabolism in the Activation of Male Behavior. *By* CHERYL F. HARDING 371

Steroid Hormone Antagonists and Behavior. *By* I. THEODORE LANDAU 379

Sex Steroids and Energy Balance: Sites and Mechanisms of Action. *By* GEORGE N. WADE 389

Steroid Receptors and Hormone Action in the Brain. *By* JEFFREY D. BLAUSTEIN 400

Noradrenergic Regulation of Progestin Receptors: New Findings, New Questions. *By* BRUCE NOCK 415

Reproductive Neuroendocrine Regulation in the Female Rat by Central Catecholamine–Neuropeptide Interactions: A Local Control Hypothesis. *By* WILLIAM R. CROWLEY 423

Hormone Effects on Serotonin-Dependent Behaviors. *By* LYNN H. O'CONNOR and CHRISTINE T. FISCHETTE 437

Hormonal Aspects of the Morphological Differentiation of Neuronal Clonal Cell Lines. *By* CHRISTIAN P. REBOULLEAU 445

Analysis of Steroid Action on Gene Expression in the Brain. *By* JOSIAH N. WILCOX 453

Index of Contributors 461
Subject Index 463

The New York Academy of Sciences believes that it has a responsibility to provide an open forum for discussion of scientific questions. The positions taken by the authors of the papers that make up this *Annal* are their own and not those of the Academy. The Academy has no intent to influence legislation by providing such forums.

Preface

This volume contains an array of reviews of reproductive processes from a behavioral and neuroendocrine perspective which were presented at the 25th Anniversary Alumni Conference of the Institute of Animal Behavior held on June 6-8, 1984 in conjunction with the 60th birthday of the Director, Professor Jay S. Rosenblatt. All the contributors have been or are members of the Institute, and all the papers address interactions among behavior, hormones, and/or reproduction. The origins of these diverse approaches toward understanding the relationships between behavior and reproduction can be traced to the seminal writings and teaching of the late Daniel S. Lehrman, the founder of the Institute of Animal Behavior. He broke the ground for the insights we have cultivated and which are presented in this volume.

We gratefully acknowledge the generous financial support for the Conference that was provided by the Graduate Dean's Office, University College, and the Institute of Animal Behavior, all of Rutgers-The State University of New Jersey, Campus at Newark. We also express our appreciation to the graduate students, technical and support staff, and faculty and alumni of the Institute of Animal Behavior for their assistance.

BARRY R. KOMISARUK
HAROLD I. SIEGEL
MEI-FANG CHENG
HARVEY H. FEDER

Jay S. Rosenblatt, 1984.

Picture credit: Donald A. Dewsbury

In Honor of Jay S. Rosenblatt

This volume is in honor of Jay S. Rosenblatt, Director of the Institute of Animal Behavior, scientist, teacher, and friend. Jay's interest in animal behavior developed while he was a student at New York University, taking classes with T. C. Schneirla and Herbert Birch. As a graduate student at N.Y.U., Jay carried out his thesis research at the American Museum of Natural History under the supervision of Lester Aronson. It was also during this time that Jay first met and worked with Daniel S. Lehrman, then a graduate student studying finches. Danny later was appointed to a position in the Psychology Department of Rutgers University in Newark, New Jersey, and subsequently founded the Institute of Animal Behavior.

Jay's research in animal behavior has spanned thirty-five years. His earlier studies of sexual behavior in male cats illustrated the relationship among hormones, copulatory behavior, and previous sexual experience. His research on cats expanded to include maternal behavior and the development of suckling and home orientation in kittens and then to the psychobiology of maternal behavior in rats. These studies began in 1959 with the techniques of observation and description, and over the years have incorporated the areas of sensory processing, development, neuroanatomy, pharmacology, and endocrinology, and addressed questions related to natural selection and molecular mechanisms of hormone action. Jay's methods, findings, and conceptual thinking have become synonymous with the current status for this area of research.

After the initial shock of the death of Danny Lehrman in the summer of 1972, Jay Rosenblatt was elected Director of the Institute of Animal Behavior by unanimous vote of the IAB faculty. His style has been different from Danny's, but his demand is the same for scholarly and scientific excellence. His critical perspective has been influential in keeping the students and faculty of the IAB on target in research strategy. He has nurtured the traditional IAB perspective of addressing problems of natural function and adaptive significance to the organism as a guiding principle in the formulation of research problems. He has encouraged and supported research initiatives of students and faculty which have led to studies involving new species, including humans, as well as new areas such as *in vitro* systems and chemistry. In this regard, he has fostered the spirit of freedom of investigation that has been a hallmark of the IAB, a tradition from which it derives its greatest strength.

Jay is also a practicing psychotherapist, and an artist and sculptor. As Director of the IAB, as a colleague, and as a friend, he is accessible, compassionate, patient, tolerant, and fair. We are deeply grateful for his help and support over the years.

BARRY R. KOMISARUK
HAROLD I. SIEGEL
MEI-FANG CHENG
HARVEY H. FEDER

Jay S. Rosenblatt (*left*) and Daniel S. Lehrman (*right*), 1965.

Historical Perspective of Daniel S. Lehrman, Founder of the Institute of Animal Behavior

COLIN G. BEER

Institute of Animal Behavior
Rutgers-The State University of New Jersey
Newark, New Jersey 07102

J. WAYNE LAZAR

Neuropsychology Division
North Shore University Hospital
Manhasset, New York 11030

CAROL DIAKOW

Biology Department
Adelphi University
Garden City, New York 11530

In a way the Institute of Animal Behavior at Rutgers-The State University of New Jersey, Campus at Newark, had its beginnings in the research Danny Lehrman had done for his doctoral dissertation. He started out with a plan to study the formation and function of brood patches of ring doves, only to find, when he got hold of some of the birds, that they lack brood patches. So he switched to a study of the control of crop milk production, and it was from this that he initially derived what became a central theme in his later work: the ways in which behavior and physiology, stimulation and hormone secretion, are linked in reciprocal influence in an "endless golden braid" leading to reproduction.

In his words, "My main orienting attitudes towards the study of animal behavior come from having been a bird watcher since I was 14 years old and if I were going to name a person in the field who had a main effect of giving me a theoretical orientation it would be T. C. Schneirla." Danny knew before he went to college that he wanted to study bird behavior, perhaps, be a warden in a bird sanctuary and study nest-building and territoriality. He said that he never thought that he would grow up working in a laboratory.

In his research, and in his theoretical and critical writing, Danny Lehrman was as tough-minded as any behaviorist: he constructed his conception of the ring dove reproductive cycle carefully and rigorously. He dug out logical and factual inconsistencies in the positions he opposed; he emphasized the point that what a science finds depends upon what its practitioners look for, which can be affected by the temperament, taste, and training of the individual scientist. He was repeatedly rough on what he called molar theories of behavior, such as those of Hull and Skinner, for failure to distinguish between questions of different sorts, and blindness to the discontinuous diversity in the animal kingdom. On this latter point he reflected the influence of his teacher T. C. Schneirla,[1,2] especially Schneirla's doctrine of psychological levels in phylogeny and ontogeny. This doctrine has some affinities with Piaget's genetic epistemology, and so it is not surprising that one of the hallmarks of a Schneirla student is preoccupation with questions of behavioral development. Lehrman was no exception.

Danny's philosophy of science is exemplified by the following quote from his last theoretical paper:

> The animal itself, and the behavior by which it lives its natural life, can seem to a human observer like a mysterious and attractive part of the world around him. The interest of the investigator in understanding the life of the animal can have a direct character—can carry an immediate emotional charge of curiosity and fascination, and of apprehension of the animal as a subject that has an existence of its own, independently of whether it is serving as an object for a human experimenter. Feelings of this kind have, for the observer, more in common with feelings involved in watching a sunset or reading a poem than they do with those involved in solving engineering problems, or in abstracting formalized general relationships from narrowly defined operations of experimenter and subject. (Reference 3, pp. 463-464)

In his concept of the Institute of Animal Behavior, Danny felt that the Institute should not be oriented by a particular theoretical orientation or technique, such as a stimulus-response orientation or a Skinner box. "We were oriented by a kind of subject matter." That subject matter was the species-specific behavior by which animals related to one another. It included a wide variety of behavior patterns, including maternal, sexual, and aggressive behavior, and each was the "kind of social interaction that actually characterizes the natural adaptation of the animal to its natural world." The methodologies of endocrinology, neurophysiology, biochemistry, ethology, and developmental psychology were to be used to look at the interactions from different points of view. They were not to offer alternative explanations, but, rather, different points of view "which stimulate each other and cooperate with each other." His job as Director of the Institute was to make sure that the various methodologies illuminated a behavioral question.

The contributions in this volume bear testimony to Danny's intellectual influence.

REFERENCES

1. SCHNEIRLA, T. C. 1949. Levels in the psychological capacities of animals. *In* Philosophy for the Future. R. W. Sellars, V. J. McGill & M. Farber, Eds. Macmillan. New York, N.Y.
2. SCHNEIRLA, T. C. 1965. Aspects of stimulation and organization in approach/withdrawal processes underlying vertebrate behavioral development. Adv. Study Behav. 1: 1-74.
3. LEHRMAN, D. S. 1971. Behavioral science, engineering and poetry. *In* The Biopsychology of Development. E. Tobach, L. R. Aronson & E. Shaw, Eds. Academic Press. New York, N.Y.

PART I. BEHAVIOR AS A REGULATOR OF HORMONAL MECHANISMS AND REPRODUCTION

Introduction

BARRY R. KOMISARUK AND MEI-FANG CHENG

*Institute of Animal Behavior
Rutgers–The State University of New Jersey
Newark, New Jersey 07102*

The recognition that environmental factors such as light and temperature stimulate specific hormone secretion opened the field of neuroendocrinology. The underlying messengers are the neuropeptides that are synthesized and released by hypothalamic neurons. They exert control over the synthesis and/or release of pituitary hormones that are in turn released into the systemic circulation. The anterior pituitary hormones in turn stimulate target endocrine glands to secrete their hormones that, via the systemic circulation, affect body morphology, metabolism, the brain, and behavior.

A major impetus to the field of neuroendocrinology was provided by Daniel S. Lehrman who emphasized that the behavior of one mate stimulates hormonal, and thereby behavioral, responses in its partner. The mate is then affected reciprocally by the altered behavior of the partner. Lehrman thus emphasized the critical role of behavior, both emitted and perceived, in mediating reproductive physiology (see, for example, Scientific American 211 (1964): 48-54).

The papers in this section reflect and extend Lehrman's influence in exploring behavioral factors mediating reproduction. *M.-F. Cheng* presents evidence that in ring doves development of the female's ovaries in response to stimuli provided by the male's courtship requires a specific behavioral response of the female. The proprioceptive and auditory self-stimulation which the female receives from her own specific behavioral response (nest-cooing) to the male, plays a major role in her own ovulation. Thus, it is active behavior rather than passive sensory reception that triggers the female's hormonal response. *C. J. Erickson* incorporates this view, emphasizing that the male's initial aggressive behavior toward the female may delay or block the female's nest-cooing response and hence her ovarian development. He points out the significance of this disruption process for sperm competition among males. *N. T. Adler* and *J. P. Toner, Jr.* focus on the role of the individual's own behavior in mediating its reproductive efficacy, in their studies on sperm competition, sperm output, sperm transport, and fertilization in rats. They show that the number of sperm entering the uterus is reduced in males that achieve fewer than six intromissions before ejaculation. Since dominant males are more effective than subordinate males in fertilizing females, and they achieve more intromissions than subordinate males prior to ejaculation, their greater fertility may be due to the increased number of sperm. *R. J. Barfield* and *D. A. Thomas* describe their studies on the function of ultrasonic vocalization by male and female rats during mating encounters, with reference to sperm competition. When a dominant and a subordinate male are placed together with a female, vocalization by the dominant male (during his mating, which occurs before the subordinate male's) may signal the subordinate male to stay away from the female, thereby inhibiting ejaculatory attempts by the subordinate male and consequently preventing sperm competition by the subordinate male. *B. L. Inman* surveys the contexts in which vocalization by avian females mediates breeding success via pair-bond formation, breeding synchrony, territoriality and mate-guarding.

Tactile copulatory stimulation plays a major role in inducing ovulation in "reflex" ovulator species and "spontaneous" ovulator species under appropriate conditions (e.g., persistent estrus). *M. J. Gibson* reports her studies in which grafts of fetal brain tissue from the preoptic area (of females or males) into the hypothalamus of adult genetically hypogonadal female mice apparently achieve functional afferent reinnervation to the extent that the females, which enter a "continual estrus" state, respond to copulation by the male by ovulating, then becoming pregnant, delivering young, and raising them successfully.

The neuroendocrine reflexes triggered by vaginal stimulation are reviewed by *B. R. Komisaruk* and *J. L. Steinman,* who also analyze the mechanism underlying the analgesia that is produced by vaginal stimulation in rats and women, and discuss its potential significance to parturition and mating behavior. Copulation-induced hormonal changes are also addressed in the paper by *J. Terkel.* He reviews his studies on the neural basis of induction of prolonged (2 weeks) diurnal and nocturnal prolactin surges following a single episode of vaginal stimulation. The prolactin surges can be triggered by the sensory stimulus even before the prolactin system matures. Another interesting feature of reflexive prolactin release is that suckling also stimulates prolactin secretion, but the episodic rhythmical pattern of prolactin secretion is absent. Similarly, stimulation-induced prolactin and oxytocin secretion is addressed in the paper by *J. M. Stern.* She discusses her findings and their possible adaptive significance with regard to birth space interval in women; prolonged (2 years) suckling maintains elevated prolactin levels and is associated with reduced coital frequency.

Tactile genital stimulation has also been shown to play an important role in the differentiation of sexual behavior in rats, thereby amplifying genetic factors, a novel perspective presented in the paper by *C. L. Moore.* She shows that the dam responds to olfactory cues resulting from androgen-induced differences in neonatal urine by increasing the amount of anogenital licking directed toward neonatal males. Rendering the dam anosmic by intubation of the nares leads to reduced vigor of masculine copulatory behavior in her male offspring, whereas increasing the dam's anogenital licking of her female offspring increases their masculine copulatory behavior as adults. Furthermore, evidence is presented that genital self-grooming in males increases their own accessory organ weight, another example of behavioral self-stimulation inducing a hormonal response.

The role of the mother as an intermediary in the differentiation of sexual behavior in the offspring is also addressed by *L. R. Herrenkohl,* but prenatally rather than postnatally. Herrenkohl reviews studies showing how stress applied to the mother, and/or perceived by the mother, can influence her offspring via hormonal changes in the prenatal environment.

The role of idiosyncratic environmental influences (e.g., prenatal stress) in generating individual differences in behavior, and the possible consequences for social structure, are addressed by *D. F. Lott* in his review.

Specific olfactory stimuli exert a major influence on reproductive physiology. The phenomenon of blockage of pregnancy by the odor of an individual male different from the impregnating male is analyzed from an adaptational perspective by *A. Storey.* Under seminatural conditions using voles she has found that when a pregnant and an estrous vole cohabit with a new male, usually only the estrous vole gives birth. The additional presence of the initial impregnating vole can protect against the pregnancy blockage. Storey discusses the adaptive significance of these and other findings in terms of limitation of pregnancy under high population density conditions, reproduction in emigrating females, protection against infanticide, and pair-bonding.

S. Lenington assesses the roles of mate selection, aggressiveness, and dominance in transmission of deleterious t-alleles in male and female wild house mice. An implication of her findings is that the genetic evolution of mouse populations is strongly influenced by specific components of reproductive behavior.

The role of chemoreception in reproduction is reviewed by *M. A. Johns,* who provides an historical review of the vomeronasal system and discusses her recent studies on the role of the vomeronasal system in stimulating ovulation in rats.

In addition to the specific sensory factors in hormone secretion reviewed in the above papers, previous behavioral experience of a reproductive cycle facilitates subsequent breeding efficiency. This is discussed in the paper by *G. F. Michel,* with special reference to prolactin secretion in ring doves. He also discusses the adaptive role of experiential factors in the synchrony of breeding cycles.

The self-selection of environmental niches by individuals provides unique sensory stimuli to which they can respond. *J. Burger* discusses ways in which hormonally based behavior (e.g., aggression) can affect selection of an environment (e.g., territory) to which an individual is exposed, which in turn can affect its reproductive capacity.

A unique problem is raised by *D. Crews* in his analysis of environmental stimulation of ovarian function in a parthenogenic species of *Drosophila.* He demonstrates the stimulus specificity of the greater effectiveness of males than females in stimulating egg production in the parthenogenic individuals, which raises the intriguing question of its adaptive significance. He raises another interesting question of the possible adaptive significance of dissociation of reproductive behavior from ovarian development in garter snakes. Understanding the nature of these special cases may shed further light on the basic underlying mechanisms.

Individual Behavioral Response Mediates Endocrine Changes Induced by Social Interaction[a]

MEI-FANG CHENG

Institute of Animal Behavior
Rutgers–The State University of New Jersey
Newark, New Jersey 07102

In 1955, G. W. Harris proposed the existence of neural control over pituitary hormone secretions,[1] a concept subsequently confirmed by successful purifications and syntheses of hypothalamic releasing hormones some 15 years later.[2,3] That this control in fact exists provides an anatomical and physiological basis for a variety of external stimuli, including in particular, stimuli emitted from other members of an animal's species to elicit different types of hormone secretion in that animal. Among the most intriguing observations made across vertebrates, including humans, are those of the promotion or suppression of ovarian activities in a social context. Menstrual cycles of college women tend to be synchronized among roommates and close friends.[4] The onset of puberty in female mice[5] and rats[6] is advanced by the presence of conspecific males, and in female prairie voles uterine growth is triggered by a conspecific male[7]; similarly, in birds, egg-laying is stimulated by songs of the male.[8] It has been the consensus that hormone secretions in all of these situations are a consequence of direct stimulations from the social partner(s). Thus, stimulus parameters of the social partner(s) are determining factors of the endocrine response.

During the past few years, at the Institute of Animal Behavior, we have conducted a series of studies testing the idea that hormone secretions induced by members of the species are passive, reflexive responses over which the individual has no control. On the basis of the evidence which I shall describe in the ensuing sections, I propose that it is, in fact, the individual's specific behavioral reaction to social stimuli, rather than the social signal as such, which promotes that individual's own hormone secretions (I have termed this a "self-feedback" or "self-stimulation" effect[9]). In this concept, the variables associated with the performance of the specific behavior and not the characteristics of the social stimuli are critical in determining the levels of hormone secretions.

The study of the follicular development of female ring doves (*Streptopelia risoria*) in response to male courtship is probably the most systematic and thorough among all the investigations of this phenomenon. Pioneered by the late Professor Daniel S. Lehrman, this line of research has been diligently pursued and expanded by his associates and students. A new interpretation of this well-studied phenomenon is therefore not valid unless it also accounts for previously found evidence. To this end, the readers are referred to an excellent paper in this volume by Dr. Carl Erickson,

[a] This paper is contribution number 427 from the Institute of Animal Behavior, Rutgers University. This work is supported by National Science Foundation Grant BNS 8121495.

one of the early defenders of the idea that male courtship directly stimulates the female's hypothalamus–pituitary–ovarian axis (HPO). In this paper, I shall limit myself to presenting the lines of evidence in support of the idea that it is the self-stimulation (or self-feedback) effect of a female's specific behavior which promotes the female's own HPO activities. It should be pointed out that the phenomenon of self-stimulation was first noted by Brockway in her study with budgerigars: testis size was found to be related to the male's own singing bouts.[10] This phenomenon takes on a new meaning in the case of the female's follicular growth which is widely thought to be regulated solely by male courtship during breeding seasons.

THE SEQUENCE OF EVENTS BETWEEN MALE COURTSHIP AND THE FEMALE'S FOLLICULAR GROWTH

Courtship in ring doves is typically initiated by the male's cooing displays (starting with bow-cooings which are soon replaced by nest-cooings) interspersed with chasing activities. Some time later (about 3 days), the female joins the male at the nest site and they perform nest-cooing in duet for 1 to 2 days. The male then leaves the female doing her solo cooings to collect materials for nest construction. A few days later, the female lays a clutch of two eggs. This whole sequence is usually completed in 7 to 10 days in a laboratory setting. This predictable and orderly sequence of events—male courtship, female nest-cooings, follicular growth, and egg-laying—is schematically represented in FIGURE 1. These orderly relationships concur with the widely held idea that male songs stimulate follicular development of the conspecific female. It is, however, not a perfect fit: unlike songbirds, female ring doves do sing (nest-cooing) during the pre-laying phase. It raises the question of why the female nest-coos at all? What role does it play in breeding success? Does it contribute to the process of follicular development? That this last question was never asked is puzzling. Considering for a moment the three events, (A) male courtship, (B) female nest-cooing, and (C) female follicular development, it should be obvious that we must analyze all possible relationships including that between B and C before a definitive causal relationship between A and C can be drawn. In identifying factors leading to egg-laying behavior, these steps were not followed; the relationship between the female's nest-cooing and follicular development was never considered a plausible research question. Two factors might have contributed to this omission: (1) strong evidence linking male courtship to female's follicular growth and, (2) the preconceived idea that follicular growth is a passive (reflexive) response to external stimuli. These pre-conceptions have dominated my research strategy as well. It was my sustained interest in female behavior that ultimately led me to the discovery of its pivotal role in follicular development.

WHAT IS THE FUNCTION OF NEST-COOING BEHAVIOR IN FEMALE RING DOVES?

Nest-cooing behavior in both the male and female consists of assuming an oblique posture with tail up, while flipping the wings (often the whole body descends into the nest-bowl as the female becomes increasingly attached to the nest); and simultaneously the birds emit two-syllable coos. A full display of nest-cooing by a female requires stimulation both by male courtship (bow-cooing, nest-cooing) and estrogen action.[11] During courtship pairing, if the female's nest-cooing behavior is reduced or absent, two events stand out: (1) male courtship, i.e., nest-cooing bouts, continues with all its original vigor, and (2) the female shows no sign of or little follicular development despite her apparent attraction to her mate (approaching him, copulating with him). In short, the proposed causal relationship between male courtship and follicular development breaks down when the female's nest-cooing behavior is removed. That is, in terms of A, B, and C in the previous section, when B does not follow A, A

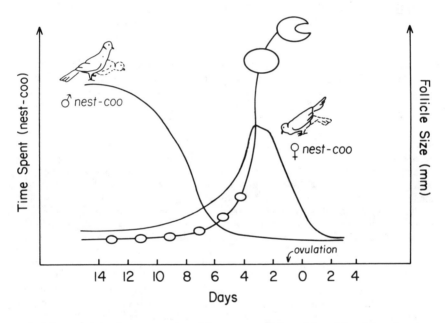

FIGURE 1. A schematic representation of the sequence of the male's and female's nest-cooing behavior in relation to the pattern of follicular development, plotted in reference to oviposition. The pattern of follicular development is based on data presented in Eleanor Sims' thesis in partial fulfillment of the requirements for the doctorate degree at Rutgers University, 1983.

does not lead to C, suggesting that B is indispensable for the manifestation of C. These basic observations have been demonstrated by studies wherein female doves were devocalized by either lesioning brain regions (the midbrain, in particular, n. intercollicularis: nICo) that mediate vocalization,[12] or by severing peripherally the motor nerves innervating the syrinx, the sound organ.[13] It is clear from these studies that the female's nest-cooing behavior is an essential component of the mechanisms underlying male courtship-induced follicular development. Assuming that this is the case, the next question is, how does the female's nest-cooing behavior influence her own ovarian activity? Elucidation of "how" in this case will also lend further support to the idea of self-feedback effects.

EFFECTS OF THE FEMALE'S OWN NEST-COOING BEHAVIOR

A female's own nest-cooing behavior may affect her follicular development by two mechanisms: (1) effects from hearing her own nest-cooings (acoustic feedback), and (2) effects from performing nest-cooings (proprioceptive feedback).

If acoustic feedback is involved in the female's nest-cooing effect on follicular growth, the following is predicted. The devocalized female will show follicular growth when she hears playbacks of her own nest-cooings, but not when hearing those of her mate. This prediction was tested in a study[14] where female doves were devocalized by inserting a 1½-in. length of polyethylene tubing (3/16-in. OD, ⅛-in. ID) into the interclavicular sac immediately surrounding the syrinx. The open end deflates the major energy source produced by the respiratory pump which is required for vocalization and hence renders the bird unable to emit sound.

Playing back the female's own nest-cooings daily for 5 days stimulated a follicular size increment significantly larger than that produced by the nest-cooing of the male; nest-cooings of other females were also effective but not as effective as the female's own. This selective endocrine response suggests that the female is capable of discriminating between the nest-cooings of males and females, a remarkable faculty considering the overall similarity between them; the male's cooings feature slightly larger intervals between some elements (see FIGURE 2).

In another study,[15] playbacks were conducted in a more "natural" setting, in contrast to the previous study where playbacks were done in soundproof chambers. Experimental doves were kept in a breeding room which also housed other pairs of the breeding colony. Intact female doves were introduced to active males for one day. On the second day, the males were removed and daily 2-hr playback sessions were begun; this continued for a total of five consecutive days. In this study, the females were exposed to the males for exactly one day whereas in the soundproof-chamber study males were present throughout the experiment. In this study, the females hearing their own nest-cooings showed significant follicular growth, suggesting that the continued presence of their mates is not essential after 24 hr. Surprisingly, the results of hearing other females' nest-cooings contradicted those of the previous study in that the nest-cooings of the other females were totally ineffective in stimulating follicular growth. This disparity may be related to the different settings in which the playbacks occurred; perhaps finely tuned discriminations are exercised only in situations where there is some necessity, such as in the breeding colony where other females' vocalizations constitute a competitive background to the female's own cooings.

It is clear from these studies that the acoustic feedback is involved in the female's nest-cooing effect. The result, however, does not preclude the importance of proprioceptive feedback.

EFFECTS OF PROPRIOCEPTIVE FEEDBACK

The procedure of removing both cochleas renders the dove unable to hear either her own or her mate's cooings and hence is an ideal condition to test if proprioceptive feedback alone is sufficient to stimulate ovarian growth, provided that the deafened female does engage in nest-cooing. The choice of the ring dove is particularly fortunate in this context. In doves, unlike songbirds, the development of cooing does not depend on acoustic input,[16] and it is likely therefore that deafening will not seriously affect the nest-cooing display. In an unpublished study, we found that about half the females subjected to deafening surgery did not recover nest-cooing during the following 3 months of repeated testing. For those that did nest-coo when paired with males, their nest-cooing behavior was indistinguishable from that of sham-operated or normal females. Nevertheless, a significantly longer duration of male courtship was necessary

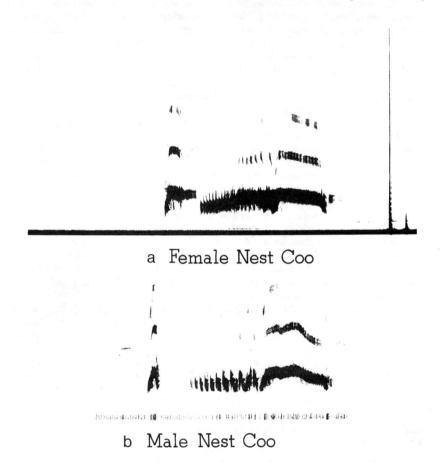

FIGURE 2. Representative sonograms of nest-cooings in the ring dove: (a) a female, (b) a male.

to induce nest-cooing behavior of deafened females. Thus, the laying latency to lay eggs for the deafened females who did nest-coo was significantly longer than that of sham-operated and intact females. Those females who did not utter nest-coos failed to lay during a 4- to 6-week period of continuous pairing with active males. This, in my view, attests to the importance of proprioceptive feedback.

Taken together, the evidence we have thus far amassed in support of the idea that the female's own nest-cooing behavior feeds back to stimulate her endocrine system is compelling. The critical nature of the female's own nest-cooings in her follicular growth does not, however, rule out entirely the role of male courtship. Male courtship is essential in the very least to initiate the female's nest-cooing behavior. To what extent this reduced role of the male's nest-cooing is significant awaits further analysis from this new perspective.

IS THERE NEUROANATOMICAL EVIDENCE LINKING THE VOCAL CONTROL SYSTEM AND THE FOLLICULAR GROWTH CONTROL SYSTEM (HYPOTHALAMUS-PITUITARY-OVARIAN AXIS)?

There is no evidence in the literature of avian neuroanatomy that I can cite to support the idea that inputs from the nICo region can be transmitted to the hypothalamus-pituitary-ovarian axis. Neural control of song is well illustrated exclusively in passerine species, by elegant work from the laboratories of M. Konishi and F. Nottebohm. The neuroendocrine basis of ovarian activities in quails and chickens, on the other hand, has been the prime research target of avian endocrinologists. The research interests of these two groups rarely overlap, and there was no a priori reason to suspect that the vocal control system was a part of the larger extrahypothalamic control of the hypothalamus-pituitary-ovarian axis until the ring dove studies mentioned above suggested the existence of just such connections.

Using electrolytic lesioning and intracranial estrogen stimulation techniques, we have identified an estrogen-sensitive part of the midbrain, the medial portion of n. intercollicularis (mICo), as one important station in the circuitry of nest-cooing production. In this region we can selectively produce nest-cooing displays with estrogen in the ovariectomized dove[12] and selectively reduce or block nest-cooing by electrolytic lesions.[12] Lesions in the other estrogen-sensitive sites, the posterior medial hypothalamus (PMH) for instance, produce a cascade of behavioral deficits of which nest-cooing is one.[17] Similarly, lesions in the vocal control nuclei, such as the robust nucleus of the archistriatum (RA), affect all vocalizations including nest-cooings.[18] We recognize therefore that the mICo provides a good strategic point from which to study the network involved in the production of this display and its connection to the hypothalamus-pituitary-ovarian axis. Modern tract tracing methods, amino acid autoradiographic tract tracing of axonal connections and horseradish peroxidase (HRP) retrograde tract tracings, provide excellent tools to determine afferents and efferents of the mICo: fibers entering the mICo may modulate or control the production of nest-cooing, and fibers leaving the mICo may mediate effects of nest-cooing on other systems, one of which is the hypothalamus-pituitary-ovarian axis.

Although this project has gotten under way only recently, we have already made an encouraging finding: the fibers leaving the region of the mICo, the same discrete site where electrolytic lesions block and estrogen stimulations induce nest-cooings, terminate at the periventricular nuclei (PVM) and the anterior hypothalamus (AM) region.[19] This is the general region where luteinizing hormone-releasing hormone (LHRH) neurons are present in ducks[20] and doves.[21] It is reasonable, therefore, to propose that this pathway mediates the self-stimulation effect of the female's nest-cooing by way of its contacts with LHRH neurons; these neurons may release gonadotropin-releasing hormone (GnRH) in response to excitatory impulses from mICo (FIGURE 3). This hypothesis, however, does not exclude the possibility that still another pathway is involved. It is quite conceivable, for instance, that auditory feedback may involve the auditory ascending and descending pathways that have been demonstrated in pigeons. Acoustic information is transmitted from the cochlea to the nucleus mesencephalicus lateralis, pars dorsalis (comparable to mammalian inferior collicular) and on to the ovoidalis of the thalamus and finally to field L in the telencephalon.[22] Detection of the female's own versus the male's coos may be processed here in the telencephalon. Descending pathways from field L may terminate in the hypothalamus

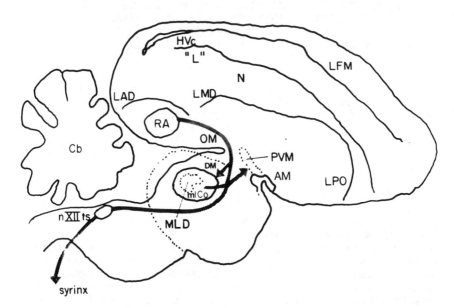

FIGURE 3. A schematic drawing showing (1) the neural controls of all vocalizations in ring doves, and (2) parts of the nest-cooing-relevant neural circuitry identified by the leucine autoradiography tract tracing method (based on data reported by Akesson et al.[19]). The projections from mICo to PVM and AM are thought to be involved in the female's self-stimulation effect on her hypothalamus-pituitary-ovarian axis. The efferent pathway RA → nXIIts is based on HRP data reported by Cohen.[18] Note at present there are not sufficient data to claim DM and mICo as two distinct regions; both are in the general medial portion of ICo. Circles indicate individual brain regions, and arrows, anatomical projections. Abbreviations: hyperstriatum ventrale, pars caudale (HVc), robust nucleus of the archistriatum (RA), medial portion of n. intercollicularis (mICo), midbrain dorsomedial nucleus of the intercollicular region (DM), hypoglossal motor neurons (nXIIts), n. periventricularis mognocellularis (PVM), n. anterior medial hypothalamus (AM).

via its projections to vocal control nuclei such as RA which sends fibers to the mICo. Alternatively, outputs from field L may descend by ansa lenticularis via its projection to paleostriatum. The observation that the female's follicular development depends on her discriminating between male and female songs reinforces the notion that part of these pathways is involved. On the other hand, for intact normal doves, the presence of proprioceptive information may be sufficient for the female to discriminate her own cooings from the male's.

IMPLICATIONS AND DIRECTIONS FOR FUTURE RESEARCH

Laboratory findings often run the risk of being peculiar to the artificial setup of the laboratory, which cannot faithfully duplicate a natural environment. It should also be stressed, however, that a laboratory finding should not automatically be tossed away as biologically insignificant. A case in point is the long-established laboratory finding of the pregnancy-blocking phenomenon in rodents (the Bruce effect), now found to prevail in the free-ranging species in the field (see the paper by A. Storey in this volume).

Likewise, it is comforting to observe that recent avian studies in the field, exemplified by field endocrinology, have not produced any major findings that contradict those previously established by studies of ring doves in the laboratory. In fact, the strength of the field findings rests on their similarity to the more rigorously controlled ring dove research. The fact remains that the ultimate significance of the self-stimulation (or self-feedback) effects of the female's song must be demonstrated by fieldwork. In this context, it is interesting to note that some field observations (see the paper by B. Inman in this volume) do suggest that such a phenomenon exists. Now that this mechanism is well established in the laboratory, more controlled observations in the field can be designed to test if it operates there.

In a wider context, the self-stimulation concept suggests a need in future research to focus on the behavior of the individual as much as on the stimulus properties of social signals to understand the dynamics of endocrine activities in a complex social environment. A response, even in the classical sense of a reflex response, is not necessarily as stereotyped as it was once thought to be. Reflex actions can be modified by the subject's past experience. This involves changes in synaptic transmission as has been elegantly demonstrated in aplasia by Kandel's group.[23] In this sense all responses, hormonal as well as behavioral, are reactions of the organism. In our analysis of socially induced endocrine changes, we need to readdress the active nature of the organism.

REFERENCES

1. HARRIS, G. W. 1955. Neural Control of Pituitary Gland. Latimer, Trend & Co. Ltd. Plymouth, England.
2. SCHALLY, A. V., A. ARIMURA & A. J. KASTIN. 1973. Hypothalamic regulatory hormones. Science **179**: 341-350.
3. BLACKWELL, R. E. & R. GILLEMON. 1973. Hypothalamic control of adenohypophysial secretions. Annu. Rev. Physiol. **35**: 357-390.
4. McCLINTOCK, M. K. 1971. Menstrual synchrony and suppression. Nature (London) **229**: 244.
5. VANDENBERGH, J. G. 1967. Effect of the presence of a male on the sexual maturation of female mice. Endocrinology **81**: 345-349.
6. VANDENBERGH, J. G. 1976. Acceleration of sexual maturation in female rats by male stimulation. J. Reprod. Fertil. **46**: 451.
7. CARTER, C. S., L. L. GESTZ, GAVISH, J. L. McDERMOTT & P. ARNOLD. 1980. Male-related pheromones and activation of female reproduction in the prairie vole (*Microtus achrogaster*). Biol. Reprod. **23**: 1038-1045.
8. LEHRMAN, D. S. 1965. Interaction between internal and external environments in the regulation of the reproductive cycle of the ring dove. *In* Sex and Behavior. F. A. Beach, Ed.: 355-380. Wiley. New York, N.Y.

9. CHENG, M.-F. 1979. Progress and prospect in ring dove research: a personal view. *In* Advances in the Study of Behavior. J. S. Rosenblatt, R. A. Hinde, C. G. Beer & M.-C. Busnel, Eds. Academic Press. New York, N.Y.
10. BROCKWAY, B. 1969. Roles of budgerigar vocalization in the integration of breeding behavior. *In* Bird Vocalization. R. Hinde, Ed.: 131-158. Cambridge University Press. London, England.
11. CHENG, M.-F. 1973. Effect of estrogen on behavior of ovariectomized ring doves (*Streptopelia risoria*). J. Comp. Physiol. Psychol. **83:** 234-239.
12. COHEN, J. & M.-F. CHENG. 1981. The role of the midbrain in courtship behavior of the female ring dove (*Streptopelia risoria*). Evidence from radiofrequency lesion and hormone implant studies. Brain Res. **207:** 279-301.
13. COHEN, J. & M.-F. CHENG. 1979. Role of vocalizations in the reproductive cycle of ring doves (*Streptopelia risoria*): effects of hypoglossal nerve section on the reproductive behavior and physiology of the female. Horm. Behav. **13:** 113-127.
14. CHENG, M.-F. Female cooing promotes ovarian development in ring doves. Physiol. Behav. In press.
15. PERRUZZI, E. 1984. Effects of Nest-Coo Vocalization on Follicular Development in Female Ring Doves (*Streptopelia risoria*). Senior Honor Project. Rutgers University. Newark, N.J.
16. NOTTEBOHM, F. & M. NOTTEBOHM. 1971. Vocalizations and breeding behavior of surgically deafened ring doves (*Streptopelia risoria*). Anim. Behav. **19:** 313-327.
17. GIBSON, M. J. & M.-F. CHENG. 1979. Neural mediation of estrogen-dependent courtship behavior in female ring doves. J. Comp. Physiol. Psychol. **93:** 855-867.
18. COHEN, J. 1983. Hormones and brain mechanisms of vocal behaviour in non-vocal learning birds. *In* Hormones and Behaviour in Higher Vertebrates. J. Balthazart, E. Pröve & R. Gilles, Eds.: 422-436. Academic Press. New York, N.Y.
19. AKESSON, T., N. DE LANEROLLE & M.-F. CHENG. Efferent pathways from the nucleus intercollicularis in the ring dove. Abstract presented at a meeting of the Society of Neuroscientists, November 6-11, 1983, Boston, Mass.
20. BONS, N., B. KERDELHUÉ & I. ASSENMACHER. 1978. Immunocytochemical identification of an LHRH-producing system originating in the preoptic nucleus of the duck. Cell Tissue Res. **188:** 99-106.
21. RODRIGUEZ-SIERRA, J. & M.-F. CHENG. Immunocytochemical identification of an LHRH-producing system in the ring dove (*Streptopelia risoria*). In preparation.
22. COHEN, D. H. & H. J. KARTEN. 1974. The structural organization of avian brain: an overview. *In* Birds: Brain and Behavior. I. J. Goodman & M. W. Schein, Eds.: 29-73. Academic Press. New York, N.Y.
23. KANDEL, E. R. 1981. Environmental determinants of brain architecture and of behavior: early experience and learning. *In* Principles of Neural Sciences. E. R. Kandel & J. H. Schwartz, Eds.: 621-632. Elsevier/North-Holland. Amsterdam, the Netherlands.

Social Induction of the Ovarian Response in the Female Ring Dove

CARL J. ERICKSON

*Psychology Department
Duke University
Durham, North Carolina 27706*

Contemporary research in behavioral endocrinology generally falls into three areas. The first and oldest of these is concerned with the activational effects of hormones on behavior, and the studies of Berthold[1] constitute the fundamental work in this area. A second field of investigation focuses on the organizing effects of hormones. The classic experiment by Phoenix et al.[2] provided the impetus for an extraordinary research effort in this area during the past quarter of a century. A third, more recent field to emerge is founded upon recognition of the fact that behavior represents more than a mere expression of neuroendocrinological activity. On the contrary, behavior is itself a powerful determinant of hormone secretion and of the ultimate activational and organizational effects that hormones may generate. Clearly, it was research conducted at the Institute of Animal Behavior of Rutgers University that was essential in establishing this third, important domain of inquiry within behavioral endocrinology.

In one of the first studies performed at the Institute, Lehrman et al.[3] showed that social stimulation from the male ring dove has a profound effect upon ovarian activity in the female. They placed female doves in breeding cages where they were maintained under a variety of stimulus conditions. Some were housed with a male mate, a nest bowl, and nesting material for 7 days; others remained in the cages for a 7-day period but the male or the male plus nesting material were removed before this interval was complete. Some females remained alone throughout the entire period. When females were killed for autopsy on the seventh day, their ovaries revealed striking differences that corresponded with the stimulus conditions with which they had been presented (FIGURE 1). When exposed to a male for 4 days or more, most females went on to ovulate 7 days later. On the other hand, when females were exposed to a male for only 2 or 3 days, their ovaries contained atretic follicles suggesting that some FSH stimulation had occurred but that it had been insufficient to produce a sustained ovarian response.

I mention this study for several reasons. First, it was one of the first large, well-controlled studies that attempted to manipulate social conditions and determine their effects on the ovarian response. Second, it suggested that although exposure to a male is important for the response, ovarian activity becomes independent of the male once a certain stage is attained. This is a point to which I will return later. Finally, this study led to a series of others at the Institute in which the nature of social input was examined more closely.

I came to the Institute in 1961, the year that Lehrman et al. published the paper I have just discussed, and I soon joined Daniel Lehrman in analyzing the nature of the male's contribution to the ovarian response of the female. Our first study[4] clearly indicated that the female responds to male behavior, not the mere presence of the male. Then later, in my doctoral dissertation, I showed that it is the androgen-dependent behavior of the male which is especially important for the effect. As the amount of androgen that I gave to a castrated male increased, the more likely it became that his female mate would ovulate within 7 days of their pairing.[5,6]

Several other graduate students and postdoctoral fellows at the Institute became interested in this problem during the mid-sixties. Lott and Brody[7] showed that female ring doves lay eggs in response to their mirror image and do so more readily when also exposed to sounds from the colony. Subsequently, Lott et al.[8] found that when

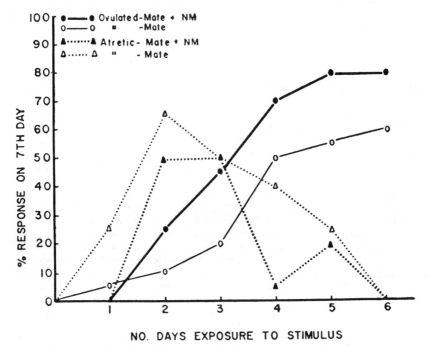

FIGURE 1. Number of birds ovulating (solid lines) and number of birds developing atretic follicles (broken lines) by the seventh day, as a function of duration of exposure to a male (light lines) or to a male and nesting material (heavy lines) during the first 0 to 6 days. (Reprinted from Reference 3 with the permission of the authors and the Society for Experimental Biology and Medicine.)

females were exposed to either castrated or intact males, the ovarian response was affected by colony sounds. The reproductive tracts from females that could see and hear a courting male but were deprived of colony sounds were no different from those of socially isolated females that could hear these sounds. During this same period Ronald Barfield joined the Institute as a postdoctoral fellow. He found that as little as 15 min of exposure to a male each day for 7 days was sufficient to produce significantly larger oviducts and ovarian follicles in colony-housed females.[9] During the early seventies other members of the Institute worked on this problem. Cheng[10] provided evidence that physical contact between male and female can contribute to the ovarian response, and Friedman[11,12] made what was, perhaps, the most intricate and ingenious attempt to understand the interaction between visual and auditory inputs in the female.

Surely the most interesting of the recent studies on these issues are those of Cohen and Cheng[13,14] indicating that feedback from the female's own vocal behavior plays a vital role in stimulating her ovarian activity. These studies are described elsewhere in this volume and will not be reviewed in detail here. Nonetheless, I wish to suggest several implications of their studies which, to my knowledge, have not been recognized.

IMPRINTING AND OVARIAN ACTIVITY

In 1913 Craig noted that when he made pets of some of his ring doves, they would direct their courtship and sexual behavior to him.[15] Other investigators, including Harth,[16] have also found this effect. Most interesting is the fact that female doves not only court their human keepers but they seem to lay eggs in response to them as well. In short, the imprinting includes neuroendocrinological changes as well as the reorientation of behavioral responses. This effect raises several interesting questions. Is there a sensitive period early in life during which the ovarian response becomes associated with the human keeper? And if so, is this period the same as or different from that in which overt behavioral responses become associated with a stimulus? If the periods for neuroendocrinological and behavioral imprinting are different, would it be possible to create a female ring dove that would court one object but lay eggs in response to another?

I think the work of Cohen and Cheng suggests answers to at least some of these questions, although they deserve experimental examination. If ovarian development is stimulated by the vocalizations associated with courtship in the female, then ovarian activity should occur in any situation in which the female nest-coos, even if she vocalizes in response to her human keeper. In other words, the neuroendocrine response is probably inextricably linked to female vocalization, and any associative learning during the developmental period should determine the stimulus for both of these responses rather similarly.

THE PRIMER EFFECT

Earlier I mentioned the study of Lehrman *et al.*[3] showing that it was not until females had received a minimum of 4 days of exposure to a male and nesting material that they would reach a stage where ovarian activity could continue in the absence of these priming conditions. In light of the studies of Cohen and Cheng it now seems likely that this initial period is necessary to establish the self-feedback phase of the cycle. Once under way, however, positive feedback can carry the system to ovulation without further stimulatory input from the male. Presumably, by the third or fourth day estrogen levels are sufficient to produce nest-cooing by the female, and this vocalization leads, through self-feedback, to still higher estrogen levels, with progesterone rising rapidly as well.[17]

MALE INHIBITION OF OVARIAN ACTIVITY

The fact that ovarian activity can continue in the absence of the male after the fourth day does not imply that the female is liberated from his influence at this time. On the contrary, I suggest that during the later stages of ovarian development, male behavior can inhibit this development. Moreover, I believe that in certain circumstances males inhibit ovarian development by preventing female vocalization and that they may benefit by doing so. Is there any evidence for this conclusion?

When I was a graduate student at the Institute I noted that if female doves were used too often for daily testing, they often engaged in nest-cooing within a few minutes of being placed with a male. I was surprised to find that this premature vocalization often elicited an attack from the male which resulted in a chase and, of course, cessation of the female display. In 1976 Patricia Zenone and I[18] confirmed this effect, showing that the same males, when exposed in turn to females that display immediately and those that do not, exhibit much more aggression toward the displaying females. What is the effect of this aggression? Two interpretations can be considered. First, the attacks may be a response to female mates that are not physiologically synchronized with the male. Or, second, the attacks may disrupt female nest-coo vocalization. This would prevent self-feedback and delay ovulation beyond the lifetime of any stored sperm the female received from earlier copulations with a competing male. These hypotheses are not mutually exclusive, and there is no reason a priori why the aggressive behavior could not contribute both to synchrony and to the defense of the male's genetic paternity. Several recent studies address these issues.

Emilie Rissman of Cornell University[19] performed an experiment to determine what would happen when male and female ring doves were both equally advanced into the cycle. In other words, she wished to find out whether a male would be more likely to attack a preexposed female than an unexposed female when he too was preexposed and presumably physiologically synchronized with the preexposed female. TABLE 1 summarizes her findings. Comparison of cells b and d confirms our earlier results; that is, unexposed males are much more aggressive toward preexposed females than toward unexposed females. Especially interesting here is her finding that, rather than being less aggressive, preexposed males are more than twice as aggressive toward preexposed females with which they are physiologically synchronized (i.e., comparison of cells a and c). (Unfortunately, Rissman provides no statistical comparison to determine whether this difference is significant, but the difference between medians is in a direction consistent with the paternity defense hypothesis.) In short, preexposed females are attacked when the male is in either physiological state.

Does male aggression delay ovulation and egg-laying? Rissman did not attempt to relate ovulation and egg-laying latencies with the aggression found in each group, but a recent unpublished study by Sims and Cheng[20] followed a similar procedure and included egg-laying latencies. They found (TABLE 2) that when an unmated male was introduced to a female that had spent 4 days with another male, egg-laying (and presumably ovulation) was significantly delayed compared to other types of male-female pairings. This was to be expected since previous studies had indicated that males would be exceptionally aggressive in such encounters. However, Sims and Cheng did not find a similar delay when preexposed females were introduced to preexposed males. These males should have behaved much like Rissman's most aggressive group (TABLE 1, cell a). But, according to Sims and Cheng, not only was there no significant delay in egg-laying, lower levels of aggression were found as well. Clearly, further research is required to resolve this discrepancy.

Zenone et al.[21,22] also found a relationship between male aggression and egg-laying latency, though their observations were performed in a context that differed from those in the studies described above. When a male ring dove discovers his fertile mate with another male, he often attacks her. Sometimes these attacks are severe and involve wing-slapping, a behavior more often seen in encounters between males than between male and female. Zenone[21] compared egg-laying latencies for females that were either slapped or not slapped by their mates and found that slapped females laid eggs more than a day later than females that were not slapped.

In 1976 Sims[23] performed a study at Duke University indicating that female doves will lay fertile eggs for up to 8 days following physical separation from a male mate. Her results were very similar to those found in a study by Riddle and Behre[24] in 1921. If we assume that egg-laying follows ovulation by about 36 to 40 hr, we can estimate maximum sperm storage to be about 5 or 6 days. Consequently, if male aggression postpones ovulation beyond this time, offspring will be fathered by the second male. Of course sperm competition may also play an important role. Several studies on birds suggest that the last male to copulate prior to ovulation fathers the

TABLE 1. Comparison of Aggressive Behavior of Male Ring Doves toward Preexposed and Unexposed Females

	Males	
	Preexposed	Unexposed
Females Preexposed	7.5^a (0-29)	6.5^b (0-28)
Unexposed	3.0^c (0-28)	2.5^d (0-22)

Note: Values given are the median number of times the male pecked the female; ranges are in parentheses. (From Rissman.[19])
  a vs. b, not significant.
  b vs. d, $p < 0.01$.

offspring.[25,26] The success of the second male is generally attributed to sperm competition, but it should be noted that the second male may also manipulate the timing of female ovulation to his own benefit. I suspect that the success of the second male depends upon the combination of delayed ovulation and sperm competition. It is quite likely that the success of fresh sperm increases with the age of prior sperm. By delaying ovulation and repeatedly inseminating the female, the second male increases his advantage. And if ovulation is delayed beyond the 5- or 6-day storage interval, his sperm will be the only viable gamete pool in the female's oviduct.

SOME CONCLUSIONS AND SPECULATIONS

During the past 25 years it has become abundantly clear that social interaction can produce changes in endocrine activity, and the ovarian response of the female ring dove has served as an excellent model for this effect. In the course of this research we have learned much about the preovulatory phase of the ring dove reproductive cycle, and from time to time our broader perspectives have changed. Fifteen years ago the central question seemed to be: What features of the male stimulate the ovarian response? Today the question takes a subtly different form. It might be put as: To what features does the female respond?

In the introduction to her talk at Bielefeld in 1982 Mei Cheng[27] argued that too much emphasis had been placed on the role of the male and that the role of the female had been too often ignored. I believe that she is correct in her view. In fact, I believe that as we continue to focus on the female, we will find several other examples of her active participation. For example, we still know almost nothing about the process whereby females begin to search for mates. Undoubtedly, physiological changes are involved in converting a female from a nonreproductive state to one in which she actively seeks out a partner. Moreover, having searched until encountering a male, we can presume that she then assesses his acceptability as a mate. Again, we know nothing about the physiological basis for this process. I suggest that it is while she is in this assessment stage that the positive qualities of the male are important. If she finds him acceptable, she enters the self-feedback stage and exhibits rapid ovarian

TABLE 2. Laying Latencies and Breeding Success of Synchronized and Nonsynchronized Pairs[a]

Group ($N = 12$)	Median Latency[b]	Successful Pairs
I. 4-day ♀ and 0-day ♂	7.7[c]	9
II. 0-day ♀ and 4-day ♂	6.1	9
III. 4-day ♀ and 4-day ♂	6.5	9
IV. 4-day ♀ and same 4-day ♂	6.1	10

[a] From Sims and Cheng.[20]
[b] Mann-Whitney U test ($p < 0.05$).
[c] Significantly different from Groups II-IV.

development. If he is unacceptable, ovarian development fails to occur. Note that this is a very different perspective than the one I held some years ago. Previously I would have said that courting males stimulate ovarian activity and that noncourting males do not. I now believe it is more correct to state that courting males are more likely to be acceptable to the female than are noncourting males—hence the difference in ovarian activity.

As Lehrman et al.[3] showed, ovarian activity and ovulation can occur in the absence of the male by about the fourth day. I suggest that the fourth day marks the end of the assessment phase and the commencement of the self-feedback phase of the cycle. During this period the male is no longer a positive influence, but through his aggression he can have a strong inhibitory effect on ovarian activity.

In summary, the female may play an active role in her ovarian activity in at least three ways: (1) by searching for a mate, (2) by assessing his qualities after he is found and then accepting or rejecting him following evaluation, and (3) by vocalizing and generating self-feedback for rapid development of her reproductive tract. Perhaps there are comparable processes in the male. In any event it is clear that even after 25 years of research on this problem many unanswered questions remain.

ACKNOWLEDGMENTS

I am grateful to M. Eleanor Sims and Mei-Fang Cheng for allowing me to present unpublished data in this paper. I am especially grateful to Jay S. Rosenblatt for many stimulating conversations over the years. The breadth of his knowledge and the depth of his understanding have had a profound impact on my education and my life.

REFERENCES

1. BERTHOLD, A. A. 1849. Transplantation der Hoden. Arch. Anat. Physiol. Wiss. Med. **16:** 42.
2. PHOENIX, C. H., R. W. GOY, A. A. GERALL & W. C. YOUNG. 1959. Organizing action of prenatally administered testosterone propionate in the tissues mediating mating behavior in the female guinea pig. Endocrinology **65:** 369-387.
3. LEHRMAN, D. S., R. P. WORTIS & P. BRODY. 1961. Gonadotropin secretion in response to external stimuli of varying duration in the ring dove *(Streptopelia risoria)*. Proc. Soc. Exp. Biol. Med. **106:** 298-300.
4. ERICKSON, C. J. & D. S. LEHRMAN. 1964. Effects of castration of male ring doves upon ovarian activity of females. J. Comp. Physiol. Psychol. **58:** 164-166.
5. ERICKSON, C. J. 1965. A Study of the Courtship Behavior of Male Ring Doves and Its Relationship to Ovarian Activity of Females. Unpublished Ph.D. Dissertation. Rutgers University. New Brunswick, N.J.
6. ERICKSON, C. J. 1970. Induction of ovarian activity in female ring doves by androgen treatment of castrated males. J. Comp. Physiol. Psychol. **71:** 210-215.
7. LOTT, D. S. & P. N. BRODY. 1966. Support of ovulation in the ring dove by auditory and visual stimuli. J. Comp. Physiol. Psychol. **62:** 311-313.
8. LOTT, D. S., S. D. SCHOLZ & D. S. LEHRMAN. 1967. Exteroceptive stimulation of the reproductive system of the female ring dove *(Streptopelia risoria)* by the mate and by the colony milieu. Anim. Behav. **15:** 433-437.
9. BARFIELD, R. J. 1971. Gonadotrophic hormone secretion in the female ring dove in response to visual and auditory stimulation by the male. J. Endocrinol. **49:** 305-310.
10. CHENG, M.-F. 1974. Ovarian development in the female ring dove in response to stimulation by intact and castrated male ring doves. J. Endocrinol. **63:** 43-53.
11. FRIEDMAN, M. B. 1977. Interactions between visual and vocal courtship stimuli in the neuroendocrine response of female doves. J. Comp. Physiol. Psychol. **91:** 1408-1416.
12. LEHRMAN, D. S. & M. FRIEDMAN. 1969. Auditory stimulation of ovarian activity in the ring dove. Anim. Behav. **17:** 494-497.
13. COHEN, J. & M.-F. CHENG. 1979. Role of vocalization in the reproductive cycle of ring doves *(Streptopelia risoria)*: effects of hypoglossal nerve section on the reproductive behavior and physiology of the female. Horm. Behav. **13:** 113-127.
14. COHEN, J. & M.-F. CHENG. 1981. The role of the midbrain in courtship behavior of the female ring dove *(Streptopelia risoria)*. Evidence from radiofrequency lesion and hormone implant studies. Brain Res. **207:** 289-301.
15. CRAIG, W. 1913. The stimulation and the inhibition of ovulation in birds and mammals. J. Anim. Behav. **3:** 215-221.
16. HARTH, M. S. 1970. The Effects of Developmental Experience on the Organization and Development of Reproductive Behavior Patterns in the Ring Dove *(Streptopelia risoria)*. Unpublished Ph.D. Dissertation. Rutgers University. New Brunswick, N.J.
17. SILVER, R., C. REBOULLEAU, D. S. LEHRMAN & H. H. FEDER. 1974. Radioimmunoassay of plasma progesterone during the reproductive cycle of male and female ring doves *(Streptopelia risoria)*. Endocrinology **94:** 1547-1554.
18. ERICKSON, C. J. & P. G. ZENONE. 1976. Courtship differences in male ring doves: avoidance of cuckoldry? Science **192:** 1353-1354.

19. RISSMAN, E. F. 1983. Detection of cuckoldry in ring doves. Anim. Behav. **31:** 449-456.
20. SIMS, M. E. & M.-F. CHENG. Unpublished study.
21. ZENONE, P. G. 1979. Protection of Genetic Paternity in the Ring Dove *(Streptopelia risoria)*. Unpublished Ph.D. Dissertation. Duke University. Durham, N.C.
22. ZENONE, P. G., M. E. SIMS & C. J. ERICKSON. 1979. Male ring dove behavior and the defense of genetic paternity. Am. Nat. **114:** 615-626.
23. SIMS, M. E. 1976. The Capacity for Sperm Storage in the Female Ring Dove *(Streptopelia risoria)*. Unpublished Honors Thesis. Duke University. Durham, N.C.
24. RIDDLE, O. & E. H. BEHRE. 1921. Studies on the physiology of reproduction in birds. IX. On the relation of stale sperm to fertility and sex in ring-doves. Am. J. Physiol. **57:** 228-249.
25. WARREN, D. C. & L. KILPATRICK. 1929. Fertilization in the domestic fowl. Poult. Sci. **8:** 237-256.
26. SIMS, M. E., G. BALL & M.-F. CHENG. Unpublished manuscript.
27. CHENG, M.-F. 1982. "Behavioral feedback" control of the endocrine state. Paper presented at the 4th Conference of the European Society for Comparative Physiology and Behavior, Bielefeld, FRG.

The Effects of Copulatory Behavior on Sperm Transport and Fertility in Rats

NORMAN T. ADLER AND JAMES PATRICK TONER, JR.

Department of Psychology
University of Pennsylvania
Philadelphia, Pennsylvania 19104

INTRODUCTION

Judging from the pervasive use of contraception in our own species, one might guess that getting pregnant is a simple process. From the point of view of biological mechanism and function, this is far from the truth. While high premiums are paid to a system which reliably produces pregnancy under the right conditions, severe penalties are paid for pregnancy under unfavorable circumstances.

The adaptive balancing of *easy* pregnancy with *prudent* pregnancy has produced a complex set of controls which regulate both the opportunity to mate and the effects of mating. These controls exist at many levels of biological organization, from molecular to social. This principle of integration at multiple levels of biological organization mandates that students of reproductive behavior adopt the corresponding strategy of integrative, multidisciplinary investigation—precisely the approach that Jay Rosenblatt has developed so elegantly, interestingly, and influentially.

In sexually reproducing species, sperm and ova must meet for fertilization to occur. In mammals, the union of egg and sperm typically requires sperm to pass from the male into the female's reproductive tract where they must pass, successively, through her cervix, uterus, uterotubal junction, and a length of oviduct. Despite the importance of these processes for fertility (and infertility), many investigators find the present understanding of the process incomplete.[1-5]

One of the ways to further our understanding of mammalian fertility is to analyze the role of copulatory behavior in this process. It is now well documented that *behavior influences sperm transport and fertility* in species spanning a broad taxonomic range.[6-22] In recent years, we have investigated the behavioral control of transcervical and transuterine sperm transport in terms of its underlying *mechanisms* and its *functional significance*. This work proceeds on three levels of biological organization: the *organismic* level, the *social* level, and the *physiological* level.

Although this narrative is something of a case study, concentrating on the ways in which behavior controls pregnancy in the rat, we think the biopsychological principles uncovered are rather general, with potential relevance for understanding problems of human fertility and infertility.

OUTLINE: SEXUAL BEHAVIOR AND THE INDUCTION OF PREGNANCY

The copulatory pattern of rats consists of a series of 10 to 15 intromissions culminating in ejaculation, during which the pair remains *immobile* for a number of seconds. At ejaculation, the male deposits sperm and a coagulating "copulatory plug" in the vagina. This plug lodges tightly up against the cervix and holds the sperm in contact with the channels in the cervix.[16,23]

The sequence of intromissions and ejaculation constitutes an *ejaculatory series*. This series is followed by a pause in the mating (the *postejaculatory interval* or PEI). The composite pattern of ejaculatory series and PEI is repeated on the average of seven times for a given pair of rats in one mating session.

The stimulation derived from this mating sequence controls the onset of subsequent pregnancy. It does so by influencing the *probability of becoming progestational* (i.e., secreting progesterone) and by *transporting sperm through the reproductive tract* into the uterus.[10,16,23–27] While both the hormonal basis of gestation and gamete transport are necessary components of pregnancy induction (and hence neither is sufficient), in the remainder of this paper we will concentrate on the latter aspect—sperm transport.

Although early workers thought that sperm were deposited directly in the uterus of the female rat at ejaculation,[23,28] their interpretation was based on qualitative estimates of sperm numbers in the uterus. Moreover, the earliest examination of the uterus made by these investigators was on the order of 1 min after ejaculation,[28] which at least leaves open the question of how many sperm were transported in the first minute following ejaculation.

More modern data, based on *quantitative* analysis of sperm transport over a more extended temporal range of the PEI, show transcervical sperm transport to be gradual.[16] In fact, several reviewers believe that convincing evidence has not been found for direct intrauterine deposition of sperm in any species.[29,30]

In the rat there are several physiological requirements for transcervical sperm transport to occur. First, the reproductive system of the female rat must be *primed* by pre-ejaculatory factors; second, the *form and placement of the copulatory plug* at ejaculation must be adequate[23]; and third, the plug's position must not be *disrupted by exogenous stimulation* during the first minutes after ejaculation.[16]

All three of these physiological requirements for successful sperm transport in the female rat are influenced by copulatory behavior. In the remainder of this paper, we will discuss the reciprocal relationship between behavior and physiology in initiating sperm output, transport, and subsequent fertilization. In the next section, we will discuss the influence of the individual behavioral components on sperm transport.

BEHAVIORAL INITIATION OF SPERM TRANSPORT

Pre-ejaculatory Behavior: Female Effects

Females require "priming" by the male rat's pre-ejaculatory intromissions if they are to transport optimal numbers of sperm following ejaculation.[10,16] Without two or more intromissions, sperm transport and pregnancy do not occur.[10] Although we do not yet understand the precise mechanism by which the intromissions exert their effect, our tentative hypothesis is that the mechanism may be humoral, since the pre-ejaculatory intromissions may be stored for a considerable period of time.

Pre-ejaculatory Behavior: Male Effects

Although we have known for some time that pre-ejaculatory intromissions affect the *female's capacity to transport sperm,* we have recently obtained evidence that the number of intromissions a male rat delivers *affects his ejaculatory performance.*[31]

To obtain these data, we allowed males and females different copulatory experiences by switching females during mating. Thirty-four virgin females were mated to sexually rested male rats. During 31 of these mating tests, the male was switched from a stimulus female to the experimental female in the course of his ejaculatory series. The switch was performed at random points in the series. Consequently, the experimental male and female experienced different numbers of pre-ejaculatory intromissions. Some females had as few as one pre-ejaculatory intromission and some males as few as three.

Reduced numbers of sperm were recovered when the *female received fewer than two intromissions* and/or when the *male had fewer than six intromissions* overall. Thus, the presence of sperm in the female rat's genital tract depended upon *both* (a) the number of intromissions the female received and (b) the number the male delivered. These relationships are illustrated in FIGURE 1.

These data confirm our previous results showing that female rats require pre-ejaculatory intromissions to transport sperm into the uterus.[10,16,17] The present data also show that there is a clear-cut threshold in females: sperm recovery was always meager when she received fewer than two intromissions and was adequate if she received three or more intromissions.

In contrast to the effect on females, the role of the stimulation the male receives from the intromissions he delivers has not been previously documented. We found that *few sperm were recovered when males had fewer than six intromissions,* even when stimulation to the female was above her threshold (i.e., she received more than two intromissions). When males had six or more intromissions, many sperm were usually recovered. This transition from poor to good sperm recovery was not, however, as abrupt as that in the female.

These self-stimulatory effects of the male rat's behavior on his own sperm output could have functional consequences. First, it is well known that the number of pre-ejaculatory intromissions among male rats is variable. Second, this variability in the number of intromissions is not a constant organismic variable between individual males but varies as a function of environmental and social contingencies. Experimentally, it is a simple matter to vary the number of intromissions by a variety of techniques.[32-34] More to the functional orientation of the present discussion, there is an ecological relationship between *social factors* and *reproductive function.* Subdominant males have fewer than half the number of pre-ejaculatory intromissions that dominant males have.[35] In a related finding, subdominant male rats seem to have depressed fertility in the presence of the dominant conspecific (but normal fertility when mating in the absence of the dominant male[36]). We suggest that in a natural context subdominant males would take fewer intromissions to ejaculate,[35] and thus would produce a suboptimal ejaculate. This reduced ejaculatory potency, whether manifested as fewer sperm, a smaller plug, or reduced ejaculate volume, could lead to reduced fertility.[31,36]

The relationship between a male's pre-ejaculatory behavior and his subsequent ejaculatory performance *may be a general feature of mammalian reproductive physiology, including humans.* In the field of animal husbandry, collection of sperm into an artificial vagina is enhanced by some type of "sexual preparation," either in the form of "false mounts" (mounts in which intromission is prevented) or by simple exposure to a female before the ejaculate is collected by the experimenter. In rabbits and bulls that were not sexually prepared, second ejaculates had more sperm than first ejaculates. Providing three false mounts to rabbits nearly quadrupled the sperm

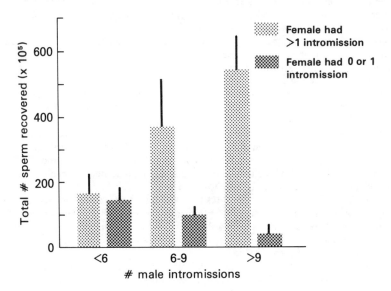

FIGURE 1. The total number of sperm recovered from the genital tract of female rats (uterine and vaginal counts) after different numbers of pre-ejaculatory intromissions. Lightly stippled bars indicate cases in which the female had two or more intromissions; darkly stippled bars indicate cases in which the female had zero or one pre-ejaculatory intromission. Thin bars indicate the SEM.

output of the first ejaculate, and increased sperm output of the second ejaculate by one-third.[37] These increases were more pronounced for the sperm fraction than the accessory gland fractions. Beef and dairy bulls also responded to three false mounts by increases in semen volume, sperm concentration, and number of sperm at the first ejaculation[38]; beef bulls also responded to one false mount and 5 min of active restraint by an increase in the percentage of sperm that were motile.[39]

There are some preliminary data to support the idea that a similar (at least analogous) phenomenon occurs in human males (A. DeCherney, personal communication to J.P.T.). If so, this finding may have application in cases of marginal male fertility as a means to enhance the chance of pregnancy.

These comparative data highlight the widespread taxonomic occurrence of the male's stimulatory effect on his own reproductive output. The phenomenon appears to be a general one as well from the point of view of behavioral dynamics for it illustrates what Cheng has called *behavioral self-feedback*.[40] This term refers to an *animal's behavior influencing its own subsequent behavior or physiology.* Other examples include the self-licking in female rats which promotes mammary development[41]; the nest-coos of female ring doves which stimulate normal follicular development[42]; and the vocalizations of young sparrows which are necessary for normal adult singing.[43]

Self-stimulation is often an efficient mechanism for an animal to regulate activities which cannot be coordinated by purely endogenous means. For example, egg-laying needs to be coordinated with the availability of an adequate nest and cannot depend solely on internal indices. Canaries check the environment directly, for tactile contact made with a completed nest allows egg-laying to occur.[44] In contrast, egg-laying in ring doves follows a certain amount of nest-building behavior, independent of the condition of the nest.[45] In both of these cases, the adequacy of the external environment for receiving the precious egg is assured by making sure that the external world has been prepared—either by checking it directly (as in the canary) or by using a "behavioral token" like nest-building activity (as in the ring dove).

Ejaculatory Behavior

Once the female is primed by the precopulatory intromissions, she is ready to receive the sperm. Mere deposition of the sperm in the vagina is, however, not sufficient—*the behavior of the pair at ejaculation influences fertility.* More sperm are transported into the uterus when the pair of rats remains immobile at ejaculation with close pelvic contact for a second or more after the final ejaculatory thrust.[16] This beneficial effect of ejaculatory immobility might be mediated by better placement of the vaginal plug against the cervix[16] since (a) the better the plug conforms to the shape of the cervix, the more sperm are transported[17] and (b) the placement of the plug is a function of the duration of the pair's ejaculatory immobility response. That is, pairs that remained immobile at ejaculation had plugs with tighter fits; and tighter-fitting plugs transport more sperm.[16-18]

Postejaculatory Behavior

The PEI is a crucial period in the reproductive life of the female rat, because intromissions delivered too soon after an ejaculation are able to reduce sperm transport, sometimes by two orders of magnitude, thereby disrupting fertility.[12,16] Postejaculatory intromissions may exert their inhibitory effects by *mechanically dislodging the vaginal plug.*[17]

The most critical postejaculatory period for this disruption is the *first 6 min* following ejaculation.[12] This length of time corresponds to the male rat's PEI. We have obtained a fair amount of evidence that this important period of reproductive quiescence in the male is assured by an active *inhibitory* state.[46-48] We conclude that one function of this state of inhibition is to prevent the male from "prematurely" resuming mating. This in turn would permit the female sufficient time without copulatory stimulation to complete sperm transport.

The periodic action of inhibition is not specific to the male rat's sexual behavior. Although the point is so basic that it is at times overlooked, a major topographic aspect of behavior is its *temporal organization*. In this sequencing of responses, behavioral inhibition is a potent and ubiquitous mechanism for "turning off" behavior at the appropriate point.

Perhaps because of the pervasiveness of behavior inhibition, we should not have been surprised to find an analogous example of lowered sexual responsiveness during the female rat's postejaculatory interval.[49,50] In the male, the postejaculatory inhibition provides a pause in the stream of copulatory behavior, helping to reconcile the *conflict* between (a) the male's producing sufficient sperm (from prolonged copulation and multiple ejaculatory series) and (b) permitting the female rat's reproductive system the quiescence to "process" his sperm from the immediately preceding ejaculation. In her copulatory behavior, the female rat also has a conflict.

The copulating female also faces a strategic conflict: her survival depends upon both her *mating and mothering*. Davis and Hall[51] have determined that approximately 50% of all pregnancies in populations of wild rats result from *postpartum* mating. (Shortly after giving birth to one litter, the female rat becomes receptive and can copulate, thus initiating a second litter.) Since female rats will spend approximately 80% of their time caring for their litters,[52-54] this presents a problem: time taken *to mate* is *time taken away from mothering*, and vice versa. To investigate how the female handles this potential conflict, we studied the occurrence and temporal organization of maternal and copulatory behavior in postpartum female rats.[49,50] We found that a strategic compromise exists in which the female exhibits both mating and mothering, alternating between the two in the manner that McFarland[55] has called "behavioral time-sharing." While time-sharing is not as elementary a mechanism as that controlling the male's PEI, it does allow for the orderly alternation of two behavior patterns, with the dominant pattern allowing the other to appear only in its wake.

In the case of postpartum mating in the rat, copulatory behavior seems to be the dominant behavior—since the female virtually always completed an ejaculatory series before returning to the next. (*Only 2 of the 32 series* were interrupted; the preponderance of maternal behavior occurred during the postejaculatory intervals.) While "sexual motivation" remains high *during* the ejaculatory series (preventing the expression of maternal behavior), it declines *after* ejaculation, during the PEI.

Thus, it seems that in both sexes of this species, the postejaculatory interval is a "sexually protected" period. Both the male and the female entered an altered (and reduced) state of copulatory readiness during the critical time when sperm must travel from the vagina into the uterus.

SOCIAL FACTORS IN THE CONTROL OF SPERM TRANSPORT

Sperm Transport: Inhibition

In the previous sections we have described the effects of copulation on sperm transport, concentrating on the *organismic level* (a single copulating pair, for a single ejaculation). As we pointed out earlier, though, behavior has consequences on other levels (both higher, social levels and lower, physiological levels). In this section, we would like to explore some of the consequences of sperm transport on the *social* level.

Of course, sexual behavior is, by definition, social since it involves at least two animals—the male and the female. The social dimension in our case study, however, applies to mating with several males (and females) that are normally present in the immediate social environment of the copulating pair.

To begin this analysis we would like to reconsider the postejaculatory phase of the mating encounter. Although the ejaculating male's own PEI may reflect an active physiological process that *protects his own sperm* from disruption by his too-quick resumption of mating, there is still the possibility that another male could interfere with the transport of the first male's sperm. We have already seen that one male *can* "cancel" the sperm transport of a male that has just ejaculated, substituting his own sperm instead.[12,46]

This inhibitory effect of one male's copulatory behavior on the reproductive success of another is an example of what has been broadly called "sperm competition."[56] One molar definition of sperm competition is: the set of behavioral and physiological mechanisms by which the probability of one male's sperm fertilizing the available ova is increased at the expense of another.

Sperm Transport: Protection

Like other dyadic relationships in biology (e.g., prey-predator relations), sperm competition between males involves pressures from both sides of the relationship. Thus, while the copulatory disruption of sperm transport is an adaptive mechanism by which one male might cancel the effects of a previous male's mating, the "disadvantaged" male might be expected to generate a set of responses (secondary adaptations) to *prevent* his sperm from being displaced, or at least to mitigate the effect of displacement. Though the mechanism is not yet clear, there is evidence that this might be occurring in rats: Following ejaculation by a dominant male, the female waits twice as long before resuming mating than if the ejaculation was delivered by a male of lower social status.[57] Thus, one way in which a male (at least a dominant male) can protect his gametic investment is to *modify the behavior of the female,* so as to protect the sperm he has just ejaculated from being dislodged by a second (subdominant) male.

In addition to protecting his sperm by modifying the behavior of the female, there is a second way in which a dominant male can ensure his sperm's success against that of a genetic adversary: by reducing the gametic output of the subordinate male, as discussed in the previous section.

A third, related strategy by which a male might protect his investment of sperm is to increase the *amount* or *quality* of copulatory contact with a given female. Lanier et al.[58] have shown that, in a competitive mating situation, male rats increase the proportion of pups that they sire when they deliver a greater number of ejaculatory series, relative to another male. They concluded that the *proportion* of sperm from each male is important in determining reproductive advantage.

In determining the final proportion of sperm that a female receives from several males, the total *number* of ejaculations is surely important. We have recently discovered that there is another major factor operating, viz., the *particular ejaculatory series* which is (are) delivered.[59]

In studying the fertility of successive ejaculatory series, we were interested in determining the basic, physiological parameters of reproduction in this species which is characterized by multiple ejaculatory series. There is another, more ecological reason for pursuing this research, though. The concepts of sperm competition and mating strategy play a central theoretical role in behavioral ecology and sociobiology. Often, these theoretical considerations are pursued with great formal clarity. In addition to this theoretical rigor, however, basic empirical data are needed to adequately evaluate competing theoretical hypotheses about the biopsychology and behavioral ecology of reproductive processes. So although it is important to consider the effects of multiple ejaculations in a social setting with several males,[20,60] it would also seem important to work out the basic pattern of fertility supported by each ejaculation in order to satisfactorily elaborate the evolutionary consequences of complex mating patterns.

Thirteen male and 276 female Sprague-Dawley CD rats (Charles River) were used in this experiment. Two groups of males were randomly selected from our pool of sexually experienced animals: an older group ($n=9$), 223 to 402 days old at the beginning of the study, and a younger group ($n=4$), 103 to 172 days old.

After 2 weeks of rest, a given male was mated successively to 6 females (in the case of the older males) or 7 females (in the case of the younger males). Within a minute after each ejaculation the female was removed and replaced with another estrous female. Females mated to the younger males were given five preliminary intromissions by a "stimulus" male before being mated to the experimental male. (These priming intromissions increased the chance that any reduction in fertility would *not be due* to a lack of progesterone secretion and would instead *depend on* reduced numbers of sperm.) One group of females was examined for numbers of sperm; another group (mated to the same males) was allowed to become pregnant, and the number of embryos was counted. The results are illustrated in FIGURES 2 and 3.

Overall, different ejaculatory series had different fertilizing capacities. The precise pattern depended on the age of the male. *Younger* male rats displayed two different patterns of sperm output, one in which initial ejaculations contained the most sperm, and another in which the second or third ejaculation was maximal for sperm. Despite these different patterns of sperm output, embryo numbers remained high until the

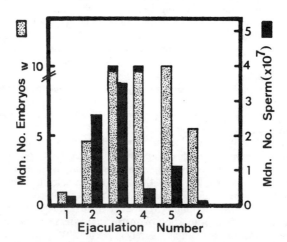

FIGURE 2. Median number of embryos (stippled bars) and sperm (black bars) resulting from successive ejaculations delivered in one mating session by older males ($n=9$). Number of observations per bar from left to right: 17, 18, 18, 16, 18, 16, 16, 16, 6, 15, 16, 10. When the median number of pups was more than 10, the corresponding bar was topped with a solid black section.

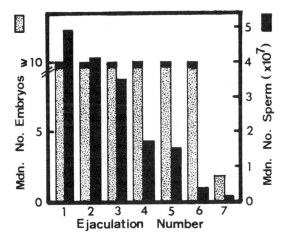

FIGURE 3. Median number of embryos (stippled bars) and sperm (black bars) resulting from successive ejaculations delivered in one mating session by younger males ($n=4$). The number of observations per bar was 7 for sperm and 4 for embryos. All males completed all seven ejaculatory series in each mating session. When the median number of pups was more than 10, the corresponding bar was topped with a solid black section.

seventh ejaculation, when the number of ejaculated sperm was smallest. Thus in young male rats, the number of ejaculated sperm was not a limiting factor in fertility until the seventh ejaculation.

Older males typically had initial ejaculations which were less potent than middle ejaculations, both in terms of sperm and embryo number. These results reflect a preliminary observation of Zucker and Wade[61] that in their one test of a male being mated to six females successively, only the first ejaculation produced no litter. Like the younger males, the last ejaculation had the fewest sperm and reduced fertility.

Modeling of optimal male strategies for distributing ejaculations among females will require evaluation of the potency of different ejaculations. The trend in recent reports, covering a range of species, suggests that sperm output varies with successive ejaculations, yet this variation is rarely reflected in variable fertility (see review in Reference 62). The maintenance of fertility in the face of reduced numbers of sperm has led to the notion that males act interested in mating (i.e., males are behaviorally potent) only when they are physiologically potent as well.[60,63] The present data at least partially contradict this principle. Older males showed a coupling of behavioral and physiological potency mainly in their middle series, and younger males mated but produced no litters with their very late series. While fertility was variable over series, males often acted more potent than they were. It should be interesting to work out the functional implications of this phenomenon.

CONCLUSION

In this paper, we have described one small component in the process by which females become pregnant: the behavioral control of sperm transport from the male, through the reproductive tract of the female, to the point where sperm and egg unite. We have concentrated on the behavioral (and social) levels of organization and have tried to show how physiological parameters underlying successful reproduction are regulated by higher levels. This process, in one species, represents a kind of microcosm reflecting the behavioral-physiological integration that occurs across a wide spectrum of behavior patterns and species and therefore has general applicability in biopsychological thought. In developing this research program, much of our thinking has been illuminated by our contacts, social and intellectual, with Jay Rosenblatt over many years. We hope we have many more years of instruction from him.

REFERENCES

1. KELLY, J. V. 1962. Myometrial participation in human sperm transport: a dilemma. Fert. Steril. **13:** 84.
2. FOX, C. & B. FOX. 1967. Uterine suction during orgasm. Br. Med. J. **1:** 300-301.
3. MOGHISSI, K. S. 1969. Sperm migration in the human female genital tract. J. Reprod. Med. **3:** 73.
4. HAFEZ, E. S. E. 1973. Transport of spermatozoa in the female reproductive tract. Am. J. Obstet. Gynecol. **115:** 703.
5. SETTLAGE, D. S. F., M. MOTOSHIMA & D. TREDWAY. 1973. Sperm transport from the external cervical os to the fallopian tubes in women: a time and quantitation study. Fertil. Steril. **24:** 655.
6. MASTERS, W. H. & W. E. JOHNSON. 1966. Human Sexual Response. Little, Brown & Co. Boston, Mass.
7. DEWSBURY, D. A. 1967. A quantitative description of the behaviour of rats during copulation. Behaviour **29:** 154-178.
8. DEWSBURY, D. A. 1981. On the function of the multiple-intromission, multiple-ejaculation copulatory patterns of rodents. Bull. Psychon. Soc. **18:** 221-223.
9. ADLER, N. T. 1968. Effects of the male's copulatory behavior in the initiation of pregnancy in the female rat. Anat. Rec. **160:** 304.
10. ADLER, N. T. 1969. Effects of the male's copulatory behavior on successful pregnancy of the female rat. J. Comp. Physiol. Psychol. **69:** 613-622.
11. ADLER, N. T. 1983. The neuroethology of reproduction. *In* Advances in Vertebrate Neuroethology. J. Ewert, R. Capranica & D. Ingle, Eds.: 1033-1061. Plenum Press. London, England.
12. ADLER, N. T. & S. R. ZOLOTH. 1970. Copulatory behavior can inhibit pregnancy in female rats. Science **168:** 1480-1482.
13. DEVINE, M. C. 1975. Copulatory plugs in snakes: enforced chastity. Science **187:** 844-845.
14. DEVINE, M. C. 1977. Copulatory plugs, restricted mating opportunities, and reproductive competition among male garter snakes. Behaviour **267:** 154-178.
15. MARTAN, J. & B. A. SHEPHERD. 1976. The role of the copulatory plug in reproduction of the guinea pig. J. Exp. Zool. **196:** 79-84.
16. MATTHEWS, M. & N. T. ADLER. 1977. Facilitative and inhibitory influences of reproductive behavior on sperm transport in rats. J. Comp. Physiol. Psychol. **91:** 727-741.
17. MATTHEWS, M. & N. T. ADLER. 1978. Systematic interrelationship of mating, vaginal plug position, and sperm transport in the rat. Physiol. Behav. **20:** 303-309.
18. MATTHEWS, M. & N. T. ADLER. 1979. Relative efficiency of sperm transport and number of sperm ejaculated in the female rat. Biol. Reprod. **20:** 540-544.
19. ROSS, P. & D. CREWS. 1977. Influence of the seminal plug on mating behaviour in the garter snake. Nature **267:** 344-345.
20. DEWSBURY, D. A. & T. G. HARTUNG. 1980. Copulatory behavior and differential reproduction of laboratory rats in a two-male, one-female competitive situation. Anim. Behav. **28:** 95-102.

21. DEWSBURY, D. A. & D. J. BAUMGARDNER. 1981. Studies of sperm competition in two species of muroid rodents. Behav. Ecol. Sociobiol. **9:** 121-133.
22. ALLEN, T. O. & N. T. ADLER. 1982. The neuroendocrine consequences of sexual behavior in mammals. *In* Reproduction. N. T. Adler & D. Pfaff, Eds. Plenum Press. New York, N.Y.
23. BLANDAU, R. J. 1945. On the factors involved in sperm transport through the cervix uteri of the albino rat. Am. J. Anat. **77:** 253-272.
24. WILSON, J. R., N. ADLER & B. LEBOEUF. 1965. The effects of intromission frequency on successful pregnancy in the female rat. Proc. Natl. Acad. Sci. USA **53:** 1392-1395.
25. ADLER, N. T., J. A. RESKO & R. W. GOY. 1970. The effect of copulatory behavior on hormonal change in the female rat prior to implantation. Physiol. Behav. **5:** 1003-1007.
26. CHESTER, R. V. & I. ZUCKER. 1970. Influence of male copulatory behavior on sperm transport, pregnancy, and pseudopregnancy in female rats. Physiol. Behav. **5:** 35-43.
27. TERKEL, J. & C. H. SAWYER. 1978. Male copulatory behavior triggers nightly prolactin surges resulting in successful pregnancy in rats. Horm. Behav. **11:** 304-309.
28. HARTMAN, C. G. & J. BALL. 1931. On the almost instantaneous transport of spermatozoa through the cervix and uterus in the rat. Proc. Soc. Exp. Biol. Med. **28:** 312.
29. FREUND, M. 1973. Mechanisms and problems of sperm transport. *In* The Regulation of Mammalian Reproduction. S. J. Segal, R. Crozier, P. A. Cofman & P. G. Condiffe, Eds. C. C. Thomas. Springfield, Ill.
30. OVERSTREET, J. W., G. W. COOPER & D. F. KATZ. 1978. Sperm transport in the reproductive tract of the female rabbit. 2. The sustained phase of transport. Biol. Reprod. **19:** 115-132.
31. TONER, J. P. & N. T. ADLER. 1986. The pre-ejaculatory behavior of male and female rats affects the number of sperm in the vagina and uterus. Physiol. Behav. **36:** 363-367.
32. BERMANT, G. 1964. Effects of single and multiple enforced intercopulatory intervals on the sexual behavior of male rats. J. Comp. Physiol. Psychol. **57:** 398-403.
33. LARRSON, K. 1959. The effect of restraint upon copulatory behaviour in the rat. Anim. Behav. **7:** 23-25.
34. SILBERBERG, A. & N. T. ADLER. 1974. Modulation of the copulatory sequence of the male rat by a schedule of reinforcement. Science **185:** 374-375.
35. MCCLINTOCK, M. K., J. J. ANISKO & N. T. ADLER. 1982. Group mating among Norway rats. 2. The social dynamics of copulation: competition, cooperation, and mate choice. Anim. Behav. **30:** 410-425.
36. COSTANZO, D. J. & R. K. ORNDOFF. 1984. Reproductive success of dominant versus subdominant male rats in a competitive mating situation. Abstract presented at Conference on Reproductive Behavior, Pittsburgh, Pa.
37. MACMILLAN, K. L. & H. D. HAFS. 1967. Semen output of rabbits ejaculated after varying sexual preparation. Proc. Soc. Exp. Biol. Med. **125:** 1278-1281.
38. ALMQUIST, J. O. 1973. Effects of sexual preparation on sperm output, semen characteristics and sexual activity of beef bulls with a comparison to dairy bulls. J. Anim. Sci. **36:** 331-336.
39. FOSTER, J., J. O. ALMQUIST & R. C. MARTIG. 1970. Reproductive capacity of beef bulls. IV. Changes in sexual behavior and semen characteristics among successive ejaculations. J. Anim. Sci. **30:** 245-252.
40. CHENG, M.-F. 1983. Behavioural "self-feedback" control of endocrine states. *In* Hormones and Behaviour in Higher Vertebrates. J. Balthazart, E. Prove & R. Gilles, Eds. Springer-Verlag. West Berlin, FRG.
41. ROTH, L. & J. S. ROSENBLATT. 1968. Self-licking and mammary development during pregnancy in the rat. J. Endocrinol. **42:** 363-378.
42. COHEN, J. & M.-F. CHENG. 1979. Role of vocalization in the reproductive cycle of ring doves (*Streptopelia risoria*): effects of hypoglossal nerve section on the reproductive behavior and physiology of the female. Horm. Behav. **13:** 113-127.
43. MARLER, P. 1970. A comparative approach to vocal learning: song development in white-crowned sparrows. J. Comp. Physiol. Psychol. **71:** 1-25.
44. HINDE, R. A. & R. P. WARREN. 1959. The effect of nest building on later reproductive behaviour in domesticated canaries. Anim. Behav. **7:** 35-41.

45. CHENG, M.-F. & J. BALTHAZART. 1982. The role of nest-building activity in gonadotrophin secretions and the reproductive success of ring doves (*Streptopelia risoria*). J. Comp. Physiol. Psychol. **96:** 307-324.
46. KURTZ, R. & N. T. ADLER. 1973. Electrophysiological correlates of sexual behavior in the male rat. J. Comp. Physiol. Psychol. **84:** 225-239.
47. ADLER, N. T. 1974. The behavioral control of reproductive physiology. *In* Reproductive Behavior. M. Montagna & W. A. Sadler, Eds. Plenum Press. New York, N.Y.
48. ANISKO, J. J., S. F. SUER, M. K. MCCLINTOCK & N. T. ADLER. 1978. Relation between 22 kHz ultrasonic signals and sociosexual behavior in the rat. J. Comp. Physiol. Psychol. **92:** 821-829.
49. GILBERT, A. N., R. J. PELCHAT & N. T. ADLER. 1980. Postpartum copulatory and maternal behaviour in Norway rats under seminatural conditions. Anim. Behav. **28:** 989-995.
50. GILBERT, A. N., R. J. PELCHAT & N. T. ADLER. 1984. Sexual and maternal behaviour at the postpartum oestrus: the role of experience in time-sharing. Anim. Behav. **32:** 1045-1053.
51. DAVIS, H. N., & O. HALL. 1951. The seasonal reproductive conditions of female Norway (Brown) in Baltimore Maryland. Physiol. Zool. **24:** 9-20.
52. GROTA, R. L. & R. ADER. 1969. Continuous recording of maternal behavior in *Rattus norvegicus*. Anim. Behav. **17:** 722-729.
53. ADER, R. & R. J. GROTA. 1970. Rhythmicity in the maternal behavior of *Rattus norvegicus*. Anim. Behav. **18:** 144-150.
54. LEON, M., P. G. CROSKERRY & G. K. SMITH. 1978. Thermal control of mother-young contact in rats. Physiol. Behav. **21:** 793-811.
55. MCFARLAND, D. 1974. Time-sharing as a behavioral phenomenon. Adv. Study Behav. **5:** 201-225.
56. PARKER, G. A. 1970. Sperm competition and its evolutionary consequences in the insects. Biol. Rev. **45:** 525-547.
57. MCCLINTOCK, M. K., J. P. TONER, J. J. ANISKO & N. T. ADLER. 1982. Group mating among Norway rats. II. The social dynamics of copulation: competition, cooperation, and mate choice. Anim. Behav. **30:** 410-425.
58. LANIER, D. L., D. Q. ESTEP & D. A. DEWSBURY. 1979. Role of prolonged copulatory behavior in facilitating reproductive success in a competitive mating situation in laboratory rats. J. Comp. Physiol. Psychol. **93:** 781-792.
59. TONER, J. P. & N. T. ADLER. 1985. Potency of rat ejaculations varies with their order and with male age. Physiol. Behav. **35:** 113-115.
60. DEWSBURY, D. A. 1982. Ejaculate cost and male choice. Am. Nat. **119:** 601-610.
61. ZUCKER, I. & G. WADE. 1968. Sexual preferences of male rats. J. Comp. Physiol. Psychol. **66:** 816-819.
62. DEWSBURY, D. A. & D. K. SAWREY. 1985. Male capacity as related to sperm production, pregnancy initiation, and sperm competition in deer mice *(Peromyscus maniculatus)*. Behav. Ecol. Sociobiol. **16:** 37-47.
63. HALLIDAY, T. & A. HOUSTON. 1978. The newt as an honest salesman. Anim. Behav. **26:** 1273-1281.

The Role of Ultrasonic Vocalizations in the Regulation of Reproduction in Rats[a]

RONALD J. BARFIELD AND DAVID A. THOMAS

*Department of Biological Sciences and Bureau of Biological Research
Rutgers-The State University of New Jersey
New Brunswick, New Jersey 08903*

To the casual observer, many rodents appear to vocalize only rarely. Usually, as with rats, one hears squeaks or cries only in response to painful or stressful stimulation. There is little audible spontaneous vocalization or vocalization directed to conspecifics. Yet, there is ample evidence that many small rodents vocalize in the ultrasonic range and that they can hear these sounds as well.[1] Perhaps it is because we cannot hear these calls that we tend to minimize the importance of acoustical communication in animals such as rats, mice, and other small rodents.

For small rodents, olfaction provides a major channel for social communication, especially in reproduction.[2] Tactile stimulation, too, is of major importance in the regulation of reproductive behavior and physiology.[3] Do ultrasonic vocalizations produced during mating activity also contribute to the coordination of reproduction in rodents? We have been investigating this question in rats and some of our findings will be presented in this review.

Adult rats emit ultrasonic calls during mating and aggressive activity, and also while investigating strange areas. Infant rats, like many other rodents, also vocalize ultrasonically.[4] In addition, lactating rats may also produce ultrasonic calls.[5] The discussion that follows will focus primarily on those calls occurring during sexual activity and their possible roles in the coordination of reproduction.

RAT MATING CALLS

Adult rats produce two main types of call, one centering on 50 kilohertz (kHz), and another at approximately 22 kHz. The 50-kHz call is produced by both male and female during the copulatory sequence. The male's call occurs primarily as he approaches and investigates the female[6] and, similarly, the female's call is associated with her solicitational and investigatory behavior.[7] A sonographic display of a 50-kHz call of a female is shown in FIGURE 1 (cf. Reference 1, p. 186). Both the vocalizations of the male and the female appear to facilitate mating interactions.

---

[a]The research reviewed here was supported by National Institutes of Health Grants HD-04484 to R. J. Barfield and NS-18526 to D. A. Thomas. Additional support was provided by the Charles and Johanna Busch Memorial Fund.

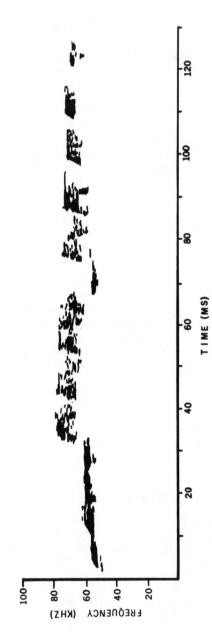

FIGURE 1. Ink tracing of a sound spectrograph of an ultrasonic vocalization of an estrous female rat during mating with a devocalized male. The sonograph is characteristic of a particular variation of female vocalization. Male mating vocalizations, although not displayed for comparison, cannot be distinguished readily from those of the female. (Reprinted with permission from Thomas and Barfield.[7])

The 22-kHz vocalization is produced by the male following ejaculation.[8] It is a loud and persistent call that lasts for approximately three-fourths of the postejaculatory refractory period. During the time that the male is vocalizing, he is inactive and, at times, shows a "tonic immobility." In addition, the male exhibits a sleep-like EEG pattern during the vocalization period.[9] In contrast, the 50-kHz call is associated with behaviorally aroused states and with a "theta" EEG pattern.[10]

Both the 50- and the 22-kHz calls also occur during aggressive behavior.[11] The 22-kHz vocalization appears to be a signal of submission or subordination and is exhibited by both males and females. The 50-kHz call appears to be emitted by both dominant and subordinate rats and its probable function is not clear[12,13] (R. L. Francis, personal communication).

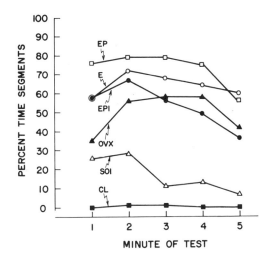

FIGURE 2. Rate of ultrasonic vocalization as a function of the hormonal condition of the female. On the ordinate is the percentage of 15-sec intervals in which vocalization occurred per minute, as indicated on the abscissa. Eighteen males were tested with each of the following: estradiol-progesterone-treated female (EP); estradiol-treated female (E); EP female given two intromissions from a stud male (EPI); ovariectomized female without hormone treatment (OVX); cage bedding soiled by an estrous female (SOI); clean cage bedding (CL). Overall, EP > E > EPI = OVX > SOI > CL, $p < 0.001$ (Sign tests). (Copyright 1978 by The American Psychological Association. Reprinted with permission from Geyer and Barfield.[14])

The 50-kHz Mating Call

The 50-kHz vocalizations of the male signal his readiness to mate. Vocalizations increase in the presence of an estrous female in response, at least in part, to olfactory stimuli (see FIGURE 2).[14] But vocalizations by the female probably affect the male as well. Latency to vocalize after the introduction of a female to the mating cage is highly inversely correlated with copulatory performance as shown in TABLE 1.

In contrast to the above (also shown in TABLE 1), the rate of vocalization is inversely correlated with the mating performance of the male.[15] This finding may

TABLE 1. Number of Vocalizations Occurring in the Introductory Period and during Copulation[a]

	Vocalizations in Introductory Period	Latency to Vocalize following Introduction of the Female (sec)	No. of Vocalizations during First 20 min of Mating
Tests resulting in ejaculation ($N=28$)			
Mean ± SEM	46.8 ± 7.8	4.9 ± 1.1	95.5 ± 16.1
Tests without ejaculation ($N=12$)			
Mean ± SEM	53.4 ± 11.6	27.3 ± 9.5	301.9 ± 43.9
p^b	N.S.[c]	<0.001	<0.001

[a] From McIntosh and Barfield.[15]
[b] Student's t test.
[c] N.S. = not significant.

reflect attempts on the part of the male to better coordinate mating activity with the female or to mate at a pace faster than that selected by the female, or it may indicate a lack of interest by the female.

As the male approaches an estrous female prior to a mount the occurrence of his 50-kHz calls increases. As shown in FIGURE 3, this is particularly apparent just before an ejaculation when the rate of vocalization is increased markedly.[15] Introduction of recorded 50-kHz calls into a cage with a solitary estrous female rat has no obvious effect on her behavior; however, if immediately after exposure to the calls she is placed with a male she shows an augmentation of both proceptive and receptive behavior.[16]

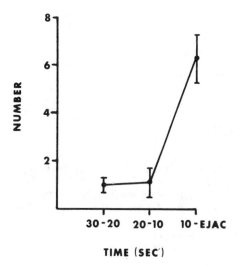

FIGURE 3. The number of 40- to 60-kHz vocalizations occurring in successive 10-sec intervals beginning 30 sec prior to ejaculation. Bars indicate the SEM ($N = 28$). (Reprinted with permission from McIntosh and Barfield.[15])

If such a female is presented with the calls while in the presence of a castrated nonmating male she shows heightened solicitation behavior and even a few lordosis responses without being mounted.[17]

It is possible to devocalize rats surgically by resection of the laryngeal nerves. This procedure has permitted us to gain greater experimental control of the vocalizations and to analyze further the functions of these calls. When normal males were

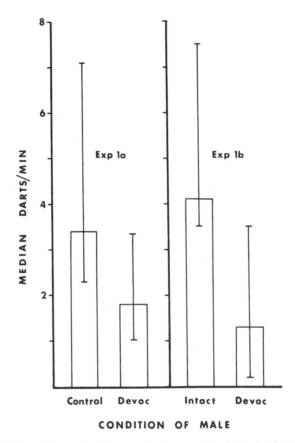

FIGURE 4. Median (and range) darts per minute during the ejaculation latency for females paired with control-operated ($N = 5$) or devocalized ($N = 8$) males in Experiment 1a (left) and females paired with intact ($N = 7$) or devocalized ($N = 7$) males in Experiment 1b (right). (Copyright 1981 by the American Psychological Association. Reprinted with permission from Thomas et al.[18])

devocalized and mated with estrous females, there was, as shown in FIGURE 4, a significant decrease in the amount of darting and ear-wiggling displayed by these animals.[18]

If given the opportunity, will females show a preference for a vocalizing male over a male that has been devocalized? Females placed in a tether apparatus with an intact male at one end and a devocalized male at the other consistently darted more frequently in the presence of the intact animal. In contrast, however, there was little difference in the amount of time spent or mating that occurred with either male.[19]

Is the vocalization itself sufficient to augment the solicitation activity of the female? Two devocalized males were placed at either end of a tether apparatus and tape-recorded vocalizations were introduced through a speaker behind one of the two animals. Each of the experimental females exhibited more darts to the male paired with the vocalizations than to the silent partner. These findings suggest that the male-produced 50-kHz vocalizations function to facilitate and orient the solicitational activity of the estrous female during mating.[19]

If the male's vocalizations augment the solicitation behavior of the female then deafening of the female should bring about the same result as devocalization of the male. Results of experiments with reversible deafening (with earplugs) have been less clear on this question. In an earlier study, deafened females showed a reduction of darting in a tether apparatus with an intact vocalizing male.[20] Current studies have yielded more complicated findings. Not only do the deafened females not dart in close temporal proximity to the male's call, they often position themselves out of the reach of the tethered male. This suggests that the call is important for the orientation of the female to the male.[21]

The 50-kHz Vocalization of the Female

Application of the devocalization procedure to males has revealed that females emit substantial amounts of vocalization during mating. As shown in FIGURE 1, these calls are structurally similar to, if not identical with, those produced by males. The occurrence of these calls is strongly associated with the estrous condition; ovariectomized females rarely vocalize in the presence of a male unless they are behaviorally estrous as a result of treatment with estrogen and progesterone.[7]

The function of the female's calls is currently under investigation. As shown in FIGURE 5 through a comparison of control matings and matings in which one member of the pair is devocalized, females emit about as many vocalizations during mating as do males. It is quite possible that the male and female actually "converse," but there is as yet no evidence for this. The calls of the females occur in association with their approaches to the male and with their hopping and darting. Rarely was vocalization observed while females were immobilized in the lordosis posture although they often vocalized during the reflexive release from this posture.[7] On the basis of our observations, it would be reasonable to propose that the female's vocalizations serve to accentuate her solicitation activity and to enhance her attractiveness to the male at times when she is most ready to mate. Recently, we have shown that devocalized females receive fewer intromissions from males yet they display an enhanced level of darting.[22] This suggests that indeed the female's call is important in the coordination of mating activity.

The 22-kHz Vocalization of the Male

Males may emit 22-kHz vocalizations throughout the mating cycle. The 22-kHz postejaculatory vocalization is most striking and predictable, but males often call at this frequency before ejaculation, especially as this culminating event becomes imminent.[23]

Males sometimes exhibit a transition from 50- to 22-kHz calls if they appear to be frustrated in their attempts to achieve ejaculation of if they are prevented from gaining access to the female by some barrier.[14] Whether this calling has any communicative function is not at all clear, but it could serve to inform the female of an impending ejaculation and thereby allow her to either accept or reject the offer. It is well established that females will pace the copulatory cycle if given the opportunity.[25] It is certainly reasonable to expect that temporal control of the ejaculatory intromission might be critical to reproductive success. It is noteworthy that McClintock[26] has reported that females in group situations are more likely to intercept the ejaculatory intromission from another female than other intromissions. Perhaps they are making use of a "preejaculatory call" in this connection.

The function or functions of the 22-kHz postejaculatory calls also remain unclear. It has been suggested that the postejaculatory vocalizations serve to maintain contact

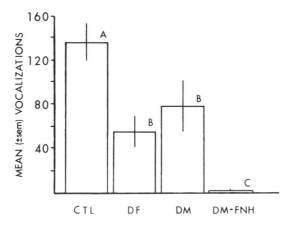

FIGURE 5. The mean (±SEM) number of 50-kHz vocalizations detected during a 5-min test period with a male-female rat dyad. Mating occurred in the first three conditions. CTL = control condition; DF = devocalized female condition; DM = devocalized male condition; DM-FNH = devocalized male with ovariectomized hormonally untreated female. (Reprinted with permission from Thomas and Barfield.[7])

with the female during the period of sexual activity and to perhaps discourage intervention by other males at this time.[20]

It has been demonstrated by Adler and Zoloth[27] that if a second male mates with a female within 5 to 6 min of an ejaculation sperm transport and therefore pregnancy may be aborted. As the period of vocalization appears to correspond closely with the time required for successful pregnancy by the female the above hypothesis has intuitive appeal. Support for this view, however, is limited.

During the refractory period of the male the female tends to reduce her activity and maintain distance from the male. When observed in a large (4 × 4-ft) enclosure, females maintained greater distance from the males during periods of their calling than when they were silent.[24] Introduction of tape-recorded 22-kHz vocalizations in the presence of mating devocalized males did not, however, affect the spacing between the pair.[28] This experiment was performed in a smaller test cage and the greater immediacy of other stimuli emanating from the male may have obscured a possible effect of the calls.

If males are observed with only a female in a test cage they appear quite "refractory" during their refractory period. They lie quietly, often almost hypnotically, on the floor rarely moving as they vocalize. In addition to the persistent calling, there is also a highly predictable pattern of urination characteristic of the male's refractory period.[29,30] Perhaps the olfactory signal provided by the urine is additive to or redundant with that inherent in the calling.

It is also reasonable to propose that the postejaculatory vocalization and urination function to keep other males away from the mating area. Although the male is refractory sexually, he is more than capable of mounting a severe attack on an intruder.[31] In fact, males attack more aggressively when they are refractory sexually than at other times.

THE SEMINATURAL ENVIRONMENT

The difficulty in decoding the acoustical communication system of rats may be aggravated by observing the animals in confined and highly unnatural test cages. We have recently begun to study the mating behavior of rats in a large enclosure and with a somewhat more complex social grouping. Groups consisting of two males and one female were observed in a 4 × 8-ft plywood enclosure and behavior was recorded on videotape from the onset of behavioral estrus until mating ceased.

As mating begins it is clear that the dominant or alpha male almost always mates first. After ejaculation the alpha male often trails after the female and may actively prevent the beta or subordinate male from mating with her. Sometimes the beta male is attacked viciously when he approaches the female.

A summary of the mating interactions observed is presented in FIGURE 6. The alpha male almost always achieves at least two ejaculations before the beta male has an opportunity to mate to ejaculation. At times the beta male cannot even gain access to the female until the alpha male has ejaculated two or three times. Under the conditions of this test situation, the refractory period is greatly attenuated and the period of vocalization is much longer and more irregular than that observed in a small test cage. Contrary to previous observations the male may emit 22-kHz vocalizations as he walks about and at times appears to direct these calls toward the other male or the female.

The observations reported here suggest that the dominant males attempt to guard the female from the second males. This would of course ensure paternity to the dominant animal, but why do they relax their vigilance only after two or three ejaculations? Is it possible that two or three ejaculations are required to ensure a fertile mating? Although it is generally thought that one ejaculation is sufficient to impregnate a female rat, there is some evidence to suggest that perhaps one ejaculation is not enough (D. A. Thomas and R. J. Barfield, unpublished results). In our laboratory, Long-Evans female rats do not normally become pregnant after only one ejaculation.[32]

On the basis of this study it is reasonable to propose that the 22-kHz vocalization of the male serves to maintain contact with the female to facilitate continued mating and also to signal to other males to keep away from the mating pair. Further study should allow us to evaluate these hypotheses.

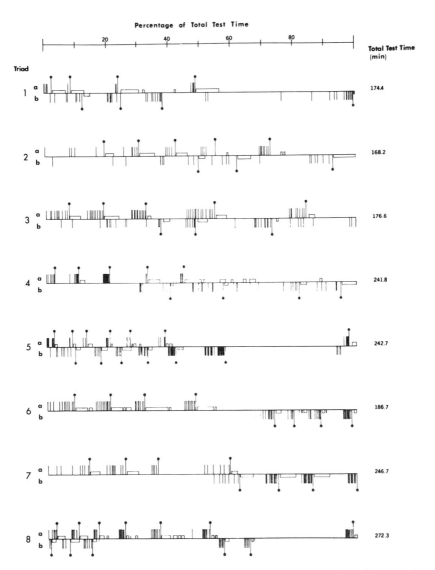

FIGURE 6. Temporal distribution of mating events and 22-kHz vocalizations with two males and one female in a large (4 × 8-ft) seminatural enclosure. The lines with dots represent ejaculations, lines without dots indicate intromissions, and horizontal bars show the occurrence of 22-kHz calls. The alpha or dominant male is indicated by "a" (the top record) and the beta or subordinate male is shown as "b."

REPRODUCTIVE INFLUENCE OF 22-kHz VOCALIZATIONS

In rodents, both tactile and olfactory stimuli affect reproductive processes.[3] Could the stimuli emanating from the male during his refractory period influence the reproductive physiology of the female? We have suggested that another function of the postejaculatory call (and perhaps the postejaculatory urination as well) is to facilitate the female's luteotrophic response and thus her pregnancy. Experiments currently in progress will assess the ability of vocalizing and devocalized males to successfully impregnate females after varying numbers of intromissions and ejaculations.

This brief review has shown that ultrasonic vocalizations occur in all aspects of rat social behavior. Although we are far from a complete understanding of the communicative functions of these calls certain general statements may be put forth:

(1) Vocalizations emitted by male and female contribute in the coordination of mating behavior.

(2) Vocalizations may serve as a territorial or dominance signal in mating contexts.

(3) As vocalizations are emitted by both mother and pups they may be significant in the mother-young interaction.

(4) Rats have a limited vocal repertoire; therefore context is probably critical in decoding the signal content conveyed by particular calls.

(5) Auditory communication is probably redundant with other lines of communication that utilize other sensory modalities such as olfaction or touch.

ACKNOWLEDGMENT

Our thanks to Jane Sherwood for her care and patience in the preparation of the final manuscript.

REFERENCES

1. SALES, G. & D. PYE. 1974. Ultrasonic Communication by Animals. Chapman & Hall. London, England.
2. BRONSON, F. H. 1971. Rodent pheromones. Biol. Reprod. **4:** 344-357.
3. ADLER, N. T. 1978. Social and environmental control of reproductive processes in animals. *In* Sex and Behavior: Status and Prospectus. T. E. McGill, D. A. Dewsbury & B. D. Sachs, Eds.: 115-160. Plenum Press. New York, N.Y.
4. NOIROT, E. 1972. Ultrasounds and maternal behavior in small rodents. Dev. Psychobiol. **5:** 371-387.
5. FRANCIS, R. L. 1977. 22-kHz calls by isolated rats. Nature **265:** 236-237.
6. SALES, G. D. (nee SEWELL). 1972. Ultrasound and mating behaviour in rodents with some observations on other behavioral situations. J. Zool. **168:** 149-164.
7. THOMAS, D. A. & R. J. BARFIELD. 1985. Ultrasonic vocalization of the female rat. (*Rattus norvegius*) during mating. Anim. Behav. **33:** 720-725.
8. BARFIELD, R. J. & L. A. GEYER. 1972. Sexual behavior: ultrasonic postejaculatory song of the male rat. Science **176:** 1340-1350.
9. BARFIELD, R. J. & L. A. GEYER. 1975. The ultrasonic postejaculatory vocalization and the postejaculatory refractory period of the male rat. J. Comp. Physiol. Psychol. **88:** 723-734.

10. McIntosh, T. K., R. J. Barfield & D. A. Thomas. 1984. Electrophysiological and ultrasonic correlates of reproductive behavior in the male rat. Behav. Neurosci. **98:** 1000-1003.
11. Sales, G. D. (nee Sewell). 1972. Ultrasound and aggressive behaviour in rats and other small mammals. Anim. Behav. **20:** 88-100.
12. Thomas, D. A., L. K. Takahashi & R. J. Barfield. 1983. Analysis of ultrasonic vocalizations emitted by the intruder during aggressive encounters among rats. J. Comp. Psychol. **97:** 201-206.
13. Takahashi, L. K., D. A. Thomas & R. J. Barfield. 1983. Analysis of ultrasonic vocalizations emitted by residents during aggressive encounters among rats. J. Comp. Psychol. **97:** 207-212.
14. Geyer, L. A. & R. J. Barfield. 1978. Influence of gonadal hormones and sexual behavior on ultrasonic vocalization in rats. I. Treatment of females. J. Comp. Physiol. Psychol. **92:** 438-446.
15. McIntosh, T. K. & R. J. Barfield. 1980. Temporal patterning of 40-60 kHz ultrasonic vocalizations and copulation in the rat (*Rattus norvegicus*). Behav. Neural Biol. **29:** 349-358.
16. Geyer, L. A., T. K. McIntosh & R. J. Barfield. 1978. Effects of ultrasonic vocalizations and male's urine on female rat readiness to mate. J. Comp. Physiol. Psychol. **92:** 457-462.
17. McIntosh, T. K., R. J. Barfield & L. A. Geyer. 1978. Ultrasonic vocalizations facilitate sexual behavior of female rats. Nature **272:** 163-164.
18. Thomas, D. A., L. Talalas & R. J. Barfield. 1981. Effects of devocalization of the male on mating behavior in rats. J. Comp. Physiol. Psychol. **95:** 630-637.
19. Thomas, D. A., S. B. Howard & R. J. Barfield. 1982. Male produced ultrasonic vocalizations and mating patterns in female rats. J. Comp. Physiol. Psychol. **96:** 807-815.
20. Barfield, R. J., P. Auerbach, L. A. Geyer & T. K. McIntosh. 1979. Ultrasonic vocalizations in rat sexual behavior. Am. Zool. **19:** 469-480.
21. Kolunie, J. M., D. A. Thomas & R. J. Barfield. 1985. Auditory regulation of sexual behavior in rats. In Proceedings, Conference on Reproductive Behavior, Asilomar, California.
22. White, N. R. & R. J. Barfield. 1985. Female devocalization increases darting and decreases intromissions. In Proceedings, Conference on Reproductive Behavior, Asilomar, California.
23. Brown, R. E. 1979. The 22-kHz pre-ejaculatory vocalization of the male rat. Physiol. Behav. **22:** 483-489.
24. Geyer, L. A. & R. J. Barfield. 1980. Regulation of social contact during the postejaculatory interval by the female rat. Anim. Learn. Behav. **8:** 679-685.
25. Krieger, M. S., D. Orr & T. Perper. 1976. Temporal patterning of sexual behavior in the female rat. Behav. Biol. **18:** 379-386.
26. McClintock, M. K. 1984. Group mating in the domestic rat as a context for sexual selection: consequences for the analysis of sexual behavior and neuroendocrine responses. Adv. Study Behav. **14:** 1-50.
27. Adler, N. T. & S. R. Zoloth. 1970. Copulatory behavior can inhibit pregnancy in female rats. Science **168:** 1480-1482.
28. Thomas, D. A., S. B. Howard & R. J. Barfield. 1982. Male-produced postejaculatory 22-kHz vocalizations and the mating behavior of estrous female rats. Behav. Neural Biol. **36:** 403-410.
29. Anisko, J. J., N. T. Adler & S. Suer. 1979. Pattern of postejaculatory urination and sociosexual behavior in the rat. Behav. Neural Biol. **26:** 169-176.
30. McIntosh, T. K., P. G. Davis & R. J. Barfield. 1979. Urine marking and sexual behavior in the rat (*Rattus norvegicus*). Behav. Neural Biol. **26:** 161-168.
31. Thor, D. H. & K. J. Flannelly. 1979. Copulation and intermale aggression in rats. J. Comp. Physiol. Psychol. **93:** 223-228.
32. Bernstein, P. L. & R. J. Barfield. 1985. Ultrasonic vocalizations and mating stimuli in the induction of pregnancy in Long-Evans rats. In Proceedings, Conference on Reproductive Behavior, Asilomar, California.

Female Vocalizations and Their Role in the Avian Breeding Cycle

BARBEE L. INMAN

Institute of Animal Behavior
Rutgers-The State University of New Jersey
Newark, New Jersey 07102

Vocalization by female birds has largely been ignored or cursorily mentioned in the avian vocalization literature. Although most female birds vocalize, the complexity of these vocalizations in some species and their resemblance to male song have been underestimated. In this paper I will review the occurrence of female "song" and its function within the breeding cycle. Possible functions for complex female vocalizations include territorial defense, pair-bond establishment and maintenance, reproductive synchronization, and nest defense. Because females of duetting and nonduetting species show different patterns of calling behavior, they will be considered separately.

Song and complex vocalizations as considered here follow Thorpe's[1] definition of song as "a series of notes generally of more than one type, uttered in succession and so related as to form a recognizable sequence or pattern in time." Thus it is distinguished from mono- and disyllabic call notes which tend to function as alarm notes, nest-defense calls, food solicitation, etc., and are given by most individuals of any species. Here, song includes vocalization by females in species outside the suborder Passeres (oscines) which comprises songbirds in the strict sense.

NONDUETTING SPECIES

Although the male is best known as the singing partner in nonduetting species, there are numerous accounts of singing females.[2-4] Most of these accounts are anecdotal, based on few observations, and have not been subject to behavioral testing.

In a discussion of song, the breeding cycle can be divided into three periods: (1) courtship and nest-building, (2) incubation and care of young, and (3) postbreeding (which usually extends into prebreeding). In most species, females tend to concentrate their singing behavior in one of these periods, with song-like vocalizations markedly less frequent at other times of the year. In some species, use of song is variable among individuals. For example, the female marsh warbler (*Acrocephalus palustris*) sings a "grated song" early in the breeding season, usually during the first days of pair establishment; some females, however, sing mostly during incubation, at nest exchange.[5] A related species, the great reed warbler (*Acrocephalus arundinaceus*), has only been recorded singing at the beginning of the breeding season.

Courtship and Nest-Building Period

Females are most often vocal during this part of the reproductive cycle. In this context, song may function to enhance a mate's territorial defense, as pair-bond reinforcement, or as reproductive self-stimulation to the hypothalamo-gonadal axis (see Cheng, this volume).

The female grasshopper sparrow (*Ammodramus savannarum*) trills from before pairing through completion of nesting.[6] The male is attracted to her call when she arrives on his territory and proceeds to court her. Once paired, she will vocalize both in response to her mate and on her own when signaling her approach. The behavioral context of her trills suggests that pair-bond establishment and maintenance are the functions of these calls, since they seem to indicate her presence to her mate.

There are anecdotal reports of female chaffinches (*Fringilla coelebs*) producing subsong, which in this species is the normal "adolescent babbling" of juvenile males.[7] The female eastern phoebe (*Sayornis phoebe*) sometimes sings the "phee-b-be-be" song early in the season, before dawn, prior to copulation.[8] Female indigo buntings (*Passerina cyanea*) and rufous-sided towhees (*Pipilo erythrophthalamus*) are reported to sing early in the breeding season.[9] The function of the female song in these species, for which there are only anecdotal reports, is obscure.

In one of the first and most thorough reports on female song, Nice[4] characterized the song of female song sparrows (*Melospiza melodia*). It is given early in the breeding season, from a perch higher than typical for the female, and it is "short, harsh, and entirely unmusical." Nice suggested that it is a "vestigial phenomenon" since other song sparrows show no response to it. Perhaps it functions in a self-stimulatory way, enhancing the female's own gonadal development, if it is not directed at conspecifics.

Song is reported at, or just subsequent to, pairing in the female white-crowned sparrow (*Zonotrichia spp.*).[10] Female cardinals also sing frequently before nesting and they often match song types with their mates.[11] This may function as pair-bond reinforcement or to support the pair's territorial defense.

Female budgerigars (*Melopsittacus undulatus*) are quite vocal and produce complex warbles similar to those of the males but never as frequently. Warbling occurs early in the breeding cycle and also when females are occupying the nest box, during early stages of incubation. Exposure to the vocalizations of males or of breeding females stimulates nest box occupation and oviposition in nonpaired females in the laboratory.[12] As these are colonial breeders, sounds of breeding behavior from other pairs may enhance and further stimulate any individual's reproductive development.

One of the more systematic studies of female vocalization has been of the Hawaiian 'elepaio (*Chasiempis sandwichensis*).[13] Both sexes are quite vocal in this territorial species. It is one of the few species for which rates of females' songs have been tabulated over the entire breeding cycle. The highest rates of singing by females occur during courtship and nest-site selection while song is least frequent during incubation. Male song peaks during nest-building and, like female song, is least frequent during incubation. The female sings throughout the breeding season and shares in territorial defense. Her relatively high rates of singing before oviposition may be a component of her territorial behavior, as playbacks stimulated female as well as male vocal responses.

Among some polygynous species, early breeding season vocalizations by females may have territorial function. Female red-winged blackbirds (*Agelaius phoeniceus*) may hold subterritories within a male's main territory. Two song types are displayed by females; one is used to answer males early in the season and may have pair-bond significance, while the other is directed mainly toward females and may be used in subterritory definition, if not defense. The second, more "territorial" song is individually identifiable, although females respond aggressively to playbacks of their own songs as well as to those of strangers.[14]

Incubation and Brood-Care Period

It would seem likely that during this period, when females are sitting on eggs or young chicks, they would be most unobtrusive. Yet some females sing on the nest or at nest exchange. Female rose-breasted grosbeaks (*Pheucticus ludovicianus*)[15] and black-headed grosbeaks (*Pheucticus melanocephalus*)[2] have been reported to sing a quiet song while incubating. They also sing before nesting and during nest-building. Ritchison[2] reports that as the brooding period progresses females and their mates sing more frequently on and around the nest. The young become quite responsive to their parents' calls. Perhaps the value of parental recognition outweighs the danger of acoustically oriented predation.

In species with precocial young, the chicks are, at hatching, in danger of wandering away from the nest or parent, and being preyed upon. Here, recognition of parental calls is especially important for survival. Precocial chicks, including guillemots (*Cepphus grylle*) and chickens (*Gallus domesticus*), readily approach playbacks of their parents' calls.[16]

The female eastern bluebird (*Sialia sialis*) has been observed to sing a song virtually identical to that of a male when disturbed from the nest by human intruders.[17] The sound of a strange male (really his female) brings her mate to the nest, which he then defends from human intruders. Ritchison[18] also reported this phenomenon in female black-headed grosbeaks. Song on these occasions has served as an indirect nest-defense mechanism by attracting the male's attention.

Postbreeding Period

The songs of species that defend feeding territories beyond the breeding season often function in territorial defense. This is one of the few situations in which there is evidence of a clear function of female song. The female American dipper (*Cinclus mexicanus*) vigorously sings a song indistinguishable from the male's between December and February, the peak time for territorial defense.[19] Female mockingbirds (*Mimus polyglottos*) hold and defend territories during fall and winter, the only time of the year at which they regularly sing.[20] Other species that defend winter feeding territories with song are the European robin (*Erithacus rubecula*), the wren tit (*Chamaea fasciata*), and the loggerhead shrike (*Lanius ludovicianus*).[3]

DUETTING SPECIES

Females are quite vocal in species that perform duets. A duet will be defined as an "overlapping bout" of vocalizations "given by a mated pair."[21] Duets can be innate or learned during pair establishment.[22] Duetting species are characterized by monogamy and year-round territoriality, and are often sexually monomorphic. An extensive list of duetting species can be found in Farabaugh[21] and in Thorpe.[23]

A duet may function in more than one capacity and a mated pair may have a repertoire of several duets serving different purposes. Duets function similarly to song in nonduetting species: in territorial defense or advertisement, pair-bond maintenance, mate-guarding, reproductive synchronization, and self-stimulation.

As there have been systematic studies of duetting birds and the role of the duet in the behavior of these species, duets will be discussed with reference to function.

Pair-Bond Establishment and Maintenance

Four characteristics of vocalizations in duetting species could be interpreted as evidence for their role in strengthening or maintaining coordination between a mated pair: (1) duetting takes place when the mate is present, (2) the duet is developed when the individuals first meet, (3) duetting continues after the breeding season if the pair remains together, and (4) the duet is used to identify the mate in situations where there is the confusion of the flock or the mate is not visible. The very nature of the duet, with shared vocal participation, suggests intrapair communication as an important function.

Evidence for the role of vocalizations in the establishment of a pair bond can be found in the Australian magpie lark (*Grallina cyanoleuca*).[24] Juveniles of this species form pair bonds and sing duets before they establish a territory. Pairs maintain a repertoire of about 11 different calls, which differ between pairs and are not sex specific. Therefore, information about the identity and sex of the caller is primarily communicated within the pair. Singing rates are highest in the period before nest-building and after the brood fledges, when song may play an important role in keeping the pair together. Changes in song frequency within the repertoire over the breeding season have not been measured. This information might lead to further understanding of a particular duet's "meaning" in behavioral and reproductive contexts.

Pair-bond reinforcement may be a secondary function in species where duetting primarily functions in territorial defense. African forest weavers (*Symplectes bicolor*) duet in unison, that is, sing the same notes at the same time.[25] While songs and repertoires vary between pairs, pair members share an identical repertoire. This suggests that the duet is learned. The learning process itself may act as a premating investment in the pair's reproductive success.

Among colonial species the duet may function in mate identification. The common grackle (*Quiscalus quiscala*) duets frequently at the nest-building phase of the breeding cycle.[26] Duets are frequently used in contact between mates when interacting within the colony or when leaving it for courtship and mating activity. The noisy miner (*Manorina melanocephala*) also breeds communally, and the duet may be important for recognition within the commune.[27] It appears that duetting for identification/recognition purposes would be more important in colonial species than in territorial species, where pairs are more widely spaced and there is less confusion in locating one's mate. In colonial species, duetting may be expected in the early stages of the breeding season; later the nest site would be a more accurate predictor of the mate's position.

Territoriality

Duetting species are often permanently territorial. If the duet is used as part of territorial defense or advertisement, one would expect to see pairs responding to intruders with increased use of the duet, calling from a prominent site, and countersinging with neighboring pairs.[28] If the duet occurs throughout the year, it may be difficult to differentiate its value in territorial maintenance from its importance in the maintenance of the pair bond.

The slate-colored boubou (*Laniarius funebris*) has four different types of duets which correspond to different situations,[28] namely, (1) mate-guarding, (2) territorial defense, (3) breeding synchronization, and (4) territorial advertisement. The territorial duets have different contexts; type 4 is used in actual territorial encounters with other pairs, whereas type 2 is associated with territorial advertisement and is primarily used in display. These duets are used year-round in support of permanent territories.

Duetting can be elicited by playbacks of conspecific foreign duets in the West African barbet (*Lybius vielilotti*)[29] and in the robin chat (*Cossypha heuglini*), another African species.[30] Harcus[31] found bokmakierie shrike (*Telophorus zeylonus*) pairs duetting within sight of other pairs, and engaging in countersinging. This species varies its location between breeding and nonbreeding seasons, and duetting upon return to breeding areas may function in territory establishment.

Female bay wrens (*Thryothorus nigricapillus*) defend their territories vigorously, initiating duets, participating in disputes, and responding to playbacks with duets. These females will maintain their territories even if the mate is removed.[32]

Duetting by a pair of West African barbets is often followed by duets of neighboring pairs. This was observed during the nonbreeding seasons, when duetting occurs only a few times daily. Thus it is unlikely that one pair's duet following another pair's duet is random,[29] which supports the role of the duet in territorial defense.

Mate-Guarding

Mate-guarding can be considered a subset of territoriality, because a mate can be considered a defendable resource. Duetting may function in mate-guarding when there is danger of cuckoldry or desertion and where there are limited mates or territories. It can be recognized by an increase in intensity of response to intruders singing solo songs rather than to territorial interactions with duetting pairs. One would expect the guarder to initiate more duets than the one being guarded. In species where polygyny sometimes occurs a female may want to use the duet to keep other females away; males may use the duet to guard their females when they are most fertile, or when there are nonmated males nearby.

Sonnenschein and Reyer[28] observed this variable response in the slate-colored boubou mentioned above, although they were unable to elicit such a response using playbacks in the field. However, the southern boubou shrike (*Laniarius ferrugineus*) did respond vigorously to playbacks of solo vocalizations.[31] Differences in environmental circumstances and the likelihood of proximate unmated individuals may account for differences in duet context in these species.

Reproductive Synchrony

Duets may function to synchronize the breeding behavior of a pair, by providing a set point from which reproduction may ensue. The duet would be directed primarily between the mates and would occur early in the breeding season. It may act as a signal within the pair, reflecting the appropriate time to initiate breeding each year, especially if environmental conditions are unpredictable.[21]

Several African species, the southern boubou shrike, the bokmakierie shrike, and the bar-throated apalis (*Apalis thoracica*), duet throughout the year, with peak frequency occurring at the very beginning of the breeding season.[31] This increase in coordinated behavior may perform a synchronizing function which further stimulates gonadal development. Conversely, it could be the result of gonadal development brought on by other environmental stimuli, including photoperiodic cues.

The slate-colored boubou discussed earlier, has a duet which may function to synchronize gonadal cycles of the pair (type 2). It is used most frequently early in the breeding season and almost disappears when nest-building starts. One season it was observed that a male sang his portion of the duet frequently, but the female's response frequency was abnormally low; no breeding occurred.[28] The female's component of the duet may be an indicator of her readiness to breed or of her reproductive state. Her vocal response may be crucial for initiation of her own breeding cycle and may be a self-feedback mechanism affecting her behavior.[33]

Further support for the role of the duet as a synchronizer can be found in the robin chat.[30] This is a monogamous species whose behavior has been studied in the laboratory. The effects of the male's vocal role in the duet on breeding behavior were examined by devocalizing males via section of the hypoglossal nerve, tracheosyringeal branch (HNS). In spite of the male's inability to perform duets, the pair remained together and performed visual displays to intruder stimuli. In another experiment, the female of a pair was deafened. Following this manipulation, she no longer participated in duets, although she continued to produce nonduet vocalizations. Under either manipulation, neither individual chose to desert when given the opportunity to do so. This implies that pair-bond maintenance is not the most important function of the duet. However, no breeding occurred. This may have been the result of the stress of the manipulations or the duet may be necessary to stimulate the hypothalamo-gonadal axis for reproduction. It is unclear whether sham-operated controls bred successfully and how important environmental cues are to the breeding cycle. The reverse experiments, devocalization of females or deafening of males, would provide insight into the female's role in the duet and to what extent her vocal participation is important in the initiation of breeding in this species.

Ring dove (*Streptopelia risoria*) pairs give a nest-coo duet specific to early pairing, when choosing a nest site. This is followed by the female's solo nest-coo through oviposition. The female will not breed if subjected to central devocalization, via lesion to the midbrain nucleus intercollicularis (ICo), rendering her unable to participate in the duet.[33] Peripheral devocalization, via HNS, produced similar disruption of the breeding system. Playback studies have established the important nature of the female's contribution to the duet in stimulating her own ovarian development. Field observations of the slate-colored boubou mentioned above suggest that the same mechanism may also operate in this species.

These laboratory experiments imply that the duet may serve as a mediator of breeding success, through which the pair agrees that the conditions are right to reproduce. It is also a means whereby each sex can exert some control over the breeding cycle.

DISCUSSION

There are several problems implicit in attempts to assign a particular function to vocalizations. A vocalization may function in several different ways depending on the context and the individual responding to it. A song or duet may mean one thing to the individual singing, something else to its mate or to an unmated conspecific of the opposite sex, and something else entirely to a neighboring territory holder.[34] Some species have only one song or a small repertoire of duets, yet these are involved in such varied contexts as definition and defense of territory, establishment or support of the pair bond, and initiation of breeding behavior each season.

A most noticeable aspect of the body of work examining female song and its role in the breeding cycle is that very little research has actually approached this question directly. There are anecdotal reports of female song which can be correlated with its occurrence in the breeding season to suggest possible functions, but field experiments involving environmental and physiological manipulations are necessary to put the correlative information to a test. Furthermore, elaborate observations of the frequency and context of song must be done for those species where only anecdotal evidence is present.

The important questions about female song have yet to be answered: What is the relationship between song in females and hormone levels? Are there important organizational factors necessary for female vocal behavior? Is the neural basis of female song the same as that for male song? The role of androgens in male song is well established[35]; how is the neural circuitry activated given the different hormonal milieu of the female? Song and singing behavior have been extensively studied in males; now, the same questions must be applied to females.

Although many of these questions have been tackled in the ring dove,[36] generalizations cannot be made until studies are performed using species in which females have more complex song. Endocrinological information can be determined from radioimmunoassay of plasma from wild birds.[37] Laboratory techniques developed using the ring dove and the canary,[35] among others, can be applied to species for which behavioral information is extensive, but physiological data are lacking. This synthesis of laboratory and field data can provide an interactive behavioral-mechanistic perspective for the singing female bird.

REFERENCES

1. THORPE, W. H. 1961. Bird Song: The Biology of Vocal Communication and Expression in Birds. Cambridge University Press. London, England.
2. RITCHISON, G. 1983. The function of singing in female black-headed grosbeaks (*Pheucticus melanocephalus*): family-group maintenance. Auk **100:** 105-116.
3. ARMSTRONG, E. A. 1963. A Study of Bird Song. Oxford University Press. London, England.
4. NICE, M. M. 1943. Studies in the life history of the song sparrow. II. The behaviour of the song sparrow and other passerines. Trans. Linn. Soc. N.Y. **6:** 1-328.
5. DOWSETT-LEMAIRE, F. 1979. Vocal behavior of the marsh warbler, (*Acrocephalus palustris*). Le Gerfaut **69:** 475-502.
6. SMITH, R. L. 1959. The songs of the grasshopper sparrow. Wilson Bull. **71:** 141-152.
7. THORPE, W. H. & P. M. PILCHER. 1958. The nature and characteristics of sub-song. Br. Birds **51:** 509-514.
8. SMITH, W. J. 1969. Displays of *Sayornis phoebe* (Aves, Tyrannidae). Behaviour **33:** 283-322.

9. NOLAN, V. 1958. Singing by female indigo bunting and rufous sided towhee. Wilson Bull. **70:** 287-288.
10. KERN, M. D. & J. R. KING. 1972. Testosterone-induced singing in female white-crowned sparrows. Condor **74:** 204-209.
11. LEMON, R. E. 1968. The relation between organization and function of song in cardinals. Behaviour **32:** 158-177.
12. HUTCHISON, R. E. 1974. Temporal patterning of external stimuli and reproductive behavior in female budgerigars. Anim. Behav. **22:** 150-157.
13. CONANT, S. 1977. The breeding biology of the Oahu 'elepaio. Wilson Bull. **89:** 193-210.
14. BELETSKY, L. D. 1983. Aggressive response to "self" songs by female red-winged blackbirds, *Agelaius phoeniceus.* Can. J. Zool. **61:** 462-465.
15. IVOR, H. R. 1944. Bird study and semi-captive birds: the rose-breasted grosbeak. Wilson Bull. **56:** 91-104.
16. IMPEKOVEN, M. 1976. Prenatal parent-young interactions in birds and their long-term effects. *In* Advances in the Study of Behavior. J. S. Rosenblatt, R. A. Hinde, E. Shaw & C. Beer, Eds. **7:** 201-253. Academic Press. New York, N.Y.
17. MORTON, E. S., M. S. GEITGEY & S. MCGRATH. 1978. On bluebird "responses to apparent female adultery." Am. Nat. **112:** 968-971.
18. RITCHISON, G. 1983. Possible "deceptive" use of song by female black-headed grosbeaks. Condor **85:** 250-251.
19. BAKUS, G. J. 1959. Territoriality, movements, and population density of the dipper in Montana. Condor **61:** 410-425.
20. MICHENER, H. & J. R. MICHENER. 1935. Mockingbirds, their territories and individualities. Condor **37:** 97-140.
21. FARABAUGH, S. M. 1982. The ecological and social significance of duetting. *In* Acoustic Communication in Birds. D. E. Kroodsma & E. M. Miller, Eds. **2:** 85-124. Academic Press. New York, N.Y.
22. WICKLER, W. 1980. Vocal dueting and the pair bond. I. Coyness and partner commitment. A hypothesis. Z. Tierpsychol. **52:** 201-209.
23. THORPE, W. H. 1972. Duetting and antiphonal song in birds. Behaviour (Suppl.) **18:** 1-197.
24. TINGAY, S. 1974. Antiphonal song of the magpie lark. Emu **74:** 11-17.
25. WICKLER, W. & U. SEIBT. 1980. Vocal dueting and the pair bond. II. Unisono dueting in the African forest weaver, *Symplectes bicolor.* Z. Tierpsychol. **52:** 217-226.
26. WILEY, R. H. 1976. Affiliation between the sexes in common grackles. II. Spatial and vocal coordination. Z. Tierpsychol. **40:** 244-264.
27. DOW, D. D. 1975. Displays of the honeyeater *Manorina melanocephala.* Z. Tierpsychol. **38:** 70-96.
28. SONNENSCHEIN, E. & H. U. REYER. 1983. Mate-guarding and other functions of antiphonal duets in the slate-coloured boubou (*Laniarius funebris*). Z. Tierpsychol. **63:** 112-140.
29. PAYNE, R. B. & N. J. SKINNER. 1970. Temporal patterns of duetting in African barbets. Ibis **112:** 173-183.
30. TODT, D. & H. HEULTSH. 1982. Impairment of vocal signal exchange in the monogamous duet-singer *Cossypha heuglini* (Turdidae): effects on pairbond maintenance. Z. Tierpsychol. **60:** 265-274.
31. HARCUS, J. L. 1977. The functions of vocal duetting in some African birds. Z. Tierpsychol. **43:** 23-45.
32. LEVIN, R. 1983. The adaptive significance of vocal duetting in the bay wren (*Thryothorus nigricapillus*). Abstract (#288) presented at the American Ornithologists' Union Conference, 1983.
33. CHENG, M. F. 1983. Behavioural "self-feedback" control of endocrine states. *In* Hormones and Behaviour in Higher Vertebrates. J. Balthazart, E. Prove & R. Gilles, Eds.: 408-421. Springer-Verlag. West Berlin, FRG.
34. SMITH, W. J. 1968. Message-meaning analysis. *In* Animal Communication. T. A. Sebeck, Ed.: 44-60. Indiana University Press. Bloomington, Ind.
35. NOTTEBOHM, F. 1980. Brain pathways for vocal learning in birds: a review of the first 10 years. *In* Progress in Psychobiology and Physiological Psychology. J. M. Sprague & A. N. Epstein, Eds. **9:** 85-124. Academic Press. New York, N.Y.

36. CHENG, M. F. 1979. Progress and prospects in ring dove research: a personal view. *In* Advances in the Study of Behavior. J. S. Rosenblatt, R. A. Hinde, C. Beer & M. C. Busnel, Eds. **9:** 97-129. Academic Press. New York, N.Y.
37. WINGFIELD, J. C. & D. S. FARNER. 1975. The determination of five steroids in avian plasma by radioimmunoassay and competitive protein binding. Steroids **26:** 311-327.

Role of the Preoptic Area in the Neuroendocrine Regulation of Reproduction: An Analysis of Functional Preoptic Homografts[a]

MARIE J. GIBSON[b]

Department of Medicine
Mount Sinai School of Medicine
New York, New York 10029

The preoptic area (POA) is intimately involved in many aspects of reproduction in rodents, most essentially as the locus of many cell bodies that contain gonadotropin-releasing hormone (GnRH).[1] This peptide hormone is vital for the normal function of the reproductive system, as it is the primary stimulus of luteinizing hormone (LH) and follicle-stimulating hormone (FSH) synthesis and secretion from the pituitary. In turn, LH and FSH are necessary for gonadal growth and development in both male and female. A portion of the GnRH cells in the POA project to the median eminence, where axons terminate at the capillaries of the pituitary portal plexus, ensuring delivery of GnRH to the gonadotrophs of the pituitary. The POA is also implicated as the location of the ovulatory mechanism that stimulates the LH surge necessary for ovulation.[2] The area is important in the mediation of male sexual behavior[3] and maternal behavior.[4,5] Many studies describe sexually dimorphic characteristics in this region that range from synaptic interactions[6] to altered nuclear density and volume.[7]

The POA has been used as a tissue source for brain grafts in two types of experiments. In one, the sexual dimorphism of the region was of primary interest.[8] Neonatal male POA was bilaterally implanted into the POA of neonatal female rats, and the effect on the females' sexual behavior in adulthood was observed. In the other series of experiments, the POA was utilized as a major source of GnRH cells. POA tissue obtained from normal fetal mouse brain was implanted into the third ventricle of brains of adult hypogonadal (hpg) mice,[9-12] which are genetically defective in hypothalamic GnRH.[13]

Of interest to researchers for decades, the study of brain tissue grafts has expanded rapidly in the last several years. Facilitated in part by the early observation that grafted tissue is not generally rejected by the brain, which some have termed an "immunologically privileged site," studies have investigated various aspects of neuronal growth and plasticity.[14,15] Others have capitalized on the neurochemical properties of certain brain regions, by transplanting specific neural tissue that contains an identified chemical

---

[a]Supported by National Institutes of Health Grants HD 19077 to the author and NS 20335 to Dorothy T. Krieger.

[b]Correspondence address: Department of Medicine, Mount Sinai School of Medicine, 1 Gustave L. Levy Place, New York, N.Y. 10029.

substance into a brain that is deficient in the substance, due either to an imposed lesion or a genetic defect. Examples include the use of substantia nigra tissue, rich in dopamine, as grafts in lesioned dopamine-deficient animals,[16] and hypothalamic grafts containing vasopressin neurons in the vasopressin-deficient Brattleboro rat.[17]

The deficiency of brain GnRH in the hpg mouse is an autosomal recessive defect resulting in a failure to show reproductive maturation after birth, while appearing to be normal otherwise. Low levels of LH and FSH are accompanied by infantile gonads and undeveloped secondary sex characteristics, with spermatogenesis absent in males and ovarian follicular development in females that rarely passes the antral stage. That the primary defect in hpg mice is the deficiency of brain GnRH has been indicated by several studies: plasma levels of LH and FSH increase in response to acute[18] or chronic[19] injections of GnRH, hpg ovaries and testes develop and function when transplanted to normal hosts,[20] and levels of GnRH in the hpg brain are either very low or absent[9] as seen by radioimmunoassay, while GnRH cells and fibers are undetectable by immunocytochemistry. Studies were designed to determine whether grafts that include the POA (and GnRH cells) from normal fetal mouse brain would function physiologically when transplanted into the brains of hpg mice.

MATERIALS AND METHODS

Mutant and normal mice of the hpg strain were housed in temperature (22-24°C) and light (LD 12:12) controlled rooms. Adult animals were anesthetized with chloral hydrate (12 mg, ip) and placed in a rat stereotaxic apparatus (David Kopf Instruments) fitted with a mouse adaptor. Either a 22-gauge needle attached via PE50 tubing to a 50-μl Hamilton syringe or a needle fitted with a plunger was used for the injections, which were placed midline at bregma; the needle was lowered 5.3 mm. These coordinates were chosen in order to place the graft in the anterior third ventricle, where grafted cells may have access to the median eminence. In most experiments, tissue for the grafts was obtained from 16- to 18-day-old fetal brains of normal animals of the same strain. For each graft, segments from two brains were pooled in a drop of 0.9% sterile saline and bisected. The total volume of the injections was 2-4 μl. In some studies similar-size segments were obtained from the fetal cortex or from the POA or cortex of mouse pups. In some experiments, representative brains from each group were reserved for GnRH immunocytochemistry and the remainder processed for radioimmunoassay of GnRH.[9] Pituitaries and plasma were stored at -20°C for radioimmunoassay of LH and FSH. Gonads and accessory sex organs were weighed and the gonads were processed for histological examination.

In studies with hpg females, the mice were examined periodically for vaginal opening. Vaginal cytology was assessed on cells obtained by vaginal lavage.

RESULTS AND DISCUSSION

The ability of POA brain grafts from normal fetuses to correct the reproductive deficiencies in adult genetically hypogonadal mice has been a consistent finding. In males[9,12] increased levels of pituitary and plasma LH and FSH were associated with testicular descent and development and increased weight of the androgen-dependent seminal vesicles. Evidence of spermatogenesis was seen (FIGURE 1). Gonadotropin levels similar to those in normal female mice were present in successfully grafted hpg females[10,11] who after vaginal opening, entered persistent vaginal estrus. Ovarian and uterine weights rose to the normal range within 2 months after graft surgery (FIGURE 2). Hypogonadal females in persistent estrus mated with normal males[11]; 7 of 10 became pregnant, and 6 of these delivered and raised healthy litters. Grafts of tissue obtained from newborn[9] or fetal[10] cortex or from the POA or cortex of 15-day-old mice[10] were ineffective in correcting any of the reproductive deficiencies in hpg mice.

Immunocytochemistry revealed that GnRH-positive cells were detectable (FIGURE 3) within the grafts in the host brains and that GnRH-containing fibers from the grafts abut capillaries in the median eminence, suggesting the feasibility of the peptide hormone being transported appropriately through the vasculature to the pituitary.

POA and the LH Surge

The POA region is regarded as the site whose integrity is essential for the ovulatory LH surge in rodents. Knife cuts that disconnect this anterior region from the medial basal hypothalamus,[21,22] or lesions that also include the suprachiasmatic nucleus region,[23,24] eliminate the estrogen-induced LH surge in rats, while electrical stimulation of the POA induces LH release.[25,26] The presumption is that the effects are through interruption or stimulation of GnRH release. There is no convincing evidence as yet that the POA grafts in hpg female mice support spontaneous ovulations since females with successful grafts generally enter continual vaginal estrus, and corpora lutea are not detectable in their ovaries. However, some of these females are capable of ovulation in response to normal male mating, as seen in the group where 7 of 10 mice became pregnant following pairing with a normal male for one night.[11] Studies continue to explore the possibility that in rare cases spontaneous cyclicity may be established.

Sexual Dimorphism in the POA

The rodent brain is particularly sensitive to its steroid environment during the perinatal period. The presence of androgens (presumably aromatized to estrogens)[27] at this time positively affects the capacity to show male sexual behavior in adulthood, and impairs the ability to produce ovulatory LH surges. Neuroanatomical differences between male and female brain have been described. Major sexual dimorphisms in the rat POA include the nature of synapses on dendritic spines[6] and the volume and cellular density of the sexually dimorphic nucleus of the preoptic area (SDN-POA).[7] These structural changes are reversible by neonatal castration of males or by androgen administration to females.[28,29] No such sexual dimorphisms have yet been clearly demonstrated in the mouse.[30,31] Nevertheless, administration of testosterone to 5-day-old female mice results in diminished fertility.[32,33] The precise timing of masculinization of the mouse brain has not yet been clarified. In the rat, a rise in fetal plasma testosterone levels at the 18th day of gestation[34] indicates that some masculine differentiation of the brain may occur at this timepoint. Although it is reported that plasma testosterone is elevated in neonatal mice[35] there is no information about such

FIGURE 1. Scanning electron micrograph of testis of hpg mouse, 2 months after fetal POA brain implant, reveals spermatozoa within the seminiferous tubule.

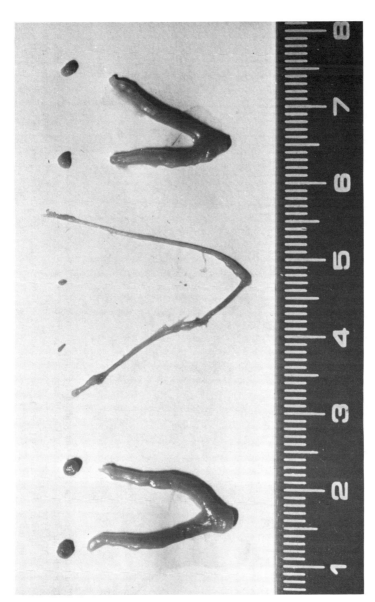

FIGURE 2. The ovaries and uterus of an hpg female with POA graft (right) are comparable to those of a normal mouse (left) as contrasted to the undeveloped ovaries and uterus of an hpg female who had received a cortex tissue graft (center).

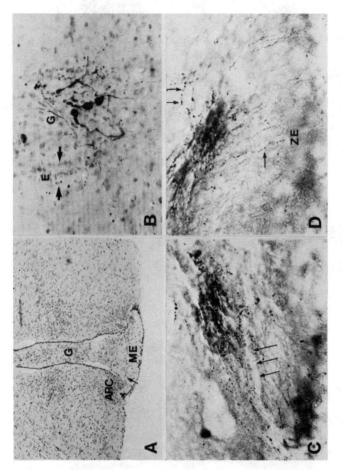

FIGURE 3. Photomicrographs of a POA brain graft in the third ventricle of the brain of an hpg mouse. A: Graft (G) is located within the third ventricle at the level of the arcuate nucleus (ARC) and appears to merge (arrow) with the median eminence (ME) of the host brain. B: GnRH cells present within the graft; (E) ependymal cells at wall of third ventricle. C: GnRH fibers abut a capillary (arrows) at lateral corner of the third ventricle. D: GnRH fibers at the lateral median eminence follow pathways observed in normal animals, for example, passing through the arcuate nucleus (upper arrows) and approaching the zona externa (ZE). (Reproduced with permission from Reference 11. Copyright 1984 by the AAAS.)

levels during gestation in this species. There remains the possibility that the mouse brain also is at least partially masculinized before birth.

In the studies with the hpg mice, however, there has been no indication that the sex of the donor of fetal POA plays any role in the effectiveness of the grafts in either male or female. Males with grafts derived from male fetuses and those with grafts derived from female fetuses showed no difference in testicular weight or testosterone-dependent seminal vesicle development[12] (TABLE 1). Similarly, four of seven female hpg mice that became pregnant had received grafts from male fetuses.[11] Tissue from neither 5-day-old neonatal males nor females has yet proven successful in supporting reproductive development in hpg females. It is thus not yet clearly established whether presumably fully differentiated tissue from males will stimulate full female development in hpg females.

Since the POA in rodents has also been implicated in the mediation of male sexual behavior,[3,36,37] the question may arise whether this behavior is deficient in the hpg mouse. In fact, the hpg male does not exhibit normal male sexual behavior following physiologically successful POA grafts, but this failure is most likely due to a secondary effect of the genetic defect in GnRH production. The hpg male brain is apparently not exposed in the perinatal period to sufficient androgens that are essential for the masculine differentiation of the brain.

TABLE 1. Weights of Testes and Seminal Vesicle in hpg Mice with POA Grafts Derived from Brain of Normal Fetal Male or Female

Sex of Donor	N	Weight of Testes (mg)	Weight of Seminal Vesicle (mg)
Male	16	87.5 ± 11.6[a]	74.3 ± 17.5
Female	17	64.8 ± 9.8	60.0 ± 13.4

[a] ± SEM.

In a test directed at the behavioral effects of male POA implants in females, Arendash and Gorski[8] placed POA grafts derived from 1-day-old male rat pups into the POA of the brains of 1-day-old female rats. When tested in adulthood, the females with bilateral grafts showed greater male mounting behavior in response to testosterone administration than did control rats or those who had received grafts of amygdala or caudate nucleus. Females with POA grafts also, however, showed unexpected increases in female lordotic behavior in response to minimal estrogen priming. This latter effect is reminiscent of that seen in tests following GnRH administration[38,39] to ovariectomized or ovariectomized, hypophysectomized rats. Since GnRH was not assessed in these grafts, the question remains whether its presence may have played a role in the effects on female sexual behavior seen in this study. It is also of interest that POA lesions facilitate sexual receptivity in female rats[40] since the introduction of foreign POA tissue into the female brain may have some lesion-like effect on the POA.

The authors[8] suggest that the increase in male sex behavior may be due to the above noted fact that male rat fetuses have elevated plasma testosterone at day 18 of gestation[34] since this may masculinize the donor tissue sufficiently to affect later behavior. Male rats that are castrated on day one appear to have bisexual capacities.

Although reductions in male sexual behavior in response to administration of testosterone in adulthood are seen,[41,42] this is most likely a peripheral effect, since neonatal castration is associated with greatly reduced penile length. When low dosages of androstenedione are administered neonatally to such males, sufficient to normalize penile length, adult male sexual behavior is normal in response to testosterone. The low dosages of androstenedione do not, however, affect the neonatal castrated male's female sexual behavior when treated with estrogen and progesterone in adulthood: lordotic responses are elicited which are not seen in males castrated at later ages.

The increase in male behavior seen in the females with male grafts may therefore be due to the origin of the grafts from male brains. However, failure to include a group that received grafts from *female* pups leaves open the question of whether some nondifferentiated characteristic of POA grafts may stimulate male behavior when bilaterally implanted in the host POA, since electrical stimulation of the POA has been shown to facilitate male behavior in the rat.[36]

POA and Maternal Behavior

The evidence to date indicates that the sole defect in the hpg mouse is that of GnRH production. As described above, when GnRH is supplied exogenously, through either injections or tissue grafts, other hormone-related deficiencies are corrected. Although many GnRH cell bodies are located in the POA region, it is important to remember that such cells are scattered, not localized to a particular nucleus, and constitute only a minute fraction of the total cell population of the region. There are at present no other known primary defects in the hpg mouse, and thus no evidence of a specific functional "POA" impairment in the mutant mice. Although the POA has been shown to be involved, for example, in mediating maternal behavior in rats,[4,5] there is no reason to believe that GnRH cells are involved in this behavior. The hpg females with successful grafts who bore young reared them to weaning. Nevertheless, analytical tests of maternal behavior have not yet been undertaken with these mice to fully examine this question.

SUMMARY

In conclusion, POA grafts that contain GnRH cell bodies are capable of correcting many of the reproductive deficiencies that are associated with a genetic failure to produce GnRH. The previously infertile animals exhibit steroidogenesis and gametogenesis. Ovulations associated with mating suggest that the grafted GnRH cells are responding to some environmental cues and may be under some regulation by the host brain. Although the grafts contain tissue that may have sexually dimorphic characteristics, the studies with the hpg mice have not yet revealed any effects of the grafts that could be related to the sex of the donor.

ACKNOWLEDGMENTS

The late Dorothy T. Krieger was the first to realize that the hpg mouse might be a good model for brain implants. I am grateful for her encouragement to join the studies in the initial planning stages and to the collaborators who have ensured their progress: Harry M. Charlton, Terry F. Davies, Michel Ferin, George Kokoris, Mark J. Perlow, Ann-Judith Silverman, and Earl A. Zimmerman.

REFERENCES

1. WITKIN, J. W., C. M. PADEN & A. J. SILVERMAN. 1982. The luteinizing hormone-releasing hormone (LHRH) systems in the rat brain. Neuroendocrinology **35:** 429-438.
2. CHRISTENSON, L. W. & R. A. GORSKI. 1978. Independent masculinization of neuroendocrine systems by intracerebral implants of testosterone or estradiol in the neonatal female rats. Brain Res. **146:** 325-340.
3. HEIMER, L. & K. LARSSON. 1966. Impairment of mating behavior in male rats following lesions of the preoptic-anterior hypothalamus continuum. Brain Res. **3:** 248-263.
4. NUMAN, M. 1974. Medial preoptic area and maternal behavior in the female rat. J. Comp. Physiol. Psychol. **87:** 746-759.
5. NUMAN, M., J. S. ROSENBLATT & B. R. KOMISARUK. 1977. Medial preoptic area and onset of maternal behavior in the rat. J. Comp. Physiol. Psychol. **91:** 146-164.
6. RAISMAN, G. & P. M. FIELD. 1971. Sexual dimorphism in the preoptic area of the rat. Science **173:** 731-733.
7. GORSKI, R. A., J. H. GORDON, J. E. SHRYNE & A. M. SOUTHAM. 1978. Evidence for a morphological sex difference within the medial preoptic area of the rat brain. Brain Res. **148:** 333-346.
8. ARENDASH, G. W. & R. A. GORSKI. 1982. Enhancement of sexual behavior in female rats by neonatal transplantation of brain tissue from males. Science **217:** 1276-1278.
9. KRIEGER, D. T., M. J. PERLOW, M. J. GIBSON, T. F. DAVIES, E. A. ZIMMERMAN, M. FERIN & H. M. CHARLTON. 1982. Brain grafts reverse hypogonadism of gonadotropin releasing hormone deficiency. Nature **298:** 468-471.
10. GIBSON, M. J., H. M. CHARLTON, M. J. PERLOW, E. A. ZIMMERMAN, T. F. DAVIES & D. T. KRIEGER. 1984. Preoptic area brain grafts in hypogonadal (hpg) female mice abolish effects of congenital hypothalamic gonadotropin-releasing hormone (GnRH) deficiency. Endocrinology **114:** 1938-1940.
11. GIBSON, M. J., D. T. KRIEGER, H. M. CHARLTON, E. A. ZIMMERMAN, A. J. SILVERMAN & M. J. PERLOW. 1984. Mating and pregnancy can occur in genetically hypogonadal mice with preoptic area brain grafts. Science **225:** 949-951.
12. SILVERMAN, A. J., E. A. ZIMMERMAN, M. J. GIBSON, M. J. PERLOW, H. M. CHARLTON, G. J. KOKORIS & D. T. KRIEGER. 1985. Implantation of normal fetal preoptic area into hypogonadal (hpg) mutant mice: temporal relationships of the growth of GnRH neurons and the development of the pituitary/testicular axis. Neuroscience **16:** 69-84.
13. CATTANACH, B. M., C. A. IDDON, H. M. CHARLTON, S. A. CHIAPPA & G. FINK. 1977. Gonadotrophin-releasing hormone deficiency in a mutant mouse with hypogonadism. Nature **269:** 338-340.
14. BJORKLUND, A., U. STENEVI & N. A. SVENDGAARD. 1976. Growth of transplanted monoaminergic neurones into the adult hippocampus along the perforant path. Nature **262:** 787-790.
15. JAEGER, C. B. & R. D. LUND. 1979. Efferent fibers from transplanted cerebral cortex of rats. Brain Res. **165:** 338-342.

16. PERLOW, M. J., W. J. FREED, B. J. HOFFER, A. SEIGER, L. OLSON & R. J. WYATT. 1979. Brain grafts reduce motor abnormalities produced by destruction of nigrostriatal dopamine system. Science **204:** 643-647.
17. GASH, D., J. R. SLADEK, JR. & C. D. SLADEK. 1980. Functional development of grafted vasopressin neurons. Science **210:** 1367-1369.
18. FINK, G., W. J. SHEWARD & H. M. CHARLTON. 1982. Priming effect of luteinizing hormone releasing hormone in the hypogonadal mouse. J. Endocrinol. **94:** 283-287.
19. CHARLTON, H. M., D. M. G. HALPIN, C. IDDON, R. ROSIE, G. LEVY, I. F. W. MCDOWELL, A. MEGSON, J. F. MORRIS, A. BRAMWELL, A. SPEIGHT, B. J. WARD, J. BROADHEAD, G. DAVEY-SMITH & G. FINK. 1983. The effects of daily administration of single and multiple injections of gonadotropin-releasing hormone on pituitary and gonadal function in the hypogonadal (hpg) mouse. Endocrinology **113:** 535-544.
20. BAMBER, S., C. A. IDDON, H. M. CHARLTON & B. J. WARD. 1980. Transplantation of the gonads of hypogonadal (hpg) mice. J. Reprod. Fertil. **58:** 249-252.
21. HALASZ, B. & R. GORSKI. 1967. Gonadotrophic hormone secretion in female rats after partial or total interruption of neural efferents to the medial basal hypothalamus. Endocrinology **80:** 608-622.
22. BLAKE, C. A., R. I. WEINER, R. A. GORSKI & C. H. SAWYER. 1972. Secretion of pituitary luteinizing hormone and follicle stimulating hormone in female rats made persistently estrous or diestrous by hypothalamic deafferentation. Endocrinology **90:** 855-861.
23. CLEMENS, J. A., E. B. SMALSTIG & B. D. SAWYER. 1976. Studies on the role of the preoptic area in the control of reproductive function in the rat. Endocrinology **99:** 728-735.
24. BROWN-GRANT, K. & G. RAISMAN. 1977. Abnormalities in reproductive function associated with the destruction of the suprachiasmatic nuclei in female rats. Proc. R. Soc. London Ser. B **198:** 279-296.
25. EVERETT, J. W. & H. M. RADFORD. 1961. Irritative deposits from stainless steel electrodes in the preoptic rat brain causing release of pituitary gonadotropin. Proc. Soc. Exp. Biol. Med. **108:** 604.
26. FINK, G., S. A. CHIAPPA & M. S. AIYER. 1976. Priming effect of luteinizing hormone releasing factor elicited by preoptic stimulation and by intravenous infusion and multiple injections of the synthetic decapeptide. J. Endocrinol. **69:** 359-372.
27. MACLUSKEY, N. J. & F. NAFTOLIN. 1981. Sexual differentiation of the central nervous system. Science **211:** 1294-1303.
28. RAISMAN, G. & P. M. FIELD. 1973. Sexual dimorphism in the neuropil of the preoptic area of the rat and its dependence on neonatal androgen. Brain Res. **54:** 1-29.
29. JACOBSON, C. D., V. J. CSERNUS, J. E. SHRYNE & R. A. GORSKI. 1981. The influence of gonadectomy, androgen exposure or a gonadal graft in the neonatal rat on the volume of the sexually dimorphic nucleus of the preoptic area. J. Neurosci. **1:** 1142-1147.
30. BLEIER, R., W. BYNE & I. SIGGELKOW. 1982. Cytoarchitectonic sexual dimorphisms of the medial preoptic and anterior hypothalamic areas in guinea pig, rat, hamster, and mouse. J. Comp. Neurol. **212:** 118-130.
31. YOUNG, J. K. 1982. A comparison of hypothalami of rats and mice: lack of gross sexual dimorphism in the mouse. Brain Res. **239:** 233-239.
32. BARRACLOUGH, C. A. & J. H. LEATHEM. 1954. Infertility induced in mice by a single injection of testosterone propionate. Proc. Soc. Exp. Biol. Med. **85:** 673-674.
33. PETERS, H. & I. N. SORENSEN. 1970. The fertility of mice after androgen injection in early infancy. Hormones **1:** 273-284.
34. WEISZ, J. & I. L. WARD. 1980. Plasma testosterone and progesterone titers of pregnant rats, their male and female fetuses, and neonatal offspring. Endocrinology **106:** 306-316.
35. PANG, S. F. & F. TANG. 1984. Sex differences in the serum concentrations of testosterone in mice and hamsters during their critical periods of neural sexual differentiation. J. Endocrinol. **100:** 7-11.
36. MALSBURY, C. W. 1971. Facilitation of male rat copulatory behavior by electrical stimulation of the medial preoptic area. Physiol. Behav. **7:** 797-805.
37. ARENDASH, G. W. & R. A. GORSKI. 1983. Effects of discrete lesions of the sexually dimorphic nucleus of the preoptic area or other medial preoptic regions on the sexual behavior of male rats. Brain Res. Bull. **10:** 147-154.

38. Moss, R. L. & S. M. McCann. 1973. Induction of mating behavior in rats by LH-RF. Science **181**: 177-179.
39. Pfaff, D. W. 1973. Luteinizing hormone-releasing factor potentiates lordosis behavior in hypophysectomized ovariectomized female rats. Science **182**: 1148-1149.
40. Powers, J. B. & E. S. Valenstein. 1972. Sexual receptivity: facilitation by medial preoptic lesions in female rats. Science **175**: 1003-1005.
41. Goldfoot, D. A., H. H. Feder & R. W. Goy. 1969. Development of bisexuality in the male rat treated neonatally with androstenedione. J. Comp. Physiol. Psychol. **67**: 41-45.
42. Stern, J. J. 1969. Neonatal castration, androstenedione, and the mating behavior of the male rat. J. Comp. Physiol. Psychol. **67**: 608-612.

Genital Stimulation as a Trigger for Neuroendocrine and Behavioral Control of Reproduction

BARRY R. KOMISARUK[a]

Institute of Animal Behavior
Rutgers-The State University of New Jersey
Newark, New Jersey 07102

JUDITH L. STEINMAN

Marine Biomedical Institute
Galveston, Texas 77550

NEUROENDOCRINE RESPONSES TO STIMULATION OF THE GENITAL TRACT

Mechanostimulation of internal and external genital structures in female mammals is essential to specific reproductive processes. This stimulation is normally provided by the male in the course of coitus or by the fetus during parturition.

Gonadotropin Secretion

Coital stimulation normally provokes ovulation in rabbits, cats, ferrets, mink, and voles,[1] which are therefore termed "reflexive ovulators." "Spontaneous ovulators" (e.g., rats, humans, mice, cattle) normally ovulate in response to cyclical changes in hormone levels,[2] but under certain conditions, such as estrogen-induced mating in diestrus,[3] barbiturate suppression of the neural activity preceding spontaneous ovulation,[4] or persistent estrus produced by constant light,[5] rats may ovulate reflexively. Thus the difference between reflexively and spontaneously ovulating species appears to be one of degree.[6]

In rabbits, coital stimulation depletes luteinizing hormone (LH) from the anterior pituitary within 15 min of mating[7] and in rats made to be in persistent estrus by exposure to continuous light, coital stimulation increases plasma LH within 10 min.[5] Artificial vaginal stimulation elevates plasma LH within 30 min in anesthetized rats.[8]

---

[a]Author to whom correspondence should be addressed.

Genital tract mechanostimulation inhibits gonadotropin secretion in chickens. The mechanostimulation of the oviduct provided by the presence of an egg provides an inhibitory timing signal for subsequent ovulation; simply inserting a thread into the oviduct inhibited ovulation for more than 25 days.[9]

Prolactin Secretion

Copulatory stimulation in rats also triggers the secretion of prolactin from the anterior pituitary (see Reference 10 and this volume). The prolactin stimulates ovarian progesterone secretion which in turn stimulates the growth of the uterus and prepares it for implantation of the ova. The amount of progesterone secreted is positively correlated with the number of intromissions that the female receives.[11] If the female rat receives too few (fewer than 3) intromissions prior to the ejaculation by the male, she does not become pregnant.[12] The reflexive prolactin secretion occurs in response to vaginal stimulation during coitus, for if the pelvic nerve (which conveys sensory input from the vagina and cervix[13]) is transected bilaterally, the prolactin surge in response to mating is markedly reduced.[14] The ova in pelvic-neurectomized rats do not implant unless exogenous prolactin[14] or progesterone[15] is administered. Pelvic neurectomy also abolishes artificial vaginal-cervical stimulation-induced pseudopregnancy (the corollary of pregnancy, during which the estrous cycle is suspended for 2 weeks, but fertilization does not occur).[16] A phenomenon of "delayed pseudopregnancy" was demonstrated by Everett.[17] He showed that if rats were anesthetized with barbiturate during the "critical period" for LH secretion on the afternoon of proestrus and allowed to mate upon recovery later the same day, they ovulated some days later and subsequently entered a delayed phase of pseudopregnancy. Everett postulated that when the corpora lutea formed subsequent to the delayed spontaneous ovulation, the persistent prolactin secretion resulting from the initial coitus was then capable of stimulating the progestational phase, thus maintaining the delayed pseudopregnancy for the characteristic 2-week period. This hypothesis was subsequently supported by the findings of others[18] who showed that vaginal stimulation triggers persistent daily surges in prolactin secretion.

Oxytocin Secretion

A third pituitary hormone that is released by vaginal stimulation is oxytocin. Copulatory stimulation in cows provokes uterine contractions (which may facilitate sperm transport) and milk letdown, both of which are responses to circulating oxytocin.[19] It has been reported that milk letdown occurs in lactating women during sexual intercourse[20] and plasma levels of oxytocin increase in women upon orgasm.[21] The Hottentot tribe uses a procedure of blowing air into the vagina in cows in order to facilitate milking.[20]

The neuroendocrine reflex in which oxytocin is released in response to mechanostimulation of the genital tract and then stimulates uterine contractions during parturition is known as the "Ferguson reflex."[22,23] Uterine contractions are stimulated by dilatation of the vaginal cervix or uterus, the most effective stimulus being stretching of the cervix. Lesioning the pituitary stalk or transecting the spinal cord at T12 abolishes the reflex.[22] Furthermore, lesions of the pituitary stalk or hypothalamic regions involved in oxytocin synthesis (e.g., paraventricular nuclei) retard delivery.[24,25] Stimulation of either the hypothalamus or posterior pituitary results in uterine contractions in rabbits[26] as well as concomitant increases in plasma oxytocin levels.[27] Vaginal distention has been shown to increase venous levels of oxytocin in ewes.[28] Similarly, oxytocin release was induced in the goat,[29] an effect potentiated by estrogen and diminished by progesterone. The levels of the enzyme oxytocinase, which metabolizes and inactivates oxytocin, were found to increase in human plasma as labor approaches.[30,31] This raises the possibility that the systemic circulation is equipped to degrade oxytocin that is released during pregnancy.

A lack of elevation in oxytocin levels at term[32] in humans casts doubt on the existence of an endogenously activated oxytocin release by uterine or vaginal distention. However, in another study, plasma oxytocin levels were found to increase during delivery in women,[33] and Mitchell et al.[34] demonstrated that short pulsatile peaks of oxytocin could be detected in the plasma of pregnant ewes at term, during labor, and at the precise delivery of each lamb. In addition, Vasicka et al.[35] clearly demonstrated that increases in plasma oxytocin levels specifically correspond to cervical dilatation and maximal vaginal distention at crowning in humans. No surges were observed at the onset of labor. Thus, negative reports may have resulted from sampling at inappropriate times. Historically, it is interesting that the initial observations by Ferguson[22] were accurate in that the "contractile-substance" was most effectively released in response to cervical dilatation.

BEHAVIORAL RESPONSES TO STIMULATION OF THE GENITAL TRACT

Vaginal mechanostimulation (VS) prolongs[36-38] and intensifies[39] the female's immobile mating stance (lordosis) which normally occurs during copulation. In unreceptive females, artificial VS induces receptivity.[40] The lordosis posture and immobilization facilitate intromission and ejaculation.[41] Furthermore, immobilization after ejaculation facilitates sperm transport in the female.[42]

VS alone has a strong immobilizing effect,[39] induces extension of the legs and digits,[41] and inhibits the righting response.[43] VS also produces a marked suppression of a variety of responses to noxious stimulation, e.g., leg flexion to foot pinch,[44,45] vibrissa retraction to ear pinch, and eye blink to corneal stimulation.[44] Vocalization in response to noxious tail shock is suppressed by VS. As the force applied to the vaginal cervix is increased using a calibrated spring-loaded probe, the vocalization threshold increases.[46] This effect is not due simply to motor inhibition, for the rat's ability to vocalize in response to non-noxious stimulation, e.g. lifting, is not suppressed by VS.[47] We further ruled out simple motor inhibition by using direct electrical stimulation of the pyramidal tract at current intensities just suprathreshold to elicit muscular movements of the face or limbs.[47] VS had no inhibitory effect on these movements, indicating that the inhibitory effect of VS acts proximal to the final common motor pathway.

These findings raised the question of the mode of action of VS-induced blockage of responses to noxious stimulation. Is the effect due to suppression of noxious sensory input? What are the loci of action of the suppression of responses to noxious stimulation? What is the neurochemical basis of the effect?

Evidence that Vaginal Stimulation is Analgesic

Neurons in the ventrobasal nuclear complex of the thalamus have been shown to respond to noxious stimulation in rats[48] and in monkeys.[49] We found that single and multiple thalamic neuronal responses to pinching the foot were markedly suppressed by VS. Furthermore, in single neurons that responded to both noxious and innocuous stimulation, VS differentially suppressed the response to noxious stimulation.[47]

A behavioral parallel is that during VS, while withdrawal responses to noxious stimulation are blocked, rats simultaneously respond to non-noxious tactile stimulation of the rump with the lordosis response.

VS-induced suppression of responses to pinch stimulation has also been reported for neurons in hypothalamus,[50] mesencephalic central gray,[51] nucleus gigantocellularis of the medulla,[52] and substantia nigra, zona compacta.[53] In addition, opposite neuronal responses to VS and foot pinch were observed in neurons in ventral tegmentum, midbrain reticular formation, and substantia nigra, although interactions were not reported in that study.[54]

Our finding[47] that VS induces differential suppression of afferent responses to noxious, but not innocuous, stimulation implies that VS is indeed analgesic, rather than that it is generally anesthetic, or that it blocks the rats' capability of performing a motor response to noxious stimulation. Furthermore, we found that rats performed an operant response which provided VS when they were confronted with noxious skin shock,[55] indicating that VS antagonizes the aversive property of skin shock.

While the above studies in rats support an interpretation that VS produces analgesia, a definitive determination requires a verbal report that VS suppresses pain. Consequently, we determined the effect of VS on pain and tactile thresholds in women.[56] Vaginal self-stimulation of calibrated force was applied using a plastic cylinder from a facial-type vibrator (motor and batteries removed) on which was mounted a pressure transducer. Pain thresholds were measured by a calibrated, steadily increasing compressive force (Ugo-Basile Analgesia Meter) applied to the fingers of the left hand. Tactile thresholds were determined on the dorsal surface of the hand using von Frey fibers. In three studies, with 10 women each, a variety of control conditions produced no significant change in the pain detection threshold (i.e., the compressive force at which pain was first perceived) or the pain tolerance threshold (force at which the pain was too uncomfortable to be increased further). That is, no significant increase in pain detection threshold occurred in response to pressing the transducer against the knee, contracting the pelvic muscles, watching a distracting film segment on a video monitor, or brushing a fur mitt on the skin.

Significant increases in pain detection threshold over control levels were observed in response to pressure applied to the anterior (but not the posterior) vaginal wall (47.4% increase), during VS that was described as pleasurable (71.4%), and during orgasm (106.7%) (FIGURE 1). Pain tolerance thresholds showed similar but less marked increases. Tactile thresholds showed no significant change under any condition. Thus, verbal reports that VS produces a differential increase in pain thresholds but not tactile thresholds support our earlier findings that VS suppresses thalamic neuronal responses to noxious but not innocuous cutaneous stimulation. On the basis of these findings we conclude that VS produces analgesia, i.e., a specific suppression of pain.

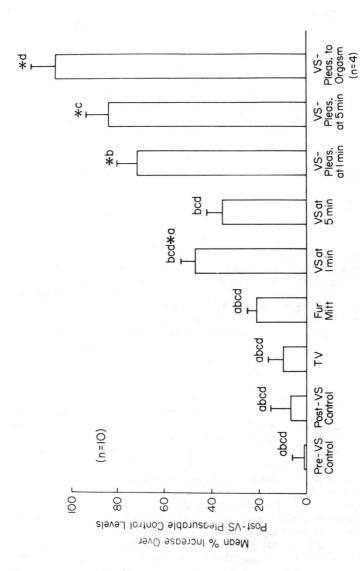

FIGURE 1. Pain detection thresholds during the various stimulus conditions. Values shown are group mean (± SEM) percentage increases over control (post-VS pleasurable control) levels. Significant differences (based upon ANOVA and subsequent Student Newman-Keuls tests, $p < 0.05$) between pairs of conditions are represented as follows: the value of any condition in which a letter is preceded by an asterisk is significantly greater than any other condition with the same letter that is not preceded by an asterisk. Example: the VS condition at 1 min is significantly greater than the Pre- or Post-VS control, TV, or Fur Mitt conditions, but not the VS-at-5 min condition. (From Whipple and Komisaruk.[56])

Further studies are necessary to relate these quantitative increases in pain threshold to subjective relief from clinical pain.

Pharmacological Bases of the Analgesic Action of Vaginal Stimulation

Role of Monoaminergic Systems

Systemic Studies. Our findings provide consistent evidence that VS activates an alpha-adrenergic mechanism which attenuates responses to noxious stimulation, and indicate a role of a serotoninergic mechanism.

Systemic administration of the tyrosine hydroxylase inhibitor, α-methyltyrosine (200 mg/kg), significantly reduced forebrain levels of norepinephrine (NE) and dopamine (DA) and significantly reduced the vocalization threshold elevation produced by VS to 60% of control levels. Specific inhibition of NE synthesis by administration of disulfiram significantly reduced the vocalization threshold elevation produced by VS to 66% of control levels. Addition of the NE precursor dihydroxyphenylserine to α-methyltyrosine-treated rats significantly and strongly potentiated the threshold-elevating effect of VS to 175% of control levels. None of these treatments significantly changed baseline (non-VS) vocalization thresholds.[57]

The role of serotonin (5-HT) was more difficult to interpret because the putative serotonin receptor blocker, cinanserin (25 mg/kg), strongly attenuated the effect of VS to 47% of control levels, but the presumed stimulation of serotonin receptors with 100 μg/kg LSD, or enhanced serotonin release with H 75/12, also attenuated the effect of VS.[57] Spinal studies (below) support a role for serotonin in the VS effect.

The effects of adrenergic agonists and antagonists are consistent with the effects of adrenergic synthesis blocker and precursor treatment. The antinociceptive effect of VS was doubled by the adrenergic agonist clonidine (100 μg/kg) and by the catecholamine releasing agent, H 77/77. Conversely, the alpha-adrenergic receptor blocking agent, phentolamine (100 or 25 mg/kg), significantly attenuated the effect of VS to 69% of control levels. The beta-adrenergic antagonist, sotalol (40 or 80 mg/kg), had no significant effect.[57]

Spinal Studies. Evidence that descending spinopetal monoaminergic systems mediate the effect of VS was obtained through the use of direct administration of monoaminergic receptor blocking agents or neurotoxins directly to the spinal cord, and by measurement of monoamines in the cord or the superfusate of the cord after VS.[58]

The alpha-adrenergic receptor blocking agent, phentolamine, or the serotoninergic receptor blocking agent, methysergide, was injected perispinally, i.e., intrathecally, into the lumbar area via chronically implanted polyethylene catheters.[59] On both the vocalization threshold test and the tail flick latency test, the effect of VS was significantly attenuated by injection of 40 μg phentolamine. Methysergide (10 μg) significantly attenuated the effect of VS on the vocalization test. Combined injection of 10 μg phentolamine and 10 μg methysergide had an additive effect, significantly attenuating the effect of VS on the tail flick test. The antagonistic effect of 10 μg methysergide on the antinociceptive effect of VS is consistent with that of systemically injected cinanserin which also antagonized the effect of VS on the vocalization threshold.[58]

As an alternative approach to the local administration of monoaminergic receptor blocking agents, monoamine neurotoxins were administered to the spinal cord. Microinjections of 6-hydroxydopamine (6-OHDA), a toxin for catecholaminergic neurons,[60] were made bilaterally into the spinal cord at the level of the cisterna magna (10 μg in 1 μl) and the rats were tested 16 days later. The effect of VS on the vocalization threshold was significantly attenuated to 60% of vehicle control levels, and the level of norepinephrine measured in the spinal cord was reduced from a mean level of 198 ng/g in the controls to a mean of 42 ng/g in the 6-OHDA rats.[57] These findings further support a role for NE mediating the antinociceptive effect of VS.

In a separate experiment 5,7-dihydroxytryptamine (5,7-DHT), a neurotoxin for serotonergic neurons,[61] was injected intrathecally and 3 days later the effects of VS on the vocalization threshold were determined. Doses of 4 or 10 μg 5,7-DHT prevented VS from significantly elevating the vocalization threshold (i.e., the increases in vocalization threshold induced by VS were 6 and 18%, respectively, compared to 91% in the saline controls).[62]

We hypothesized that if VS acts via NE and serotonin, then these transmitters would be released at neuron terminals in response to VS. We found that VS significantly increased by twofold the levels of norepinephrine and serotonin released into spinal superfusate.[58] These findings provide further evidence that the antinociceptive effect of VS is mediated at least in part by release of norepinephrine and serotonin from nerve terminals in the spinal cord, the cell bodies of which are in the lower brainstem.[63,64] In support of this, it is noteworthy that VS produces a significant increase in uptake of ^{14}C-labeled 2-deoxy-D-glucose (2-DG) in midbrain dorsal raphe.[65] Thus, VS may utilize noradrenergic and serotoninergic systems that have been shown in other contexts to exert antinociceptive effects. That is, several lines of evidence indicate that analgesia can be produced by stimulation of noradrenergic and serotoninergic neurons with cell bodies located in the lower brainstem and axons terminating in the spinal cord:

(1) Direct intrathecal application of NE to the spinal cord produces analgesia, an effect antagonized competitively by phentolamine.[66] Similarly, analgesia produced by intrathecal administration of 5-HT[67] is attenuated by methysergide, potentiated by 5-HT uptake blockers, and mimicked by other putative 5-HT agonists.[67]

(2) Analgesia is produced by electrical stimulation of lower brainstem regions (e.g., locus coeruleus) containing NE[68] or 5-HT neurons (e.g., dorsal or central nuclei of the raphe),[69,70] whose axons descend to the spinal cord.[63,64]

(3) Neuronal responses to noxious stimulation are attenuated by local iontophoretic application of adrenergic and serotoninergic agents. Thus, iontophoretic ejection of NE into the dorsal horn of the spinal cord in cats selectively inhibits responses of dorsal horn neurons to noxious, but not innocuous, sensory stimulation.[71,72] NE ejected iontophoretically also inhibits spontaneous and synaptic firing of spinal cord interneurons[73] and reduces the excitability of cutaneous primary afferent C-fibers.[74] The inhibitory effects of NE on spinal cord motoneurons in cats are due to hyperpolarization of the neuronal membrane[75,76] which is associated with a decrease in membrane conductance.[77] Increased firing rate of lamina V cells to intraarterial bradykinin was inhibited by electrical stimulation of the bulbar reticular formation (particularly the nucleus reticularis gigantocellularis) which contains cell bodies of descending noradrenergic neurons.[78] This effect was blocked by systemic administration of tetrabenazine, a monoamine depletor, and reversed by L-DOPA. 5-HT ejected into lamina II by electrophoresis reduced responses of neurons in laminae IV and V to a noxious heat stimulus.[71,72] The increased firing produced by noxious mechanical stimulation in spinothalamic tract cells identified by antidromic stimulation was diminished by the iontophoretic application of 5-HT.[79]

Role of Supraspinal Opiate Systems

There is evidence that these descending spinopetal serotoninergic and noradrenergic pathways mediate opiate-induced analgesia. Microinjection of morphine into the periaqueductal gray, which shows immunoreactive staining for enkephalin,[80–82] induces analgesia that is antagonized by phentolamine[66,83] or methysergide[83] injected intrathecally into the spinal cord. Morphine injected into the periaqueductal gray increases the release of serotonin into spinal superfusate.[84] Morphine administered locally to the locus coeruleus, which also shows immunoreactive staining for enkephalin,[80] increased nociceptive thresholds,[85] and morphine injected systemically increased the levels of metabolites of norepinephrine and serotonin in the spinal cord, but not after cord transection at C1.[86,87] Lesions of the locus coeruleus,[88] dorsal raphe,[88] or raphe magnus and pallidus[89] reduced the analgesic effect of morphine administered systemically.

These findings raise the question of whether VS activates an endogenous opiate system which in turn activates the descending noradrenergic and serotonergic spinopetal systems. We therefore investigated whether elevation of the vocalization threshold by VS would be antagonized by administration of naloxone or induction of morphine tolerance. Naloxone (10 mg/kg) did not affect the ability of VS to elevate the vocalization threshold,[90] and rats made tolerant to morphine by 35 days of morphine treatment also showed no attenuation of the VS-induced increase in vocalization threshold. Thus, on this measure, VS did not show cross-tolerance to morphine. However, Hill and Ayliffe[91] reported that 10 mg/kg naloxone did attenuate the effect of VS on different measures, i.e., elevation of the tail flick latency to radiant heat and the latency to withdraw the tail from hot water. Thus, we repeated the experiment, this time using both tail flick and vocalization threshold measures. In this study, we confirmed the results of both our earlier study and that of Hill and Ayliffe.[91] That is, naloxone did not significantly attenuate the effect of VS on the vocalization threshold, but did attenuate the effect of VS on tail flick latency.[92] In further support of an opiate component to VS-produced analgesia, we found that induction of morphine tolerance by twice-daily injections of morphine sulfate (20 mg/kg) for 10 days significantly reduced the effect of VS on the tail withdrawal latency.[92] These findings suggest that an opiate-monoaminergic system mediates the VS-produced increase in tail flick latency, but VS may bypass the opiate component in producing an increase in vocalization threshold to tail shock.

There is evidence that descending monoaminergic pathways that are opiate sensitive and mediate analgesia, course through the dorsolateral funiculus (DLF). The DLF contains descending serotoninergic axons whose cell bodies are located in the raphe nuclei,[63,93] particularly raphe magnus and also reticularis magnocellularis which together are coextensive with serotoninergic group B3[94] of Dahlstrom and Fuxe.[63] The DLF also contains descending noradrenergic axons whose cell bodies are located in nuclei A1-3.[64] Electrical stimulation of these nuclear regions produces analgesia.[68–70,78] Furthermore, lesions of the DLF have been shown to antagonize analgesia produced by electrical stimulation of the raphe nuclei,[95] morphine microinjections into the periaqueductal gray,[96] systemic morphine administration,[68] and opiate-dependent front paw shock-induced analgesia.[97,98] It is likely that VS utilizes this system, for we have found that DLF lesions significantly attenuate the increase in tail flick latency induced by VS.[99]

A role of endogenous opiates in VS-produced analgesia is also indicated by localization of immunoreactive staining for enkephalin in the dorsal raphe[80] and the medial preoptic area,[80,81,100] both of which show increased uptake of labeled 2-DG in response to VS.[65] Although no significant increase in uptake of labeled 2-DG was found in periaqueductal gray in response to VS,[65] Pert[101] stated that VS reduced endogenous opiate levels measured in periaqueductal gray, implying a stimulation-induced depletion.

Role of the Intraspinal System

In addition to the monoaminergic and opiate mechanisms that mediate the effects of VS, it is important to note that even in rats with complete spinal transection at the midthoracic level, VS can still immediately attenuate the leg withdrawal response to foot pinch.[44] In rats whose spinal cords are transected at T2, VS significantly elevates tail flick latency (TFL), but to a significantly lesser degree than in intact or decerebrate rats.[99] Thus, there are intraspinal as well as descending mechanisms that contribute to the VS effect.

There is an intrinsic spinal opiate system that may mediate the VS effect. In spinal cord, enkephalin-like immunoreactivity is most dense in laminae I (marginal zone)[102] and II (substantia gelatinosa).[80] It has been shown through autoradiographic identification with a tritiated opiate antagonist that opiate receptors are also localized in laminae I and II.[103] The opiate receptor density in these laminae was reduced by sensory deafferentation[104,105] providing evidence that the receptors are located on primary afferent terminals and suggesting that opiates may block pain input there by presynaptic inhibition.[105] Since local administration of morphine to the spinal cord induces analgesia,[106] reflexive release of intrinsic spinal opiates could produce analgesia at the spinal level.

A glycinergic component also mediates the analgesia produced by VS. We have found recently that as little as 5 μg strychnine, a glycine receptor antagonist administered locally to the spinal cord, suppresses the elevation in vocalization threshold produced by VS and antagonizes the ability of VS to block the distress vocalization response to brushing yeast-inflamed skin.[107]

Functional Significance

The analgesia produced by VS may both facilitate impregnation and reduce the stress of parturition. While moderate amounts of vaginal stimulation facilitate sexual receptivity,[40] excessive amounts are inhibitory.[108] Unless female rats receive a moderate amount of vaginal stimulation prior to the male's ejaculation, they do not become pregnant.[12] Consequently, an analgesia-producing effect of VS may make the females tolerant of a sufficient number of intromissions to adequately trigger the neuroendocrine reflexes necessary for impregnation; i.e., without such an analgesic effect, perhaps their sexual receptivity would decrease before they would tolerate the number of intromissions necessary for sufficient prolactin-progesterone secretion[11] for impregnation.

In the case of pregnancy and parturition, analgesia increases in the last week of pregnancy, reaching a peak at parturition and declining to baseline levels the following day; this analgesia is naltrexone reversible.[109] Uterine denervation reduces the increase in analgesia fivefold.[110] Perhaps the increased afferent stimulation generated by distension of the uterus, cervix, and vaginal canal by the fetuses during delivery, produces analgesia, which could decrease the stress of parturition and consequently facilitate lactation and maternal behavior. Further studies are necessary to test these hypotheses.

REFERENCES

1. ROWLANDS, I. W. 1966. In Proceedings, Zoological Society of London Symposia, No. 15. Academic Press. New York, N.Y.
2. FEDER, H. H. 1981. In Neuroendocrinology of Reproduction. N. T. Adler, Ed.: 279-348. Plenum Press. New York, N.Y.
3. ARON, C., G. ASCH & J. ROOS. 1966. Int. Rev. Cytol. **20:** 139.
4. HARRINGTON, R. E., R. G. EGGER & R. D. WILBUR. 1967. Endocrinology **81:** 877.
5. BROWN-GRANT, K., J. M. DAVIDSON & F. GREIG. 1973. J. Endocrinol. **57:** 7.
6. ZARROW, M. X. & J. H. CLARK. 1968. J. Endocrinol. **40:** 343.
7. DESJARDINS, C., K. T. KIRTON & H. D. HAFS. 1967. Proc. Soc. Exp. Biol. Med. **126:** 23.
8. BLAKE, C. A. & C. H. SAWYER. 1972. Neuroendocrinology **10:** 358.
9. HUSTON, T. M. & A. V. NALBANDOV. 1953. Endocrinology **52:** 149.
10. TERKEL, J. & C. H. SAWYER. 1978. Horm. Behav. **11:** 304-309.
11. ADLER, N. T., J. A. RESKO & R. W. GOY. 1970. Physiol. Behav. **5:** 1003.
12. ADLER, N. T. 1969. J. Comp. Physiol. Psychol. **69:** 613.
13. KOMISARUK, B. R., N. T. ADLER & J. HUTCHISON. 1972. Science **178:** 1295-1298.
14. SPIES, H. G. & G. D. NISWENDER. 1971. Endocrinology **88:** 937.
15. KOLLAR, E. J. 1953. Anat. Rec. **115:** 641-658.
16. CARLSON, R. R. & V. J. DEFEO. 1965. Endocrinology **77:** 1014.
17. EVERETT, J. W. 1967. Endocrinology **80:** 145.
18. FREEMAN, M. E. & J. D. NEILL. 1972. Endocrinology **90:** 1292.
19. VANDEMARK, N. L. & R. L. HAYS. 1952. Am. J. Physiol. **170:** 518.
20. FOLLEY, S. J. 1969. J. Endocrinol. **44:** x.
21. FOX, C. A. & G. S. KNAGGS. 1969. J. Endocrinol. **45:** 145-146.
22. FERGUSON, J. K. W. 1941. Surg. Gynecol. Obstet. **73:** 359-366.
23. HATERIUS, H. O. & J. K. W. FERGUSON. 1938. Am. J. Physiol. **124:** 314-321.
24. DEY, F. L., C. FISHER & S. W. RANSON. 1942. Am. J. Obstet. Gynecol. **42:** 459.
25. NIBBELINK, D. W. 1961. Am. J. Physiol. **200:** 1229.
26. CROSS, B. A. 1959. In Endocrinology of Reproduction. C. L. Lloyd, Ed.: 441. Academic Press. New York, N.Y.
27. BOER, K., K. CRANSBERG & J. DOGTEROM. 1980. Neuroendocrinology **30:** 313-318.
28. FLINT, A. P. F., M. L. FORSLING, M. D. MITCHELL & C. TURNBULL. 1975. J. Reprod. Fertil. **43:** 551-554.
29. ROBERTS, J. S. & L. SHARE. 1969. Endocrinology **34:** 1076.
30. FEKETE, D. 1980. Endokrinologie **7:** 364.
31. CHARD, T. 1972. J. Reprod. Fertil. Suppl. **16:** 121-138.
32. OTSUKI, Y., K. YAMAJI, M. FUJITA, T. TAKAGI & O. TANIZAWA. 1983. Acta Obstet. Gynecol. Scand. **62:** 15-18.
33. GOODFELLOW, C. F., M. G. R. HULL, D. F. SWAAB, J. DOGTEROM & R. M. BUIJS. 1983. Br. J. Obstet. Gynaecol. **90:** 214-219.
34. MITCHELL, M. D., D. L. KRAEMER, S. P. BRENNECKE & R. WEBB. 1982. Biol. Reprod. **27:** 1169-1173.
35. VASICKA, A., P. KUNARESAN, G. S. HAN & M. KUMARESAN. 1978. Am. J. Obstet. Gynecol. **130:** 263-273.
36. KUEHN, R. E. & F. A. BEACH. 1963. Behaviour **21:** 282-299.
37. BERMANT, G. & W. H. WESTBROOK. 1966. J. Comp. Physiol. Psychol. **61:** 244-250.
38. DIAKOW, C. 1975. J. Comp. Physiol. Psychol. **89:** 704-712.
39. KOMISARUK, B. R. & C. DIAKOW. 1973. Endocrinology **93:** 548-557.

40. RODRIGUEZ-SIERRA, J. F., W. R. CROWLEY & B. R. KOMISARUK. 1975. J. Comp. Physiol. Psychol. **89:** 79-85.
41. KOMISARUK, B. R. 1978. *In* Biological Determinants of Sexual Behavior. J. B. Hutchison, Ed.: 349-393. John Wiley. New York, N.Y.
42. MATTHEWS, M. & N. T. ADLER. 1977. J. Comp. Physiol. Psychol. **91:** 721-741.
43. NAGGAR, A. N. & B. R. KOMISARUK. 1977. Physiol. Behav. **19:** 441-444.
44. KOMISARUK, B. R. & K. LARSSON. 1971. Brain Res. **35:** 231-235.
45. KOMISARUK, B. R., V. CIOFALO & M. B. LATRANYI. 1976. *In* Advances in Pain Research and Therapy Vol. 1. J. J. Bonica & D. Albe-Fessard, Eds.: 439-443. Raven Press. New York, N.Y.
46. CROWLEY, W. R., R. JACOBS, J. VOLPE, J. F. RODRIGUEZ-SIERRA & B. R. KOMISARUK. 1976. Physiol. Behav. **16:** 483-488.
47. KOMISARUK, B. R. & J. WALLMAN. 1977. Brain Res. **137:** 85-107.
48. MITCHELL, D. & R. F. HELLON. 1977. Proc. R. Soc. London Ser. B **197:** 169-194.
49. FOREMAN, R. D., R. F. SCHMIDT & W. D. WILLIS. 1977. Brain Res. **124:** 555-560.
50. BARRACLOUGH, C. A. & B. A. CROSS. 1963. J. Endocrinol. **26:** 339-359.
51. PETTY, L. C. 1975. Unpublished doctoral dissertation. Virginia Commonwealth University. Richmond, Va.
52. HORNBY, J. B. & J. D. ROSE. 1976. Exp. Neurol. **51:** 363-376.
53. CHIODO, L. A., A. R. CAGGIULA, S. M. ANTELMAN & C. G. LINEBERRY. 1979. Brain Res. **176:** 385-390.
54. MAEDA, H. & G. J. MOGENSON. 1982. Br. Res. Bull. **8:** 7-14.
55. ROSS, E. L., B. R. KOMISARUK & D. O'DONNELL. 1979. J. Comp. Physiol. Psychol. **93:** 330-336.
56. WHIPPLE, B. & B. R. KOMISARUK. 1985. Pain **21:** 357-367.
57. CROWLEY, W. R., J. F. RODRIGUEZ-SIERRA & B. R. KOMISARUK. 1977. Brain Res. **137:** 67-84.
58. STEINMAN, J. L., B. R. KOMISARUK, T. L. YAKSH & G. M. TYCE. 1983. Pain **16:** 155-166.
59. YAKSH, T. L. & T. A. RUDY. 1976. Physiol. Behav. **17:** 1031-1036.
60. KOSTREZEWA, R. M. & D. M. JACOBOWITZ. 1974. Pharmacol. Rev. **26:** 199-388.
61. BREESE, G. R. 1975. *In* Biochemical Principles and Techniques in Neuropharmacology. L. L. Iversen, S. D. Iversen & S. H. Snyder, Eds.: 137-189. Plenum Press. New York, N.Y.
62. STEINMAN, J. L. & B. R. KOMISARUK. 1981. Proc. Soc. Neurosci. **7:** 583.
63. DAHLSTROM, A. & K. FUXE. 1965. Acta Physiol. Scand. Suppl. **232:** 1-55.
64. NYGREN, L.-G. & L. OLSON. 1977. Brain Res. **132:** 85-93.
65. ALLEN, T. O., N. T. ADLER, J. H. GREENBERG & M. REIVICH. 1981. Science **211:** 1070-1072.
66. REDDY, S. V. R., J. L. MADERDRUT & T. L. YAKSH. 1980. J. Pharmacol. Exp. Ther. **213:** 523-533.
67. YAKSH, T. L. & P. R. WILSON. 1979. J. Pharmacol. Exp. Ther. **208:** 446-453.
68. SEGAL, M. & D. SANDBERG. 1977. Brain Res. **123:** 369-372.
69. OLIVERAS, J. L., G. GUILBAUD & J. M. BESSON. 1979. Brain Res. **164:** 317-322.
70. OLIVERAS, J. L., F. REDJEMI, G. GUILBAUD & J. M. BESSON. 1975. Pain **1:** 139-145.
71. BELCHER, G., R. W. RYALL & R. SCHAFFNER. 1978. Brain Res. **151:** 307-321.
72. HEADLEY, P. M., A. W. DUGGAN & B. T. GRIERSMITH. 1978. Brain Res. **145:** 185-189.
73. ENGBERG, I. & R. W. RYALL. 1966. J. Physiol. **185:** 298-322.
74. JEFTINIJA, S., K. SEMBA & M. RANDIC. 1981. Brain Res. **219:** 456-463.
75. PHILLIS, J. W., A. K. TEBECIS & D. H. YORK. 1968. Eur. J. Pharmacol. **4:** 471-475.
76. ENGBERG, I. & A. THALLER. 1970. Acta Physiol. Scand. **80:** 34A-35A.
77. ENGBERG, I. & K. C. MARSHALL. 1971. Acta Physiol. Scand. **83:** 142-144.
78. TAKAGI, H., T. DOI & K. KAWASAKI. 1975. Life Sci. **17:** 67-72.
79. JORDAN, L. M., D. R. KENSHALO, JR., R. F. MARTIN, L. H. HABER & W. D. WILLIS. 1978. Pain **5:** 135-142.
80. SAR, M., W. E. STUMPF, R. J. MILLER, K.-J. CHANGE & P. CUATRECASAS. 1978. J. Comp. Neurol. **182:** 17-38.

81. ELDE, R., T. HOKFELT, O. JOHANSSON & L. TERENIUS. 1976. Neuroendocrinology **1:** 349-351.
82. GOODMAN, R. R., S. H. SNYDER, M. J. KUHAR & W. S. YOUNG III. 1980. Proc. Natl. Acad. Sci. USA **77:** 6239-6243.
83. YAKSH, T. L. 1979. Brain Res. **160:** 180-185.
84. YAKSH, T. L. & G. M. TYCE. 1979. Brain Res. **171:** 176-181.
85. YAKSH, T. L., J. C. YEUNG & T. A. RUDY. 1976. Brain Res. **114:** 83-103.
86. SHIOMI, H., H. MURAKAMI & H. TAKAGI. 1978. Eur. J. Pharmacol. **52:** 335-344.
87. SHIOMI, H. & H. TAKAGI. 1974. Br. J. Pharmacol. **52:** 519-526.
88. SASA, M., K. MUNEKIYO, Y. OSUMI & S. TAKAORI. 1977. Eur. J. Pharmacol. **42:** 53-62.
89. PROUDFIT, H. K. 1980. Pharmacol. Biochem. Behav. **13:** 705-714.
90. CROWLEY, W. R., J. F. RODRIGUEZ-SIERRA & B. R. KOMISARUK. 1977. Psychopharmacology **54:** 223-225.
91. HILL, R. G. & S. J. AYLIFFE. 1981. Pharmacol. Biochem. Behav. **14:** 631-632.
92. STEINMAN, J. L., L. A. ROBERTS & B. R. KOMISARUK. 1982. Proc. Soc. Neurosci. **8:** 47.
93. BASBAUM, A. & H. L. FIELDS. 1978. Ann. Neurol. **4:** 451-515.
94. WATKINS, L. R., G. GRIFFIN, G. R. LEICHNETZ & D. J. MAYER. 1980. Brain Res. **181:** 1-15.
95. BASBAUM, A. I., N. MARLEY & J. O'KEEFE. 1976. *In* Advances in Pain Research and Therapy Vol. 1. J. J. Bonica & D. Albe-Fessard, Eds.: 511-515. Raven Press. New York, N.Y.
96. MURFIN, R., G. J. BENNETT & D. J. MAYER. 1976. Proc. Soc. Neurosci. **2:** 946.
97. HAYES, R. L., D. D. PRICE, G. J. BENNETT, G. L. WILCOX & D. J. MAYER. 1978. Brain Res. **155:** 91-101.
98. MAYER, D. J. & L. R. WATKINS. 1981. *In* Modern Problems of Pharmacopsychiatry: The Role of Endorphins in Neuropsychiatry. H. M. Emrich, Ed. S. Karger AG. Basel, Switzerland.
99. WATKINS, L. R., P. L. FARIS, B. R. KOMISARUK & D. J. MAYER. 1984. Brain Res. **294:** 59-65.
100. JOHANSSON, O. & T. HOKFELT. 1980. J. Histochem. Cytochem. **28:** 364-366.
101. PERT, A. 1980. Acupuncture found to trigger endorphins in rat brain. Alcohol, Drug Abuse, and Mental Health Administration News 6(No. 12): 1,5.
102. SEYBOLD, V. & R. ELDE. 1980. J. Histochem. Cytochem. **28:** 367-370.
103. ATWEH, S. F. & M. J. KUHAR. 1977. Brain Res. **124:** 53-67.
104. LAMOTTE, C., C. B. PERT & S. H. SNYDER. 1976. Brain Res. **112:** 407-412.
105. FIELDS, H. L., P. C. EMSON, B. K. LEIGH, R. F. T. GILBERT & L. L. IVERSEN. 1980. Nature **284:** 351-353.
106. YAKSH, T. L. & T. A. RUDY. 1977. J. Pharmacol. Exp. Ther. **202:** 411-428.
107. ROBERTS, L., C. BEYER & B. R. KOMISARUK. 1985. Life Sci. **36:** 2017-2023.
108. HARDY, D. F. & J. F. DEBOLD. 1973. J. Comp. Physiol. Psychol. **85:** 195-202.
109. GINTZLER, A. R. 1980. Science **210:** 193-195.
110. GINTZLER, A. R., L. C. PETERS & B. R. KOMISARUK. 1983. Brain Res. **277:** 186-188.

Neuroendocrinology of Coitally and Noncoitally Induced Pseudopregnancy[a]

JOSEPH TERKEL

Department of Zoology
George S. Wise Center for the Life Sciences
Tel Aviv University
Tel Aviv, Israel

INTRODUCTION

Pseudopregnancy (PSP) is commonly referred to as a state in which the uterus undergoes a decidual reaction in response to uterine trauma.[1] It arises when luteotrophic factors activate the corpora lutea which then secrete progesterone.[2] The induction of the progestational state is requisite for the structural changes in the uterus in preparation for implantation of the fertilized egg.

The primary luteotrophic substance in rats was discovered many years ago to be prolactin (PRL).[3,4] This led to the expectation of substantial elevation of this hormone in the circulation throughout pregnancy and PSP. Yet, on the basis of once-a-day blood sampling procedures, PRL levels were found to be low throughout pregnancy with the exception of two significant elevations. The first occurs on the fourth day of pregnancy and is related to implantation, while the second occurs just prior to parturition and is associated with preparation for lactation.[5] Only when around-the-clock sampling procedures were instituted in 1972 was the daily pattern of PRL secretion during both pregnancy[6] and PSP[7] discovered. During PSP and for the first 10 days of pregnancy, PRL is secreted in two daily "surges," one nocturnal and one diurnal (for review see Reference 8). This circadian rhythm is maintained until termination of PSP by luteolysis,[9] while during pregnancy the peaks cease at midgestation, at a time corresponding to the development of the placenta.[10,11] Thus, another characteristic can be added to the classic definition of PSP—the semicircadian or twice-daily pattern of PRL secretion. The significance of this additional characteristic of PSP lies in the essential role of PRL in initiating the progestational state.[12]

In this paper, I do not attempt to provide a comprehensive overview of the role of PRL in the neuroendocrine regulation of PSP. Rather, I outline the research perspective of our laboratory, in which we explore the phenomena of nocturnal PRL secretion and the regulation of coitally induced PSP at several biological levels. A variety of noncoital factors can also induce PSP, and I compare their underlying control mechanisms with those of coitally induced PSP.

[a]The research reported in this paper was supported by the American-Israel Binational Foundation (BSF), The Israel Academy of Sciences, and The Israel Center for Psychobiology, Jerusalem, Israel.

Our interest in the regulation of PSP arose from our view that each pregnancy originates as a direct continuation of the initial pseudopregnant state triggered by the neural signal of mating. We emphasize the concept that the specific copulatory stimulus of the male initiates a "mnemonic system" (a system that assists in retaining and expressing memory for some time) which is essential for repetitive circadian PRL secretion.[13] We believe the "mnemonic system" to be the key to understanding the extension of PSP into pregnancy. Because in their initial stages PSP and pregnancy are identical in their regulation,[4] these terms are used interchangeably in this paper.

COPULATORY BEHAVIOR TRIGGERS NIGHTLY PROLACTIN SURGES

In species that are reflex ovulators (e.g., cats, rabbits, mink, ferrets, voles, and shrews), ovulation depends on the copulatory stimulation that the female receives from the male. On the other hand, in species that are spontaneous ovulators (e.g., rats, mice, hamsters, guinea pigs, cows, and primates), ovulation does not depend on the behavior of the male but is autonomously controlled by endogenous factors and occurs in every cycle.[14] Hence, one might assume that the function of the male during copulation in spontaneously ovulating species is merely to deposit sperm in the female's vagina at the appropriate time of the estrous cycle, i.e., at the time of ovulation, after which fertilization and pregnancy ensue. This assumption about the role of the male has important economic ramifications in the field of artificial insemination. For example, deposition of sperm in the reproductive tract of cows by artificial insemination has yielded significant results both in financial savings and in improved breeding and husbandry. It is surprising, then, that artificial insemination in rats, mice, and hamsters has not been successful. In hamsters, mere deposition of sperm in the vagina without additional vaginal stimulation does not result in pregnancy.[15] Why should artificial insemination be successful in some spontaneous ovulators such as cows and monkeys but unsuccessful in others such as rats and mice? The basis for this difference may be in the lengths of their estrous cycles. In spontaneous ovulators with long estrous cycles such as cows and monkeys, the corpora lutea are functional and secrete increasing amounts of progesterone for several days, thus permitting uterine implantation of the fertilized egg. In contrast, rats like mice and hamsters, have a short, incomplete estrous cycle, lacking a spontaneous luteal phase, and the corpora lutea of the estrous cycle are nonfunctional.[16] The newly formed corpora lutea of the cycling rat secrete progesterone only on the day of metestrus, with progesterone declining on the following day.[17] If the female does not mate during proestrus, and additional luteotrophic support is not provided, the corpora lutea subsequently fail and progesterone secretion wanes.[12] Hence, it is possible that one function of coital stimulation is to maintain newly formed corpora lutea and to prevent their degeneration.

The role of coital stimulation in the maintenance of pregnancy has been most thoroughly studied in rats. During the 4-day estrous cycle, the female rat accepts the male for copulation only on the night of proestrus.[18] When mating, the male mounts, intromits, and dismounts from the female several times, and after seven to nine intromissions he ejaculates. Following a short rest, the male repeats this behavior until he achieves five to seven ejaculations in a single episode of mating activity. Although there are some individual differences among males, multiple intromissions always precede ejaculation.[19] Beach[20] postulated that the stimulation that the female derives from multiple intromissions leads to secretion of progesterone which is important for the establishment and maintenance of pregnancy.

The first experimental analyses of the biological significance of this stereotyped copulatory behavior on the female's reproductive success were performed by Wilson et al.[21] and Adler.[22] These investigators manipulated the number of pre-ejaculatory intromissions that a female received. They found that following a reduction in the number of intromissions prior to ejaculation (low intromission group) only 20% of the females became pregnant or pseudopregnant. However, when males performed more than six intromissions before ejaculation, most females became pregnant or pseudopregnant. When the low intromission group females, who received three or less intromissions prior to ejaculation, were injected with progesterone, pregnancy ensued.[23] The hypothesis that male copulatory behavior initiates a progestational state was tested by measuring progesterone levels in females receiving a high or low number of intromissions. Only those females receiving a high number of intromissions had elevated progesterone levels on the third day of pregnancy and significantly higher levels on the next day.[24] Females belonging to the low intromission group showed no elevation of progesterone and normal estrous cycles continued. The authors[24] hypothesized that the multiple-intromission copulatory pattern initiates a neuroendocrine reflex resulting in secretion of progesterone in sufficient quantities to support implantation and pregnancy.

Having established that sufficient stimulation from the copulation is essential for the onset of the progestational state, the question arises as to the mechanism by which natural coital stimulation initiates this state. Together with Sawyer,[25] we directly tested whether copulatory stimulation received by the female affects the appearance of the nightly PRL surges which are essential intermediaries in the expression of the neuroendocrine reflex. We examined the correlation between nocturnal PRL surges and establishment of pregnancy or PSP in females receiving either a low or a high number of intromissions prior to ejaculation. The nightly PRL surges occurred in most females in the high intromission group. Only two out of nine females in the low intromission group exhibited PRL surges. This study revealed individual variation in thresholds for initiation of PRL secretion in response to coital stimulation. Approximately 14% of the females had a very low threshold and another 12% had such a high threshold that they did not secrete PRL in the surge pattern even after 18-20 intromissions. This natural system then demonstrates a variability in threshold which is not revealed following artificial electrical stimulation of the cervix.[26] The nightly PRL elevation is basically the first hormonal expression of the onset of PSP following copulatory stimulation.[6] This study indicates that a certain threshold level of coital stimulation must be reached to activate the series of nightly PRL surges, and that the manner in which the coital stimulation triggers PSP is an all-or-none pattern rather than a stepwise quantitative one.[25]

ADRENERGIC AND CHOLINERGIC MEDIATION OF PSEUDOPREGNANCY

The next question to which we addressed ourselves was: which neurotransmitter systems are involved in the transmission of cervical stimulation and the initiation of PSP? With Egozi[27] we tested the role of various neurotransmitters in the initiation of PSP by examining the effects of a series of drugs known to act on the adrenergic,

TABLE 1. Incidence of Pseudopregnancy when Various Drugs Were Administered 10 min before Cervical Stimulation on Day of Estrus

Treatment	Dose (mg/kg)	No. of Rats	No. Pseudopregnant	Percentage Pseudopregnant
Saline		35	30	86
Eserine	0.4	15	1	7
Prostigmine	1.0	9	7	78
Oxotremorine	0.2	8	0	0
Scopolamine HBr	1.0	4	4	100
Scopolamine HBr	8.0	5	5	100
Clonidine	0.015	7	7	100
Phenylephrine	10.0	8	8	100
Propranolol	10.0	16	11	69
WB-4101	5.0	9	2	22
Haloperidol	4.0	13	12	92
PCPA	62.0	6	5	83

cholinergic, dopaminergic, and serotonergic systems. Each drug was administered to groups of females on the morning of estrus, 10 min prior to artificial stimulation of the cervix. The initiation of PSP was blocked only by the cholinergic agonists, eserine and oxotremorine, and by the alpha-adrenergic receptor blocker WB-4101 (TABLE 1). These results strongly suggest that the induction of PSP requires activation of the adrenergic system and simultaneous inhibition of the cholinergic system. It is possible that the coital or copulomimetic stimuli lead to a depression of acetylcholine while simultaneously stimulating norepinephrine secretion.

The events in the central nervous system initiated by cervical stimulation and culminating in PSP are very short-lived since the drugs that are capable of blocking PSP 10 min prior to cervical stimulation are no longer effective when administered 10 min after the stimulation (TABLE 2) or even after a brief interval of 2-5 min post-cervical stimulation (Y. Egozi and J. Terkel, unpublished observations). This time course study establishes that the initiation of pseudopregnancy occurs during a brief interval following cervical stimulation.

ELECTROPHYSIOLOGICAL MEDIATION OF PSEUDOPREGNANCY

Considerable progress has been made in understanding the pathway by which the copulatory stimulus reaches the central nervous system (see Reference 8 for a sum-

TABLE 2. Incidence of Pseudopregnancy when Various Drugs Were Administered 10 min after Cervical Stimulation on Day of Estrus

Treatment	Dose (mg/kg)	No. of Rats	No. Pseudopregnant	Percentage Pseudopregnant
Oxotremorine	0.2	8	8	100
Eserine	0.4	7	7	100
WB-4101	5.0	6	6	100

mary). The vaginocervix acts as a transducer converting the mechanical stimulus of copulation into electrical signals. These signals are transmitted via the pelvic nerve[28,29] and ascend the anterolateral columns of the spinal cord to the brainstem. Within the brainstem the neural signals pass into the ventromedial and lateral parts of the midbrain and later into the hypothalamus via the dorsal and ventral noradrenergic bundles. The last bundle is especially important for the sensory transmission of vaginocervical stimulation which induces PSP.[30-32]

However, further mediation and processing of the incoming stimulus as well as the link between the neural events[33] and the initiation of PRL secretion characteristic of PSP still require further clarification.[34] The electrophysiological approach, among others, has been applied to unraveling the role of vaginocervical stimulation in the induction of PSP, but it has yielded ambiguous results.[8] Some studies have reported a decrease in neural activity following vaginal stimulation in the arcuate, ventromedial hypothalamus, and medial preoptic area,[35,36] while other studies have shown that cervical stimulation increases neural activity in the same areas as well as in the dorsomedial hypothalamus, anterior hypothalamic area, and suprachiasmatic nucleus.[37,38] Most of these results were obtained from acute experiments in which recordings were made at one or two brain sites.[38,39] In an attempt to achieve a more naturalistic perspective on this problem, Dafny and I (unpublished results) concentrated on recording single-multi units of electrical activity at six different brain sites which are thought to be involved in reproductive processes. The uniqueness of our approach lay in the simultaneous recording of unit activity in six brain loci in the freely behaving rat before, during, and after cervical stimulation. Recordings were made from the following areas: preoptic area, lateral hypothalamus, suprachiasmatic nucleus, ventromedial hypothalamus, anterior hypothalamus, and sensory cortex. After recovery from electrode implantation and completion of two consecutive estrous cycles, the rats received cervical stimulation on the morning of the day of estrus. Baseline neural activity levels were established for each female through recordings made for 30 min prior to stimulation and continuing for 3.5 hr following stimulation. A representative record of unit activity is shown in FIGURE 1. All of the hypothalamic areas from which we recorded simultaneously showed an increase in activity, with the exception of the preoptic area which showed an immediate decrease for about half an hour followed by an elevation above baseline.

The inhibition of neural activity in the preoptic area is of particular interest since it complements other findings implicating this neural site in the induction of PSP.[39-44] Lesions of the medial preoptic area trigger the release of nocturnal PRL surges and induce repetitive spontaneous PSP.[45-47] Conversely, electrochemical stimulation of the preoptic area reduces PRL secretion perhaps by increasing the inhibitory action of this brain site.[48] It has been proposed that the medial preoptic area may contain neurons which exert a chronic inhibitory effect on PRL surges.[8]

Measurement of the metabolic activity of different brain structures following cervical stimulation, using 2-deoxyglucose uptake as an indicator, led to the hypothesis[49] that the medial preoptic area is the early receiving area for input from cervical stimulation. Following processing of the information it may be stored and may trigger the bicircadian PRL surges.

Thus, the natural chain of events initiated by cervical stimulation may culminate in the reduction of medial preoptic area activity and consequent removal of its inhibition of the nocturnal PRL surges. Our recording studies, which demonstrated actual inhibition of ongoing unit activity in the medial preoptic area at the time of vaginal stimulation, may provide direct evidence for integration between the cervical input and the neural events in the central nervous system (CNS) leading to induction of PSP.

PHOTOPERIODICITY OF NOCTURNAL PROLACTIN SURGES

The secretion of many hormones is endogenously controlled and occurs in circadian patterns which are synchronized with the light-dark (LD) cycle.[50] Our interest in the

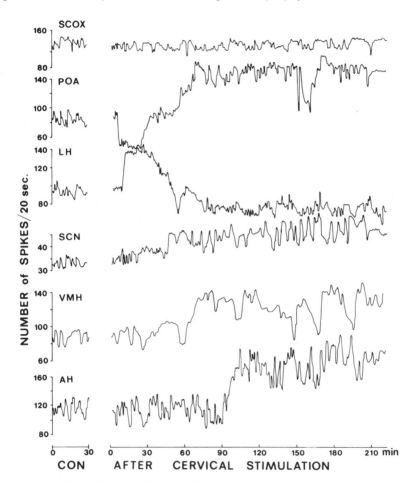

FIGURE 1. Polygraph record of neural activity at six brain loci prior to (control) and following cervical stimulation. The stimulation evoked both excitatory and inhibitory responses. Note the decrease in firing rate in the POA following stimulation. Con, control; SCOX, sensory cortex; POA, preoptic area; LH, lateral hypothalamus; SCN, suprachiasmatic nucleus; VMH, ventromedial hypothalamus; AH, anterior hypothalamus.

factors controlling the circadian nocturnal PRL surges during the first half of pregnancy and throughout PSP led us to investigate the role of the light-dark cycle in the timing of the rhythmic nocturnal surges of PRL. With Yogev,[51] we exposed groups of female rats to the following experimental photoperiods: 22L:2D, 19L:5D, 14L:10D,

8L:16D, and 3L:21D, as well as to constant light (LL) and constant darkness (DD). After a 2-month acclimation period to the light conditions, the females were mated. Chronic intraatrial cannulas were implanted in the females, thus enabling us to collect blood around-the-clock.[52] Each animal was bled 6-10 times a day on alternate days throughout the first few days of pregnancy or PSP. The results of exposure to the various lighting regimes demonstrated that the timing of the nocturnal surge can be phase-shifted by alteration of the light-dark ratio in the 24-hr light cycle. As the dark period decreases in length, the nocturnal PRL surge occurs earlier (phase advance) with respect to the onset of the dark phase. Conversely, lengthening the dark period to more than 10 hr per day caused a delay (phase delay) in the appearance of the nocturnal PRL surge (FIGURE 2).

When we analyzed these results using the midpoint of the dark or the light phase as the reference point, we found that the PRL surge occurred about an hour and a half after the midpoint of the dark phase under all but the most extreme photoperiod (3L:21D). In constant darkness the nocturnal PRL surge was synchronized among all animals. Interestingly, the magnitude of the nocturnal PRL surge and its duration were greater than under the various LD schedules.[51] Taken together, these results show that the mechanism controlling the timing of the daily nocturnal surge is flexible enough to function in extremes of daily photoregimens ranging from 22 hr of light to constant darkness. However, the condition of constant light seems to confound the regulatory mechanism since the females failed to retain a rhythmic pattern of PRL secretion. Although daily elevation of PRL still occurred following cervical stimulation, the timing was irregular and without pattern. The average PRL levels throughout the 24-hr period for the animals in constant light were less than those for all the animals in the group under constant darkness. We also found that the nocturnal PRL surge, like other circadian rhythmic phenomena, failed to occur in suprachiasmatic nucleus-lesioned females receiving cervical stimulation.[51,53]

On the basis of these and other investigations,[53-56] it is clear that during the first half of pregnancy and throughout PSP, prolactin has an endogenous rhythm with a periodicity of 24 hr occurring in phase with the light-dark cycle. Thus, once the nightly pattern of PRL secretion is established, it resembles other endogenous rhythms which are generated by the suprachiasmatic nucleus. However, it differs in that its initiation requires the exogenous trigger of cervical stimulation.

THE "MEMORY" OF CERVICAL STIMULATION

The most remarkable features of the neuroendocrine response to cervical stimulation are the retention of this information in the CNS[57] and its repeated expression for several days in the form of PRL surges[58,59]; thus, it displays the characteristics of a true "mnemonic system."[13] In this section, I shall briefly present the experimental evidence supporting the notion of a "mnemonic system" in three reproductive states: immature, cycling, and pregnant animals.

The "mnemonic system" is present in immature rats even before they are capable of PRL secretion in the surge pattern. If cervical stimulation is applied to 23-day-old females, PRL is released only 2-3 days later when the hormone response has developed.[60] Thus, although young rats are too immature to release adult-like PRL surges at the time of stimulation, they are capable of retaining the stimulus for at least 2 days until they are sufficiently developed to secrete the hormone in the adult pattern.

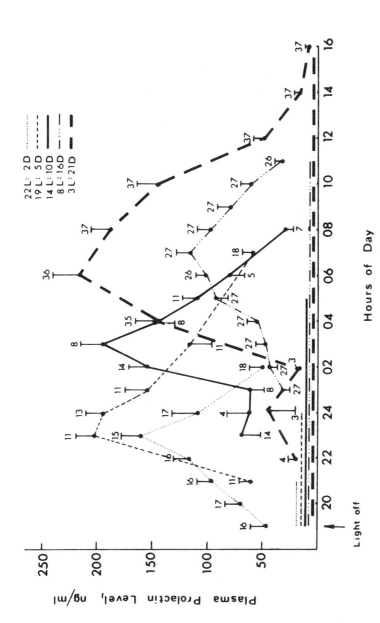

FIGURE 2. Pattern of nocturnal prolactin surges in animals exposed to five different photoperiods. The data from the first few days of pregnancy or pseudopregnancy are superimposed to show the 24-hr pattern of prolactin secretion. Each point represents the mean ± SE for samples taken at the same hour. The number of samples is indicated above the SE. The patterned lines on the abscissa indicate the dark phase per 24 hr, for the corresponding lighting shown in the key at the top right of the figure. (Reprinted with permission from Yogev and Terkel.[51])

In cycling rats, the capacity of cervical stimulation to initiate the semicircadian pattern of PRL secretion is not restricted to the specific hormonal state characteristic of proestrus or estrus. Stimulation of the uterine cervix at a time other than the day of ovulation, e.g., on the second day of diestrus, leads to the initiation of a delayed PSP which begins after the next ovulation, when the newly formed set of corpora lutea become functional.[61,62]

In the pregnant rat, administration of 1 mg ergocornine on the second day postmating resulted in termination of pregnancy followed by an estrous cycle and then a delayed PSP. Several studies have shown that spontaneous PSP is initiated if a new set of corpora lutea is formed within a 6-day period following cervical stimulation.[63,64] In the mouse, too, termination of early pregnancy by ergocornine resulted in delayed PSP.[65] During pregnancy, the last day of PRL surges is day 10 postcoitum, but if a hysterectomy is performed on day 11 or 12 of the pregnancy the PRL surges reappear 1 or 2 days later.[66] Thus, the removal of the inhibitory influence of the uterine-conceptus complex permits the reappearance of the PRL surges.

Although the initiation of the PRL surges does not require an ovarian background, the number of days that the surges are repeated depends upon the timing of the removal of the ovaries. With Yogev,[59] we found that in the majority of females ovariectomized one month before the experiment, cervical stimulation sustained nocturnal PRL elevation for 12 days (FIGURE 3). Yet when ovariectomy was performed on the same day that cervical stimulation was given, only 6 days of PRL surges were observed.[58]

Why should the duration of the neuroendocrine response, as expressed by the nocturnal PRL surges, differ under these two experimental conditions? Apparently, when the neural stimulus is given on a background of ovarian steroids, the expression of nocturnal PRL surges after the initial 6 days depends on the presence of progesterone.[26] When the neural stimulus is given in the absence of an ovarian steroid background, the PRL response remains independent of endocrine events and therefore can be repeated for a longer period. It still remains to be resolved how the same neural stimulus given against two different hormonal backgrounds produces responses of different duration.

SIGNIFICANCE OF MNEMONIC CONTROL OVER NOCTURNAL PROLACTIN SURGES

As mentioned previously, the luteal phase is initiated when the species-specific male copulatory behavior switches on the "mnemonic system" to signal repetitive PRL release which activates the corpora lutea. What might be the advantage of the involvement of a "mnemonic system" in the process of preparation for pregnancy?

For 2 days following copulation, the only hormone that shows a change in secretory pattern is PRL.[12] It is the major luteotrophic substance[3,4] which transforms the cyclic corpora lutea into the corpora lutea of pregnancy or PSP. However, the corpora lutea begin secreting progesterone only on the third day following their formation, creating an interval of about 3 days between copulation and the rise in progesterone secretion during which the PRL surges are secreted without any hormonal feedback.[12] In the absence of hormonal feedback, the initial stimulus, the multi-intromission copulatory signal, appears to be stored in the CNS and its memory is expressed in the form of daily PRL secretion. This is the proposed "mnemonic system" which is thought to underlie the unique pattern of PRL secretion that follows copulomimetic stimulation.[63]

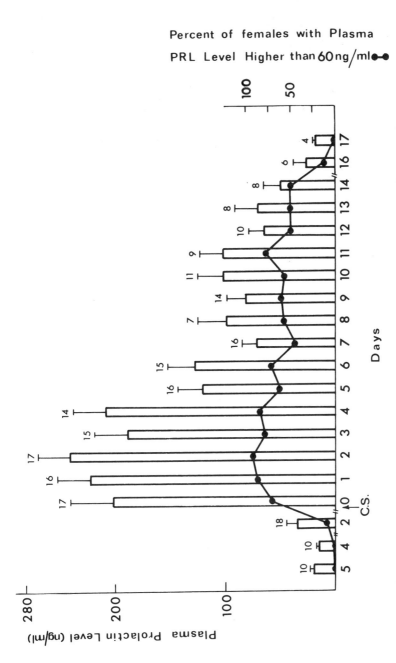

FIGURE 3. Plasma prolactin levels in ovariectomized female rats at the expected time of the nocturnal prolactin surge before and after cervical stimulation (indicated by arrow). Percentage of animals with plasma levels of prolactin higher than 60 ng/ml (●) is shown and the numbers above the error bars indicate group size. Each animal was bled twice a day at 0200 and 0400 hr and the higher level for each animal was combined to form the group mean ± SEM. (Reprinted with permission from Yogev and Terkel.[59])

The PRL surges are thus the measurable daily expression of the memory of cervical stimulation. Without this repetitive secretion of PRL, the luteal phase would not be activated. For the first 6 days following copulation, PRL surges occur autonomously, i.e., independently of progesterone secretion, even though secretion of this hormone begins on the third day postmating; only after the sixth day do PRL surges become progesterone dependent.[26]

The initial period following copulation, in which PRL is secreted autonomously, was termed by Yoshinaga[13] a state of "hypothalamic pseudopregnancy." "Hypothalamic pseudopregnancy" may be viewed as a state in which only the brain participates, responding to past cervical stimulation by producing daily PRL surges while the female reproductive system is still indistinguishable from one participating in a normal estrous cycle. In fact, the PSP and pregnancy which follow copulomimetic stimulation or mating with a fertile male are indistinguishable for the first 3 days. Implantation, which normally takes place 4-5 days after mating, thus occurs against a physiological background of a "hypothalamically pseudopregnant" female. Therefore each pregnancy could be viewed as a continuation of the hypothalamic PSP that was initiated by the copulatory stimulus. Remarkably, even if the PRL surges are experimentally blocked, the "mnemonic system" is retained and remains functional.[63,64,67] When the blocking effect on PRL ceases, the daily PRL surges resume (see the previous section). This is evidence that feedback provided by PRL is not necessary for the "memory," and that the brain is indeed the first organ to become set in the pseudopregnant state, in preparation for the ensuing pregnancy.

LACTATIONAL DIESTRUS IS NOT A PSEUDOPREGNANCY

The state of lactation in the rat is characterized by cessation of the estrous cycle and by high levels of PRL resulting from suckling stimulation. On the basis of the criteria of vaginal smears, the decidual cell response, the presence of active corpora lutea, and high levels of progesterone, many investigators have referred to lactational diestrus as a state of PSP.[68,69] However, it remained to be determined whether the autonomous nightly surges of PRL characteristic of PSP are also present during lactational diestrus. Yogev and I[70] explored this question. In order to prevent masking of the presumed nocturnal surges by high background levels of PRL induced by suckling, pups were separated from their mothers at 2100 hr and returned at 0600 hr. We collected blood daily on days 3-14 postpartum, at the expected time of the nocturnal PRL surges. We did not detect any nocturnal PRL elevation characteristic of PSP during the 2 weeks of lactation.

The observation that during lactation there are no autonomous nocturnal PRL surges indicates that during this reproductive state, PRL secretion is governed solely by the external stimulus of suckling. The well-known fact that termination of the luteotrophic phase and resumption of cyclicity occur within 3 days after cessation of suckling[70,71] may thus be attributed to the sudden reduction in suckling-induced PRL secretion and the absence of autonomous PRL secretion during lactation.

That lactational diestrus is not a PSP according to the criterion of PRL surges creates a progesterone-related paradox. Under experimental conditions, high levels of exogenous progesterone promote the initiation of PSP and daily autonomous PRL secretion for the duration of normal PSP.[72] Yet during lactation, the high levels of endogenous progesterone[73] do not trigger autonomous PRL secretion.[70] It is nonetheless possible to initiate PRL secretion in the typical surge pattern in lactating rats by providing the neural message of copulomimetic stimulation. When lactating females receive cervical stimulation during the first 2 weeks postpartum, they enter a state of PSP (L. Yogev and J. Terkel, unpublished). Furthermore, if lactating rats mate during the postpartum estrus, they exhibit both the nightly PRL surges and high levels of PRL resulting from acute stimulation by suckling pups.[10,70] Thus, during lactation, there appears to be an inhibition against superimposition of hormonally induced PSP.

We are tempted to speculate that the adaptive advantage of a lactational diestrus over a true PSP is probably related to greater reproductive efficiency for the female. Should a catastrophe befall the litter during lactation, the rat will return to the fertile state almost immediately. On the other hand, if the rat were pseudopregnant during lactation and lost her litter, she would have to wait up to 2 weeks until the termination of PSP before reentering the fertile state, a loss which would have direct impact on the female's reproductive success.

NONCOITALLY INDUCED PSEUDOPREGNANCY

Normally, we think of pseudopregnancy as a state which is initiated naturally by mating with an infertile male or artificially by copulomimetic stimulation. Yet other endocrine as well as socioenvironmental factors also may lead to this physiological state. In this section we examine some of the hormonal and social factors that can initiate PSP with the intent of comparing the underlying mechanisms controlling coitally and noncoitally induced pseudopregnancies.

Under normal husbandry conditions, rats, mice, and hamsters exhibit a 4- to 5-day estrous cycle which is repeated as long as the female does not mate and become either pregnant or pseudopregnant. However, during each estrous cycle, the female is briefly susceptible to events that may lead to cessation of cyclicity and the onset of PSP. This sensitive period occurs during the short life span of the corpora lutea, between proestrus and estrus. At this time, internal or external events may potentially activate the corpora lutea and establish PSP, even if the female did not receive cervical stimulation.

A brief treatment with any of the following hormones—prolactin, progesterone, or estrogen—will induce PSP. The tranquilizers, reserpine and chlorpromazine, which affect the endocrine system, will induce PSP as well. The pseudopregnancies that result from all these treatments are mediated by PRL secretion which in turn activates the corpora lutea (see Reference 72). The subsequent maintenance of PSP is thought to become self-sustaining when luteal progesterone levels are elevated and hence provide feedback for PRL secretion.

External factors may also lead to PSP. For example, PSP ensues when stressful treatment, which leads to elevation of plasma PRL, coincides with the sensitive period of the estrous cycle when the corpora lutea are formed. In our laboratory, we found that if routine atrial cannulation for chronic blood sampling is performed on the day of proestrus, the surgical stress often results in PSP.

Neither the hormonally induced nor the stress-induced PSP is based on an underlying "mnemonic system." When ergocornine is administered on the second day of either hormonally induced or stress-induced PSP, that state is terminated and cyclicity resumes (L. Yogev and J. Terkel, unpublished). In contrast, when coitally induced PSP is terminated by the same drug, PRL secretion is reinstated, allowing a delayed pseudopregnancy to ensue.[63,64] This suggests that the ergocornine did not affect the "mnemonic system" initiated by copulatory stimulation.[67]

Thus, from the endocrine and physiological points of view the pseudopregnancies induced by hormones, stress, and copulation would appear to be identical: they each last 12 days, are characterized by nocturnal PRL surges, and can support the decidual response. However, there is a fundamental difference in the underlying control mechanism, as shown by the response to pharmacological intervention. The humorally induced and stress-induced pseudopregnancies are self-sustaining, based on a reciprocal positive feedback between progesterone and PRL. The involvement of the brain is strictly limited to the timing of the PRL surges. In contrast, the coitally induced PSP is based on an underlying "mnemonic system" which is initially activated by male copulatory behavior.

In female mice, the social environment suppresses or enhances ovarian cyclicity. When female mice are housed in all-female groups, the social stimulation causes them to cease cycling and enter a state of prolonged diestrus.[74] This state is a PSP with a decidual response to uterine trauma, functional corpora lutea, and elevated progesterone levels.[75] The odor of a male is sufficient to terminate such group-induced PSP and to reinstate cyclicity.[76] The hormonal deficit by which PSP is naturally terminated by the male[77] is basically the same as that in the pharmacological termination of PSP by ergocornine.

Elicitation of a change in the female's response to surrounding social stimuli via a "motivational switch" in her attention to newborn pups can also induce PSP. Normally, virgin rats do not show maternal behavior when exposed to pups, but they can be induced to become maternal within several days by concaveation.[78] The concaveation-induced maternal behavior is nonhormonal in nature, since both ovariectomized and hypophysectomized virgins can be induced to become maternal by exposure to pups. This study clearly eliminates the possibility of involvement of the pituitary-gonadal axis in the initiation of maternal behavior. Once the female becomes maternal, a "motivational switch" occurs, bringing about a response to a completely different set of stimuli emanating from the pups. Jakubowski and I followed the estrous cycle of the virgin rat after the onset of maternal behavior. The change in the virgin's behavior results in concomitant endocrine changes, as evidenced by a prolonged diestrus,[79] which is a true PSP characterized by nocturnal PRL surges (FIGURE 4).[80]

The disruption of normal cyclicity is a result of tactile stimulation of the female's ventrum by the young as she crouches over them, rather than by exteroceptive stimuli (e.g., odor, sound). This was shown by separating the maternal virgin from the young by a wire mesh; under these conditions, the female continued to cycle normally. Suckling of the nipples is not essential for the establishment of PSP, since thelectomized maternal virgins also become pseudopregnant.[79] Once the pup-induced PSP has been established, it is maintained for 12 days, even if the pups are removed as early as day 4.[79] The continued maintenance of PSP is self-sustaining because of the feedback of luteal progesterone on nocturnal PRL secretion. However, ergocornine administration at the time of pup removal led to resumption of cyclicity, without establishment of delayed PSP. Thus, the response to the ergocornine treatment blocking PRL secretion is identical to that occurring in humorally induced and stress-induced PSP and contrasts with the coitally induced case.

In summary, all the cases of noncoitally induced PSP require the presence of the initial stimulus until progesterone levels are sufficiently elevated to provide positive feedback on PRL secretion. The triggering stimuli in the humorally and socially induced pseudopregnancies do not have the capacity to activate the "mnemonic system" present in coitally induced PSP.

The blockage of pregnancy in newly pregnant mice by the presence of a male (Bruce effect[81]) is worthy of particular consideration, in view of the emphasis placed on the "mnemonic system" in this paper. It is possible that the underlying mechanism of blockage of pregnancy by the social stimulation of a new male might be different than that of pharmacological termination of pregnancy or PSP. Termination of preg-

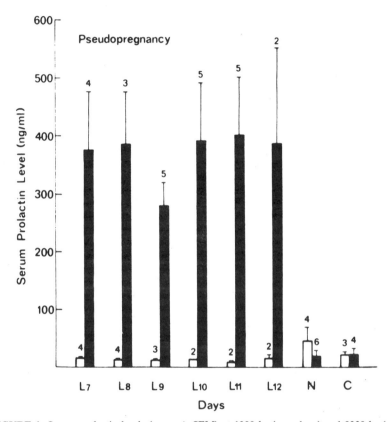

FIGURE 4. Serum prolactin levels (mean ± SEM) at 1200 hr (open bars) and 0230 hr (solid bars) in maternal virgin rats during pup-induced pseudopregnancy. The number of rats per group is shown above the SEM limits. L7-L12 are days 7-12 of pseudopregnancy. N is the day of proestrus, and C is the day of estrus. (Reprinted with permission from Yogev et al.[80])

nancy by the odor of a strange male (Bruce effect) and that by ergocornine treatment have a common end point—the suppression of PRL secretion.[63,77] Since in rats, termination of coitally induced PSP by ergocornine results in onset of delayed PSP,[63] one might also predict that a second PSP would ensue in mice following termination of pregnancy by the odor of a strange male. As this issue has not been addressed directly, it is not clear what the result might be. Both from the female's and the male's points of view it would seem advantageous that a secondary PSP not follow the termination of a pregnancy by the male, since the female would be reducing the time period during which she could become pregnant. We have considered the results of two separate studies in mice which were not originally designed to answer this question. When ergocornine was administered to mice on day 3-4 of pregnancy, approximately one-half of the small sample showed a spontaneous delayed PSP.[65] Bruce[81] reported that only 5% of the females she followed after termination of pregnancy by exposure to a strange male showed spontaneous PSP. In another rodent species, the vole (*Microtus agrestis*), Milligan et al.[67] terminated pregnancy either by injection of ergocornine or by exposure to a strange male. From their systematic study, they were able to propose that interruption of pregnancy by the strange male is due to interference with the luteotrophic "mnemonic system." The above studies suggest that the strange male induces pregnancy failure by interfering with the luteal "mnemonic system." If this is the case, it is interesting that the male's odor acts on termination of PRL secretion at a different neural level from the one on which ergocornine acts. The male's odor "erases" the "mnemonic system," whereas ergocornine temporarily blocks its expression, permitting its reexpression as a delayed PSP following the formation of fresh corpora lutea.

To obtain a more fundamental understanding of the underlying mechanism of the "mnemonic system," further experiments must be designed to directly test this unique phenomenon.

SUMMARY

In this paper, I have outlined our research perspective on coitally and noncoitally induced PSP. Coitally induced PSP requires a minimal threshold of vaginal stimulation to establish the repetitive nocturnal PRL surges which are essential for converting the corpora lutea from a nonfunctional to a functional state. The induction of PSP requires the activation of the adrenergic system and simultaneous inhibition of the cholinergic system. Reduction of neural activity in the medial preoptic area occurs at the time of coital stimulation, and is believed to be associated with the initiation of the nocturnal PRL surges. Apparently, cervical stimulation acts by inhibiting the nocturnal PRL surge inhibitory neurons at this brain site, thereby instituting PSP. It is proposed that the specific mating pattern of cervical stimulation activates a "mnemonic system" which retains information from vaginocervical stimulation permitting the expression of the repeated nightly PRL surges. In our view, each pregnancy originates as a direct continuation of the initial "hypothalamic PSP" triggered by the neural signal of mating. The mechanisms involved in the activation of the "mnemonic system" and the hypothalamically pseudopregnant state remain to be elucidated.

Pseudopregnancy can be initiated by a number of socioenvironmental factors as well, such as cohabitation in all-female groups in mice or concaveation with foster pups in virgin rats. However, unlike coitally induced PSP which activates the "mnemonic system," noncoitally induced pseudopregnancies lack a "mnemonic system," and therefore require the continued presence of the initiating stimulus until progesterone levels become sufficiently elevated to provide positive feedback on PRL secretion.

ACKNOWLEDGMENTS

The author is grateful to Esther Brezner for technical assistance, to Drs. Sarah Lenington, Herbert Hauser, and Jeff Witcher for their helpful comments, and to Dr. Amelia Terkel for help in preparing the manuscript.

REFERENCES

1. DE FEO, V. J. 1967. Decidualization. *In* Cellular Biology of the Uterus. R. M. Wynn, Ed.: 252. Appleton-Century-Crofts. New York, N.Y.
2. LU, K. H., H. T. CHEN, H. H. HUANG, L. GRANDISON, S. MARSHALL & J. MEITES. 1976. Relation between prolactin and gonadotrophin secretion in post-partum lactating rats. J. Endocrinol. **68:** 241-250.
3. ASTWOOD, E. D. 1941. The regulation of corpus luteum function by hypophyseal luteotrophin. Endocrinology **28:** 309-320.
4. ROTHCHILD, I. 1981. The regulation of the mammalian corpus luteum. Recent Prog. Horm. Res. **37:** 183-298.
5. AMENOMORI, Y., C. L. CHEN & J. MEITES. 1970. Serum prolactin levels in rats during different reproductive stages. Endocrinology **86:** 506-510.
6. BUTCHER, R. L., N. W. FUGO & W. E. COLLINS. 1972. Semicircadian rhythm in plasma levels of prolactin during early gestation in the rat. Endocrinology **90:** 1125-1127.
7. FREEMAN, M. E. & J. D. NEILL. 1972. The pattern of prolactin secretion during pseudopregnancy in the rat: a daily nocturnal surge. Endocrinology **90:** 1292-1294.
8. GUNNET, J. W. & M. E. FREEMAN. 1983. The mating-induced release of prolactin: a unique neuroendocrine response. Endocr. Rev. **4:** 44-61.
9. GOROSPE, W. C. & M. E. FREEMAN. 1981. An ovarian role in prolonging and terminating the two surges of prolactin in pseudopregnant rats. Endocrinology **108:** 1293-1298.
10. YOGEV, L. & J. TERKEL. 1978. The temporal relationship between implantation and termination of nocturnal prolactin surges in pregnant lactating rats. Endocrinology **102:** 160-165.
11. VOOGT, J., M. ROBERTSON & H. FRIESEN. 1982. Inverse relationship of prolactin and rat placental lactogen during pregnancy. Biol. Reprod. **26:** 800-805.
12. SMITH, M. S., M. E. FREEMAN & J. D. NEILL. 1975. The control of progesterone secretion during the estrous cycle and early pseudopregnancy in the rat: prolactin, gonadotropin and steroid levels associated with rescue of the corpus luteum of pseudopregnancy. Endocrinology **96:** 219-226.
13. YOSHINAGA, K. 1977. Hormonal interplay in the establishment of pregnancy. *In* Reproductive Physiology II. R. O. Greep, Ed. **13:** 201-223. University Park Press. Baltimore, Md.
14. CONAWAY, C. H. 1971. Ecological adaptation and mammalian reproduction. Biol. Reprod. **4:** 239-247.
15. DIAMOND, M. 1972. Vaginal stimulation and progesterone in relation to pregnancy and parturition. Biol. Reprod. **6:** 281-287.
16. JOHNSON, M. & B. EVERITT. 1980. Essential Reproduction. Blackwell Scientific Publications. Oxford, England.
17. SMITH, M. S., B. K. MCLEAN & J. D. NEILL. 1976. Prolactin: the initial luteotropic stimulus of pseudopregnancy in the rat. Endocrinology **98:** 1370-1377.
18. BEACH, F. 1947. A review of physiological and psychological studies of sexual behavior in mammals. Physiol. Rev. **27:** 240-307.
19. LARSSON, K. 1956. Conditioning and Sexual Behavior. Almqvist & Wiksells. Stockholm, Sweden.

20. BEACH, F. 1965. Sex and Behavior. Wiley. New York, N.Y.
21. WILSON, J. R., N. ADLER & B. LE BOEUF. 1965. The effects of intromission frequency on successful pregnancy in the female rat. Proc. Natl. Acad. Sci. USA **53:** 1392-1395.
22. ADLER, N. T. 1969. Effects of the male's copulatory behavior on successful pregnancy of the female rat. J. Comp. Physiol. Psychol. **69:** 613-622.
23. ADLER, N. T. 1968. Effects of the male's copulatory behavior in the initiation of pregnancy in the female rat. Anat. Rec. **160:** 305.
24. ADLER, N. T., J. A. RESKO & R. W. GOY. 1970. The effect of copulatory behavior on hormonal change in the female rat prior to implantation. Physiol. Behav. **5:** 1003-1007.
25. TERKEL, J. & C. H. SAWYER. 1978. Male copulatory behavior triggers nightly prolactin surges resulting in successful pregnancy in rats. Horm. Behav. **11:** 304-309.
26. FREEMAN, M. E. & J. R. STERMAN. 1978. Ovarian steroid modulation of prolactin surges in cervically-stimulated ovariectomized rats. Endocrinology **102:** 1915-1920.
27. EGOZI, Y. & J. TERKEL. 1983. Participation of the adrenergic system in the induction of pseudopregnancy. *In* Neuropeptides and Psychosomatic Processes. E. Endroczi, D. DE WIED, L. ANGENUCCI & U. SCAPANINI, Eds. Akademiai Kiadó. Budapest, Hungary.
28. CARLSON, R. R. & V. J. DE FEO. 1965. Role of the pelvic nerve versus abdominal sympathetic nerves in the reproductive function of the female rat. Endocrinology **77:** 1014-1022.
29. KOMISARUK, B. R., N. T. ADLER & J. HUTCHISON. 1972. Genital sensory field: enlargement by estrogen treatment in female rats. Science **178:** 1295-1298.
30. ARAI, Y. 1969. Effect of hypothalamic de-afferentation on induction of pseudopregnancy by vaginal-cervical stimulation in the rat. J. Reprod. Fertil. **19:** 573-575.
31. CARRER, H. F. & S. TALEISNIK. 1970. Induction and maintenance of pseudopregnancy after interruption of preoptic hypothalamic connections. Endocrinology **86:** 231-236.
32. HANSEN, S., E. J. STANFIELD & B. J. EVERITT. 1980. The role of ventral bundle noradrenergic neurones in sensory components of sexual behaviour and coitus-induced pseudopregnancy. Nature **286:** 152-154.
33. HASKINS, J. T. & R. L. MOSS. 1983. Action of estrogen and mechanical vaginocervical stimulation on the membrane excitability of hypothalamic and midbrain neurons. Brain Res. Bull. **10:** 489-496.
34. KAWAKAMI, M. & J. ARITA. 1982. Midbrain pathways for the initiation and maintenance of the nocturnal prolactin surge in pseudopregnant rats. Endocrinology **110:** 1977-1982.
35. LINCOLN, D. W. & B. A. CROSS. 1967. Effects of oestrogen on the responsiveness of neurones in the hypothalamus, septum, and preoptic area of rats with light-induced persistent oestrus. J. Endocrinol. **37:** 191-203.
36. KAWAKAMI, M. & K. KUBO. 1971. Neuro-correlate of limbic-hypothalamo-pituitary-gonadal axis in the rat: change in limbic-hypothalamic unit activity induced by vaginal and electrical stimulation. Neuroendocrinology **7:** 65-89.
37. KAWAKAMI, M. & T. IBUKI. 1972. Multiple unit activity in the brain correlated with induction and maintenance of pseudopregnancy in rats. Neuroendocrinology **9:** 2-19.
38. BLAKE, C. A. & C. H. SAWYER. 1972. Effects of vaginal stimulation on hypothalamic multiple-unit activity and pituitary LH release in the rat. Neuroendocrinology **10:** 358-370.
39. BARRACLOUGH, C. A. & B. A. CROSS. 1963. Unit activity in the hypothalamus of the cyclic female rat: effect of genital stimuli and progesterone. J. Endocrinol. **26:** 339-359.
40. PETERS, J. A. & R. R. GALA. 1975. Induction of pseudopregnancy in the rat by electrochemical stimulation of the brain. Horm. Res. **6:** 36-46.
41. BEACH, J. E., L. TYREY & J. W. EVERETT. 1978. Prolactin secretion preceding delayed pseudopregnancy in rats after electrical stimulation of the hypothalamus. Endocrinology **103:** 2247-2251.
42. ARITA, J. & M. KAWAKAMI. 1981. Role of the medial preoptic area in the neural control of the nocturnal prolactin surge in the rat. Endocrinol. Jpn. **28:** 769-774.
43. FREEMAN, M. E. & J. A. BANKS. 1980. Hypothalamic sites which control the surges of prolactin secretion induced by cervical stimulation. Endocrinology **106:** 668-673.
44. GUNNET, J. W. & M. E. FREEMAN. 1984. Hypothalamic regulation of mating induced prolactin release: effect of electrical stimulation of the medial preoptic area in conscious female rats. Neuroendocrinology **38:** 12-16.

45. GUNNET, J. W. & M. E. FREEMAN. 1985. The interaction of the medial preoptic area and the dorsomedial-ventromedial nuclei of the hypothalamus in the regulation of the mating-induced release of prolactin. Neuroendocrinology **40:** 232-237.
46. CLEMENS, J. A., E. B. SMALSTIG & B. D. SAWYER. 1976. Studies on the role of the preoptic area in the control of reproductive function in the rat. Endocrinology **99:** 728-735.
47. WIEGAND, S. J., E. TERASAWA, W. E. BRIDSON & R. W. GOY. 1980. Effects of discrete lesions of preoptic and suprachiasmatic structures in the female rat. Neuroendocrinology **31:** 147-157.
48. CLEMENS, J. A., C. J. SHAAR, J. W. KLEBER & W. A. TANDY. 1971. Reciprocal control by the preoptic area of LH and prolactin. Exp. Brain Res. **12:** 250-253.
49. ALLEN, T. O., N. T. ADLER, J. H. GREENBERG & M. REIVICH. 1981. Vaginocervical stimulation selectively increases metabolic activity in the rat brain. Science **211:** 1070-1072.
50. SAWIN, C. T. 1969. The Hormones. Churchill. London, England.
51. YOGEV, L. & J. TERKEL. 1980. Effects of photoperiod, absence of photic cues and suprachiasmatic lesions on nocturnal prolactin surges in pregnant and pseudopregnant rats. Neuroendocrinology **31:** 26-33.
52. TERKEL, J. & L. URBACH. 1974. A chronic intravenous cannulation technique adapted for behavioral studies. Horm. Behav. **5:** 141-148.
53. BETHEA, C. L. & J. D. NEILL. 1980. Lesions of the suprachiasmatic nuclei abolish the cervically stimulated prolactin surges in the rat. Endocrinology **107:** 1-5.
54. BETHEA, C. L. & J. D. NEILL. 1979. Prolactin secretion after cervical stimulation of rats maintained in constant dark or constant light. Endocrinology **104:** 870-876.
55. PIEPER, D. R. & R. R. GALA. 1979. The effect of light on the prolactin surges of pseudopregnant and ovariectomized, estrogenized rats. Biol. Reprod. **20:** 727-732.
56. NAITO, H., M. TAKAHASHI & Y. SUZUKI. 1981. Light-and-dark signal for the initiation of prolactin surges in cervically stimulated rats. Endocrinol. Jpn. **28:** 151-160.
57. EVERETT, J. W. 1968. "Delayed pseudopregnancy" in the rat, a tool for the study of central neural mechanisms in reproduction. *In* Perspectives in Reproduction and Sexual Behavior. M. Diamond, Ed. Indiana University Press. Bloomington, Ind.
58. FREEMAN, M. E., M. S. SMITH, S. J. NAZIAN & J. D. NEILL. 1974. Ovarian and hypothalamic control of the daily surges of prolactin secretion during pseudopregnancy in the rat. Endocrinology **94:** 875-882.
59. YOGEV, L. & J. TERKEL. 1980. Daily rhythm in secretion of prolactin in androgenized female rats. J. Endocrinol. **87:** 327-332.
60. SMITH, M. S. & J. A. RAMALEY. 1978. Development of ability to initiate and maintain prolactin surges induced by uterine cervical stimulation in immature rats. Endocrinology **102:** 351-357.
61. EVERETT, J. W. 1967. Provoked ovulation or long-delayed pseudopregnancy from coital stimulation in barbiturate-blocked rats. Endocrinology **80:** 145-154.
62. ZEILMAKER, G. H. 1965. Normal and delayed pseudopregnancy in the rat. Acta Endocrinol. (Kbh) **49:** 558-566.
63. DE GREEF, W. J. & G. H. ZEILMAKER. 1976. Prolactin and delayed pseudopregnancy in the rat. Endocrinology **98:** 305-310.
64. URBACH, L. 1974. Unpublished Ph.D. dissertation. Tel Aviv University. Tel Aviv, Israel.
65. CARLSEN, R. A., G. H. ZEILMAKER & M. C. SHELESNYAK. 1961. Termination of early (pre-nidation) pregnancy in the mouse by single injection of ergocornine methanesulphonate. J. Reprod. Fertil. **2:** 369-373.
66. VOOGT, J. L. 1980. Regulation of nocturnal prolactin surges during pregnancy in the rat. Endocrinology **106:** 1670-1675.
67. MILLIGAN, S. R., H. M. CHARLTON & E. VERSI. 1979. Evidence for a coitally induced 'mnemonic' involved in luteal function in the vole (*Microtus agrestis*). J. Reprod. Fertil. **57:** 227-233.
68. SELYE, H. & T. MCKEOWN. 1933. Production of pseudo-pregnancy by mechanical stimulation of the nipples. Proc. Soc. Exp. Biol. Med. **31:** 683-687.
69. PEPE, G. J. & I. ROTHCHILD. 1974. A comparative study of serum progesterone levels in pregnancy and in various types of pseudopregnancy in the rat. Endocrinology **95:** 275-279.

70. YOGEV, L. & J. TERKEL. 1982. Absence of nocturnal autonomic prolactin secretion during lactation in the rat. J. Endocrinol. **95:** 397-402.
71. OTA, K. & A. YOKOYAMA. 1965. Resumption of lactation by suckling in lactating rats after removal of litters. J. Endocrinol. **33:** 185-194.
72. DE GREEF, W. J. & G. M. ZEILMAKER. 1979. Serum prolactin concentrations during hormonally induced pseudopregnancy in the rat. Endocrinology **105:** 195-199.
73. HONDA, K., S. SUGAWARA & T. MASAKI. 1977. Serum levels of progesterone, prolactin, LH and FSH during lactation in the rat. Tohoku J. Agric. Res. **28:** 135-144.
74. LEE, S. VAN DER & L. M. BOOT. 1955. Spontaneous pseudopregnancy in mice. Acta Physiol. Pharmacol. Neerl. **4:** 442-443.
75. RYAN, K. D. & N. B. SCHWARTZ. 1977. Grouped female mice: demonstration of pseudopregnancy. Biol. Reprod. **17:** 578-583.
76. WHITTEN, W. K. & A. K. CHAMPLIN. 1973. The role of olfaction in mammalian reproduction. Handb. Physiol. Sect. F: Endocrinol. **11**(Part 1): 109-124.
77. MARCHLEWSKA-KOJ, A. & B. JEMIOLO. 1978. Evidence for the involvement of dopaminergic neurons in the pregnancy block effect. Neuroendocrinology **26:** 186-192.
78. ROSENBLATT, J. S. 1967. Nonhormonal basis of maternal behavior in the rat. Science **156:** 1512-1514.
79. JAKUBOWSKI, M. & J. TERKEL. 1980. Induction by young of prolonged dioestrus in virgin rats behaving maternally. J. Reprod. Fertil. **58:** 55-60.
80. YOGEV, L., M. JAKUBOWSKI & J. TERKEL. 1980. Two modes of prolactin secretion induced by young in maternally behaving virgin rats. Endocrinology **107:** 1808-1812.
81. BRUCE, H. M. 1962. Continued suppression of pituitary luteotrophic activity in the female mouse. J. Reprod. Fertil. **4:** 313-318.

Licking, Touching, and Suckling: Contact Stimulation and Maternal Psychobiology in Rats and Women[a]

JUDITH M. STERN

Department of Psychology
Rutgers-The State University of New Jersey
New Brunswick, New Jersey 08903

At a time when American psychology was focused on obvious physiological rewards such as food, Harry Harlow discovered the powerful role of "contact comfort" as a behavioral motivator.[2] Mother-deprived infant monkeys preferred a warm, furry, non-nutrient-bearing surrogate to a cold, wire, nutrient-bearing surrogate for clinging. In addition to its hedonic value, we now know that maternal contact stimulation also provokes beneficial physiological effects in young rats, including autonomic regulation[3] and growth hormone secretion.[4] Similarly, general body stroking of hospitalized preterm infants enhances their growth; thus, deprivation of such contact stimulation may underlie the phenomenon of psychosocial dwarfism in children.[5]

Is the physical contact with young similarly rewarding to caregivers? Specifically, is such contact necessary to induce and maintain maternal behavioral and physiological changes essential to nurturance and parental investment in the current young? In the first part of this paper, on ontogeny of maternal responsiveness, I show that onset of maternal behavior (MB) in rats through physical interaction with young (sensitization) occurs rapidly in juveniles, which is functionally significant when junior siblings are born, and is hastened in adults by prior brief contact with pups including licking and/or retrieving. Further, a parturition experience restricted to licking and handling each pup establishes a preference for pups the size of newborn rats when MB is initiated the next day. Possible human parallels to these events of the rat's maternal behavior cycle are drawn. In the second part, on offspring-induced maternal hormonal changes, I discuss recent data on suckling-induced prolactin and oxytocin secretion, suppression of fertility, and reduced sexual activity. Although the second part focuses on humans, comparisons to rats and other mammals are made.

---

[a] I began my research career by investigating the hormonal basis of incubation behavior in male ring doves with my late advisor, Daniel Lehrman.[1] Since then, the maternal behavior and physiology of both Norway rats and women have occupied much of my attention, inspired in part by Jay Rosenblatt's research program. For the study of all naturally occurring behaviors, whatever the species, the approach I learned at the Institute of Animal Behavior, encompassing proximate mechanisms, ontogenetic history, and adaptive significance, has continued to serve me well.

ONTOGENY OF MATERNAL RESPONSIVENESS

Periweaning Expression and Pubertal Repression

In Norway rats, the complex of behaviors we call maternal which are directed toward pups—retrieving, licking, and crouching over—can be thought of as present or available in the individual's genome, but generally inhibited in the adult. The report by Bridges et al.[6] that 24-day-old male and female rats become maternal rapidly (1-2 days) when housed continuously with pups dramatically revealed the availability of this behavioral repertoire. Around the time of puberty, maternal responsiveness declines and eventually reaches the typical adult latency (4-6 days).[6-8]

Maternal behavior in mammals is most relevant to the well-being of young when it occurs postpartum. What then is the adaptive significance, if any, of MB emerging in a nonlactating female, or even a male, animals incapable of fully nurturing young? Rosenblatt[9] has suggested that maternal behavior can be induced by pup stimulation in non-hormonally primed rats because such stimulation acts to maintain maternal behavior once it is established postpartum.[10,11] But nurturance emancipated from lactation is widespread among mammals,[12] and there are probably many functions served by it.

It makes good evolutionary sense for positive responses toward young to be elicited rapidly when close physical contact with young is likely to occur naturally. As much as 50% of all successful matings in the wild among Norway rats may occur during the postpartum estrus.[13] If the first litter is culled to six pups, then senior litter members typically will be 23-29 days old when the junior litter, conceived during the dam's postpartum estrus, is born.[8] Both field[14] and laboratory[8,15] observations indicate that the senior litter remains in the nest area during the birth and rearing of the junior litter. In fact, nursing will continue for the older offspring, especially if the junior litter is born prior to the senior litter's 25th day.[8,15] My students and I have seen retrieving of a junior pup by a senior pup, as well as licking, and frequent huddling together in the nest, both with and without the dam.[8] The "pubertal repression" of MB, which occurs at a time when the mother is much less likely to produce a junior litter, suggests that maturational changes inhibit the expression of rapid maternal responsiveness. Elucidation of these mechanisms may aid in understanding the reverse process, i.e., the rapid onset of MB which takes place at parturition.[10,11] The massive hormonal changes of puberty do not seem to be involved.[6-8] Because limbic system structures are implicated in the inhibition of MB (see A. Fleming, this volume), maturation of these structures may mediate the pubertal decline.

Do experiential influences modulate the decline in maternal responsiveness? In the presence of a junior litter, interaction until day 40 (but not for only 4 days after the birth of the junior litter) resulted in retention of shorter-latency MB when testing began on day 42 compared to rats remaining with the mother and same-age littermates with no junior litter present[8] (FIGURE 1). This result was not related to when the junior litter was born, or therefore to the amount of nursing the senior litter members engaged in. The failure of only 4 days of interaction to have a discernible effect in the day 42 MB latency test may be due to insufficient stimulation from pups in this situation or to a decline in previously activated nurturant behaviors. However, interaction with alien pups early in development, in the absence of the dam, does have lasting effects on maternal responsiveness of Sprague-Dawley/CD rats tested in adulthood.[16]

Physical proximity between older and younger offspring of the same mother resulting in caretaking by the former toward the latter occurs in a number of mammalian species, including African mongooses, red foxes, and African jackals, the older offspring being termed helpers.[17] A prediction from the idea concerning functional significance of pup-induced MB in Norway rats is that short-latency onset of MB among juveniles would not occur in rodent species in which there is no postpartum estrus. The golden hamster is such a species, and the unpublished results of Siegel and Rosenblatt, indicating that there is no age-related change in the latency to onset of MB in these hamsters, support this idea. In general, among nonpregnant animals, developmental changes in the readiness to nurture may be functionally related to the likelihood of close contact with young.

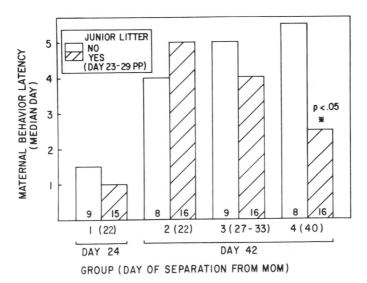

FIGURE 1. Latency to onset of maternal behavior during continuous cohabitation with alien pups (sensitization) in juvenile Long-Evans rats as a function of age and prior extent of cohabitation with junior siblings. In each litter (the number of litters is indicated at the bottom of each bar) subject 1 was tested beginning on day 24 while subjects 2-4 were tested beginning on day 42. Subject 2 was separated from the dam and home cage on day 22 and housed with age-mates until day 40; subject 3 was separated from the dam and home cage 4 days after the birth of the junior litter, or at a comparable day of age if there was no such birth (yoked controls), and then housed with age-mates until day 40; and subject 4 was separated from the dam and littermates (including junior siblings if present) on day 40. All subjects were housed individually 2 days before testing began.[8]

Maternal Behavior in Adults: Roles of Offspring Characteristics and Parturition

When MB is "turned on" in Norway rats, either as the result of the physiological changes accompanying pregnancy and parturition (see papers by A. D. Mayer and H. I. Siegel, this volume) or as a result of continuous cohabitation with pups,[10,18] it is elicited by and directed toward a particular stimulus complex. It is well known that MB is under multisensory control[19,20] and that once MB is fully established following parturition, the retrieving drive may be rather indiscriminate.[10,21] On the other hand, when choices and/or differential responding are taken into account, maternal retrieval in rats can be said to be selective in that preferences are apparent for newborn vs. older pups, own vs. alien pups, and normal vs. altered tactile/temperature characteristics of skin.[10,19,22-24]

The nature and limits of the pup-stimulus complex requisite to the *initiation* of MB have yet to be completely specified. One question to consider is whether mere exposure to exteroceptive (i.e., visual, auditory, and olfactory) pup stimuli can effect the change from indifference (or aversion) toward pups or whether actual contact (i.e., taste and tactile stimuli) is necessary for the induction or maintenance of maternal responsiveness. The smaller the cage the female is housed in with pups, the shorter the latency to onset of MB,[25] but which aspect of proximity is most effective is not clear from this finding.

On the basis of the rapid onset of MB in nulliparous rats following centrally or peripherally induced anosmia, Fleming and Rosenblatt proposed that the odor of rat pups renders the nonmaternal rat fearful; habituation to these pup odors heralds, and perhaps causes, the switch to maternal caregiving.[26] However, I found that even if such habituation plays a role in maternal sensitization, it is not sufficient in the absence of pup contact.[27] Maternally naive Long-Evans female rats were primed with either continuous exteroceptive stimulation from four rat pups (1-8 days old) placed beneath the mesh floor of their home cage or with 15 minutes per day of four rat pups placed in their home cage, in which case contact was possible; controls received no pup priming. Priming lasted 8 days, after which maternal sensitization testing began by housing all females continuously with pups. In the Contact group, those females which actually made contact, by licking the pups during the priming phase, became maternal significantly faster than females which did not lick pups during priming, or females in the Exteroceptive and Control groups. Thus, licking, not looking at, listening to, and smelling pups, is a powerful inducer of MB.[27] Pup-licking induced by tail-pinches[28] or by applications of placental fluids to the skin of pups[29] also was associated with rapid induction of MB in virgin female rats (TABLE 1).

Similar pup stimulation conditions prior to sensitization were tested with females naturally primed to be maternal by the hormonal changes of pregnancy and its termination (i.e., day 21 cesarean delivery) to assess their role in the *maintenance* of maternal responsiveness.[27] It was found that brief daily access to pups was effective in hastening the onset of MB only in those females (61.5%) which initiated MB during priming. As was true of the similarly tested virgins, exteroceptive stimulation was ineffective. Thus, physical interactions with pups seem to be critical for the maintenance, as well as the induction, of maternal responsiveness in the Long-Evans female rat.

During the natural course of events, parturition, in which the female is an active participant, provides the female with ample opportunity for physical interactions with her offspring. Indeed, MB is typically "turned on" immediately at this time, the first offspring-directed MB being extensive licking and handling. Nonetheless, parturition per se has been considered dispensable to the rapid onset of MB near term in rats because of the occurrence of MB in primiparous rats following cesarean delivery in late pregnancy as well as near term, prior to the onset of parturition (see papers by Mayer and Siegel, this volume). However, because a much smaller percentage of Long-Evans rats become maternal under these circumstances (see discussion in Reference 30) the physiological and/or behavioral events of the parturition experience may be more critical in these females.

Further, evidence has been emerging on other nurturance consequences of parturition: (1) Bridges[31] showed that experience with every other pup for the duration of parturition, after which all pups were removed, was sufficient for the retention of short-latency onset of MB 25 days later. (2) Unlike rats, sheep are highly discriminating with respect to the object of their nurturance. Ecologically, this difference makes sense: a rat mother bears and rears her litter in a separate burrow where she is unlikely to confront another mother's offspring, whereas a herd animal like the sheep will be

TABLE 1. Pup-Licking Hastens the Onset of Maternal Behavior in Virgin Rats

Condition/Group	N	Maternal Behavior Latency	Reference (Strain)
		Median Day	
Pup priming (8 days) prior to sensitization			
Contact (15 min/day)			
Pup-licking +	10	1.5^a	27
Pup-licking −	12	5.0	(Long-Evans)
Exteroceptive (24 hr/day)	16	4.0	
Control (no priming)	20	5.5	
		Mean Day	
Tail-pinches (5 min each) in presence of pups at onset of sensitization^b			
16 on days 1-2	15	3.6^a	28
8 on day 1 or 2	15	4.4^a	(Sprague-Dawley)
4 on day 1	15	5.1	
Handled controls	15	6.5	
Unhandled controls	15	6.3	
		Mean Day	
Pup treatment during sensitization			
Placenta on pups	25	3.8^a	29
Saline/saccharine on pups	14	6.7	(Long-Evans)
Untreated	18	6.3	

^a$p < 0.05$ compared to controls.
^bFifty-four out of 60 tail-pinched rats licked the pups during at least one of the tail-pinch sessions and the more vigorous the licking, the shorter the maternal behavior latency ($p < 0.001$).

exposed to many alien lambs. Within a few hours after birth, the ewe becomes olfactorily imprinted on her own lamb.[32] Apparently it is the physiological events of the parturition experience which render the mother receptive to this imprinting process. Keverne et al.[33] found that 5 min of vaginal stimulation, simulating the birth process, caused parturient ewes to accept an alien lamb at a time (2-3 hr postpartum) when such acceptance normally no longer occurs. Similarly, I found that parturition-only experienced rats are highly discriminating with respect to the range of pup stimuli that elicit their maternal attentions[30] (FIGURE 2), although individual or species-specific odors do not seem to be relevant cues.

Pups were removed from Long-Evans rats during their first parturition; 24 hr later, MB was tested with varying pup stimuli.[30] These dams were most likely to respond maternally toward newborn (0- to 2-day-old) rat pups (100%) and 6- to 8-day-old hamsters (83.3%), which are the size of newborn rats. In contrast, dams were significantly less likely to respond maternally toward newborn hamsters (50%) and 8- to 10-day-old rats (16.7%), pups which are half as large and twice as large, respectively, as newborn rats; indeed, dams were likely to attack these pups (33.3 and 25%, respectively). The rapid maternal response was similar toward dead and live newborn rats, except for reduced retrieval of dead newborns in the first minute. In contrast, primiparous cesarean-delivered dams did not display higher maternal re-

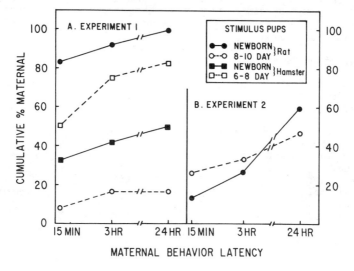

FIGURE 2. Cumulative percentage of Long-Evans primiparous females becoming maternal (A) beginning 24 hr following removal of their pups during parturition (Experiment 1; $N = 12/$ group) or (B) beginning 19-20 hr following cesarean delivery on day 21 of gestation (Experiment 2; $N = 15/$ group). Groups differed with respect to the type of pups presented. (Reprinted with permission from Stern.[30])

sponsiveness toward 0- to 2-day-old rats than toward 8- to 10-day-old rats, and 22% of them displayed infanticide toward newborn rats, compared to 0% of dams which had ever experienced parturition. These results suggest that the endogenous and/or exogenous stimuli associated with parturition enhance selective maternal responsiveness and diminish infanticide toward pups the size of newborn rats.

Thus, physical interaction with young, the readiness of which varies with age and reproductive state, leads to the establishment of MB in rats; such interaction during parturition alone has powerful consequences, including a selective bias toward pups the size of newborn rats at the later onset of maternal behavior.

A number of human parallels to these events of the rat maternal behavior cycle are evident, including the following: (1) While not foolproof, having to care for an infant or small child itself engenders nurturant behaviors and attitudes. To the extent that nurturance is more characteristic of girls and women than boys and men, it is most likely the result of cultural expectations and assignment of caretaking such as of younger siblings.[34] (2) The desirability of allowing maximal mother-infant contact shortly after parturition, in contradiction to what had become the standard practice in Western hospitals, was championed, with data, by pediatricians Kennell and Klaus,[35] who were directly inspired by the animal maternal-infant literature, including that on rats. (3) The inadequacy of the premature infant as a stimulus (small size, fragility, immature behavioral characteristics) has been addressed as a risk factor for optimal parent-infant interactions.[36]

OFFSPRING-INDUCED MATERNAL HORMONAL CHANGES

Suckling vs. Exteroceptive Control of Prolactin and Oxytocin Secretion

After birth, suckling is by far the most potent stimulus for the secretion of prolactin and oxytocin, thus ensuring milk production and release.[37] Further, the amount of prolactin and milk secretion is directly related to the amount of suckling. Hence, rats can rear a widely varying number of pups to a similar body weight at weaning and women can fully nourish twins, as well as singletons, for up to 6 months postpartum.

To what extent can non-suckling stimuli from young come to exert control of oxytocin and prolactin secretion? The conditioning of oxytocin release in response to stimuli associated with suckling is well known among mammals.[37] In nursing women, the mere thought of a baby can provoke oxytocin-induced milk let-down.[38] The adaptive significance of such conditioned release of oxytocin might be (a) in anticipation of an impending nursing episode, to provide the young with milk as soon as it begins to suck instead of after the usual delay of about a minute and (b) when the mother is physically separated from her suckling young, to ensure a rapid reunion, as in response to its cry. In fact, among women who are established lactators, oxytocin release usually occurs after the onset of suckling, and in a minority of women, shortly prior to its onset[39] (see below).

In rats, maternal behavior performance (including extensive ventrum contact), in the absence of suckling, induces modest prolactin secretion.[40] The exteroceptive control of prolactin secretion in rats was described by Grosvenor and co-workers,[41,42] who suggested that it may function to maintain lactation in the periweaning period when the suckling frequency declines. They concluded that 2 weeks of experience with a litter was required for conditioning in the primiparous dam, and that in the absence of suckling, visual and olfactory (but not auditory) pup stimuli were effective elicitors of prolactin. More recently, Terkel et al.[43] elicited marked prolactin secretion from primiparous dams 5-19 days postpartum with auditory stimulation in the form of taped pup ultrasonic distress vocalizations, but this phenomenon has not been replicated.[44,45]

Among mammals the conditioning of prolactin release to exteroceptive stimulation has been demonstrated only for rats, and only for suckling as the unconditioned stimulus, to the best of my knowledge. (The phenomenon may be common in birds: e.g., the male ring dove will secrete prolactin in response to the sight of his incubating mate following several days of his own incubating.[46]) In nursing women, prolactin secretion is tied to actual suckling, with the exception of the spontaneous nocturnal surge.[39] In 22 nursing episodes followed at 5-min intervals, oxytocin secretion preceded the onset of actual nursing in only 8 cases (36%) and prolactin did not begin to rise until at least 5 and usually 10 min after the onset of nursing[39] (FIGURE 3). This makes sense in that non-nursing stimuli from babies otherwise would disrupt the close link which exists between the amount of nursing and the amount of milk produced.

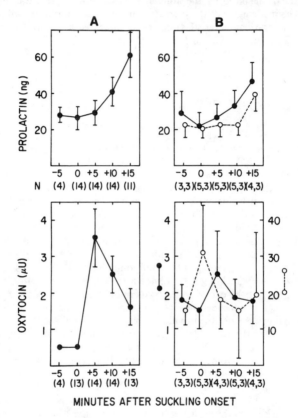

FIGURE 3. Time course of oxytocin and prolactin secretion during 22 suckling episodes followed at 5-min intervals in 19 Caucasian women between 4 and 46 months postpartum. On the left are the results from those women whose oxytocin levels were at baseline prior to the actual onset of suckling (−5 and/or 0). On the right are the results from those women whose oxytocin exceeded baseline levels prior to the onset of suckling; these seven women are further subdivided by the magnitude of their oxytocin levels.[39]

Suckling, Prolactin Secretion, and Birth-Space Intervals

Lactation not only entails an enormous physiological and behavioral investment in the current young, but it usually results in some delay in the appearance of subsequent young as well. In rats and other mammals with a postpartum estrus, there is often an implantation delay, the length of which is correlated with the number of suckling young, unless lactation ceases by removal of the current young. In many mammals, conception is not possible postpartum until the young are weaned (e.g., hamsters). In women, breast-feeding prolongs the birth-space interval by delaying the onset of menses and ovulation.[47] For example, the !Kung-San, hunter-gatherers of Botswana, have a natural birth-space interval of 3-5 years, which is probably related to their practice of frequent and prolonged nursing.[48] The duration of postpartum amenorrhea varies with the duration of lactation, the intensity and frequency of suckling, the onset of supplementary foods, and possibly the presence of night breast-feeds.[47] The suckling-induced delay in ovulation (or implantation) is primarily due to inhibition of gonadotropin secretion, although elevated levels of prolactin probably contribute to various aspects of conception delay in women.[49] Pathological levels of prolactin are associated with amenorrhea in women,[50] though menses can return while lactational levels of prolactin are elevated.[50,51]

In a study carried out in the Clinical Study Unit of Tufts University Medical Center, 20 Caucasian lactating women, between 4 and 46 months postpartum, had hourly blood samples taken and recorded their nursing episodes over a 24-hr period.[39,51] In addition, morning blood samples from a number of !Kung women were available from a previous study.[48] We confirmed the well-known finding that prolactin declines postpartum, though in contrast to textbook descriptions, prolactin does not decline to nonlactating levels in women who continue to nurse substantially until close to 2 years postpartum. In fact, the relationship of mean 24-hr prolactin to suckling duration per day is even stronger than that to time postpartum. Among the high-intensity-nursing women in our study, the mean duration of postpartum amenorrhea was 15.2 months, which would account for a birth-space interval of at least 2 years. Although the !Kung women nurse more frequently, they were comparable to the high-intensity-nursing American women in our sample in terms of number of minutes of nursing per day and in terms of morning prolactin levels.

For women in noncontracepting populations, the bottom line of this phenomenon is clear-cut: the greater the physiological investment in the current young (by prolonged and frequent nursing), the longer the delay in conception of the next young. The tragic consequence of abandoning traditional nursing behavior practices is an ever-increasing birthrate. For contracepting populations, an evolutionary perspective suggests that a natural birth-space interval of 2-3 (or more) years may well be optimal for various aspects of child development and parent-child interactions.

Maternal Behavior, Contact Comfort, and Female Sexuality

The dissection of interactions between mother and young has revealed that during the maintenance of MB, skin-to-skin thermal contact and licking serve as vital biobehavioral regulators. The temperature changes the maternal rat experiences regulate her nursing bouts[52] and maternal licking of pups' urine output achieves conservation of water.[53] Perhaps the behavioral aspects of thermal contact and licking also can be considered rewarding to the mother. Although Lehrman[54] spoke of the relief from mammary gland tensions which suckling would provide, we now know that for rats suckling can be disposed of (by removal of mammary glands and nipples) without disrupting the most obvious aspects of maternal responsiveness.[55] But the basic idea remains, namely, the hedonic pleasure of tactile stimulation or contact comfort for mothers. Most broadly, maternal nurturance, including childbirth and lactation, can be considered a vital aspect of female sexuality.[56]

Indeed, there seems to be a coordination, or a negative correlation, between the occurrence of female sexual behavior expressed with adult males and female nurturance of young. For most mammals, estrus is delayed until after weaning. In Norway rats, the postpartum estrus is abbreviated compared to the cycling estrus, and the experienced mother distributes her time appropriately between maternal and sexual behaviors during the mating interval.[57] The sexuality of the human female is most different from that of other primate females in the expression of genital sexuality during lactation.[58] But is the postpartum woman completely different in this respect from her primate relatives?

Recent evidence suggests that for both primiparous women in Scotland[59] and multiparous women in New Jersey[58] babies inhibit sexual activity. Overall, in both studies the coital rate dropped from 2.5 per week before conception to 1.25 postpartum, up to 6-9 months. There are, of course, obvious experiential bases for this phenomenon including lack of time, fatigue, and sleep disturbances, as well as satisfaction with physical closeness to baby all day. However, in the Stern and Leiblum study,[58] breast-feeders differed from bottle-feeders in resuming sexual activity later, and once resumed, there was a strikingly negative correlation between the frequency of nursing per day and the frequency of coitus per week between 13 and 24 weeks postpartum (FIGURE 4). In other words, the more a mother gives suck, the less she engages in sexual intercourse! In addition to experiential influences on this effect, we are exploring possible hormonal bases as well. Thus, hyperprolactinemia[60] and/or reduced levels of ovarian estrogens and androgens may contribute to reduced libido in nursing women.[61]

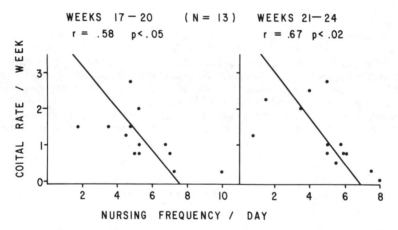

FIGURE 4. Correlations between coital rate per week and nursing frequency per day for continuing nursers only during weeks 17-20 and 21-24 postpartum. (Reprinted with permission from Stern and Leiblum.[58])

CONCLUDING REMARKS

Nurturance in mammals, multifaceted and multisensory determined, fundamentally involves physical contact between caregiver and young. The young receive nourishment, warmth, contact comfort, and other physiological benefits of contact. The caregiver—usually the natural mother—receives what in return? In terms of ultimate causation, if the caregiver is the natural mother or another relative (father, sibling), more of its genes will survive for future generations. But ultimate causes do not explain proximate mechanisms. I submit that once inhibitions concerning contact with young are lifted (which may involve such contact itself), the caregiver also receives contact comfort from the interaction, as well as physiological benefits for the service of her nurturance which may accrue from licking, touching, and suckling.

REFERENCES

1. STERN, J. M. & D. S. LEHRMAN. 1969. Role of testosterone in progesterone-induced incubation behaviour in male ring doves (*Streptopelia risoria*). J. Endocrinol. **44:** 13-22.
2. HARLOW, H. F. & R. R. ZIMMERMAN. 1959. Affectional responses in the infant monkey. Science **130:** 421-432.
3. HOFER, M. 1981. Parental contributions to the development of their offspring. *In* Parental Care in Mammals. D. J. Gubernick & P. H. Klopfer, Eds. Plenum. New York, N.Y.
4. KUNH, C. C., S. R. BUTLER & S. SCHANBERG. 1978. Selective depression of serum growth hormone during maternal deprivation in rat pups. Science **201:** 1034-1036.
5. SCHANBERG, S. Effects of tactile stimulation on maternally deprived rat pups and ICU preterm infants. Abstract. Presented at a meeting of the International Society for Developmental Psychobiology, Baltimore, Md., October 25-28, 1984.
6. BRIDGES, R. S., M. X. ZARROW, B. D. GOLDMAN & V. H. DENENBERG. 1974. A developmental study of maternal responsiveness in the rat. Physiol. Behav. **12:** 149-151.
7. MAYER, A. D. & J. S. ROSENBLATT. 1979. Hormonal influences during the ontogeny of maternal behavior in female rats. J. Comp. Physiol. Psychol. **93:** 879-898.
8. STERN, J. M., M. REINA, L. ROGERS, S. HUACUJA & B. SHEIKH. Ontogeny of maternal responsiveness in rats: maturational and experiential influences. Abstract. Presented at a meeting of the International Society for Developmental Psychobiology, Baltimore, Md., October 25-28, 1984.
9. ROSENBLATT, J. S. 1970. Views on the onset and maintenance of maternal behavior in the rat. *In* Development and Evolution of Behavior. L. R. Aronson, E. Tobach, D. S. Lehrman & J. S. Rosenblatt, Eds. W. H. Freeman & Co. San Francisco, Calif.
10. WIESNER, B. P. & N. M. SHEARD. 1933. Maternal Behaviour in the Rat. Oliver & Boyd. London, England.
11. ROSENBLATT, J. S. & D. S. LEHRMAN. 1963. Maternal behavior in the laboratory rat. *In* Maternal Behavior in Mammals. H. L. Rheingold, Ed. John Wiley. New York, N.Y.
12. SPENCER-BOOTH, Y. 1970. The relationships between mammalian young and conspecifics other than mothers and peers: a review. *In* Advances in the Study of Behavior. Vol. 3. D. S. Lehrman, R. A. Hinde & E. Shaw, Eds. Academic Press. New York, N.Y.
13. DAVIS, D. E. & E. HALL. 1951. The seasonal reproductive conditions of female Norway (Brown) rats in Baltimore, Maryland. Physiol. Zool. **24:** 9-20.
14. CALHOUN, J. B. 1962. The Ecology and Sociology of the Norway Rat. Public Health Service Publication No. 1008. Public Health Service. Bethesda, Md.
15. GILBERT, A. N., D. A. BURGOON, K. A. SULLIVAN & N. T. ADLER. 1983. Mother-weanling interactions in Norway rats in the presence of a successive litter produced by postpartum mating. Physiol. Behav. **30:** 267-271.
16. GRAY, P. & S. CHESLEY. 1984. Development of maternal behavior in nulliparous rats (*Rattus norvegicus*): effects of sex and early maternal experience. J. Comp. Psychol. **98:** 91-99.
17. BARASH, D. P. 1982. Sociology and Behavior. 2nd edit. Elsevier. New York, N.Y.
18. ROSENBLATT, J. S. 1967. Nonhormonal basis of maternal behavior in the rat. Science **156:** 1512-1513.
19. BEACH, F. A. & J. JAYNES. 1956. Studies of maternal retrieving in rats. III. Sensory cues involved in the lactating female's response to her young. Behaviour **10:** 104-125.
20. HERRENKOHL, L. R. & P. A. ROSENBERG. 1972. Exteroceptive stimulation of maternal behavior in the naive rat. Physiol. Behav. **8:** 595-598.

21. DENENBERG, V. H., G. A. HUDGENS & M. X. ZARROW. 1966. Mice reared with rats. Psychol. Rep. **18:** 455-456.
22. BEACH, F. A. & J. JAYNES. 1956. Studies of maternal retrieving in rats. I. Recognition of young. J. Mammal. **37:** 177-180.
23. STERN, J. M. & D. A. MACKINNON. 1978. Sensory regulation of maternal behavior in rats: effects of pup age. Dev. Psychobiol. **11:** 579-586.
24. MAYER, A. D. & J. S. ROSENBLATT. 1980. Hormonal interaction with stimulus and situational factors in the initiation of maternal behavior in nonpregnant rats. J. Comp. Physiol. Psychol. **94:** 1040-1059.
25. TERKEL, J. & J. S. ROSENBLATT. 1971. Aspects of nonhormonal maternal behavior in the rat. Horm. Behav. **2:** 161-171.
26. FLEMING, A. & J. S. ROSENBLATT. 1974. Olfactory regulation of maternal behavior in rats. II. Effects of peripherally induced anosmia and lesions of the lateral olfactory tract in pup-induced virgins. J. Comp. Physiol. Psychol. **86:** 233-246.
27. STERN, J. M. 1983. Maternal behavior priming in virgin and Caesarean-delivered Long-Evans rats: effects of brief contact or continuous exteroceptive pup stimulation. Physiol. Behav. **31:** 757-763.
28. SZECHTMAN, H., H. I. SIEGAL, J. S. ROSENBLATT & B. R. KOMISARUK. 1977. Tail-pinch facilitates onset of maternal behavior. Physiol. Behav. **19:** 807-809.
29. KRISTAL, M. B., J. F. WHITNEY & L. C. PETERS. 1981. Placenta on pups' skin accelerates onset of maternal behavior in nonpregnant rats. Anim. Behav. **29:** 81-85.
30. STERN, J. M. 1985. Parturition influences initial pup preferences at later onset of maternal behavior in primiparous rats. Physiol. Behav. **35:** 25-31.
31. BRIDGES, R. S. 1975. Long-term effects of pregnancy and parturition upon maternal responsiveness in the rat. Physiol. Behav. **14:** 245-250.
32. POINDRON, P. & P. LE NEINDRE. 1980. Endocrine and sensory regulation of maternal behavior in the ewe. *In* Advances in the Study of Behavior. Vol. 11. J. S. Rosenblatt, R. A. Hinde, C. Beer & M. C. Busnel, Eds.: 75-119. Academic Press. New York, N.Y.
33. KEVERNE, E. B., F. LEVY, P. POINDRON & D. R. LINDSAY. 1983. Vaginal stimulation: an important determinant of maternal bonding in sheep. Science **219:** 81-83.
34. WEISNER, T. S. & R. GALLIMORE. 1977. My brother's keeper: child and sibling caretaking. Curr. Anthropol. **18:** 169-190.
35. KENNELL, J. H., D. K. VOOS & M. H. KLAUS. 1979. Parent-infant bonding. *In* Handbook of Infant Development. J. D. Osofsky, Ed.: 786-798. John Wiley & Sons. New York, N.Y.
36. GOLDBERG, S. 1979. Premature birth: consequences for the parent-infant relationship. Am. Sci. **67:** 214-220.
37. COWIE, A. T. & J. S. TINDAL. 1971. The Physiology of Lactation. Edward Arnold. London, England.
38. CALDEYRO-BARCIA, R. 1969. Milk ejection in women. *In* Lactogenesis: The Initiation of Milk Secretion at Parturition. M. Reynolds & S. J. Folley, Eds. University of Pennsylvania Press. Philadelphia, Pa.
39. STERN, J. M., A. ROBINSON, T. HERMAN & S. REICHLIN. Oxytocin; neurophysin and prolactin: responses to suckling and 24-hour pattern during prolonged lactation in women. Manuscript in preparation.
40. STERN, J. M. & H. I. SIEGEL. 1978. Prolactin release in lactating primiparous and multiparous thelectomized, and maternal virgin rats exposed to pup stimuli. Biol. Reprod. **19:** 177-182.
41. GROSVENOR, C. E., H. MAIWEG & F. MENA. 1970. A study of factors involved in the development of the exteroceptive release of prolactin in the lactating rat. Horm. Behav. **1:** 111-120.
42. MENA, F. & C. E. GROSVENOR. 1971. Release of prolactin in rats by exteroceptive stimulation: sensory stimuli involved. Horm. Behav. **2:** 107-116.
43. TERKEL, J., D. A. DAMASSA & C. H. SAWYER. 1979. Ultrasonic cries from infant rats stimulate prolactin release in lactating mothers. Horm. Behav. **12:** 95-102.
44. STERN, J. M., D. A. THOMAS, J. RABII & R. J. BARFIELD. 1984. Do pup ultrasonic cries provoke prolactin secretion in lactating rats? Horm. Behav. **18:** 86-94.

45. VOLSOCHIN, L. M. & J. H. TRAMEZZANI. 1984. Relationship of prolactin release in lactating rats to milk ejection, sleep state, and ultrasonic vocalization by the pups. Endocrinology **114:** 618-623.
46. FRIEDMAN, M. & D. S. LEHRMAN. 1968. Physiological conditions for the stimulation of prolactin secretion by external stimuli in the male ring dove. Anim. Behav. **16:** 233-237.
47. SHORT, R. 1984. Breast feeding. Sci. Am. **250(4):** 35-41.
48. KONNER, M. & C. WORTHMAN. 1980. Nursing frequency, gonadal function, and birth spacing among !Kung hunter-gatherers. Science **207:** 788-791.
49. SCHALLENBERGER, E., D. W. RICHARDSON & E. KNOBIL. 1981. Role of prolactin in the lactational amenorrhea of the rhesus monkey (*Macaca mulatta*). Biol. Reprod. **25:** 370-374.
50. FRANTZ, A. G. 1978. Prolactin. N. Engl. J. Med. **298:** 201-207.
51. STERN, J. M., M. KONNER, T. HERMAN & S. REICHLIN. 1986. Nursing behavior, prolactin and postpartum amenorrhea during prolonged lactation in American and !Kung mothers. Clin. Endocrinol. In press.
52. LEON, M., P. G. CROSKERRY & G. K. SMITH. 1978. Thermal control of mother-young contact in rats. Physiol. Behav. **21:** 793-811.
53. GUBERNICK, D. J. & J. R. ALBERTS. 1983. Maternal licking of young: resource exchange and proximate controls. Physiol. Behav. **31:** 593-601.
54. LEHRMAN, D. S. 1961. Hormonal regulation of parental behavior in birds and infrahuman mammals. *In* Sex and Internal Secretions. W. C. Young, Ed.: 1268-1382. Williams & Wilkins. Baltimore, Md.
55. MOLTZ, H., D. GELLER & R. LEVIN. 1967. Maternal behavior in the totally mammectomized rat. J. Comp. Physiol. Psychol. **64:** 225-229.
56. NEWTON, N. 1973. Interrelationship between sexual responsiveness, birth, and breast feeding. *In* Contemporary Sexual Behavior: Critical Issues in the 1970's. J. Zubin & J. Money, Eds. Johns Hopkins Press. Baltimore, Md.
57. GILBERT, A. N., R. J. PELCHAT & N. T. ADLER. 1980. Postpartum copulatory and maternal behaviour in Norway rats under seminatural conditions. Anim. Behav. **28:** 989-995.
58. STERN, J. M. & S. LEIBLUM. 1986. Postpartum sexual behaviour of American women as a function of absence or frequency of breast feeding: a preliminary communication. *In* Primate Ontogeny, Cognition and Social Behavior, Proceedings of the International Primatological Society. J. Else & P. Lee, Eds.: 315-324. Cambridge University Press. Cambridge, England.
59. ADLER, E. & J. BANCROFT. 1983. Sexual behaviour of lactating women: a preliminary communication. J. Reprod. Infant Psychol. **1:** 47-52.
60. BUVAT, J., M. ASFOUR., M. BUVAT-HERBAUT & P. FOSSAT I. 1978. Prolactin and human sexual behavior. *In* Progress in Prolactin Physiology and Pathology. C. Robyn & M. Harter, Eds. Elsevier/North-Holland Medical Press. Amsterdam, the Netherlands.
61. SANDERS, D. & J. BANCROFT. 1982. Hormones and the sexuality of women—the menstrual cycle. Psychosom. Med. **43:** 199-214.

Interaction of Species-Typical Environmental and Hormonal Factors in Sexual Differentiation of Behavior[a]

CELIA L. MOORE

Department of Psychology
University of Massachusetts
Boston, Massachusetts 02125

It has become accepted in behavioral endocrinology that hormonal explanations of behavior are likely to be wrong or incomplete unless the roles of external stimuli are also taken into account. Although the impact of the external environment on hormonal regulation was recognized earlier (e.g., Reference 1), full appreciation of the complex interrelation of external and internal factors was closely linked with the founding and subsequent work of the Institute of Animal Behavior of Rutgers University. Lehrman's[2] analysis of reproductive behavior in ring doves and Rosenblatt's[3] analysis of sexual behavior in cats have provided provocative models that continue to generate ideas about the many ways in which external stimuli, or individual experience, can interact with hormonal effects and about the multiplicity of pathways through which hormones can alter behavior.

Gonadal hormones of vertebrates function during two major life stages: In adult stages, their actions are intimately tied to the control of reproduction; in prenatal and postnatal stages, their role is not immediately related to reproduction, but to the differentiation of male and female reproductive classes. A diagram such as that presented in FIGURE 1 could readily be used to summarize the work of Rosenblatt, Lehrman, and others on the interaction of gonadal hormones and individual experience in the control of reproduction in birds and mammals. Curiously, such interactions have not been considered in accounts of hormone action during early developmental stages.[4]

Young animals clearly do interact with their environments. Parents and other social companions respond to the individual characteristics of developing young, and the young respond selectively to stimuli that they encounter. Rosenblatt[5] has described the interactions of rat dams and litters as synchronized or mutually adjusted. The characteristics of the litter, such as age-related variations, lead to appropriate physiological and behavioral changes in the dam. Similarly, changes in the dam compel complementary changes in the young, such as those leading to independent feeding.

I began my studies of sexual differentiation in rats with the general hypothesis that whenever hormones are found to affect behavior, coaction with external stimuli is likely to be part of the process. I reasoned that if hormones interacted with external

---

[a] The research summarized in this paper was made possible by the support of the National Science Foundation through Grants BNS 77-24788, BNS 80-14669, and BNS 83-17796.

factors in adults to affect behavior and physiological condition, why not also in neonates? If maternal rats engage in mutual adjustment with whole litters, why not also with male and female variants within a litter, thus providing different stimuli to hormonally different young offspring?

SEXUALLY DIFFERENTIATED ENVIRONMENTS

Humans are keenly aware of the sex of their infants, and there is ample evidence from anthropological and psychological work of differential treatment. Despite a surprising lack of systematic study, there are some accounts of differential care of males and females among nonhuman primates.[6] Maternal rats also provide their male and female pups with different care.[7]

If looked at from the young organism's viewpoint, differential care means that males and females are exposed to systematic differences in the environments within which they develop. An infant rat is born into a nest with several littermates and a

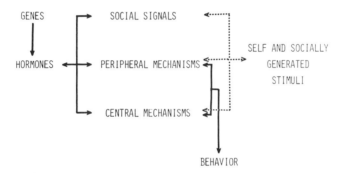

FIGURE 1. Schematic diagram illustrating the interrelated effects of hormones and environmental stimuli on organismic mechanisms that underlie behavior.

mother. These elements provide the characteristic chemical, tactile, thermal, and other stimuli that make up the species-typical environments for neonates of this species. Many of these stimuli do not differ systematically between the sexes: Males and females occupy the same nest, huddle with the same littermates, and are exposed to the same pattern of maternal nest attendance and nursing. However, handling and licking are distributed unevenly between the sexes.

Licking of the anogenital region (AGL) is a readily distinguished form of maternal licking, characterized behaviorally by particular postural adjustments on the part of both pup and dam. It is functionally related to stimulation of elimination in pups whose internal control is immature and to recycling of fluids and salts lost by dams during lactation.[8] It occurs frequently during the 2 weeks after parturition, then declines slowly.[7,8] Male pups regularly receive significantly more AGL throughout this period.[7] (See FIGURE 2.) Therefore, male and female pups are subjected unequally to the complex stimulus changes that accompany this maternal pattern. These include: thermal stimuli, both warmth from contact with forepaws and ventral surface of dam

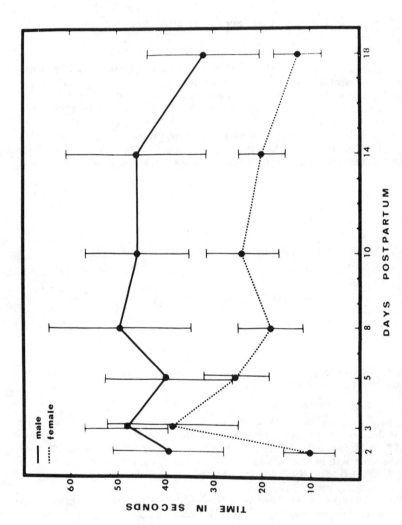

FIGURE 2. Comparison of the mean (± SE) time spent by rat dams in anogenital licking of their male and female pups throughout early development. (Adapted with permission from Moore and Morelli.[7])

and cooling from evaporation of saliva[9]; vestibular stimuli from handling and, frequently, rotation to a supine posture; and tactile stimuli from tongue, forepaws, and fur of dam. If prolonged sufficiently to elicit elimination, AGL also elicits a characteristic behavioral response from pups. This response includes extension of all limbs, spreading of the digits, deviation of the tail, and inhibition of movement. Males are not only more frequent recipients of stimulation from maternal licking, but also respond more frequently to it,[10] thus adding feedback from their own motor processes as another source of sex-related stimulus bias.

Sex-related biases in the environments of developing rats do not cease with the decline of maternal anogenital licking. When, around the time of weaning, developing pups take on the task of grooming themselves, their self-grooming reflects the same bias as that previously produced by their mothers. Males spend significantly more time than females grooming the genitalia, a sex difference that persists beyond puberty.[11] (See FIGURE 3A.) Prepubescent male and female rats also engage in social play to different extents; males are reported to participate more frequently in rough-and-tumble play.[12] The differences in behavior during this age period generate consistent and reliable sex-related differences in stimulative context, thus perpetuating and elaborating environmental biases initiated by maternal care.

SOURCES OF SEX DIFFERENCES IN SIGNALS

In order for there to be reliable sex-typical social environments for developing organisms, there must be reliable differences in the behavioral or nonbehavioral characteristics of the young organisms that will call forth different behavior from companions. In other words, young males and females must produce different signals. In young rats, a signal that is responsible, at least in part, for greater maternal attention to males is chemical and is found in the urine. Maternal discrimination of male and female pups can be disrupted by placing masking odors on the skin of the pups; dams can be lead to treat female pups like males by placing male pup urine on their skin.[13] Further, as little as 20 μl of male pup urine placed near the nest will stimulate a dam to increase anogenital licking of her pups; female pup urine does not have this effect.[14] When male and female pup urine was presented to lactating females in the absence of pups, they spent more time sniffing the urine deposits from male pups. The age of the pups from which urine was collected had no effect, within the 2-17 day postpartum range examined.[14] The difference between male and female urine is one of relative attractiveness: Both male and female pups are licked by their dams and urine produced by both sexes is attractive to dams. It is possible that the difference between the sexes is a difference in concentration of the same attractive odor, rather than qualitatively different odors. Taken together, these findings suggest that odors carried in their urine provide reliable stimuli for sex identification and may, in fact, be sufficient to account for the differential maternal treatment of male and female pups. Of course, there may also be other behavioral or nonbehavioral differences (e.g., ultrasonic vocalizations, latency to urinate) between young males and females that are detected by dams and that provide them with redundant information regarding the sex of the pup.

The sex difference in AGL-eliciting cues originates from differences in endogenous perinatal hormones. When female rat pups were injected on the day of birth with 500 μg testosterone propionate, they were treated by maternal rats as though they were males both 24 hr and 9 days later.[15] Because male pups are normally exposed to higher

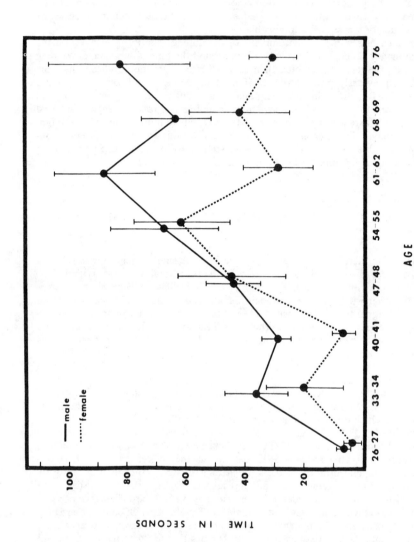

FIGURE 3A. Comparison of the mean time spent by developing male and female rats in genital self-grooming ($n = 9$). (Adapted with permission from Moore and Rogers.[11])

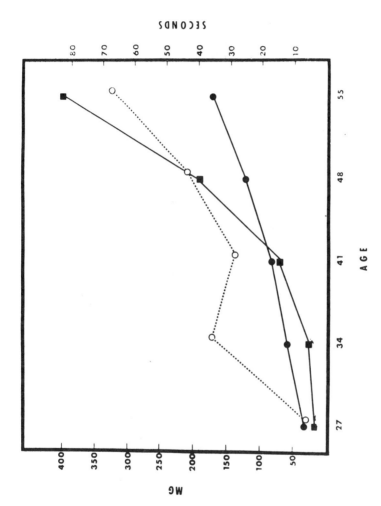

FIGURE 3B. Mean weights (in milligrams) of seminal vesicles (solid squares) and ventral prostates (solid circles) of prepubescent male rats ($n = 12$ at each age sampled). Note relationship to genital self-grooming observed in males of the same ages (open circles; data from FIGURE 3A). (Adapted with permission from Moore and Rogers.[11])

levels of testosterone than their female siblings,[16] reliable differences in the AGL-eliciting properties of males and females can be expected. It is known that sex-identifying odors in the urine of adult male rats are androgen dependent[17]; perhaps the mechanism for producing sex-identifying odors in infant male rats is similar to that in adults.

DEVELOPMENTAL CONSEQUENCES OF SEX-TYPICAL MATERNAL STIMULATION

Although maternal behavior in rats is well integrated, its underlying control is heterogeneous. It would be overly simplistic to argue that discrete behavioral patterns, like nest-building, crouching, and licking, have separate sensory control mechanisms. Nevertheless, we do know, for example, that time in the nest is largely a function of maternal temperature,[18] and maternal licking a matter of olfaction.[13,19] It is possible, therefore, to make rather selective manipulations of maternal behavior by manipulating sensory input to the dam. I have used this approach to evaluate the effect on behavioral development of those aspects of maternal stimulation provided by licking which, so far as I have been able to determine, is the only source of sexually biased maternal stimulation. Anosmic rats, deprived of the odors that elicit licking, lick their pups infrequently but otherwise have rather intact maternal behavior.[19,20] Therefore, I performed a controlled rearing study in which pups were reared by dams that differed in olfactory competence so as to determine whether level of stimulation associated with maternal licking could contribute toward the development of a sexually differentiated behavioral pattern.[21]

Anosmia was produced by lining the nasal passages with polyethylene tubing, following a procedure reported by Ruddy.[22] One group of dams was bilaterally intubated, a second group was unilaterally intubated, and a third group was untreated. The bilaterally intubated dams exhibited very low levels of anogenital and other maternal licking while the unilaterally intubated dams were intermediate to the bilateral and untreated control groups. When male and female offspring of these dams reached maturity, they were gonadectomized and treated with subcutaneous implants of testosterone propionate to ensure equivalent levels of circulating testosterone. Two weeks later, males and females were tested for masculine sexual behavior by pairing them with an estradiol- and progesterone-treated, sexually active, female partner. Males were tested twice weekly until they performed the ejaculatory pattern on two separate tests; females were tested twice weekly for 3 weeks.

All males did perform the ejaculatory pattern, but the offspring of less stimulating, intubated dams had longer ejaculatory latencies, longer interintromission intervals, and longer postejaculatory intromission latencies. They also performed a greater number of intromission patterns before the ejaculatory pattern on the first test. Interestingly, the offspring of unilaterally intubated dams, i.e., males that had experienced an intermediate level of stimulation from maternal licking, were intermediate to the other two groups on these measures of masculine copulatory performance.[21]

Females did not perform the ejaculatory pattern and could not be compared directly with their male siblings. Nevertheless, the masculine copulatory performance of the three female groups differed in a fashion analogous to that of the males. Female offspring of intubated dams performed fewer mounts and intromission patterns, and they had longer intromission latencies and interintromission intervals than controls.[21]

In a converse study,[23] female neonates were provided with additional artificial stimulation of the perineum with a soft camel's hair brush for 75 sec distributed over daily 20-min sessions from day 2 to day 14 postpartum. Female littermate controls were stimulated similarly on the shoulder. Both groups were later ovariectomized, treated with testosterone, and tested for masculine copulatory behavior as previously described. The females that had received additional genital stimulation during their early development performed more intromission patterns, probably because they were performed at a faster rate, than controls.

These studies indicate that stimulation normally provided to all developing rats by their mothers contributes to the development of masculine copulatory behavior, a species-typical behavioral pattern. In particular, it apparently contributes toward a neurophysiological mechanism that underlies the timing of copulation. Although males that received less stimulation as neonates mounted at normal levels and had no difficulty in performing either intromission or ejaculatory patterns, their intromission patterns were performed more slowly and they took longer both to reach ejaculatory threshold and to resume copulation after an ejaculation. As expected, the masculine copulatory behavior of females in these experiments was considerably more deficient than that of males. Recall that they had not been treated perinatally with testosterone. Even so, additional genital stimulation of female neonates increased their subsequent copulatory rate, and lower-than-normal levels of maternal AGL stimulation slowed it. Because masculine copulatory behavior is performed more readily and more completely in males than in females, it is possible that the greater stimulation from maternal AGL that males regularly receive contributes to this sex difference. Increased licking stimulation is one of many sequelae of perinatal testosterone and, although this is clearly not the whole story, it may indeed mediate some of the effects of this hormone on the development of sexually differentiated behavior.

DEVELOPMENTAL CONSEQUENCES OF GENITAL SELF-GROOMING

Self-grooming in rats begins before weaning and gradually develops into the adult pattern.[24] Self-grooming in juvenile males and females is very similar, with the notable exception that the genitals are groomed more extensively by males.[11,25] This sex difference depends at least in part on differences in circulating levels of testosterone during this age period. For rats observed in undisturbed social groups, gonadectomy eliminated sex differences in self-grooming, and males and females both responded to testosterone treatment with increased genital self-grooming.[25] However, when self-grooming was observed in rats that were placed alone in a novel environment, males performed more genital grooming than females even after gonadectomy.[26] Thus, differences in circulating testosterone cannot account for all sex differences in genital self-grooming. Perhaps some of the difference reflects the greater maternal anogenital licking received by males as neonates.

The sharp increase in genital self-grooming of males during the period between weaning and puberty is temporally correlated with such peripubertal physiological changes as accessory organ growth[27] (see FIGURE 3B). Mindful of a similar correlation between self-grooming of nipple line and genitalia and the growth of mammary glands in pregnant females,[28,29] Sigrid Rogers and I hypothesized that there might also be a similar causal relationship.

We[11] placed wide rubber collars around the necks of some 27-day-old males, narrow rubber collars around the necks of others, and left others uncollared. Wide collars effectively eliminated genital self-grooming; unexpectedly, the narrow collars slightly decreased genital grooming, particularly during the last week of collaring. The wide collars had no effect on the body weights of 48-day-olds, but did reduce body weights of 55-day-olds. Some animals were autopsied at 48 and others at 55 days of age and their sexual accessory organs were removed and weighed. There was a significant reduction in absolute organ weights at both ages and in relative weights at 48 days in males that wore wide collars. The narrow-collared group had accessory organ weights intermediate to the other two groups, but in only one measure (dorsal prostate weight) was the difference from uncollared control weights significant. Thus, stimulation generated during self-grooming of the genital region apparently contributes to physiological changes associated with puberty in males. Whether similar effects can be found in females has not been determined. However, the usual hallmarks of puberty in females of this species, vaginal opening and first estrus, are not correlated with particularly high levels of self-grooming. In an unpublished study, we did place wide collars on 22-day-old females and found no delay in either first estrus or vaginal opening. Perhaps other measures of physiological change (e.g., hormone levels) would have uncovered a relationship.

These data, along with the established relationship between tactile stimulation of nipples and genitals and physiological changes in adult female rats,[29] raise an interesting question for future work. Might not there also be physiological changes produced in neonates by maternally provided genital stimulation? If so, initial physiological differences between male and female neonates may be magnified by feedback from the maternal environment.

THE PROCESS OF SEXUAL DIFFERENTIATION

Although the story is far from complete, the studies reported above are sufficient to indicate that the schematic diagram presented in FIGURE 1 to illustrate the interrelations of intraorganic hormone action and input from the social and self-generated environment can usefully be applied to developing organisms as well as to adults. In the familiar terminology of current hormonal differentiation theory,[30] organizational as well as activational hormones interact with the external environment.

Male and female mammals begin their earliest developmental stage with small, but important, differences in genetic endowment. These differences are important because they determine that embryogenesis in males and females will occur in different hormonal environments. If there are other consequences to the genetic differences between males and females, they are negligible compared with those that can be traced to early hormone differences. It is apparently no exaggeration to say that all sex differences in mammalian behavior, anatomy, and physiology can be traced to an origin in different early hormones. Given that sex differences originate in early hormones, what is the process through which they account for sex differences in behavior during remote developmental periods?

First, I would like to suggest that, through their actions on various tissues, the different hormones available to males and females lead to the development of a number of separate, small, quantitative differences in male and female organisms. Originally, these differences may be quite minor, such as slight differences in muscle mass or chemical composition of urine. However, some of these differences will become elaborated and magnified through self-generated processes and through transactions with the social environment. Some of the differences between young males and females will be discriminable to caregivers and other social companions and will elicit different behavior from them. The behavior elicited by males and females may differ only quantitatively, like the differential maternal licking of male and female rat pups. Nevertheless, the result will be sex-typical environments that are reliably different for males and females. These environments may, in turn, feed back to maintain or to elaborate organismic sex differences.

Second, I would suggest that behavioral differences between the sexes may develop slowly, based on hormonally originated processes that continue between the time of early hormone availability and behavioral expression. Even though the central nervous system (CNS) surely functions differently when juvenile or adult males and females behave differently, it does not necessarily follow that there are intrinsic differences in the CNS; if there are, they need not have been differentiated during the time of early hormone availability. Sex differences in male-typical copulatory behavior, for example, originate with early hormones. Yet, male mammals become competent to perform copulatory behavior only through a rather complex developmental process that includes events, particular kinds of social experiences for example, encountered well beyond the time when these early hormones cease to be available.[5] Developmentally relevant events, such as those encountered during play or in other contact with conspecifics, may be shaped—made more likely or more salient—by residual effects of early hormones throughout the organism, including, but not limited to, the CNS.[4,23] Differential maternal attention continues beyond the early organizational period of rats, based on continuing differences in odor.[14] Similar residual effects of perinatal hormones on the signals that mediate social interaction are likely to be widespread among animals. For example, the development of a penis as a result of prenatal androgen provides infant male monkeys with a potent social signal that is attended to by the mother and other adult monkeys.[6] The anatomical differences between the sexes lead males to engage in a different pattern of sensory exploration of adult conspecifics from that found in females, which results in the maintenance of thrusting behavior and the eventual incorporation of intromission into the thrusting pattern by males but not females.[31] Young male and female rhesus monkeys are also responded to differently by their mothers as a result of differences in their prenatal hormonal conditions: Males and androgenized females receive substantially more mount solicitations from their mothers than untreated females.[32]

Finally, sexual differentiation is likely to be as heterogeneous a process as the mechanisms that underlie sexually differentiated behavior. This may be illustrated by returning to the well-studied, sexually differentiated, masculine copulatory behavior of rats. Numerous studies have demonstrated that this behavior, while functionally integrated, is not at all unitary in its underlying control.[33,34] It is entirely possible to manipulate mounting and ejaculation separately, for example, and there are a number of separate controls that contribute to each of these motor patterns. It seems reasonable to assume that the development of each of these processes may be somewhat independent. As described earlier, hormonally based differences in maternal stimulation of neonates affected copulatory rate in rats, but had no effect on a number of other aspects of copulatory behavior. Masculinization and feminization are unlikely to be unitary processes and they are unlikely to be completed during the limited perinatal period during which there are sex differences in gonadal hormones.[4]

The studies summarized here indicate that the self- and socially generated environments of developing male and female rats are different. Although each instance of differential stimulation may be small, they occur frequently and consistently, and they persist for relatively long periods of developmental time. The earliest differential stimulation by mothers is replaced after weaning by differential self- and peer-generated stimulation. The reliable presence of sex-typical biases in the environment is ensured by reliable differences in the characteristics, including signal production, of the developing males and females, which originate with differences in perinatal hormone production. This biased environmental stimulation, which can be traced through indirect pathways to an origin in hormones, loops back through equally indirect pathways to affect the development of sexually differentiated behavior. Because males and females experience different patterns of stimulation as a result of differences in their hormones, some sex differences in behavior and physiology may be accounted for by this indirect mode of hormone action.

REFERENCES

1. BEACH, F. A. 1947. A review of physiological and psychological studies of sexual behavior in mammals. Physiol. Rev. **27:** 240-307.
2. LEHRMAN, D. S. 1965. Interaction between internal and external environments in the regulation of the reproductive cycle of the ring dove. *In* Sex and Behavior. F. A. Beach, Ed.: 355-380. Wiley. New York, N.Y.
3. ROSENBLATT, J. S. 1965. Effects of experience on sexual behavior in male cats. *In* Sex and Behavior. F. A. Beach, Ed.: 416-439. Wiley. New York, N.Y.
4. MOORE, C. L. 1985. Another psychobiological view of sexual differentiation. Dev. Rev. **5:** 18-55.
5. ROSENBLATT, J. S. 1965. The basis of synchrony in the behavioural interaction between the mother and her offspring in the laboratory rat. *In* Determinants of Infant Behaviour. B. M. Foss, Ed. **3:** 3-45. Methuen. London, England.
6. FEDIGAN, L. M. 1982. Primate Paradigms. Eden Press. Montreal, Quebec, Canada.
7. MOORE, C. L. & G. A. MORELLI. 1979. Mother rats interact differently with male and female offspring. J. Comp. Physiol. Psychol. **93:** 677-684.
8. GUBERNICK, D. J. & J. R. ALBERTS. 1983. Maternal licking of young: resource exchange and proximate controls. Physiol. Behav. **31:** 593-601.
9. BARNETT, S. A. & K. Z. WALKER. 1974. Early stimulation, parental behavior, and the temperature of infant mice. Dev. Psychobiol. **7:** 563-577.
10. MOORE, C. L. & A. M. CHADWICK-DIAS. 1986. Behavioral responses of infant rats to maternal licking: variations with age and sex. Dev. Psychobiol. **19.**
11. MOORE, C. L. & S. A. ROGERS. 1984. Contribution of self-grooming to onset of puberty in male rats. Dev. Psychobiol. **17:** 243-253.
12. OLIOFF, M. & J. STEWART. 1978. Sex differences in the play behavior of prepubescent rats. Physiol. Behav. **20:** 113-115.
13. MOORE, C. L. 1981. An olfactory basis for maternal discrimination of sex of offspring in rats (*Rattus norvegicus*). Anim. Behav. **29:** 383-386.
14. MOORE, C. L. 1985. Sex differences in urinary odors produced by young laboratory rats. (*Rattus norvegicus*). J. Comp. Psychol. **99:** 336-341.
15. MOORE, C. L. 1982. Maternal behavior is affected by hormonal condition of pups. J. Comp. Physiol. Psychol. **96:** 123-129.
16. RESKO, J. A., H. H. FEDER & R. W. GOY. 1968. Androgen concentrations in plasma and testes of developing rats. J. Endocrinol. **40:** 485-491.
17. BROWN, R. E. 1977. Odor preference and urine-marking scales in male and female rats: effects of gonadectomy and sexual experience on responses to conspecific odors. J. Comp. Physiol. Psychol. **91:** 1190-1206.
18. LEON, M., P. G. CROSKERRY & G. K. SMITH. 1978. Thermal control of mother-young contact in rats. Physiol. Behav. **21:** 793-811.

19. FLEMING, A., F. VACCARINO, L. TAMBOSSO & P. CHEE. 1979. Vomeronasal and olfactory system modulation of maternal behavior in the rat. Science **203**: 372-374.
20. FLEMING, A. & J. S. ROSENBLATT. 1974. Olfactory regulation of maternal behavior in rats. I. Effects of olfactory bulb removal in experienced and inexperienced lactating and cycling females. J. Comp. Physiol. Psychol. **87**: 221-232.
21. MOORE, C. L. 1984. Maternal contributions to the development of masculine sexual behavior. Dev. Psychobiol. **17**: 347-356.
22. RUDDY, L. L. 1980. Nasal intubation: a minimally obtrusive anosmia technique applied to rats. Physiol. Behav. **24**: 881-886.
23. MOORE, C. L. 1985. Development of mammalian sexual behavior. *In* The Comparative Development of Adaptive Skills: Evolutionary Implications. E. S. Gollin, Ed.: 19-56. L. Erlbaum. Hillsdale, N.J.
24. RICHMOND, G. & B. D. SACHS. 1980. Grooming in Norway rats: the development and adult expression of a complex motor pattern. Behaviour **75**: 82-95.
25. MOORE, C. L. 1986. A hormonal basis for sex differences in the self-grooming of rats. Horm. Behav. In press.
26. MOORE, C. L. 1986. Sex differences in self-grooming of rats: effects of gonadal hormones and context. Physiol. Behav. **36**: 451-455.
27. MCCANN, S. M., S. OJEDA & A. NEGRO-VILLAR. 1974. Sex steroid, pituitary and hypothalamic hormones during puberty in experimental animals. *In* Control of the Onset of Puberty. M. M. Grumbach, G. D. Grave & F. E. Mayer, Eds.: 1-31. Wiley. New York, N.Y.
28. ROTH, L. L. & J. S. ROSENBLATT. 1967. Changes in self-licking during pregnancy in the rat. J. Comp. Physiol. Psychol. **63**: 397-400.
29. ROTH, L. L. & J. S. ROSENBLATT. 1968. Self-licking and mammary development during pregnancy in the rat. J. Endocrinol. **42**: 363-378.
30. GOY, R. W. & B. S. MCEWEN. 1980. Sexual Differentiation of the Brain. MIT Press. Cambridge, Mass.
31. HANBY, J. 1976. Sociosexual development in primates. *In* Perspectives in Ethology. P. P. G. Bateson & P. H. Klopfer, Eds. **2**: 1-67. Plenum Press. New York, N.Y.
32. GOY, R. W. Discussion following C. H. Phoenix, R. W. Goy, & A. A. Gerall, Twenty-five years later. Conference on Reproductive Behavior, Pittsburgh, Pa. (June 1984).
33. BEACH, F. A. 1971. Hormonal factors controlling the differentiation, development, and display of copulatory behavior in the ramstergig and related species. *In* The Biopsychology of Development. E. Tobach, L. R. Aronson & E. Shaw, Eds.: 249-296. Academic Press. New York, N.Y.
34. SACHS, B. D. & R. J. BARFIELD. 1976. Functional analysis of masculine copulatory behavior in the rat. *In* Advances in the Study of Behavior. J. S. Rosenblatt, R. A. Hinde, E. Shaw & C. Beer, Eds. **7**: 91-154. Academic Press. New York, N.Y.

Prenatal Stress Disrupts Reproductive Behavior and Physiology in Offspring

LORRAINE ROTH HERRENKOHL

Psychology Department
Temple University
Philadelphia, Pennsylvania 19122

Stress and anxiety affect the developing organism at any stage of the life cycle. At any stage in development, biological, psychological, and social factors influence later reproductive behavior and physiology. The present paper examines the role of prenatal stress as a disruptive factor in reproductive psychophysiology. The theses are that stress has a critical impact on the developing organism during the prenatal stage, when the neural circuitry underlying later biochemical-behavioral events is laid down, and that because of this impact, prenatal stress negatively affects later behavior in lower animals and humans as well. The orientation owes much to Schneirla's concepts[1] of similarities and differences in levels of organization in behavior and stimulus-response relationships across the phyletic scale, from lower animals to humans. It also owes much to Rosenblatt[2,3] and his conception of stages in development and reciprocity in behavior between mothers and their young. Inspired by Rosenblatt, whose training was grounded not only in experimentation but clinical theory as well, I have extended my still-ongoing interest in hard science into the clinical field. A more fully developed rationale and data appear in Mathew's book, "The Biology of Anxiety."[4]

I define stress as any real or imagined trauma, physical or psychological, that leads to the release of stress hormones associated with the adrenal glands. The range of stress-inducing stimuli is varied. It differs from organism to organism, and not uncommonly within the same organism, depending upon circumstance and time. The range of stressful stimuli examined in the psychobiological literature has also varied. Stimuli range from animal models (employing discrete changes in environmental temperature or in dosages of stress hormones exogenously administered[5-8]) to human models such as those social psychologists employ to examine the effects of life situational stress (death of loved ones, moving, marital discord) on human health and behavior.[9,10]

Anxiety is the consequence of exposure to real or imagined stress. It is a state attributed to lower animals and to humans as well. On the human level, there are two major classes of clinical anxiety (according to the "Diagnostic and Statistical Manual of the American Psychiatric Association," 3rd edit., 1980). One class includes anxiety disorders of childhood or adolescence such as overanxious disorders and separation anxiety disorders. The second class is the adult category which includes phobias, generalized anxiety reactions, panic anxiety reactions, and post-traumatic stress disorders. Some behavioral manifestations of anxiety include extremes in behavior: irritability, hyperactivity, panic, hyperattentiveness, and obsessive-compulsive reactions at one end and social withdrawal, apathy, and sadness at the other. In animal studies, anxiety has been examined in paradigms such as conditioned anxiety.[11]

CLASSIC COMPARATIVE AND PHYSIOLOGICAL PSYCHOLOGY

Prenatal maternal stress has been related to neonatal activity and irritability in lower animals and humans in documented research for the past 30 years. A classic conditioned anxiety experiment was performed by Thompson in 1957.[11] Female rats were trained in a double-compartment shuttle box first to expect strong shock at the sound of the buzzer and then to avoid the shock by opening the door between the compartments and running through to the safe side. When they had learned the task, they were mated. As soon as they were pregnant, the animals were exposed to the buzzer in the shock side of the shuttle box but with the shock turned off and the door to the safe side locked. Therefore, during gestation, the animals were exposed to "expected," not real, shock. Emotional characteristics of the offspring were compared at two later stages of development. Experimental animals showed a much higher latency of activity than control animals at both ages of testing in an open field test where measures of amount of activity and latency were recorded. Moreover, experimental animals were slower to leave the home cage ("more fearful") than controls at the first age of testing. On the basis of his findings, Thompson believed that stress hormones coursing through the mother's blood affected the developing fetus and that these hormones produced the behavioral differences. He had no direct evidence for this effect.

In the decade that followed Thompson's research,[11] many others studied the influence of maternal stress on later offspring behavior.[12–18] The implicit mechanism in all that research involved the sharing of stress hormones by the mother and fetus via a common blood supply, but no direct evidence was collected in support of that view.

CLASSIC DEVELOPMENTAL PSYCHOLOGY

Experiments in developmental psychology also illustrated the impact of maternal stress on the offspring. Lester Sontag, long an important figure in child development at the Fels Research Institute, studied pregnant women undergoing stress. Sontag reported findings of the Fels Research Institute for the study of human development in the years from 1932 to 1966.[19–21] Research was designed to explore the behavior of the human fetus, its developmental progress and individual differences, perceptions, capabilities, and responses to stimuli during the last months of pregnancy. Sontag[19] observed infants of mothers who had undergone severe emotional stress during the latter part of their pregnancies. He perceived that behavior patterns during gestation carried over into neonatal life. Sontag[19] reported that fetuses of mothers undergoing stress responded with large increases in sharp or irritable body movements, presumably the result of changes in the constituents of the mothers' blood. After birth, these infants remained irritable and hyperactive for weeks or months. They cried a great deal and slept for short periods only. Most of these infants exhibited a food intolerance and frequent or often loose stools, suggesting an autonomic or psychosomatic component of prenatal stress exposure as expressed in gastrointestinal function. Moreover, they regurgitated much of their food and frequently were switched from one formula to another without significant improvement. They failed to gain weight for a long time.

Sontag was able to correlate verbal reports by the mothers on fetal activity with kymographic records. The mothers were asked to lie on cots in the laboratory. Inflatable bags were placed over each quadrant of their abdomen. When the fetus kicked, there were pressure changes in the inflatable rubber bags which in turn were recorded by a kymograph. By this procedure and by taking advantage of spontaneously occurring natural events in the mothers' lives, Sontag was able to collect information on responses of the fetus to stress. A case is given below:

> In one instance a young woman carrying her baby, which we had been studying weekly in terms of activity and heart rate level, took refuge at the Fels Institute building one evening because her husband had just suffered a psychotic break and was threatening to kill her. She was terrified, felt alone and did not know where to turn for help. She came to the Institute, and we gave her a bed and room for the night. When she complained after a few minutes conversation that the kicking of her fetus was so violent as to be painful, we proceeded to record the activity level. It was more than 10-fold what it had been in the weekly sessions prior to this incident. Another case came to our attention when a woman we had been studying lost her husband in an automobile accident. Again, the violence of the activity and the frequency of movement of the fetus increased by a factor of more than 10. During the period of 10 years, we managed to collect 8 such dramatic incidences, all showing the same phenomena of extreme increase in fetal activity in response to grief, fear and anxiety. Children of such mothers, who suffered their emotional trauma late in pregnancy and not early, showed, of course, no congenital defect. In general, they were, however, irritable, hyperactive, tended to have frequent stools and 3 of them had marked feeding problems. (Sontag, 1966, Reference 21, p. 784.)

Later Stott[22] in Scotland reported the findings of a follow-up study from birth to the fourth year of life of the effects of prenatal stress on a sample of 200 infants. By using medical records and interviews by health nurses in the home, Stott[22] drew impressive profiles of the impact of prenatal stress on later health and behavioral development.

The types of child morbidity Stott[22] associated with personal tensions in pregnancy included physical illness (twice as much eczema and middle ear infections, somewhat more bronchitis, and severe respiratory trouble); minor physical and functional abnormalities (small size, profuse sweating, flushing, or choking); developmental difficulties (twice the incidence of late walking or poor walking such as flat-footed or clumsy, some speech defects); and behavioral abnormalities (twice as many entries for fretful, whimpering, restless, or clinging behaviors). Ten of the 14 cases had one or more indications of behavioral disturbances characteristically associated with congenital hyperactivity.

Stott's[22] most striking finding was that marital crisis during pregnancy produced the highest child morbidity scores. A case involving marital discord during pregnancy which produced a high child morbidity score is described below:

> Marital relationship is not good. There has been frequent quarreling all through the marriage. The husband is reputed to be a heavy drinker, particularly at weekends when he is very abusive and often puts the wife out of the home. He is, however, a good worker (on constant night shift) and supports his wife and family fairly well. There is nevertheless,

signs that the family may break down completely. The night before the birth there had been a violent quarrel with husband. (Stott, 1973, Reference 22, p. 776.)

By the mid-1970s therefore, a body of evidence had developed that maternal stress negatively influences subsequent health and behavior of offspring. Significant parallels in maternal stress effects on offspring were apparent in laboratory animals and humans.

CONTEMPORARY PRENATAL STRESS RESEARCH

Recent advances in prenatal stress research have been marked by the quest for an underlying mechanism. In interdisciplinary research, psychologists studying hormones and behavior have collaborated with biologists, endocrinologists, and biochemists to attempt a unified approach to the study of maternal stress on offspring.

One current model of prenatal stress research employs sexual differentiation of the brain. The sexual differentiation model of the brain states that the hormonal milieu, not the genome, determines sexual dimorphism in an inherently female brain.[23] Normal patterns of sexual behavior and gonadotropin secretion are established by adulthood as a function of the presence or absence of androgens during critical perinatal sexual differentiation stages. Two lines of evidence support this view: chemical or surgical castration of genetic males during perinatal life feminizes and demasculinizes reproductive functions; and exposure of genetic females to androgens during a critical developmental stage masculinizes and defeminizes reproductive physiology, morphology, and behavior. The severity of the masculinizing and defeminizing action of perinatal androgens (particularly its potent metabolite estradiol) depends upon the amount and timing of the hormone.[23]

The possibility that maternal stress can influence sexual differentiation was stimulated by Ward's discovery[24] that prenatal stress feminizes and demasculinizes sexual behavior in males. She exposed rats in late pregnancy to the stress of heat, restraint, and bright lights three times daily during the last trimester of gestation and reported that by adulthood the offspring showed a significant reduction in the percentage of stressed males that copulated and ejaculated compared to control males, and a significant increase in lordotic performance. She believed that the prenatal stress syndrome in males, characterized by diminished copulatory patterns and increased lordotic behavior potentials, developed from diminished exposure of fetal males to gonadal androgens, presumably as a result of increased exposure to stress steroids.[24] She also believed that parallel influences might be a cause for homosexuality in men.[25]

Moyer et al.[26,27] have suggested that prenatal stress may modify the neuroanatomical and biochemical organization of the brains of both males and females and turn the direction of male fetal brain development toward that of the female sex. They combined the microdissection procedure of Palkovits[28] for removing individual brain nuclei with sensitive radioisotopic enzymatic assays for norepinephrine (NE) and dopamine (DA). In pregnant mothers, Moyer et al.[26] discovered that stress during pregnancy reduced steady-state levels of NE in brain regions associated with gonadotropic secretion. The major noradrenergic pathway that underwent change during stress was the ventral ascending bundle (i.e., the medial preoptic nucleus, anterior hypothalamus, and median forebrain bundle). They also reported that the locations of DA decreased as a function of prepartal stress and overlapped with those brain regions in which catecholamine (CA) depletions have been implicated in functional affective disorders in humans.

They postulated that the relatively high incidence of certain mental disorders when sex steroids and CA fluctuate widely (as during diestrus, the postpartum period, and at menopause) suggests an interrelationship among female hormones, CA, and psychological state. Stress during pregnancy with corresponding changes in brain monoamines may set the stage for postpartum disorders.

Moyer et al.[27] also examined the effects of prenatal stress on CA concentrations in the brains of the male and female offspring as adults. The major pattern of brain change in male offspring was similar to that in stressed mothers. The major system that underwent change involved NE, the major direction of change was a decrease, and the major brain regions that underwent change were those associated with gonadotropic secretion from the anterior pituitary gland and with the regulation of sexual behavior. They postulated that decreases in brain NE may be the basis for the feminized and demasculinized sexual behavior of males.[24]

Among the most interesting findings was the observation that prenatal stress markedly affected CA concentrations in female offspring. Prenatal stress increased the steady-state concentration of DA in the hypothalamic arcuate nucleus of stressed female offspring by 153%! Because marked alterations in arcuate DA have been associated with abnormalities in the release of gonadotropic hormones from the anterior pituitary gland, it was predicted and ultimately observed that prenatal stress would produce reproductive dysfunctions in female offspring.[5,29]

Herrenkohl[5] exposed pregnant rats to restraint-heat stress during the last trimester of pregnancy and examined female offspring for reproduction function in adulthood. Pregnant rats were restrained in 18 × 8-cm semicircular Plexiglas cages under four bright incandescent lights. The procedure caused a surface illumination of 400 ft-cd and a surface temperature of 34°C. This was the standard restraint-heat stress procedure in all of Herrenkohl's experiments. Stress sessions occurred from day 14 to day 21 of gestation three times daily for 45 min starting at 10 AM with alternating 45-min periods of rest in the home cage. Nonstressed mothers remained unhandled. When compared to nonstressed female offspring, stressed female offspring exhibited higher incidences of estrous cycle disorders, spontaneous abortions, and vaginal hemorrhages.[5,29] They also had higher incidences of stillbirths and neonatal mortalities in a subsequent generation of progeny (TABLE 1). Cross-fostering procedures between and within treatment groups ruled out the possibility that prepartal-stress-induced disturbances during the postnatal period were the primary causes of the reproductive deficits in the offspring.[5,30] Prenatal stress by itself markedly reduced fertility and fecundity in female offspring.

Recent experiments have found that maternal stress changes testicular (Leydig cell) and brain steroid aromatase activity in rat fetuses in patterns similar to those alterations in fetal testosterone levels.[31,32] It has also been reported that maternal stress differentially alters pituitary, gonadal, and adrenal function in rats and mice.[6-8,33,34]

Critical issues remaining in the field involve answers that reconcile apparent contradictory results. For example, with regard to stressed male offspring, Ward[24] has found both a demasculinization and feminization of sex behavior (decreases in copulatory performance and increases in lordosis), whereas Whitney and Herrenkohl[35] have reported only behavioral feminization. Also in a "replication" of Herrenkohl's experiments,[5] Beckhardt and Ward[36] found that reproductive functioning in prenatally stressed female offspring remained intact. Such inconsistencies may be explained by differences in strain of animal, rearing conditions, and/or experimental procedures. Ward, for example, characteristically uses only 200 ft-cd of light stress,[24,36] whereas Herrenkohl uses twice as much.[5,29]

A final issue remaining is the mechanism of prenatal stress. In an experiment with the endocrinologist Weisz, Ward presented evidence that maternal stress altered plasma testosterone in fetal males.[31] They employed radioimmunoassay to measure plasma testosterone in cesarean-delivered normal and stressed fetal males during the last trimester of pregnancy. In normal males a surge of plasma testosterone characteristically occurred on the 18th or 19th gestational day. This testosterone surge occurred prematurely in the stressed male, on gestational day 17. On this basis, Ward and Weisz[31] concluded that the central nervous system (CNS) of fetal males becomes demasculinized and feminized, not as a result of decreased exposure in absolute amount of circulating testosterone during the late gestational stage, but because there is a desynchrony between the maturational stage of the CNS and patterns of testosterone secretion during fetal life.

The most parsimonious explanation with respect to female offspring is that the prenatal stress syndrome in females may result from exposure to increased androgens. The source of the androgens might include: (a) the fetal adrenals: restraint-heat stress elevates maternal body temperature and thereby may affect the fetus directly[37]; (b) the maternal adrenals: restraint-heat stress produces maternal pathology and adrenocortical response; and/or (c) the fetal testis: this intriguing possibility is suggested by the observation that masculinization of fetal females can be mediated by proximity to fetal males.[38,39]

TABLE 1. Summary of Effects of Prenatal Stress on Fertility and Fecundity in Female Rat Offspring in Five Experiments[a]

Female Offspring	N	Percentage Pregnant	Percentage Giving Birth	Percentage of Stillbirths	Percentage of Neonatal Deaths	Percentage of Litters Intact
Prenatally stressed	274	39 **	22 *	21 **	37 **	42 **
Nonstressed	303	89	86	5	2	93

[a] Some of the data presented were taken from Herrenkohl.[5]
*$p < 0.01$.
**$p < 0.001$.

CURRENT SOCIOCLINICAL RESEARCH

In one kind of newly emerging research strategy, psychobiologists are adapting socioclinical research to issues of stress and reproductive dysfunction in women. In one of these studies, Herrenkohl is examining the relationships among stress, personality, mood, and menstrual distress. Preliminary findings support the hypothesis that prenatal stress may affect personality, mood, and menstrual activity in women. The following observations pertain to several samples of women questioned retrospectively about the incidence of life situational change (as measured by the Holmes-Rahe Social Stress Scale) over four stages of the life cycle (Prenatal Stage, Early Childhood, Adolescence and Young Adulthood, Adulthood):

1. The higher the Prenatal Stress score (death of loved one, moving, marital discord), the higher the Menstrual Distress score (Moos Menstrual Distress Questionnaire).

2. The higher the Prenatal Stress score, the more masculine the social self-perception (Bem Androgyny Scale).
3. The higher the Early Childhood Stress score, the more irregular the menses.
4. The higher the Adolescence and Young Adulthood Stress score, the more masculine the social self-perception.
5. The higher the premenstrual distress, the higher the anxiety, depression, and neurotic symptomology (Beck Anxiety Checklist, Beck Depression Inventory, Hopkins Self-checklist).

Currently we are gathering evidence on the interrelationship of prenatal stress, premenstrual syndrome, and postpartum depression. We are also formulating hypotheses on mechanisms by which psychosocial stress may affect the developing reproductive substrate and later behavior.

We are also trying to tap the information bank that exists at the National Institute of Neurological Diseases and Strokes. It has a massive compilation of data from a collaborative perinatal study on over 10,000 women and their pregnancies. The study examines the relationship between prenatal conditions (personal circumstances, drugs, health) and outcomes for the offspring (neonatal health, neurological development, school performance). From some of the findings to date,[40] it is already possible to note that relationships exist between marital status during pregnancy and survival and health of offspring. The incidence of neonatal deaths and neurological impairment among infants is extraordinarily high among widows (Reference 40, pp. 54, 57), suggesting that the prenatal trauma associated with the death of the husband had deleterious and far-reaching consequences for the offspring. Also, the incidence of low-birth-weight young was high among nonmarried women (Reference 40, p. 55). Is there a relationship between prenatal maternal stress and reproductive dysfunction in the daughters?

Laukaran and van den Berg[41] have already found that negative maternal attitudes toward pregnancy deleteriously influence offspring outcome and obstetrical complications. Tapping an already-existing data bank on 8000 or so women at the Kaiser Permanente Foundation in the San Francisco Bay area, they controlled for such factors as socioeconomic class, maternal health, and nutrition. They concluded that the single major factor associated with postpartum infection and hemorrhages, and deaths and congenital abnormalities in newborns, was a negative maternal attitude toward having the baby. This was probably due to "stress-mediated change in hormones." There is also evidence that prenatal loss of father elevates the incidence of psychiatric disorders in adulthood.[42]

SUMMARY AND CONCLUSIONS

Prenatal maternal stress has been related to neonatal activity and irritability in both lower animals and humans in documented research for at least the past 30 years. Contemporary animal research demonstrates that prenatal stress feminizes and demasculinizes the sexual behavior of males and reduces fertility and fecundity in females, producing estrous cycle disorders, spontaneous abortions, or vaginal hemorrhaging and high neonatal mortality. Mechanisms of stress are being sought in the maternal-fetal blood exchange, hormonal alterations in the hypothalamus-pituitary-gonads and adrenals, and in brain catecholamines.

Contemporary human research demonstrates that negative maternal attitudes toward pregnancy are related to high incidences of congenital abnormalities and infant deaths. Severe psychosocial stress is related to high incidences of neonatal deaths and neurological impairments in infants, and a high incidence of psychiatric disorders in adulthood. Data derived from both animal and human research may help explain the etiology and mechanisms of prenatal-stress-induced reproductive dysfunctions as well as some forms of human psychopathology.

REFERENCES

1. SCHNEIRLA, T. C. 1965. Aspects of stimulation and organization in approach-withdrawal processes underlying vertebrate behavioral development. In Advances in the Study of Behavior. D. S. Lehrman, R. A. Hinde & E. Shaw, Eds. I: 1-74. Academic Press. New York, N.Y.
2. ROSENBLATT, J. S. 1983. Olfaction mediates developmental transition in the altricial newborn of selected species of mammals. Dev. Psychobiol. 16: 347-375.
3. ROSENBLATT, J. S. & H. I. SIEGEL. 1980. Maternal behavior in the laboratory rat. In Maternal Influences and Early Behavior. R. W. Bell & W. P. Smotherman, Eds.: 155-199. Spectrum Publications. New York, N.Y.
4. HERRENKOHL, L. R. 1982. The anxiety-prone personality: effects of prenatal stress on the infant. In The Biology of Anxiety. R. J. Mathew, Ed.: 51-86. Brunner-Mazel, Inc. New York, N.Y.
5. HERRENKOHL, L. R. 1979. Prenatal stress reduces fertility and fecundity in female offspring. Science 206: 1097-1099.
6. POLITCH, J. A. & L. R. HERRENKOHL. 1984. Effects of prenatal stress on reproduction in male and female mice. Physiol. Behav. 32: 95-99.
7. POLITCH, J. A. & L. R. HERRENKOHL. 1984. Prenatal ACTH and corticosterone: effects on reproduction in male mice. Physiol. Behav. 32: 135-137.
8. POLITCH, J. A. & L. R. HERRENKOHL. 1984. Postnatal ACTH and corticosterone: effects on reproduction in mice. Physiol. Behav. 32: 447-452.
9. RAHE, R. H., M. MEYER, M. SMITH, G. KJAER & T. H. HOLMES. 1964. Social stress and illness onset. J. Psychosom. Res. 8: 34-44.
10. HOLMES, T. H. & R. H. RAHE. 1967. The social readjustment rating scale. J. Psychosom. Res. 11: 213-218.
11. THOMPSON, W. R. 1957. Influence of prenatal maternal anxiety on emotionality in young rats. Science 125: 698-699.
12. THOMPSON, W. R. & L. W. SONTAG. 1956. Behavioral effects in the offspring of rats subjected to audiogenic seizure during the gestational period. J. Comp. Physiol. Psychol. 49: 454-456.
13. HOCKMAN, C. H. 1961. Prenatal maternal stress in the rat: its effects on emotional behavior in the offspring. J. Comp. Physiol. Psychol. 54: 679-684.
14. KEELEY, K. 1962. Prenatal influence on behavior of offspring of crowded mice. Science 135: 44-45.
15. LIEBERMAN, M. W. 1963. Early developmental stress and later behavior. Science 141: 824-825.
16. MORRA, M. 1965. Level of maternal stress during two pregnancy periods of rat offspring behaviors. Psychon. Sci. 3: 7-9.
17. DEFRIES, J. C., M. W. WEIR & J. P. HEGMANN. 1967. Differential effects of prenatal maternal stress on offspring behavior in mice as a function of genotype and stress. J. Comp. Physiol. Psychol. 63: 332-334.
18. ADER, R. & S. M. PLAUT. 1968. Effects of prenatal maternal handling and differential housing on offspring emotionality, plasma corticosterone levels, and susceptibility to gastric erosions. Psychosom. Med. 30: 277-286.

19. SONTAG, L. W. 1944. Differences in modifiability of fetal behavior and physiology. Psychosom. Med. **6:** 151-154.
20. SONTAG, L. W., E. L. REYNOLDS & V. TORBET. 1944. Status of infant at birth as related to basal metabolism of mothers in pregnancy. Am. J. Obstet. Gynecol. **48:** 208-214.
21. SONTAG, L. W. 1966. Implications of fetal behavior and environment for adult personalities. Ann. N.Y. Acad. Sci. **134:** 782-786.
22. STOTT, D. H. 1973. Follow-up study from birth of the effects of prenatal stresses. Dev. Med. Child Neurol. **15:** 770-787.
23. GORSKI, R. A. 1980. Sexual differentiation of the brain. *In* Neuroendocrinology. D. T. Krieger & J. C. Hughes, Eds.: 215-222. Sinauer Associates. Sunderland, Mass.
24. WARD, I. L. 1972. Prenatal stress feminizes and demasculinizes the behavior of males. Science **175:** 82-84.
25. WARD, I. L. 1977. Sexual diversity. *In* Psychopathology: Experimental Models. J. D. Masur & M. E. P. Seligman, Eds.: 387-403. Freeman & Co. San Francisco, Calif.
26. MOYER, J. A., L. R. HERRENKOHL & D. M. JACOBOWITZ. 1977. Effects of stress during pregnancy on catecholamines in discrete brain regions. Brain Res. **121:** 385-393.
27. MOYER, J. A., L. R. HERRENKOHL & D. M. JACOBOWITZ. 1978. Stress during pregnancy: effect on catecholamines in discrete brain regions of offspring as adults. Brain Res. **144:** 173-178.
28. PALKOVITS, M. 1973. Isolated removal of hypothalamic and other brain nuclei of the rat. Brain Res. **59:** 449-450.
29. HERRENKOHL, L. R. & J. A. POLITCH. 1978. Effects of prenatal stress on the estrous cycle of female offspring as adults. Experientia **34:** 1240-1241.
30. HERRENKOHL, L. R. & J. B. WHITNEY. 1976. Effects of prepartal stress on postnatal nursing behavior, litter development and adult sexual behavior. Physiol. Behav. **17:** 1019-1021.
31. WARD, I. L. & J. WEISZ. 1980. Maternal stress alters plasma testosterone in fetal males. Science **175:** 82-84.
32. WEISZ, J. 1983. Influence of maternal stress on the developmental pattern of the steroidogenic function in Leydig cells and steroid aromatase activity in the brain of rat fetuses. *In* Drugs and Hormones in Brain Development. M. Schlumpf & W. Lichtensteiger, Eds.: 184-193. Karger. New York, N.Y.
33. HERRENKOHL, L. R. 1983. Prenatal stress may alter sexual differentiation in male and female offspring. *In* Drugs and Hormones in Brain Development. M. Schlumpf & W. Lichtensteiger, Eds.: 176-183. Karger. New York, N.Y.
34. HERRENKOHL, L. R. & S. SCOTT. 1984. Prenatal stress and postnatal androgen: effects on reproduction in female rats. Experientia **40:** 101-103.
35. WHITNEY, J. B. & L. R. HERRENKOHL. 1977. Effects of anterior hypothalamic lesions on the feminized sexual behavior of prenatally-stressed male rats. Physiol. Behav. **19:** 167-169.
36. BECKHARDT, S. & I. L. WARD. 1983. Reproductive functioning in the prenatally stressed female rat. Dev. Psychobiol. **16:** 111-117.
37. CHAPMAN, R. H. & J. M. STERN. 1978. Maternal stress and pituitary-adrenal manipulations during pregnancy in rats: effects on morphology and sexual behavior of male offspring. J. Comp. Physiol. Psychol. **92:** 1974-1083.
38. VOM SAAL, F. S. & F. H. BRONSON. 1980. Sexual characteristics of adult female mice are correlated with their blood testosterone levels during prenatal development. Science **208:** 597-599.
39. MEISEL, R. L. & I. L. WARD. 1981. Fetal female rats are masculinized by male littermates located caudally in the uterus. Science **212:** 239-242.
40. NISWANDER, K. R. & M. GORDON. 1972. The Women and Their Pregnancies. U.S. Department of Health, Education and Welfare. Washington, D.C.
41. LAUKARAN, V. H. & B. J. VAN DEN BERG. 1980. The relationship of maternal attitude to pregnancy outcomes and obstetric complications. Am. J. Obstet. Gynecol. **136:** 374-379.
42. HUTTUNEN, M. O. & P. NISKANEN. 1978. Prenatal loss of father and psychiatric disorders. Arch. Gen. Psychiatry **35:** 429-431.

Hormonal and Metabolic Factors Underlying Intraspecific Variation in Reproductive Social Behavior

DALE F. LOTT

Wildlife and Fisheries Biology
University of California
Davis, California 95616

Recently we have become increasingly aware of the number of species that manifest different social systems at different times or in different places. Lott[1] has reviewed more than 150 instances of intraspecific variation in social systems (IVSS) in wild vertebrates, even when primates are excluded. Many of these variations are in reproductive social systems. Some species alternate between polygyny and monogamy. Examples are: the Savannah sparrow, *Passerculus sandwichensis*[2]; the beaver, *Castor canadensis*[3]; and the yellow-bellied marmot, *Marmota falviventris*.[4,5] Other species shift from polygyny to promiscuity, e.g., the deer mouse, *Peromyscus maniculatus*.[6] Still others alternate between monogamy and promiscuity. An example is the acorn woodpecker, *Melanerees formicivourus*.[7] A few species alternate between monogamy and simultaneous polyandry; for example, Harris's hawk, *Parabuteo unicinctus*.[8] Parental care, too, varies. Sometimes it varies from one-parent to two-parent care, as in deer mice, *Peromyscus maniculatus*,[6] and the long-billed marsh wren, *Telmatodyte palustris*.[9] Other species alternate between communal care and parental care; for example, the speckled mousebird, *Colis strictus*,[10] and the red fox, *Vulpes vulpes*.[11,12] Finally, several species vary between parental care and helpers at the nest. Examples are the black-backed jackal, *Canis mesomelas*,[13] and the pied kingfisher, *Ceryle rudis*.[14]

Sometimes reproductively oriented social systems vary. Topi antelope, *Damiliscus korrigum*, sometimes breed in large, all-purpose territories, encompassing much or all of the home range of a small, stable group of females. In other areas they lek, and the females live in large, unstable groups.[15]

IVSS raises several questions. The two most salient are: (1) What is the functional significance of expressing alternative social systems? That is, is it adaptive, and if so, in what way? (2) What mechanism(s) produces each of the observed alternatives? The functional significance of the presence or absence of alternative social systems can be discovered by conventional socioecological analysis. But discovery of the mechanisms that produce or preclude social system variability will require a different approach. In some cases IVSS will be due to genetically based polymorphism, but in many cases it is already clear that selection has selected a property or mechanism that can respond at some point in the animal's life in a way that modifies the social interactions and emergent social systems. That is, selection has selected for some degree of flexibility.

The study of this variation, and the flexibility that underlies it, will take place in an area of the discipline of animal behavior that is presently little occupied. Wilson[16] predicted that studies of animal behavior would fall more and more into one of the two ends of a continuum. One end of that continuum is studies of neural and biochemical mechanisms. The other end is field studies of evolutionary questions, such as the effects of ecology or kinship on social organization and social behavior. He forecast that the middle of the continuum, which includes the more molar study of interactions between individuals, and an analysis of the psychological mechanisms that underlie them, would atrophy. This forecast has proven generally correct, partly due to the increased specialization of the training of the current crop of scientists, but also partly due to the perception that there is no work to be done in the middle. However, part of the study of IVSS will take place in that middle ground. That part will be the discovery of how proximate mechanisms operate to produce intraspecific variation in social systems.

These mechanisms must be studied because a complete understanding of the expression of two alternative social systems requires two kinds of analysis. The first is a socioecological analysis to determine if the alternative can be related to ecological circumstances. As an example of this approach, Savannah sparrows are more often monogamous at high latitudes where the breeding season is short than at lower latitudes where more complex breeding relationships would have longer to develop and there would be more time to bring off the young produced.

In general, IVSS has been analyzed in terms of ecological adaptation or ultimate cause. This is an important analytic approach, and one that has contributed greatly to our understanding of animal social systems. But it is also essential to complete our understanding of these phenomena by trying to conceptualize them in terms of proximate mechanisms. An important set of variables that act proximately on animal social behavior, and thus on animal social systems, are the direct and indirect effects of reproductive hormones.

In the year-round cycle of animal social behavior we observe tremendous shifts in behavior related to seasonal fluctuations in the reproductive hormones of seasonally reproducing animals. Perhaps these fluctuations will prove to be a useful model for studying the variations in the mating and parental care systems we observe within species.

This paper attempts to bridge the present gap between IVSS and hormones by identifying possible (though not proven) hormonal mechanisms for IVSS. Not many instances of IVSS have been analyzed from the perspective of the proximate mechanism that produces them. Consequently, many of the proposed mechanisms by which hormonal changes might produce intraspecific variations in animal social systems are necessarily more the product of ingenuity than of scholarship. In short, this paper contains a lot of speculation. This amount of speculation is necessary to call attention to the potential link between hormonal mechanisms and intraspecific variation in reproductive social systems. The links proposed vary considerably in their plausibility and hence the probability of proving correct. A primary function of each of them, however, is to identify the conceptual necessity of the missing mechanism that translates a change in circumstances into a change in social systems.

HORMONAL MECHANISMS THAT MIGHT PRODUCE IVSS

One possible source of variation in reproductive social systems of adult mammals is the influence of different hormonal regimes in prenatal life. Rats born of mothers stressed during pregnancy are more anxious as adults. This could alter their later, adult social system by making them more prone to join conspecifics in anxiety-provoking situations. Their greater anxiety might make the reduction of anxiety produced by the proximity of conspecifics more reinforcing. This would influence the outcome of experiential effects on social behavior. Desert wood rats, *Neotoma lepida*, sometimes defend territories, yet in other places live in undefended home ranges.[17] The territorial pattern is most common where there are no avian predators. The more anxiety an individual experiences, the more aversive the travel away from cover that territorial defense requires in this species would be. Consequently, one possible determinant of living in an undefended home range would be the endocrinological consequences of being born to a mother who had experienced the higher stress associated with higher predation pressures. This social system change could also affect the breeding system. Animals living in territories controlled by a single male are more likely to have monogamous or polygynous than promiscuous breeding systems.

Stress experienced by the mother might also affect the parental care expressed by males. More anxious individuals might be more inclined to be reinforced by the presence of conspecifics. Some mice are known to show a two-parent parental system when in unusually dense aggregations,[6] presumably because the males are exposed to young and become parental through the process of "sensitization."[18] In some species communal, rather than one-parent or two-parent, systems could be produced by the same mechanism.

Testosterone production is altered in the male fetuses of female rats stressed during the last days of pregnancy.[19] The testosterone pulse comes earlier than usual. Consequently, the fetuses are not exposed to testosterone at the normal stage of prenatal development. Their nervous systems undergo less masculinization and defeminization than those of unstressed mothers. Such males show less sex-typical behavior as adults.[20] This may have broad effects on their social behavior and consequently on the social system they produce. Various forms of aggressive behavior have been shown to be related to the level of testosterone in males[21] and the establishment of such reproductively oriented social systems as territoriality could be affected by this condition.

The behavior of female mice as adults is also a function of the exposure of their nervous system to testosterone as fetuses. Prenatal exposure to androgen increases the readiness of females to be aggressive as adults.[22] Those positioned between male siblings inside the uterus are partially masculinized and defeminized by them.[23] As adults they are more aggressive, less attractive to males, and more inclined to urine mark their environment, all behavioral differences that seem likely to have social system consequences. This change has a complicated history. In the first place it is due to the testosterone levels produced by their male siblings during intrauterine development. Since the testosterone level of these males is determined by the stress their mothers are exposed to, the adult phenotypes of female mice seem likely to be determined in part by their mother's exposure to stress.

This in turn could act as a determinant of breeding systems. Armitage[24] has concluded that whether or not a particular male yellow-bellied marmot is in a monogamous or a polygynous system is determined to a great extent by the aggressiveness of the first female he mates with. A single aggressive female that attacks all immigrant females will keep the territory free of other females, and hence produce a monogamous breeding system. Less aggressive or more sociable females that tolerate the presence of additional females in the territory create the potential for a polygynous breeding system. Armitage has proposed that the differences in phenotype observed in female yellow-bellied marmots are genetically determined, but it is possible that in this or other species, the aggressive phenotypes could be partially determined by masculinization of the females as fetuses.

NUTRITIONAL EFFECTS

By no means are all hormonal changes that might affect adult social behavior due to early hormone effects. The level of protein in the blood determines the blood's testosterone carrying capacity.[25] Poor physical condition or poor diet could indirectly lower the level of blood testosterone, thus possibly changing social behavior. With lower testosterone levels, several reproductive social system changes could occur. Less aggressive males would be more likely to aggregate. In aggregations, males are more likely to be exposed to young and therefore, to become parental by "sensitization."[18] By this same mechanism, increased patchiness of food resources (which might be produced by a drought) could sharply increase the differences between males in nutrition, and thus the blood's testosterone carrying capacity. Consequently, well-fed animals could be considerably more aggressive than those less well fed. Such an uneven distribution of aggressive behavior among the males could produce a shift from territoriality (often associated with monogamy) to despotism (which is almost always associated with polygyny).

Poorly nourished females frequently do not go through the endocrine changes associated with reproduction. If food were scarce and unevenly distributed (for example, because of dominance in females), the upshot would be that only some would reproduce. Some of those that did not might remain and function as helpers. This could produce a shift from a parental care system to a system of helpers at the nest.

On the other hand, well-nourished female birds might recover the endocrinological characteristics of breeders quickly after producing the first clutch. This potential for rapid second clutch production might combine with an endocrinological set appropriate to initiation of mating. That, in turn, might produce a shift from monogamy to serial or even simultaneous polyandry in exceptionally well nourished females.[26]

In addition to general nutritional effects on the endocrinological state of females, there is a strong possibility that ingested hormones will have a marked effect on reproductive social behavior. Gonadotropic hormone production has been shown to be modified in several herbivorous vertebrates by the contents of their food plants. Green plants enhance reproduction in some rodents and rabbits[27,28] and possibly white-crowned sparrows,[29] but inhibit reproduction in voles.[30] High levels of phytoestrogens in the plants eaten by sheep severely disrupt female reproductive functioning, although specific behavior changes have not been reported.[31,32] California quail reproduction is inhibited in years of poor growth of their food plants.[33] There is some evidence that the inhibition of at least the egg production component of reproduction in this species is due to the concentration of phytoestrogens in the food plants.[34] The slower the plant growth, the higher the concentration of phytoestrogens. Stunted plants in drought years have very high phytoestrogen concentrations. Male Japanese quail (*Coturnix coturnix*) fed biochanin A, an important phytoestrogen, show sharply depressed levels of courtship and copulatory behavior.[35] In species in which these behaviors are important for pair-bonding, their absence might produce a general breeding failure, or a shift from monogamy to promiscuity. De Man[35] did not observe any reduction in the number of eggs produced by treated subjects. The conditions of testing (a male and a female together in a small cage) did not permit observation of any mating system effects.

SOCIAL EXPERIENCE EFFECTS ON BEHAVIOR-DETERMINING HORMONES

Stress experienced as an adult, such as losing a fight, can also reduce testosterone levels.[36,37] There are several situations in which the number of fights lost differs in different social systems. Topi in an all-purpose territory have many fewer border encounters than topi in a lek.[15] Each border encounter involves both a winner and a loser. Therefore, topi in lekking systems have more losses and tolerate much closer neighbors, at least after the system has matured. It is a far leap to suggest that topi are able to settle into low levels of aggression in a lekking system because the losses they have experienced have reduced their level of testosterone and hence testosterone-motivated aggression. After all, the basic mechanism has only been shown to operate in monkeys and mice. But some mechanism is clearly operating, and some such hypothesis is necessary.

A shift in reproductively oriented social systems might also occur if crowding at a resource creates a situation in which normally territorial animals must aggregate. When that happens some individuals will suffer many losses of fights to the dominant individual. Under these circumstances in such species, a despotism system tends to develop.[1] In such a system the lower-level individuals would be likely to produce much less testosterone. This can set up a shift from a breeding system of territorial monogamy to an aggregated polygyny.

I have proposed some speculative answers to some of the questions raised by IVSS. I hope that I have identified some fruitful research possibilities. I am certain that I have shown that IVSS raises a rich set of questions. The answers to some of them will be found in the middle ground of animal behavior that Wilson[16] accurately predicted is being abandoned. This will benefit the discipline of animal behavior; not because the middle ground deserves attention to achieve some sort of cosmic balance, but because it offers a rich opportunity to advance our knowledge of the naturally occurring social behavior of animals.

REFERENCES

1. LOTT, D. F. 1984. Intraspecific variation in the social systems of wild vertebrates. Behaviour **88:** 266-325.
2. WEATHERHEAD, P. J. 1979. Ecological correlation of monogamy in tundra-breeding savannah sparrows. Auk **96:** 391-401.
3. BUSHER, P. E., R. J. WARNER & S. H. JENKINS. 1983. Population density, colony composition, and local movements in two Sierra Nevada beaver populations. J. Mammal. **64:** 314-317.
4. ARMITAGE, K. B. & J. F. DOWNHOWER. 1974. Demography of yellow-bellied marmot populations. Ecology **55:** 1233-1245.
5. JOHNS, D. W. & K. B. ARMITAGE. 1979. Behavioral ecology of alpine yellow-bellied marmots. Behav. Ecol. Sociobiol. **5:** 133-157.
6. MIHOK, S. 1979. Behavioral structure and demography of subarctic *Cleithrinomys gapperi* and *Peromyscus maniculatus*. Can. J. Zool. **57:** 1520-1535.
7. STACY, P. B. & C. E. BOCK. 1978. Social plasticity in the acorn woodpecker. Science **202:** 1298-1300.
8. MADER, W. J. 1975. Biology of Harris' hawk in southern Arizona. The Living Bird **14:** 59-84.
9. VERNER, J. 1964. Evolution of polygamy in the long-billed marsh wren. Evolution **18:** 252-261.

10. DECOUX, J. P. 1982. Les particularities demographiques et sociologiques du *Coliu strie* dans le nord-est du Gabon. I. Donnes d'observation. Rev. Ecol. (Terre Vie) **36:** 37-78.
11. MACDONALD, D. W. 1979. "Helpers" in fox society. Nature **282:** 69-71.
12. MACDONALD, D. W. 1980. Social factors affecting reproduction amongst red foxes, *Vulpes vulpes. In* The Red Fox: Symposium on Behaviour and Ecology. E. Zimen, Ed.: 123-175. Biogeographica, Vol. 18. Dr. W. Junk B.V. The Hague.
13. MOEHLMAN, P. D. 1979. Jackal helpers and pup survival. Nature **277:** 382-383.
14. REYER, H. U. 1980. Flexible helper structure as an ecological adaptation in the pied kingfisher (*Ceryle rudis rudis* L.). Behav. Ecol. Sociobiol. **6:** 219-227.
15. MONFORT-BRAHM, N. 1975. Variations dans la structure sociale du topi, *Damaliscus korrigum* Ogilby, au Parc National de l'Akagera, Rwanda. Z. Tierpsychol. **39:** 332-364.
16. WILSON, E. O. 1975. Sociobiology: The New Synthesis. Belknap Press. Cambridge, Mass.
17. VAUGHN, T. A. & S. T. SCHWARTZ. 1980. Behavioral ecology of an insular wood rat. J. Mammal. **61:** 205-218.
18. NOIROT, E. 1964. Changes in responsiveness to young in the adult mouse. II. The effect of external stimuli. J. Comp. Physiol. Psychol. **57:** 97-99.
19. WARD, I. L. & J. WEISZ. 1980. Maternal stress alters plasma testosterone in fetal males. Science **207:** 328-329.
20. CLEMENS, L. G. & B. A. GLAUDE. 1978. Feminine sexual behavior in rats enhanced by prenatal inhibition of androgen aromatization. Horm. Behav. **11:** 190-201.
21. CHRISTIE, M. H. & R. J. BARFIELD. 1979. Effects of castration and home cage residency on aggressive behavior in rats. Horm. Behav. **13:** 85-91.
22. SAAL, F. S. VOM. 1979. Prenatal exposure to androgen influences morphology and aggressive behavior of male and female mice. Horm. Behav. **12:** 1-11.
23. SAAL, F. S. VOM & F. H. BRONSON. 1980. Sexual characteristics of adult female mice are correlated with their blood testosterone levels during prenatal development. Science **208:** 597-599.
24. ARMITAGE, K. B. 1977. Social variety in the yellow-bellied marmot: a population-behavioral system. Anim. Behav. **25:** 585-593.
25. LLOYD, C. W. & J. WEISZ. 1975. Hormones and aggression. *In* Neural Bases of Violence and Aggression. W. S. Fields & W. H. Sweet, Eds.: 92-127. Warren H. Green. St. Louis, Mo.
26. LENINGTON, S. 1984. The evolution of polyandry in shorebirds. *In* Shorebirds: Breeding Behavior and Populations. J. Burger & B. L. Olla, Eds.: 149-167. Plenum. New York, N.Y.
27. FRIEDMAN, H. M. & G. S. FRIEDMAN. 1939. Gonadotrophic extracts from green leaves. Am. J. Physiol. **124:** 486-490.
28. BRADBURY, J. T. & D. E. WHITE. 1954. Estrogens and related substances in plants. Vitam. Horm. **12:** 207-233.
29. ETTINGER, A. E. & J. R. KING. 1981. Consumption of green wheat enhances photostimulated ovarian growth in white-crowned sparrows. Auk **98:** 832-834.
30. BERGER, P. J., E. H. SANDERS, P. D. GARDNER & N. C. NEGUS. 1977. Phenolic plant compounds functioning as reproductive inhibitors in *Microtus montanus*. Science **195:** 575-577.
31. BENNETS, H. W., E. J. UNDERWOOD & F. L. SHIER. 1946. A specific problem of sheep on subterranean clover pastures in western Australia. Aust. Vet. J. **22:** 2-12.
32. ADAMS, N. R., H. HEARNSHAW & C. M. OLDMAN. 1981. Abnormal function of the corpus luteum in some ewes with phytoestrogenic infertility. Aust. J. Biol. Sci. **34:** 61-65.
33. MCMILLAN, I. I. 1964. Annual population changes in California quail. J. Wildl. Manage. **28:** 702-711.
34. LEOPOLD, A. S., M. ERVIN, J. OH & B. BROWNING. 1976. Phytoestrogens: adverse effects on reproduction in California quail. Science **191:** 98-100.
35. DE MAN, E. 1982. Dietary ferulic acid, biochanin A, and the inhibition of reproductive behavior in Japanese quail (*Coturnix coturnix*). Pharmacol. Biochem. Behav. **17:** 405-411.
36. BRONSON, F. H. & C. DESJARDINS. 1971. Steroid hormones and aggressive behavior in mammals. *In* The Physiology of Aggression and Defeat. B. E. Eleftheriou & J. P. Scott, Eds.: 43-63. Plenum. New York, N.Y.
37. HARDING, S. F. 1981. Social modulation of circulating hormone levels in the male. Am. Zool. **21:** 223-231.

Advantages to Female Rodents of Male-Induced Pregnancy Disruptions[a]

ANNE STOREY

Department of Psychology
Memorial University
St. John's, Newfoundland, Canada

The Institute of Animal Behavior of Rutgers University stimulated my interest in the integration of different levels of behavioral analysis, from the physiological to the ecological. I became particularly interested in questions about the adaptiveness of behavior patterns in which the underlying mechanisms had been studied. Male-induced pregnancy disruptions, called the Bruce effect, are a good example of a response that had been studied at the sensory and physiological levels with more recent concern focusing on its function or survival value. In 1959 Bruce reported[1] that exposure to strange males causes pregnancy disruptions in some recently mated female mice. Other researchers have since found preimplantation pregnancy disruptions in several genera of rodents including both spontaneous and induced ovulators (see reviews in References 2 and 3) and recently postimplantation disruptions have also been observed in two microtine rodents (prairie voles, *Microtus ochrogaster*[4]; and meadow voles, *Microtus pennsylvanicus*[5,6]).

Pregnancy disruptions appear to be mediated by olfactory cues since female mice exposed only to soiled bedding from a strange male will reabsorb pregnancies.[7] At the neural level, 6-OHDA lesions to the vomeronasal pathway (medial olfactory stria and accessory olfactory bulb) cause females to reabsorb pregnancies when reunited with the original male whereas controls are blocked only by a new male.[8] Exposure to the new male facilitates gonadotropin release which is incompatible with the hormonal events surrounding implantation. Injections of either progesterone or prolactin[9] or the naturally high prolactin levels during postpartum estrus[10] protect females from the pregnancy disrupting effects of a new male.

There has been some controversy about whether pregnancy disruptions are important in natural populations or whether they are merely laboratory artifacts. Many species show the response, and in some species, females are susceptible to disruptions virtually throughout pregnancy. That the response is so widespread suggests that pregnancy disruptions may greatly affect the reproductive success of the individuals involved. There is evidence in at least one species, the meadow vole, that pregnancy disruptions do occur in natural populations.[11] If pregnancy disruptions function in natural populations, it is of interest to consider how they affect the reproductive success of the new male, the original sire, and the female.

---

[a] This research was funded by operating grants from the Natural Sciences and Engineering Research Council of Canada.

Pregnancy disruptions would clearly benefit the new male since the disrupted female quickly returns to the receptive state and mates with the new male, thus allowing him to sire a litter earlier than if the female completed the pregnancy and mated with him during postpartum estrus. This timing advantage for new males has also been hypothesized in species of polygamous mammals in which new males practice infanticide on unweaned young following a group takeover (see review in Reference 12). In contrast, pregnancy disruptions do not appear to benefit the original male; I will discuss later how his reproductive options appear to be curtailed by the female's potential to reabsorb his litter.

In 1975 Wilson[13] noted that male-induced pregnancy disruptions are obviously advantageous to new males but then added, "It is less easy to see why it is advantageous to the female and therefore how the response could have been evolved by direct natural selection." Several researchers have since started to analyze the possible advantages for females. Male-induced pregnancy disruptions may benefit females by (a) limiting reproduction in high-density populations, (b) allowing pregnant emigrating females to quickly become receptive to new males, (c) preventing infanticide, and (d) keeping the original male with the female. I will discuss these potential advantages to females in the next section.

DISRUPTIONS LIMIT REPRODUCTION IN A HIGH-DENSITY POPULATION

A high-density population may be a signal that there are insufficient resources or too much social disruption for successful reproduction and it might be advantageous for some females to reabsorb pregnancies. Lloyd and Christian[14] found that with increasing population size in house mice, there was a decreasing proportion of females, particularly young ones, that produced litters. Examination of females in the high-density phase of this study indicated that many females without litters showed anatomical evidence of ovulation and pregnancy suggesting that reabsorption had occurred. One mechanism whereby higher density would facilitate reabsorption would be if females were more likely to reabsorb as the number of strange males increased, as has been shown by Chipman and Fox.[15] On the other hand, other individuals that would also be present at high densities, such as additional females[16] and the original male,[7] have been shown to reduce the incidence of disruptions.

DISRUPTIONS FACILITATE EMIGRATION

Rogers and Beauchamp[2] have suggested that pregnancy disruptions may be advantageous to males in minimizing interdeme gene mixing. Alternatively, disruptions may help females enter a new group or territory by allowing her to rapidly become receptive to a new male. Studies on grouped prairie[17] or meadow voles,[18] housed in seminatural enclosures, showed that in most cases only one male and female per enclosure had a litter. I have established several groups of meadow voles each consisting of two females in different reproductive conditions housed with one male. The most receptive female, that is the estrous female rather than the pregnant one, was usually the only female to give birth. In no instance did a pregnant female give birth to a litter sired prior to her introduction to the group. These findings suggest that being able to become receptive quickly may be important in female-female competition, and thus it would be advantageous for emigrating females to be able to reabsorb a pregnancy quickly and become receptive to a new male. This potential advantage for females should be investigated further.

DISRUPTION PREVENTS INFANTICIDE BY THE NEW MALE

Several researchers have found that polygamous mammals, including species of rodents (collared lemmings, *Dicrostonyx groenlandicus*[19]; house mice[20,21]; and meadow voles[22]), kill the offspring sired by another male, thus quickly bringing the female into estrus. While several authors[3,23] have suggested that infanticide prevention appears to be the most important factor making pregnancy disruptions advantageous to females, I suggest that infanticide should be viewed as the most extreme form of differential treatment of pups by related and unrelated males. Differences in paternal care provided by related and unrelated males should also determine whether a female should reabsorb a pregnancy.

DISRUPTIONS MAINTAIN PRESENCE OF THE ORIGINAL MALE

Pregnancy disruptions may function to keep the original male near the female since the presence of the original male lessens the blocking potential of the new male (this is seen in house mice,[7] prairie deer mice,[24] and meadow voles[6]). In meadow voles, I have found that whether the original male protects the female from disruption depends on the degree of proximity to the female and the time in pregnancy that the new male is introduced. To test these effects of the new male-old male combination, I placed primiparous females with either the original or a new male, and the second of the two males was placed on the other side of a wire partition. This change in the social environment was made on either day 4 or day 12 of the 21-day pregnancy. On day 4, few females with the new male only (6%) retained the original pregnancy whereas having the original male in the adjacent compartment (60%) resulted in a significantly higher incidence of pregnancy retention.[6] Thus early in pregnancy the presence of the original male partially protects the female from disruption and the female's ability to reabsorb pregnancies limits the potential of the original male to desert her and associate with other females.

In contrast, on day 12, there were significantly more retained pregnancies if the female was with the new male only (53%) than when the original male was in the compartment adjacent to the new male-female pair (15%). Two questions are suggested by these results. First, why does the presence of the original male not entirely protect the female from disruption when the new male is also present? Second, why is the proximity of the original male so ineffective in preventing disruptions late in pregnancy when he is not as close to the female as is the new male? Answering these questions brings us to a consideration of why the male's continued presence would benefit the female in terms of his contribution to pup care.

Schwagmeyer[3] has suggested that pregnancy disruptions may allow females to reabsorb pregnancies when the original sire is no longer present but a new male has come into the territory. Reabsorption after desertion or replacement of the original male would be most advantageous to females in species where males make some postcopulatory investment in the young, and furthermore if males make more parental investment in their own pups than in unrelated ones.[3] Males apparently assess the probability of paternity by the duration of their past exposure and recent copulatory experience with females.[20-22] Males of many rodent species show paternal behavior in the laboratory (e.g., California mice, *Peromyscus californicus*[25]; house mice[26]; and several species of muroid rodents[27]) and evidence is accumulating for long-term bonding in natural populations of some species (e.g., oldfield mice, *Peromyscus poliinotus*[28]; prairie voles[29]; and pine voles, *Microtus pinetorum*[30]). It is important to realize that the high level of paternal behavior seen in the laboratory may not occur in nature, although it is difficult to imagine why paternal responsiveness would have evolved in the male if the social systems were not sufficiently flexible for it to be one of the options shown in the field.

We have examined the extent of paternal care when female meadow voles and their litters are exposed to different combinations of the original and new males.[31] Females were placed with various combinations of new and old males late enough in pregnancy so that pregnancy disruptions did not occur, thus making it possible to determine what happens to pups if reabsorption does not occur. Not surprisingly, original males were with pups on significantly more hourly spot checks (27%) than were males in any new male-female condition (6%). It is also of interest to note that the original male spent significantly more time with the pups when he was alone with the pups and the female than when the new male was on the other side of the partition (17%, Reference 31). This may explain why pregnancy disruptions occur in some females when the original male is present in addition to the new male. Perhaps the presence of the new male signals the original male that the paternity of the litter is uncertain and hence the original male may limit his paternal care. When the new male is on the other side of the partition from the pair, paternal care is reduced but infanticide does not occur. This supports the hypothesis that probable paternal care, as well as potential for infanticide, influences whether females reabsorb pregnancies.

Almost all females reabsorbed pregnancies when they were exposed on day 12 to the new male when the old male was in the adjacent compartment. In the presence of the original male, new males provide little pup care and the new male is often hostile to the pups which the female attempts to guard closely. In this case, the presence of the original male on the other side of the partition appears to actually decrease the survival of his pups and these observations may explain why late in pregnancy females so readily reabsorb litters when they are closer to the new male than the original male.

One important line of future research should be to learn more about how a female responds to different combinations of individual males as a function of how much she has already invested in the pregnancy. In most experimental groups in which pregnancy disruptions occur, not all females show disruptions. We should examine whether there are individual differences in the behavior of the new male that might signal to the female whether that male is likely to show parental care or, on the other hand, perhaps harm pups if reabsorption does not occur.

Future research should also more closely examine the male's role in pup care and whether his care actually increases the females' reproductive success. Because of this laboratory-field difference in the amount of paternal care shown, further attempts should be made to observe behavior in larger, seminatural enclosures (as in References 17, 31, and 32). In order to get 24-hr records of parental care of pups we have developed an apparatus which contains microswitches in the nest box tunnel. These microswitches are linked to a camera and an event recorder, thus providing a record of how much time each marked individual spends with pups. Males have also been observed in the company of older pups and so we should also be able to observe the response of males to mobile pups in terms of socialization and defense.

SUMMARY

Several, not mutually exclusive, hypotheses have been discussed to account for the advantages of pregnancy disruptions for female rodents. Those hypotheses focusing on differential male behavior to related and unrelated pups may, with additional research, provide the most plausible explanation for the changes throughout pregnancy in the probability that a female will reabsorb a pregnancy when exposed to a new male.

ACKNOWLEDGMENT

Gerard Martin provided helpful comments on the manuscript.

REFERENCES

1. BRUCE, H. M. 1959. An exteroceptive block to pregnancy in the mouse. Nature **184**: 105.
2. ROGERS, J. G. & G. K. BEAUCHAMP. 1976. Some ecological implications of primer chemical stimuli in rodents. *In* Mammalian Olfaction, Reproductive Processes, and Behavior. R. Doty, Ed.: 181-195. Academic Press. New York, N.Y.
3. SCHWAGMEYER, P. L. 1979. The Bruce effect: an evaluation of male/female advantages. Am. Nat. **114**: 932-938.
4. STEHN, R. & M. RICHMOND. 1975. Male-induced pregnancy termination in the prairie vole, *Microtus ochrogaster.* Science **75**: 1211-1213.
5. KENNEY, A. M., R. L. EVANS & D. A. DEWSBURY. 1977. Postimplantation pregnancy disruptions in *Microtus ochrogaster, M. pennsylvanicus,* and *Peromyscus maniculatus.* J. Reprod. Fertil. **49**: 365-367.
6. STOREY, A. E. 1986. Influence of sires on male-induced pregnancy disruptions in meadow voles *(Microtus pennsylvanicus)* differs with stage of pregnancy. J. Comp. Psychol. **100**: 15-20.
7. PARKES, A. S. & H. M. BRUCE. 1961. Olfactory stimuli in mammalian reproduction. Science **134**: 1049-1054.
8. KEVERNE, E. B. & C. DE LA RIVA. 1982. Pheromones in mice: reciprocal interactions between the nose and brain. Nature **296**: 148-150.
9. DOMINIC, C. J. 1966. Observations on the reproductive pheromones of mice. II. Neuroendocrine mechanisms involved in the olfactory block to pregnancy. J. Reprod. Fertil. **11**: 415-421.
10. BRUCE, H. M. & A. S. PARKES. 1961. The effect of concurrent lactation on olfactory block to pregnancy in the mouse. J. Endocrinol. **22**: 6-7.

11. MALLORY, F. F. & F. V. CLULOW. 1977. Evidence of pregnancy failure in the wild meadow vole, *Microtus pennsylvanicus.* Can. J. Zool. **55:** 1-17.
12. HRDY, S. B. 1979. Infanticide among animals: a review, classification, and examination of the implications for the reproductive strategies for females. Ethol. Sociobiol. **1:** 13-40.
13. WILSON, E. O. 1975. Sociobiology. Harvard Press. Cambridge, Mass.
14. LLOYD, J. A. & J. J. CHRISTIAN. 1969. Reproductive activities of individual females in three experimental freely growing populations of house mice. J. Mammal. **50:** 49-59.
15. CHIPMAN, R. K. & K. A. FOX. 1966. Factors in pregnancy blocking: age and reproductive background of females: numbers of strange males. J. Reprod. Fertil. **12:** 399-403.
16. BRUCE, H. M. 1963. Olfactory block to pregnancy in grouped mice. J. Reprod. Fertil. **6:** 451-460.
17. THOMAS, J. A. & E. C. BIRNEY. 1979. Parental care and mating system of the prairie vole, *Microtus ochrogaster.* Behav. Ecol. Sociobiol. **5:** 171-186.
18. STOREY, A. E. Manuscript in preparation.
19. MALLORY, F. F. & R. J. BROOKS. 1978. Infanticide and other reproductive strategies in the collared lemming, *Dicrostonyx groenlandicus.* Nature **273:** 144-146.
20. LABOV, J. B. 1980. Factors influencing infanticidal behavior in wild male house mice (*Mus musculus*). Behav. Ecol. Sociobiol. **6:** 297-303.
21. HUCK, U. W., R. L. SOLTIS & C. B. COOPERSMITH. 1982. Infanticide in male laboratory mice: effects of social status, prior sexual experience, and basis for discrimination between related and unrelated young. Anim. Behav. **30:** 1158-1162.
22. WEBSTER, A. B., R. G. GARTSHORE & R. J. BROOKS. 1981. Infanticide in the meadow vole, *Microtus pennsylvanicus*: significance in relation to social system and population cycling. Behav. Neurol. Biol. **31:** 342-347.
23. LABOV, J. B. 1981. Pregnancy blocking in rodents: adaptive advantages for females. Am. Nat. **118:** 361-371.
24. TERMAN, C. R. 1969. Pregnancy failure in prairie deermice related to parity and social environment. Anim. Behav. **17:** 104-108.
25. DUDLEY, D. 1974. Contributions of paternal care to the growth and development of the young in *Peromyscus californicus.* Behav. Biol. **11:** 155-166.
26. PRIESTNALL, R. & S. YOUNG. 1978. An observational study of caretaking behavior of male and female mice housed together. Dev. Psychobiol. **11:** 23-30.
27. HARTUNG, T. G. & D. A. DEWSBURY. 1979. Paternal behavior in six species of muroid rodents. Behav. Neurol. Biol. **26:** 466-478.
28. FOLTZ, D. W. 1981. Genetic evidence for long-term monogamy in a small rodent, *Peromyscus polionotus.* Am. Nat. **117:** 665-675.
29. GETZ, L. L., C. S. CARTER & L. GAVISH. 1981. The mating system of the prairie vole, *Microtus ochrogaster*: field and laboratory evidence for pair-bonding. Behav. Ecol. Sociobiol. **8:** 189-194.
30. FITZGERALD, R. W. & D. M. MADISON. 1983. Social organization of a free-ranging population of pine voles, *Microtus pinetorum.* Behav. Ecol. Sociobiol. **13:** 183-187.
31. STOREY, A. E. & D. T. SNOW. Male identity and enclosure size affect paternal attendance of meadow voles. Anim. Behav. In press.
32. MCGUIRE, B. & M. NOVAK. A comparison of the maternal behavior in the meadow, pine, and prairie voles (*Microtus pennsylvanicus, M. pinetorum,* and *M. ochrogaster*). Paper presented at the meeting of the Animal Behavior Society, Lewisburg, Pa., 1983.

Reproductive Behavior as a Phenotypic Correlate of T-Locus Genotype in Wild House Mice: Implications for Evolutionary Models

SARAH LENINGTON

Institute of Animal Behavior
Rutgers-The State University of New Jersey
Newark, New Jersey 07102

For the past 30 years there has been considerable debate among evolutionary biologists over factors controlling the frequency of alleles at the T-locus in wild house mice (*Mus musculus*) (reviewed in Reference 1). About 25% of wild mice carry a recessive t-allele at this locus, while the remaining mice are homozygous for the wild-type +-allele.[2] Many different t-alleles have been identified in natural populations, all of which have deleterious effects when homozygous. Some are lethals, producing death during fetal development, and others produce sterility in homozygous males.[3] On the basis of these phenotypic effects alone, one would expect that t-alleles would be rare in natural populations, but in fact they are fairly common. A major reason for their presence in wild mice is that they have very large effects on the fertilizing ability of sperm. When a heterozygous male reproduces he often transmits his t-allele to 90-100% of his offspring.[4]

In the 1950s and early 1960s theoretical models were constructed attempting to estimate the frequency of heterozygotes among wild mice on the basis of the effects of t-alleles on embryonic mortality and male sterility counteracted by the transmission ratio of t-bearing sperm.[5-7] These models predicted that the frequency of t-carrying animals should be in the range of 70-90%, a value much higher than the actual frequency of 20-25%. The controversy that developed surrounded attempts to determine what additional factors could be responsible for lowering the frequency of t-alleles among wild mice. This controversy has provided an unusual opportunity to integrate behavior, genetics, ecology, and evolution.

It is a truism of evolutionary theory that although natural selection operates on phenotypes, evolutionary change comes about through changes in gene frequency. However, only rarely have changes in gene frequency associated with phenotypic changes been documented. In recent years sociobiologists have studied the evolution of complex behavioral traits and often have been successful in elucidating the adaptive nature of behavior under natural conditions. Yet virtually nothing is known about the genetics of the behaviors under study. Behavior geneticists have gathered considerable information about the genetic control of behavior. However, many studies in behavior genetics are done with inbred laboratory strains, often carrying mutant alleles not found in natural populations. As a consequence, this research can shed considerable light on the organization of genetic systems but often has little relevance to the genetic basis of behavior of animals living under natural conditions.

A few years ago it seemed to me that it might be possible to use the T-locus to integrate these diverse approaches. There were strong theoretical reasons to think the T-locus might be associated with behavioral effects. If behavioral effects were found, it might be possible to shed some light on the genetic control of a behavioral phenotype in wild animals. In addition, ideally it would be possible to devise theoretical models incorporating all known phenotypic effects (including behavior) as parameters, which would be able to account for the frequency of alleles at this genetic locus in natural populations. Although the findings arising from such a study would be specific to processes occurring with respect to a particular genetic locus, they might also serve as a model for more general interactions between genetic and phenotypic evolution.

The hope that behavioral (and other sociobiologically relevant) effects would be associated with T-locus genotype has been more than amply fulfilled. However, the solution to the second half of the problem (that of accounting for the frequency of alleles in natural populations) seems increasingly difficult. The reason for this is that genotype at the T-locus in mice seems to be correlated with almost every possible phenotypic characteristic of interest to sociobiologists and evolutionary biologists. Furthermore, these phenotypic correlates of T-locus genotype do not all have the same consequences for the frequency of alleles at the T-locus. The abundance of phenotypic correlates also has a number of implications for population genetics models attempting to explain gene frequency and sociobiological models attempting to predict optimum phenotype. The data presented below can be viewed as an evolutionary cautionary tale for any who hope that the relationship between selection and gene frequency will be either simple or straightforward.

It is highly likely that not all phenotypic correlates of genotype at the T-locus are actually *due* to the T-locus itself. We have attempted to randomize genetic background effects by using wild mice derived from many, widely separated populations. In the studies carried out in our laboratory, we have used mice from 15 different populations, caught in five states. However, the T-locus is linked to other genetic loci.[8] Some of the characters we have studied are affected by these other loci rather than the T-locus *per se* (e.g., odor difference between +/+ and +/t females).[9] Furthermore, we strongly suspect that some characters are influenced by the experience of animals with parents or siblings of particular genotypes rather than by the animal's own T-locus genotype (e.g., the effect of parental genotype on odor preferences of males and females).[9,10] However, regardless of the basis for the phenotypic characters, all are *correlated* with T-locus genotype. Therefore all will have consequences for the frequency of alleles at the T-locus. Conversely, the distribution of genotypes at the T-locus will have consequences for the distribution of these correlated phenotypes.

Initially, it seemed probable that there should be very strong selection for mice to be able to discriminate +/+ from +/t individuals and to avoid mating with individuals who carried t-alleles. Such nonrandom mating might, by itself, account for the lower than expected frequency of t-alleles in natural populations. An earlier study[11] had indicated that among mice paired in laboratory cages, females seemed to have a mating preference for +/+ as opposed to +/t males. In addition, strong social preferences were found for +/+ animals among mice tested in arenas.[12] When a female was placed in an arena with two males, one of each genotype, females approached +/+ males more frequently than +/t males. Furthermore, males were more likely to mount +/+ than +/t females. We have also found that mice, given a choice of odors of +/+ and +/t animals, prefer odors of +/+ animals.[9,10,12] This ability of mice to use odor to discriminate +/+ from +/t individuals could provide a cue for mating preferences. However, for nonrandom mating to have an effect on the frequency of t-alleles in natural populations, it cannot only be manifest under controlled conditions but also must be present when mice are freely interacting. Consequently, we looked at the behavior of mice in a seminatural environment where very little was controlled except the external conditions of the room.[13]

We placed 8-10 mice, marked with fur dye for individual recognition, in a room for 20 days. During the 20-day period, we recorded all dominance interactions for 1 hour per day. In addition, the room was checked several times each day to see if any females were in behavioral estrus. If females were in behavioral estrus, all mating behavior was recorded. At the end of 20 days, we removed all animals from the room and put in a new group of animals. This procedure was repeated for a total of 20 trials so that the entire study was based on 108 females and 68 males. After removal from the room, mice were housed individually in cages for 20 days to see if females produced litters. If there was any question about who fathered the litter, the female, her progeny, and all potential fathers were killed for electrophoretic analysis.

In this setting, mating behavior is highly constrained by dominance interactions and we found no statistically significant indication of nonrandom mating. However, we did see trends suggesting mating preferences that might have been statistically significant if the study had involved a larger number of trials and animals. More females went into behavioral estrus in trials in which a $+/+$ male was dominant as compared with trials in which a $+/t$ male was dominant. In addition, there were 40% more matings between two $+/+$ individuals and 30% fewer matings between two $+/t$ individuals than would be expected if all mating were random with respect to T-locus genotype.[13]

Thus there is some evidence that nonrandom mating may occur with respect to T-locus genotype under naturalistic conditions, the pattern of which would serve to reduce the frequency of t-alleles. However, the effect of nonrandom mating on the population genetics of t-alleles is probably weak, particularly in comparison with the effects of selection on heterozygotes (see below) and, therefore, not sufficient to account for the frequency of t-alleles in natural populations.

When carrying out the original studies of behavior of mice in arenas I noted a phenomenon that promised to cause considerable difficulty for any attempt to explain the low frequency of t-alleles. Specifically, males carrying t-alleles seemed to be much more likely to be dominant than $+/+$ males when paired in aggressive encounters in arenas.[12] Since dominance rank is an important component of fitness, these data suggested that heterozygous males might have a higher fitness than $+/+$ males, a possibility consistent with results published a number of years ago indicating that $+/t$ males might have greater viability, fertility, and levels of aggression than $+/+$ males.[14-16] However, in the seminatural environment, in contrast to arenas, male genotype was unrelated to male dominance rank.[13] Despite the absence of genotypic effects on male dominance rank, when $+/t$ males became top-ranking they were twice as likely to kill all their subordinates than were top-ranking $+/+$ males. Furthermore, as subordinates, $+/t$ males were more likely to father young than were subordinate $+/+$ males. The net result of these differences was that overall, $+/t$ males fathered 35% more young than did $+/+$ males.[13]

Because any fitness advantage for $+/t$ males is augmented by segregation distortion of t-alleles, even a small increase in fitness for $+/t$ males would result in a large increase in the frequency of t-alleles in mouse populations. Thus, if our finding of a higher fitness for heterozygous males in the seminatural environment generalizes to natural populations of mice, the task of accounting for the frequency of t-alleles in natural populations becomes considerably more complex.

Possession of a t-allele had very clear-cut consequences for female fitness in the seminatural environment.[13] Females who carried t-alleles were significantly less likely to be dominant than were +/+ females. Although female dominance rank was an important predictor of which females would go into behavioral estrus, genotype was also a significant predictor of female estrous behavior. Furthermore, among females who went into behavioral estrus, +/t females were less likely to produce litters when controlling for dominance rank than were +/+ females. The ultimate outcome was that the fitness of +/t females was only about one-third that of +/+ females.

The effect of selection at the T-locus therefore appears to differ between the sexes. Although the fitness of +/t males is no lower and may well be higher than that of +/+ males, heterozygous females had a much lower fitness than did +/+ females. We used the mathematical model developed by Hartl[17] to predict the equilibrium frequency of t-alleles given differential selection on males and females.[13] The consequences of a reduced fitness for +/t females for the frequency of t-alleles in natural populations depend primarily on the magnitude of the relative fitness of males of the two genotypes. If the fitness of +/t males is equal to that of +/+ males, the reduced fitness of +/t females will lower the frequency of t-alleles to a value close to that observed in natural populations. If, however, selection favors +/t males, negative selection on females will have virtually no effect on the frequency of t-alleles.

Given that the effects of selection are sex specific, it is important to examine the sex ratio of progeny. Biases in sex ratio will have consequences for the frequency of t-alleles when selection differs between the sexes. Matings in which the male was +/t had sex ratios biased toward male progeny, whereas matings in which the female alone was +/t had female-biased sex ratios (TABLE 1). Pairs in which the female alone is +/t produce equal numbers of +/+ and +/t progeny. However, pairs in which the male alone is +/t produce a very high proportion of +/t males, males who may have a high relative fitness. Thus the effect of this sex ratio bias should be to further increase the frequency of t-alleles in natural populations (thus increasing the difficulty of explaining the actual low frequency of t-alleles in the wild).

Thus, to summarize: selection against homozygotes, negative selection against females, and nonrandom mating would be relatively weak forces in decreasing the frequency of t-alleles compared with the effect of positive selection on males, augmented by segregation distortion and sex ratio bias. However, there is a final factor that may have a crucial role in producing the low frequency of t-alleles in natural populations. As mentioned earlier, the mice used in our laboratory were derived from a large number of wild populations. Sometimes these mice were bred to mates from their own population and sometimes they were given mates caught from another population. When we examined the transmission ratio of +/t males in our laboratory (transmission ratio being defined as the proportion of all progeny that are +/t), we noted that the transmission ratio of +/t males mated to females from their own population

TABLE 1. Mean Sex Ratio (Number of Males/Number of Females) of Progeny of Wild-Caught and F_1 Pairs

Genotype of Sire	+/+	+/+	+/t	+/t
Genotype of Dam	+/+	+/t	+/+	+/t
$\bar{x}$	1.06	0.72	1.66	1.15
SD	0.54	0.31	1.32	0.57
Number of Pairs	25	13	19	8
Number of Animals	46	24	31	16
Number of Young	407	249	308	114

TABLE 2. Mean and Standard Deviation of Transmission Ratios of Males Mated to Females from Their Own Population (Within Population Crosses) and Females from Another Population (Between Population Crosses)

	Within Population Crosses	Between Population Crosses
$\bar{x}$	0.54	0.81
SD	0.25	0.24
Number of pairs	15	19
Number of young	254	191

was considerably lower than the transmission ratio of males mated to females from another population (TABLE 2).[18] In fact, the segregation distortion of +/t males that has been a seminal characteristic of t-alleles was only found among those males mated to females from another population. When males were mated to females from their own population, the proportion of +/t progeny did not differ significantly from 50%. This finding suggests that t-alleles may be maintained only under conditions of high migration. Support for this possibility is provided by the finding that commonly in closed laboratory populations of mice, male transmission ratio tends to decline over time.[19] Furthermore, island populations of mice rarely have t-alleles,[20] a phenomenon that may result from restricted migration on islands.

Although a vast oversimplification genetically (see Reference 8 for a recent map of the T-locus), from an operational point of view the T-locus can be treated as a single locus with two alleles, the kind of genetic locus favored by population geneticists when constructing models for predicting gene frequency. Despite this apparent simplicity, however, the task of accounting for the distribution of t-alleles in natural populations is formidable. The problems arise not from an absence of phenotypic correlates, but rather from a plethora of phenotypic correlates, with differing effects on fitness. Even if transmission ratio reduction turns out to be an important phenomenon in natural populations, it will not be a simple matter to use this information to estimate the expected distribution of t-alleles. No analytical model will be able to incorporate all the effects on fitness we have so far identified; instead, complex computer simulations will be required. Furthermore, it will be necessary to have estimates for the rates of migration between populations to even begin to evaluate the potential impact of transmission ratio variability. Rates of gene flow among wild mouse populations have, in themselves, been a topic fraught with considerable controversy.[21-29]

In addition, the large number of phenotypic characters correlated with T-locus genotype has substantial implications for optimization models constructed by sociobiologists. When, for example, selection for genes correlated with increased fertility or aggression in males (i.e., selection for +/t males) may produce a correlated response in terms of decreased aggression and fertility of females, changes in the sex ratio of progeny, and increased embryonic mortality, the optimum phenotype becomes impossible to achieve.

Most of the controversy in recent years in population genetics has surrounded factors controlling the frequency of allozymes: enzymatic variation that, as yet, has not been shown to have obvious effects on fitness. Attempts to account for the frequency of allozymes have spawned much debate regarding the possibility of selective neutrality of alleles and the role of random genetic drift in natural populations. In contrast, major gene complexes that produce marked phenotypic effects, have been regarded as posing fewer analytical and theoretical problems. However, major gene complexes which occupy large segments of chromosomes, could typically be expected to be associated with a large array of phenotypic effects. For example, the H-2 locus, a major gene linked to the T-locus, is associated with an extremely large number of phenotypic effects.[30-32] Such genes may ultimately create greater difficulties for the empirical study of gene frequency than enzymatic variation. At this point it seems likely that the debate over the control of gene frequencies at the T-locus may well last another 30 years.

REFERENCES

1. LACY, R. 1978. Dynamics of t-alleles in *Mus musculus* populations: review and speculation. The Biologist **60:** 41-67.
2. BENNETT, D. 1978. Population genetics of T/t complex mutations. *In* Origins of Inbred Mice. H. Morse, Ed.: 615-632. Academic Press. New York, N.Y.
3. BENNETT, D. 1975. The T-locus of the mouse. Cell **6:** 441-454.
4. DUNN, L. C. 1939. The inheritance of taillessness (anury) in the house mouse. III. Taillessness in the balanced lethal line 19. Genetics **24:** 728-731.
5. BRUCK, D. 1957. Male segregation ratio advantage as a factor in maintaining lethal alleles in wild populations of house mice. Proc. Natl. Acad. Sci. USA **43:** 152-158.
6. DUNN, L. C. & H. LEVENE. 1961. Population dynamics of a variant t-allele in a confined population of wild house mice. Evolution **15:** 385-393.
7. LEWONTIN, R. C. 1968. The effect of differential viability on population dynamics of t-alleles in the house mouse. Evolution **22:** 262-273.
8. ARTZT, K., P. MCCORMICK & D. BENNETT. 1982. Gene mapping within the T/t complex of the mouse. II. Anomalous position of the H-2 complex in t haplotypes. Cell **28:** 471-476.
9. EGID, K. & S. LENINGTON. 1985. Responses of male mice to odors of females: effects of T and H-2 locus genotype. Behav. Genet. **15:** 287-295.
10. LENINGTON, S. & K. EGID. 1985. Female discrimination of male odors correlated with male genotype at the T-locus in *Mus musculus*: a response to T-locus or H-2 locus variability? Behav. Genet. **15:** 53-67.
11. LEVINE, L., R. C. ROCKWELL & J. GROSSFIELD. 1980. Sexual selection in mice. V. Reproductive competition between $+/+$ and $+/t^{w5}$ males. Am. Nat. **116:** 150-156.
12. LENINGTON, S. 1983. Social preferences for partners carrying "good genes" in wild house mice. Anim. Behav. **31:** 325-333.
13. FRANKS, P. & S. LENINGTON. 1986. Dominance and reproductive behavior of wild mice in a seminatural environment correlated with T-locus genotype. Behav. Ecol. Sociobiol. In press.
14. DUNN, L. C. & J. SUCKLING. 1955. A preliminary comparison of the fertilities of wild house mice with and without a mutant at locus T. Am. Nat. **89:** 231-233.
15. DUNN, L. C., A. B. BEASLEY & H. TINKER. 1958. Relative fitness of wild house mice heterozygous for a lethal allele. Am. Nat. **92:** 215-220.
16. MARTIN, P. G. & H. G. ANDREWARTHA. 1962. Success in fighting of two varieties of mice. Am. Nat. **96:** 375-376.
17. HARTL, D. L. 1970. A mathematical model for recessive lethal segregation distorters with differential viabilities in the sexes. Genetics **66:** 147-164.
18. LENINGTON, S. & I. L. HEISLER. Effect of behavior, on transmission ratio distortion of t-haplotypes in wild house mice. Submitted for publication.
19. BENNETT, D., A. K. ALTON & K. ARTZT. 1983. Genetic analysis of transmission ratio data by t haplotypes in the mouse. Genet. Res. **41:** 29-45.
20. DOOHER, G. B., R. J. BERRY, K. ARTZT & D. BENNETT. 1981. A semi-lethal t-haplotype on the Orkney Islands. Genet. Res. **37:** 221-226.
21. ANDERSON, P. K. 1964. Lethal alleles in *Mus musculus*: local distribution and evidence for isolation of demes. Science, N.Y. **145:** 177-178.

22. PETRAS, M. L. 1967. Studies of natural populations of *Mus.* II. Polymorphism at the T-locus. Evolution **21**: 466-478.
23. SELANDER, R. K. & S. Y. YANG. 1969. Protein polymorphism and genetic heterogeneity in a wild population of the house mouse (*Mus musculus*). Genetics **63**: 653-667.
24. SELANDER, R. K. 1970. Biochemical polymorphisms in populations of the house mouse and old-field mouse. Symp. Zool. Soc. London **26**: 73-91.
25. SELANDER, R. K. 1970. Behavior and genetic variation in natural populations. Am. Zool. **10**: 53-66.
26. BERRY, R. J. & M. E. JACOBSON. 1974. Vagility in an island population of the mouse. J. Zool. Soc. London **173**: 341-354.
27. POOLE, T. G. B. & H. D. R. MORGAN. 1976. Social and territorial behaviour of laboratory mice (*Mus musculus* L.) in small complex areas. Anim. Behav. **24**: 476-480.
28. BUTLER, R. G. 1980. Population size, social behaviour and dispersal in house mice: a quantitative investigation. Anim. Behav. **28**: 78-85.
29. MILLER-BAKER, A. E. 1981. Gene flow in house mice: introduction of a new allele into free living populations. Evolution **35**: 243-258.
30. IVANYI, P. 1978. Some aspects of the H-2 system, the major histocompatibility system in the mouse. Proc. R. Soc. London Ser. B **202**: 117-159.
31. YAMAZAKI, K., G. K. BEAUCHAMP, J. BARD, L. THOMAS & E. A. BOYSE. 1982. Chemosensory recognition of phenotypes determined by the *Tla* and *H-2k* regions of chromosome 17 of the mouse. Proc. Natl. Acad. Sci. USA **79**: 7828-7831.
32. YAMAZAKI, K., G. K. BEAUCHAMP, C. J. WYSOCKI, J. BARD, L. THOMAS & E. A. BOYSE. 1983. Recognition of H-2 types in relation to the blocking of pregnancy in mice. Science **221**: 186-188.

The Role of the Vomeronasal Organ in Behavioral Control of Reproduction

MARGARET A. JOHNS[a]

Division of Endocrinology
Mt. Sinai School of Medicine
New York, New York 10029

A careful drawing by the Dutch anatomist, Frederic Ruysch[1] in 1703, of the left side of the nasal septum of a young man, provided the first known illustration of a vomeronasal organ (VNO). Almost a hundred years later, in 1809, another anatomist, Samuel Soemmerring,[2] also drew a human VNO and described it.

No suggestion that this organ might play a role in human reproductive physiology was made by either anatomist. This is not surprising, considering that nothing at all was known about the rest of the vomeronasal (VN) system. After the Danish physician Ludwig Jacobson,[3] in 1811, demonstrated the presence of the VNO (also called Jacobson's organ or the "accessory" olfactory organ) in a wide range of animals—from monkeys and elephants to horses and cats—some type of physiologic function for this small, inconspicuous organ was occasionally postulated. In 1975 Powers and Winans[4] reported a role for the VNO in mediating sexual behavior of male hamsters. It was only in 1978, however, that the first experimental evidence for any physiologic function for the VNO in mammals (as distinct from behavioral function) appeared.[5]

Studies conducted at Rutgers University between 1973 and 1977 suggested the possibility that the olfactory epithelium of the VNO provides an important link between mammalian reproductive behavior and physiologic responses.[5-7] Until that time the many known or suspected effects of chemical (pheromonal) communication between mammals on reproductive neuroendocrinology were widely thought to be mediated by "primary" olfactory epithelia.

The mammalian VNO is a bilateral, tubular, blind, olfactory structure, lined with mucosa, that is found in the anterior ventral part of the nasal septum. It usually opens via a duct in front of the region of the nasopalatine canal, and it is connected to the VN bulb (or "accessory" olfactory bulb) by nerves that pass through the cribriform plate. In contrast, the mucous membrane of the "primary" olfactory organ is located in the upper part of the superior concha and in the corresponding portion of the septum. Although nerves that connect the "primary" olfactory organ to the "main" olfactory bulb also pass through the cribriform plate, there appears to be no connection between the two olfactory systems at any level. It has long been assumed that effective stimuli must be conveyed to the VN epithelium in a liquid vehicle.[8]

This paper will briefly summarize our behavioral and physiologic research and the research of others that influenced our thinking, provide an overview of speculations about the functions of the VNO in mammals up to the time of our discovery, and present subsequent experimental evidence that supports our hypothesis:

[a] Present address: Mountainville, N.Y. 10953.

Under natural conditions the primary olfactory system merely serves to orient females toward contact with chemical factors that act on the vomeronasal system. The vomeronasal system then mediates gonadotropin release responsible for reproductive maturation, pregnancy block, estrus synchronization, and reflex ovulation. (Johns, Reference 7, p. 111)

In 1980 this hypothesis was extended to include VN system mediation of pheromonally induced gonadotropin release in males, to add another role for females (successful implantation), and to substitute the more general "estrous cycle modification" (e.g., cycle lengthening, cycle shortening, estrus synchronization) for "estrus synchronization."[9]

Because of a long-standing interest in ways the environment might affect reproduction in women (including reflex ovulation), I wanted to conduct experiments that would isolate environmental factors in reproduction through the use of data obtained from laboratory animals. The model I chose was the light-induced persistent-estrus (LLPE) rat. The LLPE rat provides a good model for the study of environmental effects on a species that normally ovulates spontaneously, but is prevented from doing so by an environmental influence.

Beginning in the early 1970s, Barry Komisaruk and I investigated the nature of the stimulus (or stimuli) that triggers reflex ovulation (i.e., ovulation occurring shortly after brief exposure to appropriate stimuli) using LLPE Sprague-Dawley rats. These LLPE rats resemble naturally occurring reflex ovulators (such as female cats and rabbits) in that they tend to ovulate following copulation rather than spontaneously.

We also investigated effects of adrenalectomy. Brown-Grant *et al.*, in 1973,[10] had suggested that adrenal progesterone, released in response to stress, might be involved in induced ovulation. We conducted three studies involving concurrent testing of adrenalectomized (ADX) and adrenal-intact females. Ovulatory responses, occurring within 19 hr, following brief (1/2 hr) exposure to a variety of environmental stimuli were assessed.

In the first study we explored responses of ADX and adrenal-intact rats to vaginal taping, male-soiled bedding, novel cages, and/or male mounts without intromissions. Adrenalectomy did not reduce the percentage of females that ovulated reflexly in response to male mounts without intromissions (ADX=48% vs. intact=54%). This demonstrated that adrenal progesterone release is not a requirement for induced ovulation. However, ADX females tended to ovulate more often after vaginal taping than adrenal-intact females (ADX=33% vs. intact=22%), and less often after exposure to male-soiled bedding (ADX=25% vs. intact=54%). A parallel situation exists within at least one strain of laboratory mice. Adrenalectomy gives complete protection from pregnancy block resulting from exposure to a strange male.[11]

Among adrenal-intact females, we found that exposure to vaginal taping or novel cages had no apparent effect on ovulation. To our surprise, however, exposure to male-soiled bedding in the home cage induced reflex ovulation in about half the females. The same percentage ovulated in response to male mounts without intromissions.

Although all females tested were in vaginal estrus not all were in behavioral estrus. Their sexual behavior ranged from nonreceptive, through receptive (permitting mounting), to highly proceptive (showing soliciting behavior such as hopping, darting, and ear-quivering in the presence of the male). We found that the degree of proceptivity/receptivity was highly correlated with the ovulatory responses to male mounts among adrenal-intact animals. For example, *all* females that were proceptive ovulated in response to male mounts without intromissions. This demonstrated that penile intromissions are not the primary stimulus for reflex ovulation in adrenal-intact proceptive LLPE rats.[12,13] A similar correlation between degree of proceptivity/receptivity and ovulatory response to male-soiled bedding remains to be tested.

The soiled bedding used in the above experiments contained ejaculatory plugs from male rats along with excreta of both males and ovariectomized females (obtained from nonrelated studies) injected with estrogen and progesterone to bring them into heat. We hypothesized that exposure to the male urine was adequate to induce ovulation. In the second study, we tested this hypothesis by giving females 1/2 hr of contact with laboratory bedding that had been sprayed with either 9-10 cm^3 or 20 cm^3 of pooled male urine collected from rats of the same strain as the females, but from a different animal supplier (i.e., the same males used throughout all our experiments, with the exception of Study 3).

We thought we were on the wrong track when no female ovulated after being exposed first to what we considered to be a "huge" amount of urine, 9-10 cm.^{3} Our hypothesis was confirmed, however, when we discovered that 38% ovulated when exposed to 20 cm^3 of male urine. Indeed, the latter condition was almost as effective as exposure to soiled bedding in Study 1. In Study 3, urine collected from male rats of the same strain *and* same animal supplier was found to be far more effective than the original urine. Urine from castrated males or other females was ineffective.[6]

At this point it seemed reasonable to assume that we had found a new response to chemical stimuli (reflex ovulation) via the only olfactory system with which we were familiar—the "primary" olfactory system. (Note: I have placed the terms "primary," "main," and "accessory" within quotation marks throughout this paper. These terms no longer seem appropriate to describe either of the two important olfactory systems.) There had been no previous report of any chemically mediated effect on mammalian reproductive physiology via any other olfactory system. The consensus was that effects of chemical stimuli (e.g., urine) on reproductive physiology are mediated by small quantities of volatile, low-molecular-weight substances that act via the distance receptors of the "primary" olfactory system.

Almost all research in the field of "primer" pheromones (those affecting reproduction in contrast to "signaling" pheromones that affect behavior) from the time of the important pioneer experiments beginning in the 1950s to the present involved *direct* physical contact of the test animal with chemical substances. These experiments were important because they all demonstrated that external chemical stimuli alone are sufficient to produce physiologic changes in male and female rodents. Contact with another animal is not necessary for these changes. Endocrine effects of stimulating olfactory receptors were distinguished from effects of stimulating nonolfactory (tactile, auditory, and visual) receptors.

For example, most investigations that have demonstrated effects of exposure to urine of (or bedding soiled by) adult male or female rodents on ovarian function,[14-18] pregnancy blockage,[19,20] or puberty of immature male and female rodents[21-24] have allowed *direct* exposure of test animals to the material.

There have been exceptions: two studies have found that relatively indirect exposure to male mouse urine (or extract of urine) placed in small holders blocks pregnancies in females[25] and a similar type of *indirect* exposure to urine of pregnant or lactating mice affects vaginal estrous smears of other females.[26]

Indirect chemical stimuli are usually administered via air currents. In 1968 the first air current study was published. Whitten *et al.*[27] assessed effects of chemical stimuli derived from male mice on estrous cycles of grouped females caged beneath the males or upwind or downwind of them. The responses varied with the cage locations. Whitten and his associates concluded that chemical stimuli from the males were volatile and that they almost certainly acted through the "primary" olfactory receptors. Their influential paper has been widely cited.

A later study by Kranz and Berger[28] showed that chemical stimuli carried on air currents can be sufficient to significantly increase the pregnancy rates among recently inseminated females compared with other females deprived of the additional male stimuli. Other air current experiments demonstrated effects of chemical stimuli of female rats on the estrous cycles of other females.[29,30] McClintock thought it more likely that the communication system involved in shortening and regulating ovarian cycles depends on volatile chemicals acting on the "primary" olfactory system rather than on the VN system.

It is true that all of the above experiments provide good evidence that *indirect* exposure to external chemicals can affect mammalian reproduction. However, none was designed to rule out involvement of the VN system and none did so. This applied to our critical next experiment. In all our previous pheromone experiments female rats had been in *direct* physical contact with urine or male-soiled bedding. To justify the use of the term "olfactory" to describe effects of male urine on ovulation, LLPE females should ovulate when not in *direct* physical contact with urine. Thus, in what we thought would be our final experiment, groups of adrenal-intact LLPE females were permitted either *direct* or *indirect* contact with bedding soiled by male rats. *Indirect* contact was arranged by placing the soiled bedding approximately 3 cm beneath the wire mesh screen of the cage floor. The results were unexpected: *no female separated from the soiled bedding ovulated*. In marked contrast (and as we had found in previous experiments), half (49%) of the females allowed direct contact with the soiled bedding ovulated.

We repeated the experiment, hoping for data we were better prepared to interpret. This time we used larger cages in the home room of male rats maintained for mating. The results were essentially the same. Of 19 females tested, 47% ovulated following contact with soiled bedding but only 10% of 20 ovulated when *direct* contact was prevented by a wire mesh floor. We considered setting up a new type of experiment to determine whether the effects previously found were gustatory rather than olfactory. The females might be ingesting the pheromone and responding physiologically—like bees presented with queen substance. Vandenbergh and his associates had suggested such a possibility.[31]

On the other hand, Raisman and Field[32] in 1971 had pointed out that the mammalian VN system communicates with neural systems that are important for gonadotropic release as well as sexual behavior. This information clearly challenged the traditional belief that gonadal as well as behavioral effects of chemical stimuli act exclusively via the "primary" olfactory system and suggested a more likely interpretation of our findings. Moreover, a role for the VN system, in combination with the "primary" olfactory system, had recently been found for male hamster mating behavior.[4]

Anne Mayer of the Institute of Animal Behavior of Rutgers University was at this time in the process of investigating possible roles of the VN system in rat maternal behavior. To do so she had devised an elegant method for disrupting the rat VN system, peripherally, without disturbing the "primary" olfactory system. With her specially designed electrode, she could selectively lesion the mucosal lining of the floor of each nasal vestibule (4-6 mm from the surface of the external nares) approximately over the opening and anterior portion of the vomeronasal canal.

Collaborating with us, Mayer used electrocautery to occlude the canals leading to each VNO of test females and cauterized a region of comparable size elsewhere in the nasal mucosa of control females.

In the first experiment, 54% of LLPE females with control lesions ovulated following half an hour of *direct* contact with male-soiled bedding. In contrast, ovulation occurred in only 9% of females that had received vomeronasal occlusions. Further experiments confirmed these initial results. (Parallel experiments indicated that the deficit did not impair the ability of VNO-occluded rats to respond to mating with normal reflex ovulation.) We concluded that

> the inability to respond to male urine, in the absence of other sensory stimuli, is related to the inability of the primary olfactory system to compensate for the loss of input to the vomeronasal system, at least during the limited time allowed. Our findings suggest that the vomeronasal system may have an important role in urine-induced ovulatory responses. (Johns et al., Reference 5, p. 448)

Why did it take so long for experimental evidence to replace largely untested speculations about VNO function (TABLES 1 and 2)?[2,3,33-54] One possible answer is that most investigators in the field of "primer" pheromones did not expect to find evidence for VNO function in mammalian reproductive physiology and, therefore, did

TABLE 1. The Function of the Vomeronasal Organ in Mammals: Speculations (1809-1956)

Year	Description
1809	Organ of smell.[2]
1811	Lubricates the nasal passages. Possibly a sex chemoreceptor.[3]
1845	Receptor of sexually stimulating odors.[33]
1861	An olfactory sense organ.[34]
1877	Tests animal's own juices.[35]
1893	Rudimentary accessory olfactory organ.[36]
1899	Receptor of sexually stimulating odors.[37]
1904	A precise and sensitive olfactory specialization.[38]
1912	No apparent function in man and the primates. Only retrogressive remnant.[39]
1920	Device for smelling liquid-borne odors.[8]
1934	"The function of Jacobson's organ remains a puzzle to this day."[40]
1954	"The sexual behavior of the male guinea pig is not modified by VN nerve section and Jacobson's organ is not specialized in the perception of genital odors only." (Translated from French.)[41]
1955	"The organ does not react to airborne smells.... Its precise function in the mammal is unknown."[42]
1956	"The bilateral organ in question is so beautifully designed that one cannot fail to ascribe a purposive function to it."[43]

not look for it. Two exceptions are Whitten's[45] unpublished study in 1963, which produced inconclusive results, and Barber and Raisman's[52] 1977 study (on estrous cycle shortening in female mice after exposure to male mice), which suggested that the VN system is not essential for male-induced estrus in mice. Both studies involved VN nerve section. (Subsequently, Barber[55] discovered that his and Raisman's conclusions had been based on the mistaken assumption that VN nerves do not regenerate after transection.) For other possible answers, see the 1980 review article dealing with the role of the VN system in mammalian reproductive physiology.[9]

Our hypothesis that the VNO is the only mediator of chemically induced changes in reproductive physiology was not limited to reflex ovulation. It included *all* changes in reproductive physiology. It was almost a year after our two discoveries were reported in *Nature* (March 1978)[5] that the first support for this hypothesis appeared. A paper by Reynolds and Keverne[56] (1979) reported VNO involvement in estrus suppression in grouped mice. This report was soon followed by others equally supportive (TABLE

TABLE 2. The Function of the Vomeronasal Organ in Mammals: Speculations (1961-1979)

Year	
1961	Flehman may have something to do with VNO function. Related to sexual activity.[44]
1963	Perhaps relevant receptor for the shortening of the estrous cycle in female mice after exposure to male mice.[45]
1964	Flehman may have something to do with the organ.[46]
1970	No connection between the VNO and Flehman.[47]
1971	Involved in perception of low-molecular-weight or highly volatile compounds.[48]
1972	"The close connection of the accessory olfactory system with the medial hypothalamus and preoptic area brings up the question of its possible involvement in endocrine control mechanisms."[49]
1972	"Considering how little is presently known on the whole subject, it can only be said that evidence is lacking to prove or disprove the involvement of the female vomeronasal organ in the perception of male urinary pheromones."[50]
1975	"Despite the fact that no one has succeeded in obtaining physiologic evidence concerning the functional specificities of the mammalian vomeronasal receptor, the proposition that it has an olfactory-like function has never been seriously questioned."[51]
1977	Induction of estrus in mice by male urine is not mediated by the VN system. Transection of the VN nerves failed to prevent estrous synchrony in grouped females.[52]
1978	"Evidence suggests the VNO has a particular role in the control of mating behavior in mammals."[53]
1979	"A periscope from the diencephalon, the VN system may monitor exogenous hormones, 'pheromones.'" Also, "... the function of Jacobson's organ remains unknown...."[54]

3).[56-71] By now there is considerable (and growing) evidence that the mammalian VNO does indeed mediate all physiologic responses to "primer" pheromones.

What is the likelihood that there are one or more functional human pheromones,[72] just as there seem to be functional prosimian primate pheromones?[73] More than 20 years ago Wilson[74] noted that human "primer pheromones might be difficult to detect since they can affect the endocrine system without producing overt specific behavioral

TABLE 3. Some Physiologic Functions of the Vomeronasal System in Mammals: Experimental Findings (1978-1984)

Year	
1978	Mediates male urine-induced reflex ovulation in anovulatory rats.[5]
1979	Mediates suppression of estrus in grouped mice.[56]
1979	Involved in male odor regulation of the estrous cycle in the female rat.[57]
1979	Involved in luteinizing hormone (LH) release and ovulation in rats.[58]
1979	Mediates male-induced precocious puberty in the female rat.[59]
1980	Involved in activation of female reproduction in the vole.[60]
1980, 1981, 1982	Mediates puberty acceleration in mice.[61,62,63]
1980, 1982	Mediates induction of delayed implantation in mice.[64,65]
1981	May be involved in activation of female reproduction (e.g., changes in concentrations of luteinizing hormone-releasing hormone (LHRH) in accessory olfactory bulb tissue) in prairie voles.[66]
1982	Involved in prolactin level changes in mice.[67]
1983	Mediates release of LH in the female rat following olfactory stimuli from male or female rats.[68]
1983	Mediates female-induced testosterone surges in mice.[69]
1984	May be involved in male-induced failure of implantation in newly mated female mice (the Bruce effect) but not the Whitten effect.[70]
1984	Involved in pheromonally induced release of LH in male mice.[71]

responses." He added, "certain observations suggest that the relation of odors to human physiology can bear further examination." The first studies examining this relationship have produced encouraging results.[75]

In 1971 Comfort[76] wrote: "Humans have a complete set of organs which are traditionally described as non-functional, but which, if seen in any other mammal, would be recognized as part of a pheromone system." He was thinking of certain apocrine sweat glands, among other organs, when he wrote this. He might well have been referring to the VNO. For example, Eggston and Wolff wrote in 1947 that in humans the VNO "does not function in the adult."[77] The terms "nonfunctional," "rudimentary," and "vestigial" are still commonly used in anatomy texts to describe the VNO in humans.[78] Wysocki,[54] in an important 1979 review article, made the following point: "Since the function of Jacobson's organ remains unknown, testing the functional state of the system in humans or other primates said to possess a 'rudimentary' or 'vestigial' VN apparatus is not yet possible." When Wysocki submitted his article for publication, experimental evidence for VNO function in physiology rested on one study only: ours.[5] Quite reasonably, he used the word "equivocal" to describe evidence that had not yet been replicated.

Observations cited in this paper[1-3,40,77,78] support the argument that the VNO occurs widely in mammals ranging from rodents to humans. In addition, functions have been found for this olfactory organ in behavioral (or social) control of reproductive physiology in rodents (see TABLE 3). It is time to look for related functions in non-rodent species, including humans.

ACKNOWLEDGMENTS

Professor Jay S. Rosenblatt, Director of the Institute of Animal Behavior of Rutgers University made it possible for me to pursue these studies. I am most grateful to him for (to quote my dissertation) "much stimulating discussion, for his active interest in this research, his generous help on numerous occasions, his patience and his wise suggestions."

I also wish to thank my two advisors, Barry R. Komisaruk and Harvey H. Feder, for their guidance and advice and Anne Mayer and Sarah S. Winans for their substantial contributions to this work.

REFERENCES

1. RUYSCH, F. 1703. Thesaurus Anatomicus III. Table IV, Figure V, p. 49. Joannem Wolters. Amsterdam, the Netherlands.
2. SOEMMERRING, S. T. V. 1809. Abbildungen der menschilichen Organe des Geruches. Table 111, Figures 1-9. Varrentrapp, Wenner. Frankfurt, Germany.
3. JACOBSON, L. L. 1811. Description anatomique d'un organe de secretion dans le nez inconnu jusqu'ici. *In* Ouvrages Sur l'Organe Vomero-nasal (Part 1); avec preface et notes par O. C. Hollnagel-Jensen et E. Andersen. Munksgaard. Copenhagen, Denmark. (Translated from Danish, 1948.)
4. POWERS, J. B. & S. S. WINANS. 1975. Vomeronasal organ: critical role in mediating sexual behavior of the male hamster. Science **187**: 961-963.

5. JOHNS, M. A., H. H. FEDER, B. R. KOMISARUK & A. D. MAYER. 1978. Urine-induced reflex ovulation in anovulatory rats may be a vomeronasal effect. Nature **272**: 446-448.
6. JOHNS, M. A., H. H. FEDER & B. R. KOMISARUK. Male urine provokes ovulation in light-induced constant estrus rats. Abstract presented at the Eastern Conference on Reproductive Behavior, Storrs, Conn., 1977.
7. JOHNS, M. A. 1979. Reflex Ovulation in Light-Induced Persistent-Estrus Rats: Roles of the Vomeronasal System, the Adrenal and Ovarian Steroids. University Microfilms International. Ann Arbor, Mich.
8. BROMAN, I. 1920. Das organon vomero-nasale Jacobsonein Wassergeruchsorgen! Anat. Hefte **58**: 143-188.
9. JOHNS, M. A. 1980. The role of the vomeronasal system in mammalian reproductive physiology. *In* Chemical Signals. D. W. Muller-Schwarze & R. M. Silverstein, Eds.: 341-364. Plenum. New York, N.Y.
10. BROWN-GRANT, K., J. M. DAVIDSON & F. GREIG. 1973. Induced ovulation in albino rats exposed to constant light. J. Endocrinol. **57**: 7-22.
11. BRUCE, H. M. 1970. Pheromones. Br. Med. Bull. **26**(1): 10-13.
12. JOHNS, M. A. & B. R. KOMISARUK. The role of sensory stimuli and the adrenals in mediating ovulation and receptivity in persistent-estrus rats. Abstract presented at the Eastern Conference on Reproductive Behavior, Nags Head, N.C., 1975.
13. JOHNS, M. A., H. H. FEDER & B. R. KOMISARUK. 1980. Reflex ovulation in light-induced persistent estrus (LLPE) rats: role of sensory stimuli and the adrenals. Horm. Behav. **14**: 7-19.
14. RICHMOND, M. E. & C. H. CONAWAY. 1969. Induced ovulation and oestrus in *Microtus ochrogaster*. J. Reprod. Fertil. Suppl. **6**: 357-376.
15. HOOVER, J. E. & L. C. DRICKAMER. 1979. Effects of urine from pregnant and lactating female house mice on oestrous cycles of adult females. J. Reprod. Fertil. **55**: 297-301.
16. CHAMPLIN, A. K. 1971. Suppression of oestrus in grouped mice: the effects of various densities and the possible nature of the stimulus. J. Reprod. Fertil. **27**: 233-241.
17. MARSDEN, H. M. & F. H. BRONSON. 1964. Estrus synchrony in mice: alteration by exposure to male urine. Science **144**: 1469.
18. CHATEAU, D., J. ROOS, S. PIAS-ROSER, M. ROOS & C. ARON. 1976. Hormonal mechanisms involved in the control of oestrous cycle duration by the odor of urine in the rat. Acta Endocrinol. **82**: 426-435.
19. BRUCE, H. M. 1965. The effect of castration on the reproductive pheromones of male mice. J. Reprod. Fertil. **10**: 141-143.
20. DOMINIC, C. J. 1965. The origin of the pheromones causing pregnancy block in mice. J. Reprod. Fertil. **10**: 469-472.
21. COWLEY, J. J. & D. R. WISE. 1972. Some effects of mouse urine on neonatal growth and reproduction. Anim. Behav. **20**: 499-506.
22. DRICKAMER, L. C. 1974. Contact stimulation, androgenized females and the accelerated maturation in female mice. Behav. Biol. **12**: 101-110.
23. VANDENBERGH, J. R. 1969. Male odor accelerates female sexual maturation in mice. Endocrinology **84**: 658-660.
24. LOMBARDI, J. R. & J. G. VANDENBERGH. 1977. Pheromonally induced sexual maturation in females: regulation by the social environment of the male. Science **196**: 545-546.
25. MONDER, H., C. LEE, P. J. DOVONICK & R. G. BURRIGHT. 1978. Male mouse urine extract effects on pheromonally mediated reproductive functions of female mice. Physiol. Behav. **20**: 447-452.
26. HOOVER, J. E. & L. C. DRICKAMER. 1979. Effects of urine from pregnant and lactating female house mice on oestrous cycles of adult females. J. Reprod. Fertil. **55**: 297-301.
27. WHITTEN, W. K., F. H. BRONSON & J. A. GREENSTEIN. 1968. Estrus-inducing pheromone of male mice: transport by movement of air. Science **161**: 584-585.
28. KRANZ, L. K. & P. G. BERGER. 1975. Pheromone maintenance of pregnancy in *Microtus montanus*. *In* Proceedings, 55th Annual Meeting of the American Society of Mammologists. P. 66.
29. MCCLINTOCK, M. K. 1978. Estrous synchrony and its mediation by airborne chemical communication (*Rattus norvegicus*). Horm. Behav. **10**(3): 264-276.

30. McClintock, M. K. 1981. Social control of the ovarian cycle and the function of estrous synchrony. Am. Zool. **21**: 243-256.
31. Vandenbergh, J. G., J. M. Whitsett & J. R. Lombardi. 1975. Partial isolation of a pheromone accelerating puberty in female mice. J. Reprod. Fertil. **43**: 515-523.
32. Raisman, G. & P. M. Field. 1971. Sexual dimorphism in the preoptic area of the rat. Science **173**: 731-733.
33. Gratiolet, L. P. 1845. Recherches sur l'orgone de Jacobson. These (No. 164) Faculte de Medecine de Paris, Paris, France. (Cited in Reference 50, p. 316.)
34. Balogh, C. 1860. Ueber das Jacobson'sche Organ des Schafes. Untersuch. Z. Natur. d. Mensch. u.d. Thiere. Giessen. F: 595-597.
35. Kollicker, A. 1877. Ueber das Jacobson'sche Organ des Menschen. Gratulationschrift d. Wurzburger mediz. Fakultat (Festschrift) fur Rinecker, Leipzig **4**: 1-12.
36. Rose, C. 1893. Ueber das Jacobsonsche Organe von Wombat u. Oppossum. Anat. Anz. **8**(16): 766-768.
37. Mihalkovics, V. V. 1898. Nasenhohle und Jacobson'sches Organ. Anat. Hefte, Wiesb. 1 Abt. **11**: 1-107.
38. Ramon Y. Cajal, S. 1899. Textura del sistema nervioso del hombre y de los vertebrados, Vol. 2. Madrid, Spain.
39. Frets, G. P. 1912-1913. On the Jacobson organ of primates. K. Akad. v. Wetenschappen, Amsterdam **15**: 134-137.
40. Pearlman, S. J. 1934. Jacobson's organ (Organon vomero-nasale, Jacobsoni): its anatomy, gross, microscopic and comparative, with some observations as well on its function. Ann. Otol. Rhinol. Laryngol. **43**: 739-768.
41. Planel, H. 1954. Etudes sur la physiologie de l'organe de Jacobson. Extrait des Archives d'Anatomie, d'Histologie et d'Embryologie Tome XXXVI. Fasc. 4/8: 199-205. Alsatia, Colmar, France.
42. Adrian, E. D. 1955. Synchronized activity in the vomeronasal nerves with a note on the function of the organ of Jacobson. Pfluegers Arch. Gesamte Physiol. **260**: 188-192.
43. Negus, V. E. 1956. The organ of Jacobson. J. Anat. **90**: 515-519.
44. Mann, G. 1961. Bulbus olfactorius accessorius in Chiroptera. J. Comp. Neurol. **116**: 135-144.
45. Whitten, W. K. Is the vomeronasal organ a sex receptor in mice? *In* Proceedings, Second Asia and Oceanic Congress of Endocrinology, Sidney, Australia, 1963.
46. Knappe, H. 1964. Zur funktion des Jacobsoncschen organs. Zool. Gart. (N.F.) **28**: 188-194.
47. Dagg, A. L. & A. Taub. 1970. Flehmen. Mammalia **34**(4): 686-695.
48. Muller, W. 1971. Vergleichende elektrophysiologische Untersuchungen an den Sinnesepithelien des Jacobsonschen Organs und der Nase von Amphibien (*Rana*), Reptilian (*Lacerta*) and Saugetieren (*Mus.*). Z. Vgl. Physiol. **72**: 370-385.
49. Raisman, G. 1972. An experimental study of the projection of the amygdala to the accessory olfactory bulb and its relationship to the concept of a dual olfactory system. Brain Res. **14**: 395-408.
50. Estes, R. D. 1972. The role of the vomeronasal organ in mammalian reproduction. Extrait de Mammalia **36**: 315-341.
51. Scalia, F. & S. S. Winans. 1975. The differential projections of the olfactory bulb and accessory olfactory bulb in mammals. J. Comp. Neurol. **161**: 31-56.
52. Barber, P. C. & G. Raisman. 1977. Experimental investigation of the possible involvement of the accessory olfactory system to primer pheromones in female mice. Proc. Int. Un. of Physiol. Sci. **13**: 51.
53. Moulton, D. C. 1978. Olfaction. *In* Handbook of Behavioral Neurobiology. R. B. Masterton, Ed. **1**: 91. Plenum. New York, N.Y.
54. Wysocki, C. J. 1979. Neurobehavioral evidence for the involvement of the vomeronasal system in mammalian reproduction. Neurosci. Biobehav. Rev. **3**: 301-341.
55. Barber, P. C. 1981. Axonal growth by newly-formed vomeronasal neurosensory cells in the normal adult mouse. Brain Res. **216**(2): 229-238.
56. Reynolds, J. & E. B. Keverne. 1979. The accessory olfactory system and its role in the pheromonally mediated suppression of oestrus in grouped mice. J. Reprod. Fertil. **57**: 31-35.

57. SANCHEZ-CRIADO, J. E. 1979. Blockade of the pheromonal effects in rats by central deafferentation of the accessory olfactory system. Rev. Esp. Fisiol. **35**(2): 137-141.
58. BELTRAMINO, C. & S. TALEISNIK. 1979. Effect of electrochemical stimulation in the olfactory bulbs on the release of gonadotropin hormones in rats. Neuroendocrinology **28**(5): 320-328.
59. SANCHEZ-CRIADO, J. E. & A. GALKEGO. 1979. Male induced precocious puberty in the female rat. Role of the vomeronasal system. Acta Endocrinol. Suppl. **225**: 255. (Abstract.)
60. CARTER, C. S., L. L. GETZ, L. GAVISH, J. L. MCDERMOTT & P. ARNOLD. 1980. Male-related pheromones and the activation of female reproduction in the prairie vole (Microtus ochrogaster). Biol. Reprod. **23**: 1038-1045.
61. KANEKO, N., E. A. DEBSKI, M. C. WILSON & W. K. WHITTEN. 1980. Puberty acceleration in mice. 11. Evidence that the vomeronasal organ is a receptor for the primer pheromone in male mouse urine. Biol. Reprod. **22**: 873-878.
62. DRICKAMER, L. C. & S. M. ASSMANN. 1981. Acceleration and delay of puberty in female housemice (Mus musculus): methods of delivery of the urinary stimulus. Dev. Psychobiol. **14**(5): 487-497.
63. LOMAS, D. E. & E. B. KEVERNE. 1982. Role of the vomeronasal organ and prolactin in the acceleration of puberty in female mice. J. Reprod. Fertil. **66**(1): 101-107.
64. BELLRINGER, J. F., H. P. M. PRATT & E. B. KEVERNE. 1980. Involvement of the vomeronasal organ and prolactin in pheromonal induction of delayed implantation in mice. J. Reprod. Fertil. **59**: 223-228.
65. LLOYD-THOMAS, A. & E. B. KEVERNE. 1982. Role of the brain and accessory olfactory system in the block to pregnancy in mice. Neuroscience **7**(4): 907-913.
66. DLUZEN, D. E., V. D. RAMIREZ, C. S. CARTER & L. L. GETZ. 1981. Male urine changes luteinizing hormone-releasing hormone and norepinephrine in female olfactory bulb. Science **212**: 573-575.
67. KEVERNE, E. B. 1983. The accessory olfactory system and its role in pheromonally mediated changes in prolactin. *In* Olfaction and Endocrine Regulation. W. Briepohl, Ed.: 127-140. IRL Press Ltd. London, England.
68. BELTRAMINO, C. & S. TALEISNIK. 1983. Release of LH in the female rat by olfactory stimuli: effect of the removal of the vomeronasal organs or lesioning of the accessory olfactory bulbs. Neuroendocrinology **36**(1): 53-58.
69. WYSOCKI, C. J., Y. KATZ & R. BERNHARD. 1983. Male vomeronasal organ mediated female-induced testosterone surges in mice. Biol. Reprod. **28**(4): 917-922.
70. GANGRADE, B. K. & C. J. DOMINIC. 1984. Studies of the male-originating pheromone involved in the Whitten effect and Bruce effect in mice. Biol. Reprod. **31**: 89-96.
71. COQUELIN, A., A. N. CLANCY, F. MACRIDES, E. P. NOBLE & R. A. GORSKI. 1984. Pheromonally induced release of luteinizing hormone in male mice. Involvement of the vomeronasal system. J. Neurosci. **4**(9): 2230-2236.
72. CUTLER, W. B., G. PRETI, G. R. HUGGINS, B. ERICKSON & C. R. GARCIA. 1985. Sexual behavior frequency and biphasic ovulatory type menstrual cycles. Physiol. Behav. **34**: 805-810.
73. SCHILLING, A., M. PERRET & J. PREDINE. 1984. Sexual inhibition in a prosimian primate: a pheromone-like effect. J. Endocrinol. **102**: 143-151.
74. WILSON, E. O. 1963. Pheromones. Sci. Am. **208**: 100-114.
75. JARETT, L. R. 1984. Psychosocial and biological influences on menstruation: synchrony, cycle length, and regularity. Psychoneuroendocrinology **9**(1): 21-25.
76. COMFORT, A. 1971. Likelihood of human pheromones. Nature **230**: 432-433, 479.
77. EGGSTON, B. S. & D. WOLFF. 1947. Histopathology of the Ear, Nose and Throat. P. 554. The Williams & Wilkins Co. Baltimore, Md.
78. GRAY, H. Gray's Anatomy of the Human Body. 36th British edit., 1980. P. L. Williams & R. Warwick, Eds. Pp. 149, 994, 1142. Churchill Livingstone, New York, N.Y.

Experiential Influences on Hormonally Dependent Ring Dove Parental Care

GEORGE F. MICHEL[a]

Developmental Psychobiology Unit
Psychiatric Research
Children's Hospital Medical Center
Boston, Massachusetts 02115

Differences in breeding success between reproductively naive and experienced animals are reported frequently and have been used as suggestive evidence that experience plays a role in the development of the mature form of species-typical patterns of reproductive behavior.[1] However, because naive animals are often younger than experienced animals, it is possible that differences between them may result from some one or more factors correlated with age rather than experience itself. In field studies relating reproductive success to breeding experience, one such factor might be ability to hold a suitable nesting site against competition from others.[2] Even if this and similar factors are eliminated in a controlled laboratory setting, it is possible that age-related maturational differences in neural and endocrine functioning could account for differential breeding success. To control for some age-related differences, Lehrman and Wortis[3] compared age-matched naive and experienced ring doves in a laboratory setting, and found that experienced doves bred more efficiently. Previous breeding experience also affects the way age-matched ring doves behave in response to injection of hormones.[4-8]

These findings suggest that breeding experience, independently of age-related factors, can contribute to the organization of a reproductive cycle and to its success as measured by efficient production of viable offspring. However, participation in a breeding cycle exposes the animal to a variety of different events, each potentially providing the conditions by which experience has its impact. Therefore, there is no reason to assume that previous breeding experience exerts its effects through a single mechanism. Because so much is known about the behavioral, neural, and endocrine characteristics of its breeding cycle, the ring dove is a suitable model for experimental identification of the mechanisms by which experience operates.

Although social experience before sexual maturity can affect reproduction in ring doves,[9,10] the study of experience during this age period in doves has attracted little attention; thus, the present paper will review the roles of previous breeding experience in the expression of behaviors that are characteristic of the hormonally distinct phases of the ring dove breeding cycle. Both earlier experiences within a cycle and experiences

---

[a] Correspondence address: Developmental Psychobiology Unit, Psychiatric Research, Fifth Floor-Gardner House, Children's Hospital Medical Center, 300 Longwood Avenue, Boston, Mass. 02115.

that are carried over between cycles will be considered. Contributions of experience to each phase of the cycle will be described separately, and hypotheses about the processes by which experience operates will be considered.

HORMONES, EXPERIENCE, AND THE RING DOVE BREEDING CYCLE

In ring doves, as in many other species, the expression of specific behaviors during the progression of a breeding cycle is associated with the secretion of specific hormones. Of course, hormones do not simply cause the expression of associated behaviors. Rather, hormones augment or potentiate[11] behavioral expression. In doing so, a given hormone operates in conjunction with other internal and external factors, usually including very specific social and environmental stimulation[12] and the previous and concurrent presence of certain other hormones.[13] Thus, behavioral expression involves a complex causal network in which hormonal condition plays a role.[14] Furthermore, although the expression of behavior may be augmented by hormones in some circumstances, in other circumstances the same behavior may be expressed independently of the dove's hormonal condition.[15,16] Therefore, a question may be raised about the function of hormonal augmentation of behavioral expression in the progress of a reproductive cycle.

Part of the answer to this question resides in the fact that reproduction depends on a temporal organization of physiological processes that underlie gamete formation, fertilization, and subsequent development of progeny. In birds, these processes and their temporal organization are controlled in large part by the pattern of hormone secretion in parents. Successful reproduction in birds also involves the behavioral participation of one or both parents. This participation occurs in a temporal pattern in which specific behaviors predominate at different times in ways that are appropriate for the physiological progression of reproduction. For example, copulation occurs predominantly during the period when eggs have developed to a state appropriate for fertilization. Thus, the breeding cycle of birds can be thought of as three rather tightly coupled cycles: the physiological progression of reproduction from gamete to fledgling, the hormonal cycle of parents, and the behavioral cycle of parents. Since the same hormones that regulate the physiological progression of reproduction also augment the expression of specific behaviors, hormones can coordinate the parent's behavior to appropriate phases of the physiological cycle.

Successful reproduction in many avian species, including ring doves, requires the behavioral participation of both parents. Parents in this category must be able to synchronize their behavioral expression not only with the physiological events of reproduction, but also with one another so that the behavior of each either complements or supplements that of the other. Antagonistic behavior or phase differences in expression could disrupt the progression of the cycle and reduce breeding efficiency.[17] Because hormones secreted by parents help to regulate the temporal organization of both the physiological and behavioral aspects of a breeding cycle, behavioral synchrony between mates can be assured by having the secretion of at least some hormones sensitive to stimuli provided by the behavior of the mate.[16] Stimuli provided by a mate sometimes even affect the hormonal status of an individual indirectly, through feedback from the behavior performed by the individual in response to mate-produced stimuli.[18] Thus, calls and displays, or presentation of food or nesting material, or construction of a nest by the mate may provide the stimuli that induce changes in hormonal status.

Experience is likely to be a relevant factor in conditions where the secretion of a hormone is influenced by external stimuli.

A ring dove in its first reproductive cycle must establish coordination of behavioral changes with the hormonally governed progress of its reproductive physiology and, because both males and females engage in parental care, each dove must also establish hormonal and behavioral synchrony with its mate. Because hormones help to synchronize the behavior of mates and coordinate behavior to the physiological progress of reproduction, breeding experience can operate in two different but related contexts: experience provided by one cycle might make subsequent cycles more likely to succeed (intercyclic experience) and experience provided by earlier phases of each cycle (intracyclic experience), including the first, might contribute to successful progression of that cycle.[3,19] Thus, a first cycle may differ from subsequent cycles in a number of respects, including the degree of dependence on intracyclic hormonal or stimulative events and the degree of synchrony of the mates. Furthermore, in view of the different physiological contributions of males and females to reproduction, breeding experience may affect the sexes differently.

If the difference in breeding efficiency of reproductively naive and experienced ring doves resides in the degree of behavioral synchrony established between the mates, then this synchrony may be affected, in part, by the relation that exists between the presence of a hormone and the expression of specific behaviors. For example, breeding experience could make the expression of behavior in subsequent cycles less dependent on the animal's hormonal condition and more sensitive to the social and environmental circumstances in which the behavior was expressed during the initial cycle. This could produce greater behavioral synchrony between mates because advances in the endocrine and behavioral status of one could be matched by rapid changes in the behavior of the other. Since the female's physiological condition determines the progress of reproduction through egg-laying, it might be expected that previous breeding experience would make the male's rather than the female's behavior more independent of the hormonal conditions characteristic of the pre-laying phases of a cycle.[20] Thus, experienced males should be more sensitive than experienced females to the social and environmental stimuli associated with the pre-laying phases. With this increased sensitivity, the experienced male's behavior can be adjusted to the peculiarities of his mate's progression to egg-laying which would account for the experienced male's ability to increase the breeding success of his naive mate.[3]

Breeding experience may also consolidate the relation between behavior and hormonal condition. Behavior that is closely tuned to the animal's hormonal condition can serve as a more valid indicator of the physiological progress of the cycle. It might be expected that during the pre-laying phase experience with a previous breeding cycle would make a female's behavior more dependent on her hormonal condition. The male's sensitivity to her behavior would allow her to act as a clock that could correct any drift in phase between his behavior and her physiological condition. To ensure greater synchrony, it might be expected that the female's hormonal condition during this period will be very sensitive to the displays, calls, etc., directed specifically to her by the male[21] and that experience will further tune this sensitivity.

In monogamous species, like ring doves, continued breeding experience with the same mate may increase dependence on the specific characteristics of the mate for the maintenance of synchrony, such that if they are paired with unfamiliar mates they may breed less efficiently than naive pairs.[2,22,23] Thus, mate familiarity is one important component of breeding experience that might play quite different roles in monogamous and polygamous species.

The differences in breeding efficiency between naive and experienced doves are significant but modest in contrast to the greater differences obtained in experiments on the behavioral effects of injected hormones. Because the experience of the dove early in its first cycle could provide the background required to orient it appropriately to the objects (nest, eggs, young) that appear later in the same cycle, doves that are put, by injection, into a hormonal state typical of one phase of the cycle, are missing many of the psychobiological conditions appropriate for that phase.[19] The use of this technique can make any effects of previous breeding experience on behavioral performance more evident. This means that experiments using hormonal induction of behavior will provide a more sensitive measure of the contributions of previous breeding experience to the organization of the reproductive cycle than will comparison of first and subsequent cycle performance.

EXPERIENCE AND COURTSHIP

The behavioral characteristics of the courtship phase of a breeding cycle include male bow-cooing, female receptive crouching, and male and female kah-calling, courtship preening, and nest-site solicitation (wing-flipping and nest-cooing). In general, the level of expression of these behaviors depends on the presence of testosterone or estrogen.[13,24,25] The level of behavioral expression by the mates affects the speed of their transition to the nest-construction and egg-laying phases of the cycle through the effect the behavior has, either directly or indirectly, on the female's hormonal secretion.[18]

A bow-cooing male elicits withdrawal and/or preening from a female. To the extent that bow-cooing delays or reduces male nest-site solicitation, it will delay egg-laying and incubation even in experienced and receptive females.[17] Reproductively experienced males begin bow-cooing more quickly than naive males when encountering a companion (personal observation). In addition, experienced males are quicker than naive males to engage in high levels of nest-site solicitation.[4] This attracts the female and elicits the mutual nest-site solicitation that stimulates the egg production process and facilitates the transition to nest construction and egg-laying.[18]

Because the female's hormones control the order and timing of egg production, breeding experience ought to tighten the association between her hormonal state and the specific behaviors expressed. Such tightened association should reduce the chances of inappropriately timed behavioral expression, making her behavior more tuned to the progress of the cycle. To the degree that a female's behavior influences that of her mate, the male's behavior should be more synchronized to the cycle of an experienced female. Experience ought to make the female's endocrine secretions much more tuned to the stimuli that are directed specifically to her by the mate. In this way, the physiological progression that leads to laying eggs can be better adjusted to the specific characteristics of the male.

In the male, breeding experience ought to allow greater dissociation between hormone condition and behavioral expression. This would allow the male to rapidly adjust his behavior to the specific characteristics of the female. Experienced males might also be better than naive males in stimulating hormonal changes in the female. Thus, experienced males ought to be better at stimulating unreceptive females into a receptive state and they ought to adjust their behavior more rapidly to that of receptive females.

Although it has been learned that experienced males can increase the reproductive success of naive females and experienced females lay more quickly than naive females,[3] detailed study of the effects of experience on the courtship phase of the cycle has been lacking. A recent study[4] found that experienced males engage in more courtship behavior (nest-site solicitation) than naive males. Although both experienced and naive males seem to discriminate the receptive condition of the female, experienced males not only engage in more courtship but seem to adjust their courtship more appropriately to the receptivity of the mate. Unfortunately, there has been no study of experiential influences on female courtship.

By controlling the levels of circulating testosterone through castration and injection, Cheng et al.[4] showed that experienced males perform more courtship behavior than naive males, even when levels of circulating testosterone are the same. Since courtship behavior can be elicited by estrogen and estrogen injections remove the behavioral differences between experienced and naive males, Cheng et al. proposed that breeding experience somehow facilitates the aromatization of testosterone into estrogen by the neural cells involved in courtship behavior (cf. Reference 26). Similar experiential influences on aromatization have been reported for rat sexual behavior.[27,28] Thus, to understand the neurobiochemistry of reproduction, the effects of breeding experience must be studied.

EXPERIENCE AND THE HORMONAL INITIATION OF INCUBATION

If experienced doves receive a 7-day series of injections of progesterone while visually isolated from other doves, they exhibit incubation within minutes of being paired and placed in a breeding cage. No other hormone produces such a rapid onset of incubation behavior,[29,30] but progesterone induces a rapid onset of incubation in females only if they have certain levels of estrogen.[16] Similarly, progesterone-induced incubation in males occurs only if they have appropriate levels of testosterone.[31,32]

During a cycle, progesterone levels sharply increase in females during the nest-construction phase, immediately preceding egg-laying, and lead to the characteristic expression of incubation that occurs before the appearance of the eggs.[16,33] Thus, in females, increased progesterone secretion is closely tied to the normal onset of incubation. No clear pattern of change in progesterone secretion has been detected in males. Therefore, it has been argued that the initiation of incubation in males occurs in response to the incubation behavior of the mate.[20,34]

Male doves deprived of endogenous progesterone by dexamethasone (DEX), an ACTH blocking agent, continued to incubate throughout the 7 days of treatment[34] and castrated males, given testosterone propionate before (but not after) pairing with an intact female, began to incubate in synchrony with the female.[20] Unfortunately, the DEX treatment was begun only after the first egg was laid (some 3 days after the initiation of incubation) and both studies used reproductively experienced males. Therefore, Cheng[15,16] has argued that although progesterone levels do not change, it is still likely the hormone responsible for the onset of incubation in males.

Cheng proposed that the onset of male incubation occurs either as a result of changes in the ratio of progesterone to other hormones that are rising and/or falling during this period, or as a result of a change in the male's sensitivity to progesterone as a result of the changing androgen levels during courtship.[15,16] Indeed, the rise in testosterone secretion that occurs during the first 3 days of courtship[35] can increase the number of progestin receptors in areas of the hypothalamus specifically involved in progesterone-induced incubation.[36] Thus, the neural mechanisms involved in progesterone-induced incubation[31] are more sensitive to progesterone as a result of the testosterone secreted during courtship. It has also been shown that reproductively experienced males readily establish incubation after progesterone injections even if their naive mates do not.[37] At least under these conditions, the incubation of males is not dependent upon the incubation of the mate but rather on their current hormonal state and their previous experience.

Although Lehrman and Wortis[6] reported that the rapid expression of incubation in response to progesterone injections was not present in reproductively naive doves, it was found subsequently that the previous breeding experience need only include experience with the pre-laying, nest-construction phase of the cycle.[7] Experience with phases of the cycle subsequent to nest construction, including egg-laying and incubation, was not necessary for the rapid expression of incubation after progesterone injections. However, previous experience with courtship and nest-site soliciting behavior was not sufficient to make the dove behave as an experienced bird when injected with progesterone. These latter results are especially interesting because the endogenous rise in circulating testosterone or estrogen levels that makes the dove's nervous system more sensitive to progesterone likely occurred in these courtship-experienced doves. Nevertheless, they showed no obvious increased sensitivity to progesterone.

Two explanations may be proposed for why previous experience with the nest-construction phase of a breeding cycle is sufficient to produce rapid progesterone-induced incubation. Either the experience of nesting behavior prepares the dove's nervous system to respond to progesterone with rapid expression of incubation, or the progesterone secreted during the nest-construction phase prepares or primes the nervous system so that it is more sensitive to progesterone administered several weeks later. Recent data suggest that these two explanations are not mutually exclusive.[38] Previous nesting experience is the predominant factor leading to the rapid expression of incubation after progesterone injections. However, progesterone priming, particularly if combined with courtship experience, will facilitate the expression of nest-site-related behaviors during testing. This facilitation could, in a sufficiently large sample, reveal an advantage of progesterone-primed over nonprimed naive doves in latency to establish incubation. Although there were no sex differences in the establishment of incubation, time spent nest-soliciting was affected more by progesterone priming in females than in males.

Of particular interest in the results of this study was the discovery that previous experience with nest-building activity was not a necessary factor for progesterone-induced incubation. Rather, nesting experience (settling in a nest, nest-site soliciting in a nest, and/or forming an attachment to a nesting place) was sufficient to prepare the dove's nervous system to be responsive to subsequent progesterone. Interestingly, such nesting experience, rather than the actions of nest-building, is also sufficient self-stimulation for the secretion of gonadotropic hormones necessary for egg-laying in experienced female doves.[39]

Thus, participation in the courtship and nest-construction phases of an initial breeding cycle exposes doves to major changes in the amounts and relative ratios of circulating gonadal, pituitary, and hypothalamic hormones as well as changes in patterns of neurotransmitter secretion in diverse areas of the nervous system. Each of these changes reflects either the dove's actions or the consequences of its exposure to social and environmental stimuli that are associated with these phases of a cycle. All of these changes can have "priming" effects and may require coaction with progesterone to facilitate the transition to incubation. Thus, the dove's experience during the courtship and nest-construction phases of the breeding cycle is not separable from changes in its physiology at that time. Perhaps this is why Lehrman[19] considered intercycle experience to be similar to intracycle experience. That is, the experientially related changes in the dove's physiology that occur during the courtship and nest-construction phases can promote the conditions favoring behavioral changes in subsequent phases of the same cycle (cf. Reference 39) as well as alter breeding efficiency in subsequent cycles.

EXPERIENCE AND THE MAINTENANCE OF INCUBATION

Lehrman and Brody[40] showed that prolactin secretion was important not only for preparing the dove's crop for feeding the young squabs but also for maintaining incubation behavior during the period before hatching. To some extent this secretion of prolactin is stimulated by contact with the eggs during incubation.[40,41] Thus, incubation was thought to be initiated by progesterone secretion and maintained by prolactin secretion as a result of contact with the eggs. However, if reproductively experienced males had established incubation before being separated from their incubating mates by a glass plate, then they would continue to secrete prolactin, despite their lack of contact with eggs, presumably in response to seeing their mates incubate.[42] Thus, stimuli other than contact with eggs can stimulate prolactin secretion (cf. References 41, 43, and 44).

The ability to secrete prolactin during observation of an incubating mate is also affected by previous breeding experience. Michel and Moore[45] paired either reproductively naive or experienced males with experienced females and allowed them to progress through various phases of a breeding cycle before the male was separated by a glass plate from his incubating mate. Naive males required 8 days of incubation before observation of the incubating mate was sufficient to stimulate prolactin secretion (assessed by crop sac weight), whereas experienced males required only 3 days.

These results indicate that the stimuli associated with those directly involved in stimulating prolactin secretion can come, during the course of the first breeding cycle, to also stimulate prolactin secretion. Therefore, during a second cycle, these associated stimuli provide assurance that the male's physiological condition will be appropriate for brooding and feeding the squab despite disruptions of his incubation behavior. The role of preparatory sequences or coaction of other hormones in the operation of this experiential effect remains to be examined.

THE TRANSITION FROM INCUBATION TO BROODING

Although the behavior of the parents during the first 3-4 days after hatching is little different from that during incubation, Moore[46] found that brooding young is a different motivational state from incubation. When given a choice of sitting on a nest with a 2-day-old squab or one with eggs, birds that are incubating in their own cycle are more likely to choose the egg nest, whereas birds that have squabs of their own are more likely to choose the squab nest. This difference between incubating and brooding doves is more marked in reproductively naive than experienced doves. Moore[8] proposed that first-cycle doves rely more on their internal state and second-cycle doves rely more on stimuli with which they are familiar.

Either previous experience with sitting on a young squab or injection of prolactin is capable of making an incubating naive dove much more likely to choose to sit on a squab nest.[8] It is likely that the experience of sitting on the squab affects the naive dove's choice of nest to sit on by stimulating the endogenous secretion of prolactin. Sitting on a squab does result in prolactin secretion even early in incubation.[43,44,47] Thus, increasing levels of prolactin, either by injection or through exposure to a squab, are responsible for the change in motivational state from incubation to brooding during the naive dove's first breeding cycle. For experienced doves, the increasing levels of prolactin secreted during incubation seem to be sufficient to encourage this change in state. Hence, experienced doves should be able to make the transition from incubation to brooding more efficiently than naive doves.[46]

EXPERIENCE AND CARE OF SQUABS

Early in his career, Lehrman[5] found that reproductively experienced, but not naive, doves will feed 7-day-old squabs when injected with prolactin. However, when 2-day-old squabs are placed in the nests of incubating naive doves they are fed within a day even if it is at a point during incubation when the adults have little, if any, circulating prolactin.[47,48] Indeed, there is some evidence of overresponsiveness to the squab in naive as compared to experienced doves.[44,49] The naive dove's behavioral response to the squab may result, in part, from the ability of stimuli from squabs to rapidly induce prolactin secretion.[41,50]

In a preliminary study, Lott and Comerford[51] found that 50% of naive doves would feed young squabs if injected with a combination of progesterone and prolactin for a week prior to testing. Neither prolactin alone nor the control vehicle resulted in squab feeding during the 50-hr test period. However, 10% of the progesterone-injected doves did feed the squab, perhaps because the squab stimulated endogenous prolactin secretion. Lott and Comerford concluded that progesterone can prime the dove's nervous system (making it sit on the nest) so that increasing levels of prolactin (either exogenously supplied or endogenously generated by exposure to squabs or by prolonged incubation) make the dove more responsive to squabs. Once a dove has experienced this hormone-stimulus regime during a normal breeding cycle, exposure to just one of the hormones (prolactin) and some of the stimuli (from squabs) is sufficient to induce feeding the squabs. Neither of these hormone or stimulus conditions alone will be sufficient in reproductively naive doves. This study demonstrates again that the dove's breeding experience provides it with the appropriate combination of changes in hormone events and exposure to stimuli that makes experienced doves more likely to respond appropriately when tested for hormonal induction of behavior.

CONCLUSIONS

During the first breeding cycle, the reproductively naive dove's nervous system is exposed to changing levels, ratios, and temporal patterns of hormonal secretions. It is exposed also to temporal patterns of social and environmental stimulation. This dual exposure normally co-occurs in a dynamic composite which includes the dove's nervous system as a major participant through its control of behavior and hormone secretion. Identifying the components of this composite and unraveling their relative contributions to the dynamics of the breeding cycle have occupied the efforts of many researchers during the last three decades. In general, the results of these efforts show:

1. The secretion of many of the gonadal, pituitary, and hypothalamic hormones that control the progression of a cycle is determined, in part, by the social and environmental stimuli that characterize a breeding cycle.[16,52]

2. These stimuli can affect hormonal secretion either directly, through neuroanatomical connections between sensory structures of the brain and hypothalamic structures that control pituitary hormone secretion, or indirectly, by altering the observing dove's own behavior which, through special feedback mechanisms, will stimulate its own hormone secretion.[18]

3. Because particular hormones have an affinity for specific cells, circuits, and components of the nervous system, they can selectively affect neural functioning. Thus, changes in circulating levels of a hormone can alter the relative frequency of expression of specific behaviors, increasing some, decreasing others, and leaving some unaffected, and/or alter the dove's sensitivity to specific social[8,43,46,47] and environmental[53] stimuli.

4. Hormones alter synaptic density, the number of hormone-specific receptors on neural cells, the number of neural cells with receptors specific to a particular hormone, and even the cellular metabolism of hormones.[54] Similar mechanisms may operate in ring doves and account for how changes in a hormone's level can affect the dove's sensitivity to other hormones that will appear later in the same cycle or to the same hormone when it occurs again in a subsequent cycle.

All of these processes form the dynamics of the breeding cycle and create the conditions that Lehrman called intracyclic experience. That is, the dove's behavioral participation in antecedent phases of the cycle, or its exposure to the social and environmental stimuli characteristic of those phases, effects changes in its hormonal and neural condition that prepare its behavior, its perceptual sensitivity to stimuli, and its physiological condition so that subsequent phases of the cycle can occur.

For the most part, the consequences of breeding experience represent the retention of these intracyclic experiences across cycles. The effects of breeding experience do not represent the operation of fortuitous or unique mechanisms. Rather intercyclic experience operates through the same complex of mechanisms that governs the organization of a breeding cycle. Continued investigation of intercyclic experiential effects can significantly extend our understanding of the processes by which the ring dove breeding cycle is organized.

The research presented in this review has shown that intercyclic experience can consolidate a hormone's influence on the expression of behavior, making the behavior more dependent on the presence of the hormone. Experience also makes stimuli more effective in eliciting hormonal secretion. Intercyclic experience enables a hormonal condition and a stimulus situation to substitute for each other in controlling the expression of behavior. Experience affects the ability of a neural cell to alter a hormone into a behaviorally more active agent so that the experienced dove will appear to be more sensitive than the naive dove to the same level of a circulating hormone. Also, by exposure to the hormones secreted during the acquisition of the experience of a first cycle, the dove's nervous system will be more sensitive to subsequent exposure to those same hormones. Although conditioning may play a role in some of the experiential effects, this would not imply that the organization of a breeding cycle is learned. Indeed, this research extends the concept of experience beyond the constraints of learning theory into domains more appropriate for understanding the development of species-typical behavior.

REFERENCES

1. LEHRMAN, D. S. 1961. Hormonal regulation of parental behavior in birds and infrahuman mammals. *In* Sex and Internal Secretions. 2nd edit. W. C. Young, Ed. **2**: 1268-1382. Williams & Wilkins. Baltimore, Md.
2. COULSON, J. C. 1966. The influence of the pair bond and age on the breeding biology of the kittiwake gull (*Rissa tridactyla*). J. Anim. Ecol. **35**: 269-279.
3. LEHRMAN, D. S. & R. P. WORTIS. 1967. Breeding experience and breeding efficiency in the ring dove. Anim. Behav. **15**: 223-228.
4. CHENG, M.-F., T. KLINT & A. JOHNSON. 1986. Breeding experience modulating androgen dependent courtship behavior in male ring doves (*Streptopelia risoria*). Physiol. Behav. **36**: 625-630.
5. LEHRMAN, D. S. 1955. The physiological basis of parental feeding behaviour in the ring dove (*Streptopelia risoria*). Behaviour **7**: 243-286.
6. LEHRMAN, D. S. & R. P. WORTIS. 1960. Previous breeding experience and hormone-induced incubation in the ring dove. Science **132**: 1667-1668.
7. MICHEL, G. F. 1977. Experience and progesterone in ring dove incubation. Anim. Behav. **25**: 281-285.
8. MOORE, C. L. 1976. Experiential and hormonal conditions affect squab-egg choice in ring doves (*Streptopelia risoria*). J. Comp. Physiol. Psychol. **90**: 583-589.
9. HARTH, M. S. 1970. The Effects of Developmental Experience on the Organization and Development of Reproductive Behavior Patterns in the Ring Dove (*Streptopelia risoria*). Unpublished doctoral dissertation. Rutgers-The State University of New Jersey.
10. HUTCHISON, J. B. & R. E. HUTCHISON. 1983. Hormonal mechanisms of mate choice in birds. *In* Mate Choice. P. Bateson, Ed.: 389-405. Cambridge University Press. Cambridge, England.
11. BEACH, F. A. 1974. Behavioral endocrinology and the study of reproduction. Biol. Reprod. **10**: 2-18.
12. LEHRMAN, D. S., P. N. BRODY & R. P. WORTIS. 1961. The presence of the mate and nesting material as stimuli for the development of incubation behavior and for gonadotrophin secretion in the ring dove. Endocrinology **68**: 507-516.
13. CHENG, M.-F. & D. S. LEHRMAN. 1975. Gonadal hormone specificity in the sexual behavior of ring doves. Psychoneuroendocrinology **1**: 95-102.
14. LEHRMAN, D. S. 1965. Interaction between internal and external environments in the regulation of the reproductive cycle of the ring dove. *In* Sex and Behavior. F. A. Beach, Ed.: 355-380. Wiley. New York, N.Y.
15. CHENG, M.-F. 1975. Induction of incubating behavior in male ring doves: a behavioral analysis. J. Reprod. Fertil. **42**: 267-276.
16. CHENG, M.-F. 1979. Progress and prospects in ring dove research: a personal view. *In* Advances in the Study of Behavior. J. S. Rosenblatt, R. A. Hinde, C. Beer & M.-C. Busnel, Eds. **9**: 97-129. Academic Press. New York, N.Y.
17. LOVARI, S. & J. B. HUTCHISON. 1975. Behavioural transitions in the reproductive cycle of Barbary doves (*Streptopelia risoria L.*). Behaviour **63**: 126-150.
18. CHENG, M.-F. 1983. Behavioural "self-feedback" control of endocrine states. *In* Hormones and Behaviour in Higher Vertebrates. J. Balthazart, E. Prove & R. Gilles, Eds.: 408-421. Springer-Verlag. West Berlin, FRG.
19. LEHRMAN, D. S. 1971. Experiential background for the induction of reproductive behavior patterns by hormones. *In* The Biopsychology of Development. E. Tobach, L. R. Aronson & E. Shaw, Eds.: 297-302. Academic Press. New York, N.Y.
20. SILVER, R. & H. H. FEDER. 1973. Reproductive cycle of the male ring dove. II. Role of gonadal hormones in incubation behavior. J. Comp. Physiol. Psychol. **84**: 464-471.
21. FRIEDMAN, M. B. 1977. Interactions between visual and vocal courtship stimuli in the neuroendocrine response in female doves. J. Comp. Physiol. Psychol. **91**: 1408-1416.
22. ERICKSON, C. J. 1973. Mate familiarity and the reproductive behavior of ringed turtle doves. Auk **90**: 780-795.
23. SIMS, E. & M.-F. CHENG. 1986. Reproductive synchrony and reproductive success. I. The importance of breeding experience and mate retention. Anim. Behav. In press.

24. HUTCHISON, J. B. 1978. Hypothalamic regulation of male sexual responsiveness to androgen. *In* Biological Determinants of Sexual Behaviour. J. B. Hutchison, Ed.: 277-318. Wiley. New York, N.Y.
25. KLINT, T., A. L. JOHNSON & M.-F. CHENG. 1984. Factors in the dose-response effect of testosterone in the male ring dove. Physiol. Behav. **32:** 1037-1040.
26. STEINER, T. & J. B. HUTCHISON. 1981. Androgen increases formation of behaviourally effective oestrogen in dove brain. Nature **292:** 345-347.
27. LUPO DI PRISCO, C., N. LUCARINI & F. DRESSI-FULGHERI. 1978. Testosterone aromatization in rat brain is modulated by social environment. Physiol. Behav. **20:** 345-348.
28. DRESSI-FULGHERI, F., L.-G. DAHLOF, K. LARSSON, C. LUPO DI PRISCO & S. TOZZI. 1980. Anosmia differentially affects the reproductive hormonal patterns in sexually experienced and inexperienced male rats. Physiol. Behav. **24:** 607-611.
29. LEHRMAN, D. S. 1958. Effects of female sex hormones on incubation behavior in the ring dove. J. Comp. Physiol. Psychol. **51:** 142-145.
30. LEHRMAN, D. S. & P. BRODY. 1961. Does prolactin induce incubation behavior in the ring dove? J. Endocrinol. **22:** 269-275.
31. KOMISARUK, B. R. 1967. Effects of local brain implants of progesterone on reproductive behavior in ring doves. J. Comp. Physiol. Psychol. **64:** 219-224.
32. STERN, J. M. & D. S. LEHRMAN. 1969. Role of testosterone in progesterone-induced incubation behaviour in male ring doves. J. Endocrinol. **44:** 13-22.
33. SILVER, R., C. REBOULLEAU, D. S. LEHRMAN & H. H. FEDER. 1974. Radioimmunoassay of plasma progesterone during the reproductive cycle of male and female ring doves (*Streptopelia risoria*). Endocrinology **94:** 1547-1554.
34. SILVER, R. & J. D. BUNTIN. 1973. Role of adrenal hormones in incubation behavior of male ring doves (*Streptopelia risoria*). J. Comp. Physiol. Psychol. **84:** 453-463.
35. FEDER, H. H., A. STOREY, D. GOODWIN, C. REBOULLEAU & R. SILVER. 1977. Testosterone and "5α-dihydrotestosterone" levels in peripheral plasma of male and female ring doves (*Streptopelia risoria*) during the reproductive cycle. Biol. Reprod. **16:** 666-677.
36. BALTHAZART, J., J. D. BLAUSTEIN, M.-F. CHENG & H. H. FEDER. 1980. Hormones modulate the concentration of cytoplasmic progestin receptors in the brain of male ring doves (*Streptopelia risoria*). J. Endocrinol. **86:** 251-261.
37. MICHEL, G. F. 1976. Role of mate's previous experience in ring dove hormone-induced incubation. J. Comp. Physiol. Psychol. **90:** 468-472.
38. MICHEL, G. F. & C. L. MOORE. 1985. Contribution of nesting experience to ring dove progesterone-induced incubation. J. Comp. Psychol. **99:** 259-265.
39. CHENG, M.-F. & J. BALTHAZART. 1982. The role of nest-building activity in gonadotrophin secretion and the reproductive success of ring doves (*Streptopelia risoria*). J. Comp. Physiol. Psychol. **96:** 307-324.
40. LEHRMAN, D. S. & P. BRODY. 1964. Effect of prolactin on establishing incubation behavior in the ring dove. J. Comp. Physiol. Psychol. **57:** 161-165.
41. BUNTIN, J. D. 1977. Stimulus requirements for squab-induced crop sac growth and nest occupation in ring doves (*Streptopelia risoria*). J. Comp. Physiol. Psychol. **91:** 17-28.
42. FRIEDMAN, M. C. & D. S. LEHRMAN. 1968. Physiological conditions for the stimulation of prolactin secretion by external stimuli in the male ring dove. Anim. Behav. **16:** 233-237.
43. HANSEN, E. W. 1971. Responsiveness of ring dove foster parents to squabs. J. Comp. Physiol. Psychol. **77:** 382-387.
44. HANSEN, E. W. 1973. A further analysis of the responsiveness of experienced and inexperienced ring dove foster parents to squabs. Dev. Psychobiol. **6:** 557-565.
45. MICHEL, G. F. & C. L. MOORE. 1986. Contribution of reproductive experience to prolactin dependent crop growth in ring doves. Anim. Behav. **34:** 790-796.
46. MOORE, C. L. 1976. The transition from sitting on eggs to sitting on young in ring doves, *Streptopelia risoria*: squab-egg preferences during the normal cycle. Anim. Behav. **24:** 36-45.
47. HANSEN, E. W. 1966. Squab-induced crop growth in ring dove foster parents. J. Comp. Physiol. Psychol. **62:** 120-122.
48. KLINGHAMMER, E. & E. H. HESS. 1964. Parental feeding in ring doves (*Streptopelia roseogrisea*): innate or learned? Z. Tierpsychol. **2:** 338-347.

49. HANSEN, E. W. 1971. Squab-induced crop growth in experienced and inexperienced ring dove (*Streptopelia risoria*) foster parents. J. Comp. Physiol. Psychol. **77**: 375-381.
50. BUNTIN, J. D., M.-F. CHENG & E. W. HANSEN. 1977. Effect of parental feeding activity on squab-induced crop sac growth in ring doves (*Streptopelia risoria*). Horm. Behav. **8**: 297-309.
51. LOTT, D. F. & S. COMERFORD. 1968. Hormonal initiation of parental behaviour in inexperienced ring doves. Z. Tierpsychol. **25**: 71-75.
52. LEHRMAN, D. S. 1959. Hormonal responses to external stimuli in birds. Ibis **101**: 478-495.
53. WHITE, S. J. 1975. Effects of stimuli emanating from the nest on the reproductive cycle in the ring dove. II. Building during the prelaying period. Anim. Behav. **27**: 869-882.
54. BEYER, C., K. LARSSON & M. L. CRUZ. 1979. Neuronal mechanisms probably related to the effect of sex steroids on sexual behavior. *In* Endocrine Control of Sexual Behavior. C. Beyer, Ed.: 365-387. Raven. New York, N.Y.

The Role of Perception in Habitat Selection in Colonial Birds

JOANNA BURGER

*Department of Biological Sciences and
Bureau of Biological Research
Rutgers-The State University of New Jersey
Piscataway, New Jersey 08854*

INTRODUCTION

The history of ethology, or indeed of any science, involves a progression from observation and description of events or behavior, through controlled observation, to manipulation of the subjects or phenomena being studied. Prior to the mid-1930s, the study of animal behavior was largely limited to descriptions of some aspects of behavior of only a few species. In the 1930s, Uexkull, Lorenz, and others[1] began the systematic description of specific aspects of behavior in closely related species[2] as well as the repertoire of behavior of individual species. A major step in ethological research was the shift to examination of behavior from the perspective of the animal (physiologically, environmentally, and socially), and to manipulation of the animal and its internal hormonal and external physical and social environment. Ethological research involving these approaches usually combines the paradigms and methodologies of both biology and psychology. Habitat selection is an important aspect of behavior directly affecting an animal's fitness and our understanding of it has benefited greatly from this approach. In this paper I discuss the factors affecting habitat selection in colonial birds and the role perception plays. Factors affecting perception involve internal mechanisms (often mediated by hormonal shifts) as well as external environmental cues.

PERCEPTION AND ETHOLOGY

Perception involves selecting or responding to selected information from the vast array of stimuli available.[3] The selection process clearly is influenced by internal hormonal state. This process of information extraction allows adaptation to the world.[4] Philosophers characterize perception "as the sort of event in which we acquire direct knowledge of the physical world."[5] The cognitive processes that deal with knowledge include perception, learning, and thinking. All three affect one another, although Forgus[4] has argued that perception subsumes learning and thinking.

Nonetheless, animals in nature are exposed to competing stimuli and they select or pay attention to only some of them.[6] Scobey and Harrington[7] have even gone so far as to question the "real" world of things, pointing out that different species perceive different properties of concrete objects. Further, depending on internal hormonal state, age, or other factors, the same animal may perceive different aspects of the same stimulus. As early as 1934 Uexkull wrote about the Umwelt of animals, made up of the perceptional world and the effector world.[8] He noted that animals select stimuli from those available, and respond accordingly. "If we still cling to the fiction of an all-encompassing universal space, we do so only because this conventional fable facilitates mutual communication."[8] Thus, we must identify the perceptual cues each animal selects from all the stimuli in the environment.

The internal mechanisms whereby perceptual cues are selected by a given animal may be hormonally mediated and related to age or experience. For example, an incubation response to an egg may be related to brood patch development which is hormonally controlled.[9]

Animals respond differently to cues about time, space, and distance. Indeed, some animals use form and motion as two different perceptual cues.[8] For example, predators often respond to prey only when the prey are moving, and do not respond to the mere shape of the prey organism. When faced with a predator, many species feign death to avoid being eaten. This behavior persists only because it is successful, at least with some predators.

Perception in animals has been studied from several viewpoints, including sensory abilities,[10-12] genetic perceptual preferences,[13-15] observer state,[16] perceptual development or sharpening,[16-18] and the role of factors such as movement,[19] redundancy,[20] and frequency[21] in perception or perceptual learning. Perception involves all sensory modalities and not just visual cues, although many authors limit themselves to the latter.[22]

Perception studies have largely been conducted in the laboratory under controlled conditions. However, perception clearly affects how animals respond in nature even though it may be difficult to study. The difficulties of distinguishing the perceptual cues actually used from those available have frequently led to studies where several environmental factors have been examined with little discussion of the stimuli used by the animals, the internal hormonal state affecting perception, or the functional analysis of the outcome of the animals' selection of particular stimuli. Habitat selection has frequently been examined in terms of physical parameters alone,[23,25] without regard to the "real" cues used by the animal for selection or to the animal's internal state.

ON THE IMPORTANCE OF HABITAT SELECTION

Habitat selection is critical for all animals since it determines the environment where an animal will live and reproduce. Choices involved in habitat selection directly relate to survival and longevity, and to individual fitness. Habitat selection can be divided into selection at different ages (juvenile, adult), for different stages (breeding season, nonbreeding season), or for different functions (mating, nesting or breeding, foraging, roosting). Hormone levels directly affect the reproductive phase an animal is in, and thus affect the type of habitat selection. In all cases, the critical selection factors affecting habitat choice include considerations of how to avoid predators,[26,27] increase foraging efficiency,[28-31] and adapt to adverse climatic or weather conditions.[32,33]

Habitat selection directly affects fitness. Fitness, as measured by reproductive success in a given reproductive season,[34] can be seen to vary among individuals breeding in different habitats. For example, habitat and nest site selection affect reproductive success in fish (*Gasterosteus aculeatus*)[35] and turtles (*Malaclemys terrapin*)[36] and in such diverse birds as doves and robins (*Zenaida macroura, Turdus migratorius*)[37] and terns and gulls (*Sterna* sp., *Larus* sp.).[38–41] Further, reproductive success, or the lack thereof, results in habitat shifts for the next reproductive effort. For example, stonechats (*Saxicola torquata*)[42] and black skimmers (*Rynchops niger*)[43] both shift nesting sites the next year if they are unsuccessful. In both species predation is the greatest cause of shifts in nest sites.

Habitat selection during the breeding season involves general habitat selection, territory acquisition, and specific nest site selection.[44] In many species these three types of selection may occur simultaneously. However, in colonial birds they occur at different times, may involve different individuals, and may involve different proximate or ultimate cues. Potential selection parameters include environmental features (substrate, vegetation, slope), social factors (presence of conspecifics, colony members, or predators), hormonal state, and timing consideration. I believe that birds respond to some of the above parameters based on their perceptions of these factors, and that it is their perceptions that require examination before reliable predictions about habitat selection can be made. Finally, careful study of the hormonal mechanisms underlying an animal's choice of stimuli to respond to will enable us to understand the variability in the type of habitat selection.

HABITAT SELECTION IN COLONIAL BIRDS

Types of Selection and Timing

Birds initially choose a general habitat type for breeding, such as a marsh, forest, field, or grassland. This type of habitat selection may occur only once during the lifetime of a bird, just prior to the first breeding experience, or annually. Subsequently, birds select a particular territory to be used for breeding. In colonial birds, territories are used only for reproductive activities, and all foraging is done elsewhere. Depending on the degree of philopatry, territory selection may occur once or many times during the life of an individual colonial bird. The internal factors affecting territory acquisition have been partially examined in birds. Motivation, partially determined by hormone levels, affects both the quantity and outcome of aggressive interactions.[45,46] Aggression levels directly affect territory size.[47] Depending on their internal hormonal state, birds choose to perceive or respond to particular aggressive displays by neighbors. Even the ability to give particular aggressive calls may be directly influenced by hormone levels. For example, Terkel *et al.*[48] showed that presence and frequency of long calling in gulls are influenced by hormones.

Following territory acquisition, birds select particular spots within their territories for nest sites. In some colonial species the same nest site is used in successive years.[49]

Two aspects of the process of habitat selection bear examination, particularly as they relate to the role of perception: (1) timing, and (2) individual roles.

In some colonial species the three types of selection occur every year, whereas in others, selection occurs only once in the lifetime of each individual. Most colonial birds exhibit a pattern in between these two extremes. For example, in gulls general habitat selection usually occurs only once, territory acquisition occurs once to many times depending upon how environmental factors change, and nest site selection occurs many times since specific habitat features (e.g., vegetation) change. Presumably the gull's perception of the suitability of the environmental and social features varies during its lifetime, depending on its internal hormonal state and on its own previous experience.

However, even when all three types of selection occur in one breeding season timing is an important factor, for general habitat selection occurs early, territory acquisition occurs shortly thereafter, and nest site selection may occur days, weeks, or even months later. In most habitats environmental features change during the selection process. For example, gulls may arrive when the nesting islands are covered with ice and snow and there is no live vegetation. However, by the time the gulls select nest sites the nesting colony may be covered with vegetation.

For other colonial birds the three types of habitat selection may occur in rapid succession. For example, in weaverbirds (*Quelea quelea*) the onset of the breeding season is stimulated by their perception of rains.[50] With the rains there is a greening of the African savannah (with increasing food supplies), and one or more of these changes provide the stimulus for internal hormonal changes that bring them rapidly into breeding condition. Such hormonal changes in turn affect perception and the birds immediately select colony sites and nest sites and begin to build nests. I have seen thousands of weaverbirds in South Africa all with half-constructed nests on the same day. Thus, for the weavers, internal hormonal changes result in perceptual changes leading to immediate performance of all three types of habitat selection.

Perceptually, then, birds must either respond to the cues of the physical environment (dead vegetation, snow, and ice), or use these features as an indication of future nesting conditions. Ice, snow, and bare ground are all hard surfaces on which to stand, but ice and snow may indicate low pockets that will later be too wet for nest sites. Similarly, tall, dead vegetation may be a cue to where the tallest, live vegetation will grow.

In addition to changes in the physical environment, the factors important for the three types of habitat selection vary, and different aspects of the environment may be perceived. For example, proximity to adequate food reserves is critical to overall habitat selection. During territory acquisition, open areas for displaying to females and ability to defend the territory against intruders are critical. However, in nest site selection, protection from predators and inclement weather may be the most important characteristic. Thus, during general habitat selection vegetation may be ignored, during territory acquisition sparse vegetation may be selected, and during nest site selection the nest may be placed under the vegetation. Birds in the same physical space perceive and respond to the physical features depending on their stage in the reproductive cycle.

Further, pairs and not individuals engage in reproductive activities. Thus it is important for the investigator to identify which sex or individual performs each of the three types of selection. Both sexes select a general habitat type. In some species only males select the territory, in others both sexes do.[44] In some species nest sites are selected exclusively by the female, in others by the male or by both members of the pair. Clearly if one member of the pair selects the territory, and the other (or both together) selects the nest site, the cues used to select each may differ.

Physical Factors

Physical factors used in nest site selection might include vegetation characteristics, substrate characteristics (sand, mud, rocks), proximity to physical features (such as trees, rocks, water), and slope. Such physical factors are often used in nest site selection studies because they are easy to measure, and lend themselves to sophisticated statistical analysis. However, the characteristic of a feature used as a cue in nest site selection may not be the physical parameter measured.

Vegetation

Vegetation is one of the most obvious and easily measured features in most habitats. It can be measured (e.g., height, percent cover) and classified physiognomically (e.g., few branches, many branches, bush type, understory tree, canopy tree, etc.). Yet colonial birds may perceive not the vegetation height or physiognomy, but other cues such as strength, cover, or visibility. For species, such as herons or egrets, that nest on branches, the strength of the branch may be the salient feature. McCrimmon[23] reported that herons on the North Carolina coast select nest sites on the basis of environmental features such as diameter of nest tree or branch, number of supporting branches, and other vegetational aspects. I suggest, however, that the herons may be perceiving strength or ability of the site to support the nest, young, and adults. Birds could test strength merely by landing on the branches and perceiving the resistance to deflection. Strength would surely vary as a function of branch diameter (as McCrimmon measured), but would also vary by species of tree, and perhaps even among branches of the same size. Strength could be measured directly by investigators by attaching weights to branches and noting the downward movement of each branch.

As well as nesting on vegetation, colonial species nest under vegetation. Such vegetation provides cover from inclement weather (rain, hail), predators, and conspecifics.[32] Traditionally, vegetation height or density has been measured as an important cue in nest site selection. However, the critical cue perceived by birds selecting nest sites may be visibility or the inverse (i.e., cover; FIGURE 1). I have tested this directly in several species of gulls.[51,52] Visibility takes into account vegetation height, density, and placement (FIGURE 1), and can be measured easily with the use of a fish-eye lens. In gulls, visibility rather than any of the vegetation characteristics distinguishes between nest sites and random locations in the colony. In herring gulls (*Larus argentatus*) internest distance is directly related to visibility, and not to vegetation density or height (FIGURE 1).

Visibility affects internest distance in several species of gulls nesting in marshes (FIGURE 2A-C) and dry land (FIGURE 2D). Beyond a certain distance, visibility ceases to affect internest distance. Aggression toward neighbors is the mechanism whereby gulls establish and defend territories. At some distance, variable for individuals and species, gulls no longer attack intruders. Even within one species (herring gulls), habitat determines how visibility affects internest distance, but the relationship still exists (FIGURE 3).

Similarly, cover (the inverse of visibility) may be the critical cue in nest site selection where predation or inclement weather are important factors that decrease fitness. No single physical measurement of vegetation (other than visibility) adequately characterizes the critical feature, visual protection from predators or the sun.

Elevation

Vantage points or lookout points are exceedingly important to nesting birds because they allow them to see predators before they are themselves seen (FIGURE 4A). In

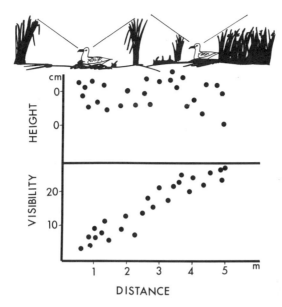

FIGURE 1. Relative importance of vegetation height compared to visibility, as measured by a visibility index (after Burger[52]), in affecting distances between nearest neighbors in herring gulls (Captree, N.Y., 1981).

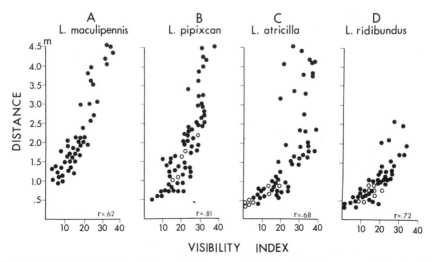

FIGURE 2. Relationships of internest distance and visibility index (0 = completely closed, 40 = completely open) for four species of gulls. A = brown-hooded gull, B = Franklin's gull, C = laughing gull, D = black-headed gull. Open circle = 5 nests, solid circle = 1 nest.

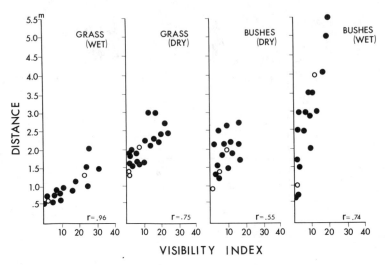

FIGURE 3. Relationships of internest distance and visibility index for herring gulls nesting in four habitats.

New Jersey salt marshes ditches were dug for mosquito control, and the dirt was piled alongside (called spoil). Spoil is slightly higher in elevation than the surrounding marsh, providing higher nest sites that allow good visibility of approaching predators and additional protection from flood tides. Several marsh-nesting species studied selected more spoil sites for nesting than random sites (FIGURE 5). In all cases the differences were significant. Their choice of nest sites on spoil is adaptive since high flood tides sufficient to wash out over half of the nests occur in Great and Barnegat Bay marshes once every 4 or 5 years. Often during exceptionally high tides only the eggs in nests placed on spoil survive. Further, high tides during the early chick phase are particularly destructive since chicks are vulnerable to cold stress.

In most nest site selection studies elevation is the physical parameter measured. Yet two birds nesting at the same relative elevation have different perceptual distances depending upon other nearby physical features and the surrounding elevations. I have found that least terns (*Sterna antillarum*) select certain vantage points for nests even though they are not significantly higher than surrounding areas. The differences between ridges and troughs may be only 6-10 cm, yet the terns select ridges and slopes (TABLE 1). Least terns nest on open beaches above the high-tide line, but in front of the dunes. Recently these areas have been extensively developed for condominiums and the population of people, dogs, and cats has increased dramatically. Ridges provide a suitable vantage point for early detection of approaching predators.

Spatial versus Real Barriers

Birds nesting in colonies may nest close together when there are physical barriers between them (e.g., tall rocks, vegetation; FIGURE 4A). Investigators often measure the distance between nests without regard to such barriers. Where they are measured, they are usually obvious, elevated objects (trees, rocks). However, I would like to suggest that the cue perceived by nesting birds may also relate to spatial barriers or discontinuities of habitat. That is, birds may nest 0.5 m apart when there is a sheer cliff between them (FIGURE 4B). Thus, the perceived cue may be the number of steps it would take to get to the nearest nest (in our above example it might take 20 steps down, and 20 steps up the cliff before reaching the nest). In other words, under some conditions, space can be a barrier as well. In Culebra, Puerto Rico, laughing gulls (*Larus atricilla*) nest on boulders that rise above the vegetation. Where there are adjacent boulders with a drop between them the gulls will nest as close as 0.3 m.

Social Factors

Perception plays a critical role in how social factors affect habitat, territory, and nest site selection. The choice of a colony site itself is dependent upon the presence

FIGURE 4. Importance of visual (A) and spatial (B) barriers in determining internest distances. In A, clear visibility to see predators or nearby neighbors affects the closeness of nests, and their relative placement. In B, a cliff between two nearby rock ledges makes it possible for birds to nest relatively close, whereas when there is no spatial barrier birds nest farther apart.

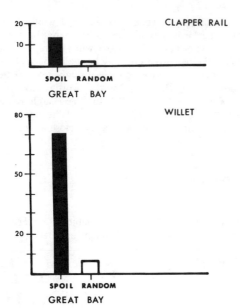
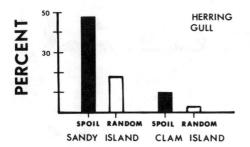
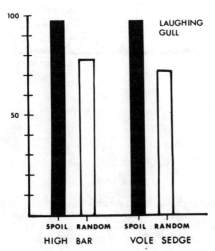

FIGURE 5. Percentage of birds nesting on spoil (elevated areas) compared to random sites on salt marshes. Clapper rail = *Rallus longirostris*. Willet = *Catoptrophorus semiplamatus*.

TABLE 1. Nest Site Selection in Least Terns at Brigantine Beach, New Jersey (1983-1984)[a]

	Used by Terns (%)	Available on the Beach (%)
Ridge	28	12
Slope	29	13
Trough	12	17
Flat areas	31	58

[a] Based on 400 early nests and 600 random points ($\chi^2 = 15.8$, $df = 3$, $p < 0.01$).

of conspecifics, or of closely related species. For many species the presence of a colony of birds in the appropriate habitat is sufficient to stimulate nesting. Thus, in egret colonies it is not unusual to find only one pair of a particular species among several pairs of another species even though one may never find a solitary-nesting egret.[53] Similarly, most colonially nesting species are known to nest both in monospecific and mixed-species colonies.[54] Indeed some species may nest in colonies with a particularly aggressive species that would provide protection from predators.

Vegetation also affects nest site selection with respect to social interactions since it provides visual barriers between close-nesting individuals. Close neighbors are disadvantageous if closeness results in increased territorial aggression or mortality of eggs and chicks. Visibility directly relates to internest distances for several colonial species (see above). Since the importance of neighbors decreases as the distance between them increases, visibility ceases to be important beyond a critical nearest-neighbor distance. The bird's perception of its visual field, however, is "real" in that with decreased visibility of neighbors there will be decreased territorial encounters, and lowered rates of predation or cannibalism on their eggs and young.

Nest sites are often chosen to minimize territorial aggression while optimizing territory size and location.[55] Just as vegetation size and shape affect visibility and nest site choice, relative size and shape of neighbors influence aggression rates. Perception is particularly critical in territorial defense. Birds often orient themselves or display to appear larger to their opponents.[56-58] Similarly, size affects how a defending bird responds. For example, in herons and egrets territory space (or internest distance) relates directly to body size of the defending bird (FIGURE 6). Larger species of ardeids nest higher, and require more space around their nests. Thus nest or territorial space is related to body size. Larger birds require more space for landing, courtship, and for the chicks to move about. If that were the only critical cue, then all intruders would be chased at the same distance regardless of their species identification. However, larger species are attacked when they are farther away than smaller species (TABLE 2). Thus snowy egrets attack great egrets when they are farther from the nest than glossy ibises. This is adaptive since a great egret intruder would require more space for a nest site and territory than would a glossy ibis. It appears that when the defending bird perceives an opponent of a particular size, it attacks—whether it is a small bird that is close, or a large bird that is farther away. Comparative psychologists have often noted the importance of size, shape, and movement cues in eliciting responses of aggressive behavior.[19,58] However, there have been few applications to habitat selection in nature.

FIGURE 6. Relationships of body size of herons, egrets, and ibises to nest placement and relative nesting space. The relative distance from the center of one bird's body to the next indicates the relative territory size per body size.

TABLE 2. Distances from which Snowy Egrets Will Initiate a Chase against an Intruder as a Function of the Species of Intruder[a]

Intruder	Size of Intruder (cm)	Distance Chased[b] (m)
Glossy ibis, *Plegadis falcinellus*	47	0.81 ± 0.52
Snowy egret, *Egretta thula*	50	1.23 ± 0.65
Little blue heron, *Florida caerulea*	55	1.36 ± 0.42
Louisiana heron, *Hydranassa tricolor*	55	1.43 ± 0.32
Great egret, *Egretta alba*	80	2.41 ± 0.56

[a]Islajo Island, New Jersey, 1975.
[b]Values given are means ± one standard deviation.

In territorial encounters with neighbors when territories and nest sites are first being established it is often the male that first enters the encounter or engages in the higher level of aggression (see references in Burger[47]). In most of these cases the male is larger, and provides the greatest perceptual cues. Such territorial aggression rarely results in contact fights where size might actually make a difference. Further, in some colonial seabirds, the females begin to defend the territories once they have been firmly established. At this stage contact fights rarely occur, and size is not important.

Hormonal Factors

In the above sections I discussed how perception affects the physical and social factors used in habitat selection. Perception, however, seems to be profoundly affected by the internal hormonal state of the bird. Most importantly, circulating hormone levels affect whether a bird is in breeding or nonbreeding condition, thus affecting whether it will select a breeding habitat or not. Circulating hormone levels vary seasonally and daily (even within the breeding season),[60,61] and hormone sensitivity also varies seasonally and daily.[62] Such variations affect many behavior levels such as courtship,[63] receptivity,[64] and aggression.[47] Thus perception of suitable stimuli also varies since in many cases the stimuli are present all day, but are responded to only at particular times of the day.

If hormone levels can affect perception of social stimuli, then perception of physical factors should be similarly influenced. For example, circulating levels of testosterone affect aggression levels, and aggression levels are usually highest early in the reproductive season when birds are selecting territories. When aggression is high, birds may select nest sites with visual barriers between themselves and close neighbors. Later in the season, when circulating hormone levels are lower, birds may select sites with fewer visible barriers. Thus hormone levels may indirectly be affecting perception of suitable sites. Since both social and physical factors are used in habitat selection, hormonal shifts should affect daily and seasonal perceptions of suitable habitats. Such an effect has been shown for weaverbirds (see above), and remains to be investigated with other birds. The role of hormones in mediating habitat choice is an important new aspect of habitat selection requiring intensive study by ethologists and neuroendocrinologists.

OPTIMIZING CONFLICTING CUES

In nature birds must make choices with respect to nest site selection, and this entails balancing the social and environmental cues. Further, the behavioral activities of each stage in the reproductive cycle affect the appropriateness of their choices.

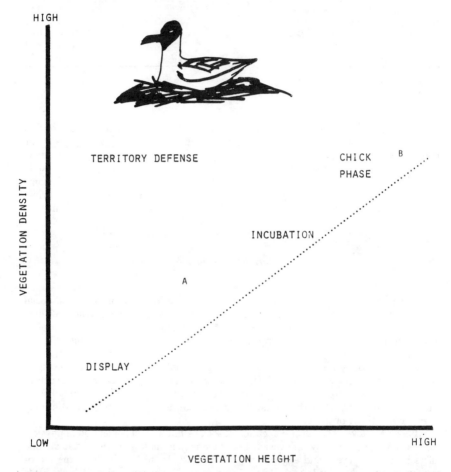

FIGURE 7. Relative importance of vegetation height and density during the different stages of the reproductive cycle of gulls. The space below the dotted line indicates the combinations of characteristics that are not optimal for any of the reproductive stages. See text for further explanation.

Presumably, over time, selection will favor those individuals that choose habitats that result in raising the most young. Thus, whichever stage (displaying to potential mates, defending territories, incubating, caring for chicks) in the reproductive effort is in the most jeopardy will have the greatest effect on their habitat selection. For example, in gulls, as in most other birds, vegetation characteristics are critical habitat variables. Vegetation height and density affect behavioral outcomes during all stages of the reproductive cycle (FIGURE 7). But, in general, high vegetation height and low vegetation density are not optimal for any of the reproductive activities. Thus I would not expect birds to nest under the conditions found below the dotted line in FIGURE 7. Displaying birds require open space (low vegetation density), territorial birds have less difficulty defending areas with few open spaces, and incubating birds prefer intermediate density and height to maximize their ability to see predators while still being partially hidden. During the chick phase tall and dense vegetation provides cover and protection from sun, rain, predators, and cannibals. With such a model we can map where gulls actually do nest, and infer the relative importance of particular stages to the overall fitness of the birds. Thus, if a particular species of gull nests at location A in FIGURE 7, we might infer that territorial defense, display activities, and incubation are all critical. Yet if it nests at B, then the chick phase might be where vegetation plays the critical role. It is important to remember that vegetation grows during the season. Thus vegetation at the optimum height for displaying birds may grow to optimum height for the chick phase. But the relationship is not passive since gulls could select habitats where the vegetation does not change in height during the nesting season.

Similar analyses can be performed with any characteristics that seem to be used by birds to select habitats, territories, and nest sites. It is critical that we determine the features the birds perceive and act upon, rather than the characteristics that are easy to measure. Ultimately we need to develop models to account for habitat selection over a continuum of perceptual cues, relating differences in cues selected to fitness. Further, we can test the perceptual abilities of birds with different hormone levels under controlled conditions. The role of perception in habitat selection of birds clearly requires further study by scientists grounded in the methodologies, paradigms, and tools of both comparative psychology and biology.

ACKNOWLEDGMENTS

Over the last 12 years I have been studying animals in groups, primarily birds nesting in colonies. I have been particularly interested in how environmental and social factors affect aspects of coloniality such as habitat selection, antipredator behavior, foraging behavior, and interspecific and intraspecific interactions. All of these subjects have benefited greatly from the application of the rigorous principles of experimental design and approaches of comparative psychology, and the evolutionary and ecological approach of biology. The Institute of Animal Behavior of Rutgers University (Newark, N.J.) provides a rare opportunity for comparative psychologists and biologists to work and interact. My own research has been profoundly affected by my experience at the Institute, and I would like to take this opportunity to thank Colin Beer, Jay Rosenblatt, and the other faculty and students for their valuable insights and discussions.

I also thank M. Gochfeld and C. Rovee-Collier for valuable comments on the manuscript and for useful references. Good colleagues are a pleasure.

REFERENCES

1. SCHULLER, C. H. (Ed.). 1957. Instructive Behavior. International Universities Press, Inc. New York, N.Y.
2. SCHLEIDT, W. M. 1973. Ethology. **3:** 119-138. Academic Press. New York, N.Y.
3. THORNDIKE, E. L. 1911. Animal Intelligence. Macmillan. New York, N.Y.
4. FORGUS, R. H. 1969. Perception: The Basic Process in Cognitive Development. McGraw-Hill. New York, N.Y.
5. YOST, R. M. 1974. Some Philosophical Problems of Perception. Pp. 19-40 (p. 21). Academic Press. New York, N.Y.
6. BOLLES, R. C. 1973. The comparative psychology of learning: the selective association principle and some problems with "general" laws of learning. *In* Perspectives on Animal Behavior. G. Bermant, Ed.: 280-306. Scott, Foresman and Company. Glenview, Ill.
7. SCOBEY, R. & T. HARRINGTON. 1973. Sensory systems. *In* Perspectives on Animal Behavior. G. Bermant, Ed.: 129-170. Scott, Foresman and Company. Glenview, Ill.
8. UEXKULL, J. VON. 1934. A stroll through the world of animals and men. (As reprinted in SCHULLER, C. H. (Ed.). 1957. Instructive Behavior. Pp. 5-80 (p. 29). International Universities Press, Inc. New York, N.Y.)
9. JONES, R. E. 1969. Hormonal control of incubation patch development in the California quail *Lophortyx californicus.* Gen. Comp. Endocrinol. **13**(1): 1-13.
10. BLAKE, R. 1981. Strategies for assessing visual deficits in animals with selective neural deficits. *In* Development of Perception: The Visual System. R. N. Aslin, J. R. Alberts & M. R. Peterson, Eds. **2:** 95-112. Academic Press. New York, N.Y.
11. STEIN, B. E. & B. GORDON. 1981. Maturation of the superior colliculus. *In* Development of Perception: The Visual System. R. N. Aslin, J. R. Alberts & M. R. Peterson, Eds. **2:** 157-196. Academic Press. New York, N.Y.
12. WADE, N. J., C. M. M. DEWEERT & M. T. SWANSTON. 1984. Binocular rivalry with moving patterns. **35**(2): 111-122.
13. KOVACH, J. K. 1980. Visual information and approach behavior in genetically manipulated quail chicks: preference hierarchies and interactions of flash rate, flash amplitude, luminance, and color. J. Comp. Physiol. Psychol. **94**(1): 178-199.
14. KOVACH, J. K. 1983. Constitutional biases in early perceptual learning. I. Preferences between colors, patterns, and composite stimuli of colors and patterns in genetically manipulated and imprinted quail chicks (*C. coturnix japonica*). J. Comp. Psychol. **97**(3): 225-239.
15. KOVACH, J. K. 1983. Constitutional biases in early perceptual learning. II. Visual preference in artificially selected, visually naive and imprinted quail chicks (*C. coturnix japonica*). J. Comp. Psychol. **97**(3): 240-248.
16. HANSEN, R. S. & A. D. WELL. 1984. The effects of stimulus sequence and probability on perceptual processing. Percept. Psychophys. **35**(2): 137-143.
17. WIENS, J. A. 1972. Anuran habitat selection: early experience and substrate selection in *Rana cascadae* tadpoles. Anim. Behav. **20**(2): 218-220.
18. CAMPBELL, B. A. & V. HAROUTUNIAN. 1983. Perceptual sharpening in the developing rat. J. Comp. Psychol. **97**(1): 3-11.
19. WADE, N. J. & M. T. SWANSTON. 1984. Illusions of size change in dynamic displays. Percept. Psychophys. **35**(2): 286-290.
20. ENNS, J. T. & W. PRINZMETAL. 1984. The role of redundancy in the object-line effect. Percept. Psychophys. **34**(1): 22-32.
21. SINNOTT, J. M., M. B. SACHS & R. D. HIENZ. 1980. Aspects of frequency discrimination in passerine birds and pigeons. J. Comp. Physiol. Psychol. **94**(3): 401-406.
22. JACKSON, F. 1977. Perception—A Representative Theory. Cambridge University Press. Cambridge, England.
23. MCCRIMMON, D. A., JR. 1978. Nest site characteristics among five species of herons on the North Carolina coast. Auk **95**(2): 267-280.
24. MACKENZIE, D. I. & S. G. SEALY. 1981. Nest site selection in eastern and western kingbirds: a multivariate approach. Condor **83**(4): 310-321.

25. COLLINS, S. L., F. C. JAMES & P. G. RISER. 1982. Habitat relationships of wood warblers (*Parulidae*) in northern central Minnesota. Oikos 39(1): 50-58.
26. TINBERGEN, N. 1963. On adaptive radeation in gulls tribe *Larini*. Bonn. Zool. Beitr. 39: 209-223.
27. ALEXANDER, R. D. 1971. The search for an evolutionary philosophy of man. Proc. R. Soc. Victoria 84(1): 99-120.
28. CROOK, F. H. 1964. The evolution of social organization and visual communication in the weaver birds (*Ploceinae*). Behaviour Suppl. 10: 1-178.
29. MURTON, R. K. 1971. The significance of specific search image in the feeding behavior of the woodpigeon. Behaviour 39: 10-42.
30. WARD, P. & A. ZAHAVI. 1973. The importance of certain assemblages of birds as 'information centres' for food finding. Ibis 115(4): 517-534.
31. KREBS, J. R. 1974. Colonial nesting and social feeding as strategies for exploiting food in the great blue heron (*Ardea herodias*). Behaviour 51: 99-134.
32. AUSTIN, O. L., JR. 1933. The status of Cape Cod terns in 1933. Bird-Banding 4(12): 190-198.
33. FINCH, D. M. 1983. Seasonal variation in nest placement of Albert's towhees. Condor 85(1): 111-113.
34. TRIVERS, R. L. 1972. Parental investments and the descent of man. In Sexual Selection and the Descent of Man. B. Campbell, Ed.: 136-207. Aldene. Chicago, Ill.
35. SARGENT, R. C. & J. B. GEBLER. 1980. Effects of nest site concealment on hatching success, reproductive success, and paternal behavior of the threespine stickleback, *Gasterosteus aculeatus*. Behav. Ecol. Sociobiol. 7(2): 137-142.
36. BURGER, J. 1977. Determinants of hatching success in diamond-back terrapin, *Malaclemys terrapin*. Am. Midl. Nat. 97(2): 444-464.
37. YAHNER, R. H. 1983. Site-related nesting success of mourning doves and American robins in shelterbelts. Wilson Bull. 95(4): 573-580.
38. DEXHEIMER, M. & W. E. SOUTHERN. 1974. Breeding success relative to nest location and density in ring-billed gull colonies. Wilson Bull. 86(3): 288-290.
39. MONTEVECCHI, W. A. 1978. Nest site selection and its survival value among laughing gulls. Behav. Ecol. Sociobiol. 4(2): 143-161.
40. BURGER, J. & F. LESSER. 1978. Selection of colony sites and nest sites by common terns *Sterna hirundo* in Ocean County, New Jersey. Ibis 120(4): 433-449.
41. PIEROTTI, R. 1982. Habitat selection and its effect on reproductive output in the herring gull in Newfoundland. Ecology 63(3): 854-868.
42. GREIG-SMITH, P. W. 1982. Dispersal between nest-sites by stonechats *Saxicola torquata* in relation to previous breeding success. Ornis Scand. 13(3): 232-238.
43. BURGER, J. 1982. The role of reproductive success in colony-site selection and abandonment in black skimmers (*Rynchops niger*) 1982. Auk 99(1): 109-115.
44. BURGER, J. 1985. Nest site selection in temperate marsh nesting birds. In Nest Site Selection. M. Cody, Ed. Academic Press. New York, N.Y.
45. CROOK, J. H. & P. A. BUTTERFIELD. 1968. Effects of testosterone propionate and luteinizing hormone on agonistic and nest building behavior of *Quelea quelea*. Anim. Behav. 16(2): 370-384.
46. DAVIS, D. E. The hormonal control of aggressive behavior. In Proceedings of the 13th International Ornithological Congress, 1963. Pp. 994-1003.
47. BURGER, J. 1984. Adaptive significance of territoriality in herring gulls (*Larus argentatus*). Am. Ornithologist Union Monogr. 34: 1-92.
48. TERKEL, A. S., C. L. MOORE & C. G. BEER. 1976. The effects of testosterone and estrogen on the rate of long-calling vocalization in juvenile laughing gulls, *Larus atricilla*. Horm. Behav. 7(1): 49-57.
49. VERMEER, K. 1970. Breeding Biology of California and Ring-Billed Gulls: A Study of Ecological Adaptation to the Inland Habitat. Canadian Wildlife Service Report Series No. 12, pp. 1-52. Ottawa, Ontario, Canada.
50. LACK, D. 1966. Population Studies of Birds. Oxford University Press. London, England.
51. BURGER, J. 1976. Nest density of the black-headed gull in relation to vegetation. Bird Study 23(1): 27-32.

52. BURGER, J. 1977. Role of visibility in nesting behavior of *Larus* gulls. J. Comp. Physiol. Psychol. **91**(6): 1347-1358.
53. BURGER, J. 1979. Resource partitioning: nest site selection in mixed species colonies of herons, egrets and ibises. Am. Midl. Nat. **101**(1): 191-209.
54. BURGER, J. The advantages and disadvantages of mixed species colonies of seabirds. *In* Proceedings of the 18th International Ornithological Congress, Moscow, 1982. In press.
55. EWALD, P. W., G. L. HUNT, JR. & M. WARNER. 1980. Territory size in western gulls: importance of intrusion pressure, defense investments, and vegetation structure. Ecology **61**(1): 80-87.
56. GALUSHA, J. G. & J. F. STOUT. 1977. Aggressive communications by *Larus glaucescens*. IV. Experiments on visual communication. Behaviour **62**: 221-235.
57. AMLANER, C. J., JR. & J. F. STOUT. 1978. Aggressive communication by Larus glaucescens. VI. Interactions of territory residents with a remotely controlled, locomotory model. Behaviour **66**: 223-251.
58. BRONSTEIN, P. M. 1981. Commitments to aggression and new sites in male *Betta splendens*. J. Comp. Physiol. Psychol. **95**(3): 443-449.
59. BAGNARA, S., F. SIMION & C. UMILTA. 1984. Reference patterns and the process of normalization. Percept. Psychophys. **35**(2): 186-192.
60. MEIER, A. H., J. T. BURNS & J. W. DUSSEAU. 1969. Seasonal variation in the diurnal rhythm of pituitary prolactin content in white-throated sparrow, *Zonotrichia albicollis*. Gen. Comp. Endocrinol. **12**(1): 282-289.
61. MEIER, A. H. & R. MACGREGOR III. 1972. Temporal organization in avian reproduction. Acta Zool. **12**(1): 257-277.
62. MEIER, A. H., J. T. BURNS, K. B. DAVIS & T. M. JONES. 1971. Circadian variations in sensitivity of the pigeon cropsac to prolactin. J. Intern. Cycle Res. **22**: 161-170.
63. BURGER, J. 1976. Daily and seasonal activity patterns in breeding laughing gulls. Auk **93**(2): 308-323.
64. ADKINS, E. K. & N. T. ADLER. 1972. Hormonal control of behavior in the Japanese quail. J. Comp. Physiol. Psychol. **81**(1): 27-36.

Comparative Behavioral Endocrinology[a]

DAVID CREWS

Institute of Reproductive Biology
Department of Zoology
University of Texas
Austin, Texas 78712

INTRODUCTION

While the diversity in reproduction has always been acknowledged, what it can tell us about the evolution of reproductive controlling mechanisms has not received the attention it deserves. This is now being addressed by comparative behavioral endocrinology, a newly emerging discipline. Three fundamental principles are involved in comparative behavioral endocrinology research (FIGURE 1). The first is the study of diverse species under both laboratory and field conditions. Species that differ in one or more components of reproduction reveal the end result of ecological and evolutionary processes. The study of these "natural"[1] experiments thus provides clues to the fundamental nature of the morphological, physiological, and behavioral aspects of reproduction. To this end, comparative behavioral endocrinology investigations frequently are conducted in the laboratory as well as in the field: while the laboratory is necessary for determining the physiological bases of reproductive behavior, the field is the only possible arena for establishing the adaptive function of the mechanism(s). The utilization of a variety of morphological, physiological, and behavioral measures is the second basic principle of comparative behavioral endocrinology. Only in this way can the relations between the different levels of biological organization be illuminated. Finally, the distinct but related questions first articulated by Niko Tinbergen for the study of behavior are the third criterion of comparative behavioral endocrinology. For example, studies of the development and causal mechanisms of reproductive behavior are more meaningful if they also consider the adaptive function and evolution of the behavior pattern in question.

In this paper I first discuss the diversity of neuroendocrine mechanisms regulating mating behavior. I next consider the diversity of mechanisms by which behavior may act to facilitate reproduction. I end by speculating on the evolution of reproductive controlling mechanisms. Most of the examples will be drawn from research on four species, three of them reptiles. These are the green anole lizard (*Anolis carolinensis*), the red-sided garter snake (*Thamnophis sirtalis parietalis*), and the whiptail lizard (*Cnemidophorus uniparens*). The fourth species to be discussed is the fruit fly,

---

[a] The research reported here was supported by grants from the National Institute of Child Health and Human Development, the National Institute of Mental Health, the National Science Foundation, the Whitehall Foundation, and the Kinsey Institute.

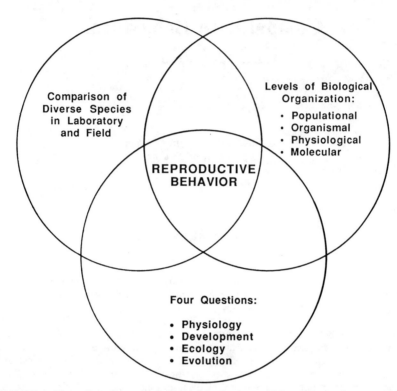

FIGURE 1. Three dimensions of behavioral investigations which, when combined as in comparative behavioral endocrinology, shed new light on the causes and functions of reproductive behavior.

Drosophila mercatorum; the latter research was conducted in the laboratory of H. L. Carson.

DIVERSITY IN NEUROENDOCRINE MECHANISMS REGULATING MATING BEHAVIOR

In a large number of vertebrates, as the time of mating approaches, gonadal activity increases (gonadal activity being defined here both as gamete production and sex hormone secretion). The consequence, of course, is that mature gametes are available at the time of mating. Mating behavior in these species is brought about by the hormones produced by the developing gonads. An example of such a functional association among gamete production, sex hormone secretion, and mating behavior comes from work on the green anole lizard.[2] Sexually mature male green anoles perform a species-typical courtship display to resident and transient females. This courtship behavior is dependent on testicular androgens. An intact male performing at a high level of sexual activity will have a high level of testosterone in his systemic circulation. Castration leads to a drop in blood testosterone to basal levels. The male's sexual behavior also drops precipitously. If androgen replacement therapy is given, the circulating level of testosterone increases and the male's sexual behavior is restored to its preoperative level.

A similar causal relationship exists between ovarian activity and sexual receptivity in the female green anole lizard.[3] As the female enters into the breeding season, a single ovarian follicle will begin to accumulate yolk until it reaches about 8 mm in diameter. The circulating level of estrogen also increases during this period, peaking around the time of ovulation. After ovulation, estrogen levels fall and progesterone levels increase. The estrogen-progesterone ratio is reversed again as an enlarging follicle in the opposite ovary becomes preovulatory. During the course of the 4- to 5-month breeding season, the ovaries alternate in the maturation and ovulation of a single ovum every 14-17 days. The female green anole lizard exhibits distinct cycles of sexual receptivity during the breeding season that are related precisely to the growth of the ovarian follicle. The female is receptive to male courtship only when the follicle is between 6 and 8 mm in diameter. Removal of the ovaries abolishes sexual receptivity, whereas replacement therapy with estrogen followed by progesterone will reinstate sexual receptivity in the ovariectomized female.

There is an abundance of other examples in vertebrates of hormone dependency of mating behavior. Because representative species of every vertebrate class have been found to exhibit hormonal mediation of mating behavior, it has been assumed that there is an intrinsic functional association among gamete production, sex steroid hormones, and mating behavior. Comparative behavioral endocrinological studies show this assumption is not warranted. While mating behavior serves the same function in different vertebrate species, the mechanisms underlying the mating behavior among the different species need not be similar.

Although it is true that almost all of the species studied to date show a functional association among gamete production, sex hormone secretion, and mating behavior (a pattern I have termed an *associated reproductive tactic* (FIGURE 2)[4]), the number of species for which we know the state of gonadal activity at the time of mating is remarkably small. Further, an alternative reproductive tactic, first described by Cour-

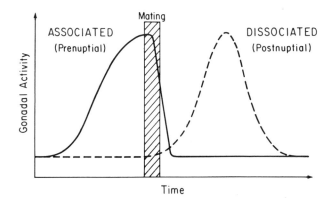

FIGURE 2. Associated and dissociated reproductive tactics exhibited by vertebrates. Gonadal activity is defined as the maturation of eggs or sperm and/or gonadal steroid hormone secretion. In individuals exhibiting the associated reproductive tactic, the gametes are mature and circulating levels of sex steroid hormones are maximal at the time of mating. In individuals exhibiting the dissociated reproductive tactic, the gonads are not producing gametes and sex steroid hormone levels are low at the time of mating. (Reprinted with permission from Crews.[4])

rier in 1927,[5] exists. In some vertebrates, the gonads are not producing gametes and sex hormone levels are lowest at the time of mating. This pattern of gonadal activity in relation to mating I have suggested calling a *dissociated reproductive tactic* (FIGURE 2). Thus, in species exhibiting a dissociated reproductive tactic, gametogenesis and its accompanying increase in the circulating level of sex hormones occur only after all mating behavior has ceased.

What stimulates sexual behavior in species that exhibit a dissociated reproductive tactic? How is it that mating behavior can be displayed at a time when circulating levels of sex steroid hormones are low and the gonads are collapsed? Some answers are provided by research on the Canadian red-sided garter snake, the most thoroughly investigated species exhibiting the dissociated reproductive tactic.[6]

Garter snakes are the most abundant of all snakes in North and Central America, ranging from Mexico and South America to the northwestern United States and Canada. In Manitoba, Canada, the red-sided garter snake spends most of its time underground in limestone caverns that serve as subterranean hibernacula. The snakes emerge in the spring (April), at which time the majority of matings occur. The breeding season can last from as little as 3 days to no longer than 3 weeks. Immediately after mating, the snakes disperse to feeding grounds up to 15 km away. In the fall (October), the snakes return to the hibernaculum from which they had emerged the previous spring.

Male red-sided garter snakes emerge from hibernation first; the females emerge about 1 week later. Also, unlike the males who emerge *en masse*, the females come out singly or in small groups. Because each female must pass through the aggregated males in order to leave the den area, as many as 100 males may simultaneously court the female, resulting in the formation of "mating balls." The female mates with only one male and after mating, she leaves the den area. The males, including the males that succeed in mating, remain at the den until all of the females have emerged.

Testicular activity in male red-sided garter snakes at the time of emergence is minimal.[7] It is only 8-10 weeks later, after the males have left the den sites, that the testes grow. A similar pattern exists with regard to androgen secretion. Androgen levels in the blood are relatively low at the time of emergence and increase at the time of spermatogenesis. The sperm produced during this period are then stored in the epididymides and vasa deferentia where they remain until the next spring.

Courtship behavior in the male red-sided garter snake is activated by the increase in temperature following winter dormancy.[7-10] Males castrated before entering into hibernation, or even on emergence from hibernation the preceding year, continue to court females on emergence[7,11]; there is no difference in the level or the nature of courtship behavior of castrated males, sham-operated males, and castrated males given androgen implants. So long as males have gone through a period of low-temperature dormancy, castration and androgen replacement will neither inhibit nor prolong the display of sexual behavior. Indeed, the pituitary can be removed on emergence from hibernation or even before the animal enters hibernation, and the males will still exhibit intense courtship behavior on emergence.

Male red-sided garter snakes will not show intense sexual behavior again unless they hibernate (TABLE 1). A number of experiments designed to induce courtship in noncourting males maintained under summer-like conditions have been conducted[11,12] (TABLE 1). Administration of testosterone, dihydrotestosterone, or estradiol in various concentrations and in various combinations has only occasionally, and then inconsistently, induced sexual behavior. A variety of administered hypothalamic hormones and neurotransmitters have also failed to elicit courtship behavior. Even implantation of minute amounts of testosterone directly into the anterior hypothalamus-preoptic area (AH-POA) will not induce sexual behavior in males maintained under summer-like conditions.[13]

Recent research has indicated that placement of electrolytic lesions in the AH-POA of males as they emerge from hibernation will terminate the males' sexual behavior.[14] The AH-POA is a major integrative area of temperature-sensitive neurons, and I hypothesize that the decline in courtship behavior in males having lesions in this area is not due to the destruction of hormone-sensitive neurons, but rather to the destruction of temperature-integrating neurons. One possibility is that male garter snakes with lesions in the AH-POA are unable to integrate information derived from the temperature change.[15]

Although sex hormones are not necessary for the exhibition of courtship behavior in the adult male red-sided garter snake, these hormones do appear to be involved in the expression of courtship behavior. Treatment with testosterone will cause males castrated as neonates or as yearlings to court attractive females.[16] Further, only those

TABLE 1. Experimental Manipulations and Their Effects on the Courtship Behavior of the Male Canadian Red-Sided Garter Snake (*Thamnophis sirtalis parietalis*)[a]

Manipulation	Response	Authority
Hibernation	+	Aleksiuk and Gregory[8]; Hawley and Aleksiuk[9,10]; Garstka et al.[12]
Castration		
On emergence	−	Camazine et al.[7]
Before hibernation	−	Garstka et al.[12]; Crews et al.[11]
Breeding	−	Crews et al.[11]
Adrenalectomy and castration	−	Crews et al.[11]
Hypophysectomy		
On emergence	−	Crews et al.[11]
Before hibernation	−	Crews et al.[11]
Systemic hormone treatment		
Sex steroids	−	Camazine et al.[7]; Garstka et al.[12]; Crews et al.[11]
Hypothalamic hormones	−	Garstka et al.[12]
Neurotransmitters	−	Garstka et al.[12]
Cyproterone acetate	−	Crews et al.[11]
Hypothalamic hormone treatment	−	Friedman and Crews[13]
Hypothalamic lesions	+	Friedman and Crews[14]

[a] + signifies an effect on courtship behavior; − signifies no effect on courtship behavior.

snakes that had been treated with androgen courted females on emergence from hibernation. Because male red-sided garter snakes probably do not reach sexual maturity until the second summer after emergence, I suspect the androgen produced during the first testicular recrudescence programs the brain to respond to the upcoming shift in temperature.

The female red-sided garter snake exhibits a pattern of gonadal activity similar to that of the male.[6] When a female emerges from hibernation, her ovarian follicles are small. Yolking begins only after the female has left the den area and ovulation does not occur for 6 weeks. Mating is known to be one of the stimuli responsible for ovarian growth; indeed, in some females, the act of mating initiates ovarian growth. Experiments have established that when the female mates, there is a very rapid but transient surge in estradiol secretion.[17] This estradiol surge not only facilitates transport of recently deposited sperm to sperm storage tubules in the oviduct,[18] but it stimulates the mobilization of phospholipids so that yolking can occur.[17] In females that do not mate, the concentration of circulating estradiol continues to be basal.

Not only does the study of species exhibiting a dissociated reproductive tactic reveal profound differences in the neuroendocrine mechanisms subserving sexual behavior, it also provides a framework within which to regard the great diversity in reproduction found among sexually reproducing vertebrates. Since most individuals can only exhibit one reproductive tactic at a time, either an associated reproductive tactic or a dissociated reproductive tactic, one would predict there would be at least four combinations of reproductive tactics found among vertebrates. Examples of each of these categories exist in nature (TABLE 2).[4] In many species both the male and female exhibit mating behavior in close association with maximum gonadal activity. There also are a number of species in which both the male and female show a dissociation between gonadal activity and mating behavior. Finally, there are species having "mixed" reproductive strategies in which one sex exhibits one tactic and the other sex exhibits the opposite tactic; that is, there exist species in which the male exhibits the associated reproductive tactic and the female the dissociated reproductive tactic as well as species in which the male exhibits the dissociated reproductive tactic and the female the associated reproductive tactic.

TABLE 2. Relationship between Gonadal Activity and Mating Behavior in Vertebrates[a]

Female	Male	
	Associated	Dissociated
Associated	Many laboratory and domesticated mammals; most birds, temperate and tropical lizards, crocodilians	Most temperate turtles (e.g., *Chrysemys picta; Gopherus polyphemus*), *Crotalus h. horridus, Virginia striatula, Ambystoma tigrinum; Esox lucius*
Dissociated	*Suncus murinus; Fulmarus glacialis; Hemiergis peronii, Eumeces egregius, Phyllodactylus marmoratus, Sceloporus g. microlepidotus, Vipera aspis, Micrurus fulvius; Cymatogaster aggregata, Trachycorystes striatulus*	Vespertilionid and rhinolophid (hibernating) bats, *Sceloporus a. bicanthalis*, several crotalid, elapid, and colubrid snakes, *Carphophis vermis; Desmognathus* spp.; *Pleuronectes platessa, Cyprinus carpio*

[a]Adapted from Reference 4.

The research with the garter snake shows that the neuroendocrine mechanisms underlying reproductive behavior in animals exhibiting the dissociated reproductive tactic are fundamentally different from those found in animals exhibiting an associated reproductive tactic. It follows that the neuroendocrine mechanisms of each sex will differ in species exhibiting a "mixed" strategy. Indeed, in the spring-breeding painted turtle (*Chrysemys picta*), the male requires high body temperatures (28°C) for testicular growth whereas the female requires low body temperatures (17°C) to undergo ovarian recrudescence.[19] In the male Asian musk shrew, *Suncus murinus*, mating behavior coincides with testicular growth and is dependent on testicular androgens; in the female, mating behavior induces growth and eventually ovulation of the ovarian follicle.[20]

DIVERSITY IN THE BEHAVIORAL MECHANISMS FACILITATING REPRODUCTION

Morphological, physiological, and behavioral aspects of reproduction co-evolve under different selection pressures, resulting in different evolutionary strategies. One example is the behavioral facilitation of reproduction of animals. It is well established in vertebrates that the presence and behavior of the male facilitate ovarian growth in the conspecific female. In the green anole lizard, the courtship behavior of the male is critical for normal levels of pituitary gonadotropin secretion in the female.[2]

A "natural" experiment that contributes to our understanding of the evolution of the behavioral facilitation of reproduction is presented by the whiptail lizard *Cnemidophorus uniparens*. Many species of *Cnemidophorus* reproduce parthenogenetically, consisting only of female individuals.[18,21] In at least five parthenogenetic lizard species individuals perform complementary male-like and female-like displays similar to those seen during courtship and copulation in closely related sexual species.[22-25]

We have established experimentally that in *C. uniparens* the individual's behavioral role is precisely correlated with its reproductive state.[24] Individuals behaving as females have large yolking follicles, whereas individuals behaving as males have small ovaries. We have also shown that the female alternates her behavior as she progresses through her ovarian cycle. As the follicles enlarge the dominant behavior observed is female-like until ovulation, at which time the dominant behavior displayed is male-like. A female will interact sexually only with other individuals in a complementary ovarian and behavioral condition. That is, if one female has large preovulatory follicles, the other individual is likely to have small ovarian follicles. This is paralleled by female-like and male-like behavioral roles being performed respectively.

There is evidence that behavioral facilitation of reproduction occurs in unisexual, as well as in sexual, lizards. Experiments have demonstrated that the presence and behavior of conspecifics facilitate reproduction in *C. uniparens*.[26] Thus, each female stimulates and is stimulated by its conspecifics. Since all individuals are genetically identical to one another, the benefits derived will be shared equally.

A comparative study was afforded when Carson and his colleagues[27] reported that individuals of a parthenogenetic strain of *Drosophila mercatorum* are less likely to mate with males than are females of the sexual ancestral strain. This loss of sexual receptivity was interpreted as evidence for the loss of a trait (sexual receptivity) due to the absence of a selective force (the need for males for reproduction). We decided to study the influence of different social conditions on reproduction in both sexual and parthenogenetic *D. mercatorum*.[28] There already was some evidence to suggest that behavioral facilitation of egg production may also occur in *Drosophila*.[29,30] However, in our experiment, individual females of a sexual strain and the parthenogenetic strain were placed separately in one of the following social conditions ($n=10$/group): female housed in isolation, female housed with another female, female housed with one sterile (XO) male, female housed with two sterile males, female housed with one fertile (XY) male, and female housed with two fertile males. Eggs produced by each female were counted on a daily basis for 4 days.

In the sexual strain, females housed with males laid the most eggs; egg production by females housed with males was 13 times that by isolated females and 6 times that by females housed in pairs (FIGURE 3). The ovaries were developed in isolated females, but they laid only sporadically. Females housed together laid a significantly greater number of eggs than isolated females.

It is likely that the courtship behavior, and not copulatory stimuli, provided by the male is responsible for this facilitation of reproduction. Egg production of females with sperm in the terminal receptacles was not different from the egg production of unfertilized females. Further, there was no difference among females housed with sterile males vs. fertile males. In addition, research with other species of *Drosophila* indicates the behavior of sterile males is identical with that of fertile males. Egg production of females housed with two males, fertile or sterile, was significantly less than that of females housed with one male. This may be due to aggression between males inhibiting ovarian activity in the female as in the green anole lizard.[2]

In contrast, there was no effect of social condition on egg production in the parthenogenetic strain. Egg production was not significantly different among the

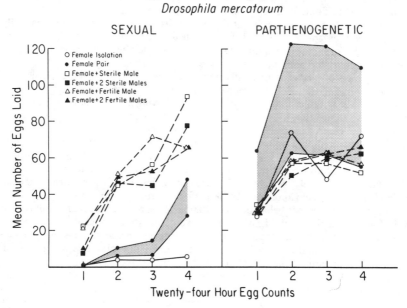

FIGURE 3. Egg production over four consecutive days by sexual and parthenogenetic strains of *Drosophila mercatorum* placed as virgins on day 1 into various social conditions. Because it was not possible to determine the extent to which each female contributed to the total egg count in the female pair category, both the total egg count (the maximum each female could lay) and one-half that amount (assuming both females were contributing equally) are presented (shaded area). (Reprinted with permission from Crews *et al.*[28] Copyright 1985 by the AAAS.)

different social conditions. Parthenogenetic female isolates laid approximately 15 times the number of eggs laid by sexual female isolates. When compared to measures obtained when the stock was derived in 1961, the average number of eggs produced by a female has increased by sevenfold. It is significant also that the parthenogenetic rate of the unisexual stock has remained constant since 1961. Although not proof, this is strong evidence for a loss of behavioral facilitation of reproduction coincident with a loss of sexual receptivity in parthenogenetic *D. mercatorum*.

Of great interest was the unexpected finding that despite the loss of genetic variability in the parthenogenetic strain,[31] the egg production of individual females was normally distributed. Furthermore, the variability both within and between parthenogenetic females was not different from that exhibited by females of the sexual strain.

This accords with our observations that individual *C. uniparens* differ markedly in their behavior.

EVOLUTION OF REPRODUCTIVE CONTROLLING MECHANISMS

Evolution selects for outcomes, not mechanisms.[32] It has been said also that it is not the structure of steroid hormones that has changed in evolution but rather the uses to which the steroid hormones have been put.[33] There are many instances in which internal signals such as hormones have been co-opted, as have external factors such as photoperiod and temperature, to serve a signaling function for synchronizing physiological processes. One can speculate on how this may have evolved. There appears to be an ancient functional relationship between gamete production and gonadal steroid secretion.[34] If this is true, it is likely that the ability of the sex steroid hormones to activate sexual behavior in the various vertebrate species has evolved independently many times. How might this have occurred? I suggest that consequent to the development of hypothalamic feedback control of pituitary gonadotropin secretion was the development of a functional association between gonadal hormone secretion and reproductive behavior. This could have been achieved by slight expansion of the steroid-sensitive hypothalamic system involved in the feedback control of pituitary function, thereby recruiting behavioral integrative (limbic) areas in the central nervous system (CNS). A further step in the evolution of hormone dependency of sexual behavior is seen in some species in which specific sexual signals elicit the secretion of reproductive hormones.[35] However, it should be emphasized that the presence of sex steroid hormone-concentrating neurons in behavioral integrative areas is not evidence a priori of a functional association between sex steroid hormones and sexual behavior. Neither garter snakes[6,11] nor female Asian musk shrews[20] exhibit hormone-dependent sexual receptivity, yet sex steroid hormone-concentrating neurons have been identified in the anterior hypothalamus-preoptic area of both species[36,37]; in the garter snake these areas are involved in the feedback control of pituitary gonadotropin secretion.[13]

The co-evolution of signal and receptor has been noted in numerous neuroethological investigations.[38] The same parallel can be drawn between social signals and physiological and/or behavioral responses.[39] In crickets, the template for receiving the song information has a common chromosomal basis with the expression of the song.[40] Using different sensory mutant and mosaic *Drosophila,* the genetic bases of the production and reception of social signals critical for normal courtship and copulation have been examined.[41] These studies reveal that different parts of the brain specify different sex-specific behaviors as well as the sensitivity to social signals responsible for the elicitation of these behaviors. The finding of single-gene products that regulate the expression of visual, olfactory, and auditory cues, thereby influencing a number of behavior patterns independently as well as interdependently, has opened new vistas for the investigator interested in the evolution of neuroendocrine controlling mechanisms.

In this regard comparisons of sexual and parthenogenetic species represent a powerful "natural" experiment for understanding the evolution of at least one reproductive controlling mechanism, namely, the behavioral facilitation of reproduction. Parthenogenetic whiptail lizards exhibit "sexual" behavior as well as behavioral facilitation of reproduction. Parthenogenetic *D. mercatorum,* on the other hand, has lost sexual receptivity as well as the behavioral facilitation of egg-laying characteristic of its sexual ancestor. Thus, the parthenogenetic fruit fly has lost both gonadal sex (gonochorism) and behavioral sex (mating behavior), whereas the parthenogenetic whiptail lizard has lost only gonadal sex.

The difference between parthenogenetic whiptail lizards and fruit flies may lie in the nature of the chromosomal origin of the respective species. The parthenogenetic whiptail lizard is a triploid and arose from the hybridization of two closely related sexual species. The parthenogenetic strain of *D. mercatorum,* on the other hand, arose from isolated virgin females of a sexual strain. It is perhaps significant that in the former species, which arose from a sexual act, the neuroendocrine mechanisms subserving the behavioral aspects of reproduction have been retained, whereas in the latter species, which arose from isolated virgin females, those mechanisms have been lost. Alternatively, the mode of parthenogenesis may be critical for the heritability of neuroendocrine controlling mechanisms. In *C. uniparens,* triploidy is restored by endoduplication prior to meiosis resulting in fixed heterozygosity,[42] whereas in *D. mercatorum,* diploidy is restored largely by duplication of single pronuclei resulting in completely homozygous clones.[31]

Neurogenetic studies have shown how the linkages in the multilocus system subserving a process as complex as reproduction can be manipulated so as to examine the component parts of reproduction.[41] It is reasonable to assume that the linkage between loci controlling mating behavior and the loci controlling fecundity might also be to some extent independent of one another. Finally, the observation that some all-female species engage in sexual behavior questions the commonly held assumption that gonadal sex and behavioral sex are intrinsically functionally associated. That the functional outcome of "sexual" behaviors in unisexual species is similar to that found in sexual vertebrate species leads one to ask whether behavioral sex is not more fundamental than gonadal sex.[43] That is, many lineages have lost males, yet behaviors facilitating reproduction have persisted in unisexual and asexual organisms. This suggests that behavioral facilitation of reproduction may be ancestral to the existence of separate sexes.

ACKNOWLEDGMENTS

I would like to thank Eric Bittman, Harvey Feder, and Ernst Mayr for comments on the manuscript.

REFERENCES

1. DIAMOND, J. M. 1983. Laboratory, field and natural experiments. Nature **304:** 586-587.
2. CREWS, D. 1975. Psychobiology of reptilian reproduction. Science **189:** 1059-1065.
3. CREWS, D. 1980. Interrelationships among ecological, behavioral and neuroendocrine processes in the reproductive cycle of *Anolis carolinensis* and other reptiles. Adv. Study Behav. **11:** 1-74.
4. CREWS, D. 1984. Gamete production, sex hormone secretion, and mating behavior uncoupled. Horm. Behav. **18:** 22-28.
5. COURRIER, R. 1927. Etude sur le determinisme des caracteres sexuels secondaires chez quelques mamiferes a activite testiculaire periodique. Arch. Biol. **37:** 173-334.

6. CREWS, D. & W. R. GARSTKA. 1982. The ecological physiology of reproduction in the Canadian red-sided garter snake. Sci. Am. **247:** 158-168.
7. CAMAZINE, B., W. GARSTKA, R. TOKARZ & D. CREWS. 1980. Effects of castration and androgen replacement on male courtship behavior in the red-sided garter snake (*Thamnophis sirtalis parietalis*). Horm. Behav. **14:** 358-372.
8. ALEKSIUK, M. & P. T. GREGORY. 1974. Regulation of seasonal mating behavior in *Thamnophis sirtalis parietalis*. Copeia **1974:** 681-689.
9. HAWLEY, A. W. L. & M. ALEKSIUK. 1975. Thermal regulation of spring mating behavior in the red-sided garter snake (*Thamnophis sirtalis parietalis*). Can. J. Zool. **53:** 768-776.
10. HAWLEY, A. W. L. & M. ALEKSIUK. 1976. The influence of photoperiod and temperature on seasonal testicular recrudescence in the red-sided garter snake (*Thamnophis sirtalis parietalis*). Comp. Biochem. Physiol. **53A:** 215-221.
11. CREWS, D., B. CAMAZINE, M. DIAMOND, R. MASON, R. TOKARZ & W. R. GARSTKA. 1984. Hormonal independence of courtship behavior in the male garter snake. Horm. Behav. **18:** 29-41.
12. GARSTKA, W. R., B. CAMAZINE & D. CREWS. 1982. Interactions of behavior and physiology during the annual reproductive cycle of the red-sided garter snake, *Thamnophis sirtalis parietalis*. Herpetologica **38:** 104-123.
13. FRIEDMAN, D. W. & D. CREWS. 1985. Role of the anterior hypothalamus-preoptic area in the regulation of courtship behavior in the male Canadian red-sided garter snake (*Thamnophis sirtalis parietalis*): intracranial implantation studies. Horm. Behav. **19:** 122-136.
14. FRIEDMAN, D. W. & D. CREWS. 1985. Role of the anterior hypothalamus-preoptic area in the regulation of courtship behavior in the male Canadian red-sided garter snake (*Thamnophis sirtalis parietalis*): lesion studies. Behav. Neurosci. **99:** 942-949.
15. SATINOFF, E. 1983. A reevaluation of the concept of the homeostatic organization of temperature regulation. *In* Handbook of Behavioral Neurobiology, Vol. 6: Motivation. E. Satinoff & P. Teitlebaum, Eds.: 443-472. Plenum Press. New York, N.Y.
16. CREWS, D. 1985. Effects of early sex hormone treatment on courtship behavior and sexual attractivity in the red-sided garter snake, *Thamnophis sirtalis parietalis*. Physiol. Behav. **35:** 569-575.
17. GARSTKA, W. R., R. R. TOKARZ, A. HALPERT, M. DIAMOND & D. CREWS. 1985. Social and hormonal control of yolk synthesis and deposition in the red-sided garter snake, *Thamnophis sirtalis parietalis*. Horm. Behav. **19:** 137-153.
18. HALPERT, A., W. R. GARSTKA & D. CREWS. 1982. Sperm transport and storage and its relation to the annual sexual cycle of the female red-sided garter snake, *Thamnophis sirtalis parietalis*. J. Morphol. **174:** 149-159.
19. GANZHORN, D. & P. LICHT. 1983. Regulation of seasonal gonadal cycles by temperature in the painted turtle, *Chrysemys picta*. Copeia **1983:** 347-358.
20. DRYDEN, G. L. & J. N. ANDERSON. 1977. Ovarian hormone: lack of effect on reproductive structures of female Asian musk shrews. Science **197:** 782-784.
21. COLE, C. J. 1975. Evolution of parthenogenetic species of reptiles. *In* Intersexuality in the Animal Kingdom. R. Reinboth, Ed.: 340-355. Springer-Verlag. West Berlin, FRG.
22. COLE, C. J. & C. R. TOWNSEND. 1983. Sexual behavior in unisexual lizards. Anim. Behav. **31:** 724-728.
23. CREWS, D. & K. T. FITZGERALD. 1980. "Sexual" behavior in parthenogenetic lizards (*Cnemidophorus*). Proc. Natl. Acad. Sci. USA **77:** 499-502.
24. MOORE, M. C., J. M. WHITTIER, A. J. BILLY & D. CREWS. 1985. Male-like behaviour in an all-female lizard: relationship to ovarian cycle. Anim. Behav. **33:** 284-289.
25. WERNER, Y. L. 1980. Apparent homosexual behavior in an all-female population of a lizard, *Lepidodactylus lugubris* and its probable interpretation. Z. Tierpsychol. **54:** 144-150.
26. GUSTAFSON, J. E. & D. CREWS. 1981. Effect of group size and physiological state of a cagemate on reproduction in the parthenogenetic lizard, *Cnemidophorus uniparens* (Teiidae). Behav. Ecol. Sociobiol. **8:** 267-272.
27. CARSON, H. L., L. S. CHANG & T. W. LYTTLE. 1982. Decay of female sexual behavior under parthenogenesis. Science **218:** 68-70.

28. CREWS, D., L. T. TERAMOTO & H. L. CARSON. 1985. Behavioral facilitation of reproduction in sexual and parthenogenetic *Drosophila*. Science **227:** 77-78.
29. COOK, R. M. 1970. Control of fecundity in *D. melanogaster*. Drosoph. Inf. Service **45:** 128.
30. MULLER, H. J. 1944. Use of males with defective Y's to promote the laying of unfertilized eggs. Drosoph. Inf. Service **18:** 58.
31. CARSON, H. L. 1973. The genetic system in parthenogenetic strains of *Drosophila mercatorum*. Proc. Natl. Acad. Sci. USA **70:** 1772-1774.
32. LEHRMAN, D. S. 1970. Semantic and conceptual issues in the nature-nurture problem. *In* Development and Evolution of Behavior. L. R. Aronson, E. Tobach, D. S. Lehrman & J. S. Rosenblatt, Eds.: 17-52. W. H. Freeman and Co. San Francisco, Calif.
33. BARRINGTON, E. J. W. 1964. Hormones and Evolution. Robert E. Krieger Publishing Co. Inc. Huntington, N.Y.
34. KANATANI, H. & Y. NAGAHAMI. 1980. Mediators of oocyte maturation. Biomed. Res. **1:** 273-291.
35. CREWS, D. 1986. Functional associations in behavioral endocrinology. *In* Masculinity/Femininity: Concepts and Definitions. J. Reinisch, Ed. Oxford University Press. London, England. In press.
36. HALPERN, M., J. MORRELL & D. W. PFAFF. 1982. Cellular ^{3}H-estradiol and ^{3}H-testosterone localization in the brains of garter snakes: an autoradiographic study. Gen. Comp. Endocrinol. **46:** 211-224.
37. KEEFER, D. A. & G. L. DRYDEN. 1982. Nuclear uptake of radioactivity by cells of pituitary, brain, uterus, and vagina of the Asian musk shrew (*Suncus murinus*) following [^{3}H] estradiol administration. Gen. Comp. Endocrinol. **47:** 125-130.
38. INGLE, D. & D. CREWS. 1985. Vertebrate neuroethology: definitions and paradigms. Annu. Rev. Neurosci. **8:** 457-494.
39. THIESSEN, D. D. 1983. Thermal constraints and influences on communication. Adv. Study Behav. **13:** 147-171.
40. HOY, R. R. 1974. Genetic control of acoustic behavior in crickets. Am. Zool. **14:** 1067-1080.
41. QUINN, W. G. & R. J. GREENSPAN. 1984. Learning and courtship in *Drosophila*: two stories with mutants. Annu. Rev. Neurosci. **7:** 67-93.
42. CUELLAR, O. 1971. Reproduction and the mechanism of meiotic restitution in the parthenogenetic lizard *Cnemidophorus uniparens*. J. Morphol. **133:** 139-166.
43. CREWS, D. 1982. On the origin of sexual behavior. Psychoneuroendocrinology **7:** 259-270.

PART II. HORMONES AS REGULATORS OF REPRODUCTIVE BEHAVIOR
A. Parental Behavior

Introduction

JAY S. ROSENBLATT

Institute of Animal Behavior
Rutgers-The State University of New Jersey
Newark, New Jersey 07102

Maternal behavior is among the most interesting behavior patterns in the field of hormones and behavior. Apart from its importance in the life of animals and as a central behavior pattern around which are organized many other patterns of behavior, maternal behavior has inherently intriguing features. In the rat, maternal behavior consists of a rich repertoire of behavior including nursing, nest-building, retrieving pups back to the nest and licking them to eliminate. Considerable progress has been made in analyzing the causal factors underlying the appearance of these behaviors and the function they play in the interaction between the mother and the young. During its course there are times when maternal behavior is largely based upon physiological processes associated with gestation and delivery and other times when behavioral and psychological processes play the more important roles.

Maternal behavior arises during late pregnancy rather than, as formerly believed, during parturition and passes through several phases before it gradually wanes with weaning of the pups between 20 and 25 days of age. In his paper *H. I. Siegel* examines the experimental models that have been developed for studying the hormonal basis of the initiation of maternal behavior. The evidence indicates that the rise in circulating levels of estrogen at the end of pregnancy preceded by the fall in progesterone levels and primed by the earlier high levels of progesterone and, perhaps, low levels of estrogen, prior to the fall, is responsible for the onset of maternal behavior. R. W. Bridges in a recently published article (Science **227** (1985): 783-784) has proposed that prolactin, released in response to estrogen, may, at least under certain circumstances, be the ultimate hormonal stimulus of maternal behavior. If this is the case, then previous studies showing a failure to prevent the onset of maternal behavior using agents that inhibit the release of pituitary prolactin will have to be reexamined. Prolactin has already been shown to play an important role in the maternal behavior of the rabbit and as *J. D. Buntin* shows in his review of the avian literature, prolactin plays an important role in incubation and parental care in the ring dove and a variety of birds.

The complexity of maternal behavior and of the endocrine changes accompanying the onset of maternal behavior may have discouraged many investigators during the late 1940s and most of the 1950s from studying this behavior pattern. On the other hand, this complexity may serve as a challenge and as a continual source of new and exciting ideas. Nowhere in the study of maternal behavior is the complexity as evident as in the study of the prepartum onset which *A. D. Mayer* has reviewed, much of the work being her own, with great clarity. While the hormonal changes underlying the last phase of pregnancy play the major role in the onset, there may be a contribution from the uterine contractions that precede actual parturition and both of these may act via a change in olfactory responsiveness to newborn pups making them attractive whereas earlier they evoked avoidance responses. The onset of nest defense appears to share with the onset of maternal care, common underlying hormonal mechanisms but the two differ in their degree of dependence upon pup stimulation for initiation.

Behavioral and endocrine approaches to the study of maternal behavior in the rat have been supported by studies of the neural sites of organization of the pattern and of the action of hormones, chiefly, estrogen, mediating the behavior. The neural networks underlying maternal behavior have been studied by *M. Numan* and *A. Fleming*, among others, and their studies and syntheses are presented in their papers in this volume. The complexity of the behavior pattern is matched, understandably, by an intricate neural organization.

Several disciplines converge in the study of maternal behavior as, for example, the study of individual and social (i.e., mother-young) behavior, endocrinology, neuroendocrinology, neurophysiology, and neuroanatomy. There has been, nevertheless, an unusual degree of communication among the disciplines and the integration of their findings is evidenced by the contributions to this volume. Certain general themes in the study of maternal behavior have therefore emerged and these provide a set of common problems across the disciplines.

It is generally acknowledged that there are two general phases in the regulation of maternal behavior, a hormonal phase during which maternal behavior is initiated prepartum, and a nonhormonal phase which has its beginning during parturition but receives strong impetus in the initial interactions postpartum between the mother and her young. The hormonal stimulus has been shown to persist for variable lengths of time in different strains of rats, when pups have been removed prior to female contact with them and females have been tested periodically with foster pups. The period during which this hormonal stimulus persists and the new nonhormonal regulation is being established has been called the transition period. This period has been studied extensively in humans and in sheep where it is often called the "critical" period for the formation of the mother-young relationship.

The nonhormonal phase of maternal behavior normally arises only in the context provided by the hormonal onset but it can be isolated from the hormonal phase in nonpregnant cycling females exposed to pups by a procedure that has been labeled "sensitization." These females begin to show maternal behavior after about 5 to 7 days of continuous exposure to pups that have been renewed every 24 hr. Avoidance of pups either because of their novelty, but more likely because they are olfactorily repugnant to nonmaternal females, needs to be overcome before nonpregnant females can respond to their attractive properties and exhibit maternal behavior toward them. The ability to study the onset of maternal behavior by a method other than hormone stimulation has enabled developmental studies to be done, in both males and females, before hormones play a role. Also, it has introduced into the study of maternal behavior the role of sensory stimuli (chiefly olfactory, but other sensory stimuli as well) with respect to their attractive and repellent effects, the corresponding approach and avoidance tendencies of females and males, and the accompanying emotional responses. How these change under the influence of the hormones which stimulate the rapid onset of maternal behavior is a challenging problem and how this, in turn, is influenced by previous experience as a mother, adds to the challenge.

The complexity of maternal behavior reflects the special role it plays in the ontogenetic and phylogenetic processes among species. It is subject to mutual natural selection during ontogeny by the offspring and the resources in the habitat and by the need to be coordinated with the physiological processes underlying all other reproductive functions (e.g., mating, pregnancy, lactation, etc.). Phylogeny has imposed certain limits on maternal care based upon structural, physiological, and behavioral capacities but it has also favored this form of ensuring survival of young.

Among all animal lines we can note the evolution of maternal care from a more generalized dispersal of the products of reproduction without further care by either parent. Since to be instrumental in evolution all of the adaptations must be transmitted from parent to offspring, the addition of parental care to other means of transmission places this behavior centrally in phylogeny as well as ontogeny.

Hormonal Basis of Maternal Behavior in the Rat[a]

HAROLD I. SIEGEL

Institute of Animal Behavior
Rutgers-The State University of New Jersey
Newark, New Jersey 07102

Just about 50 years ago, Riddle and his colleagues wrote, "Prolactin is the anterior pituitary hormone specifically concerned in the activation of maternal behavior in virgin rats" (Reference 1, p. 734). On the basis of further studies in which different groups of intact and ovariectomized nulliparous rats were injected for a series of days with a variety of substances, Riddle *et al.* wrote:

> Recurrent maternal behavior in the reproducing animal (which the rats in these tests were not) probably does not depend upon the several hormones here found to be directly or very indirectly active, but upon that one of the group which *a*, is released in increased amount at the right time; *b*, exerts an antigonad action; and *c*, then directly or indirectly increases the excitability of the sensorimotor mechanism specifically involved in this instinctive behavior. Though present information is inadequate the hormone which apparently best fits these requirements is prolactin. (Reference 2, p. 316)

The conclusions of Riddle *et al.* pertain to the two major issues I wish to address in this paper. The first concerns the identity of the hormonal factors that stimulate maternal behavior. Despite five decades of research since the work of Riddle *et al.*, the precise role of prolactin remains unclear. I will summarize the results of those studies showing an effect of prolactin and the approximately equal number of studies that have failed to demonstrate such an effect. In addition, I will review the more recent evidence supporting a role for other hormones, specifically, estrogen, progesterone, and oxytocin.

The second issue is a methodological one. The studies of Riddle *et al.* and a number of other investigators have used the naive virgin animal as a model to understand the role of hormones (and other sensory, developmental, and nonhormonal factors) in the display of maternal behavior in the parturient rat. While this approach has been quite valuable, attempts must be made to relate the findings in nulliparous animals to the more relevant question of the mechanisms by which a mother becomes maternal toward her own litter. Although important, the use of the "natural" mother is difficult because of experimentally induced complications involving pregnancy and parturition. These difficulties will be pointed out and the currently available models will be assessed.

---

[a] Institute of Animal Behavior Publication No. 423.

THE EARLY STUDIES

A critical role for prolactin in the activation of maternal behavior cannot be supported by the data of Riddle et al.[1,2] Treatment with prolactin resulted in maternal care by 71% and 79% of intact and ovariectomized rats, respectively. Other treatments stimulated the behavior in similar or greater percentages of animals: in intact females, by estrone withdrawal (78%) and progesterone (68%); in ovariectomized animals, by testosterone (85%) and intermedin (77%). It was perhaps sensitization (see next section) that accounted, at least in part, for these results. The animals in these studies were exposed to pups on a daily basis both before and during treatment, and it is now known that such exposure can induce maternal behavior in nulliparous animals.

Attempts to confirm the effectiveness of prolactin or that of other hormones did not appear for 20 years. None of the several studies in the early 1960s on prolactin,[3,4] progesterone,[5] or estrogen[4] could demonstrate a hormonal stimulation of maternal behavior in virgin animals. Thus, at this time, there was no reliable treatment to induce maternal care in nulliparous rats.

METHODOLOGICAL ADVANCES

Several important research advances during the late 1960s and early 1970s have contributed greatly to the current knowledge of the hormonal basis of maternal behavior. These advances, which have formed the basis of virtually all subsequent experiments in this field, include the systematic investigation of sensitization, the development of radioimmunoassay procedures, and pregnancy termination as a model for the study of parental care.

Sensitization, also known as concaveation, involves the presentation of test pups to animals not expected to be immediately maternal. Ordinarily, the pups are exchanged daily for freshly nourished young until the experimental animals display maternal care. Rosenblatt[6] showed that intact, gonadectomized, and hypophysectomized virgin female and male rats could be sensitized after several days of continuous exposure (24 hr/day) to young pups. These animals build nests, lick and retrieve pups, and crouch over the young in a lactating posture although they are unable to produce milk.

An understanding of sensitization has proven useful in several ways. Sensitization latencies in ovariectomized animals provide a baseline, nonhormonal measure of maternal responsiveness. The latencies of animals given exogenous hormones and then exposed to pups can be used to assess the effectiveness of these hormones. Treatments that result in latencies significantly shorter than those of controls (4 to 7 days in our Charles River Sprague-Dawley rats) are assumed to be related to endogenous factors operating in the pregnant and parturient animals. In addition, females show shorter latencies as gestation advances suggesting an increasing responsiveness to pups during pregnancy (see Reference 7). Sensitized animals have also been used to separate the effects of pregnancy from the display of maternal care. For example, the maternal aggression of lactating animals has been compared to that of virgin animals induced to be maternal through pup stimulation (see A. D. Mayer, this volume).

Clearly, the parturient animal cannot rely on her sensitization capacity. The newborns would not survive the several days required for sensitization to occur. What then is the role of this nonhormonal responsiveness to pups? Rosenblatt[8] has suggested that the onset of maternal behavior at parturition is mediated by the endogenous changes during pregnancy. Once hormonally stimulated, the mother's behavior is maintained by her contact with the litter. The endocrine profile of lactation is very different from that of the prepartum period that stimulated maternal behavior and further, ovarian hormones are not necessary for the postpartum maternal care. In addition, it is the changing characteristics of the pups that determine the level of the mother's litter-related behavior from shortly after birth until weaning. Substituting different-aged pups can either increase or decrease the intensity and frequency of maternal care.

While sensitization provided both an understanding of the overall regulation of maternal behavior and a methodology to test the effectiveness of various agents, it was the application of radioimmunoassay procedures that directed the search for the hormones involved in the onset of maternal behavior. Although a number of hormones are present and fluctuate during gestation, especially the prepartum period, most research attention has focused on the ovarian steroids and prolactin. For reference, FIGURE 1 shows the levels of these hormones during pregnancy in the rat.

It had been known that surgical termination of pregnancy by cesarean delivery near the time of expected parturition resulted in short-latency maternal care.[9,10] Lott and Rosenblatt[11] and later Rosenblatt and Siegel[12] terminated pregnancy by hysterectomy at different stages of gestation. Hysterectomy performed at advancing stages of gestation resulted in progressively shorter latencies to the onset of maternal behavior (FIGURE 2). (This finding is similar to that mentioned earlier for intact females tested with pups on different days of pregnancy.) Further, females hysterectomized and ovariectomized displayed longer latencies than those hysterectomized only.

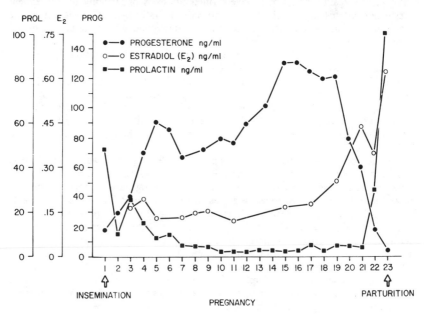

FIGURE 1. Circulating levels of progesterone, estradiol, and prolactin during pregnancy in the rat. (Reprinted from Rosenblatt et al.[7] with permission of Academic Press, Inc.)

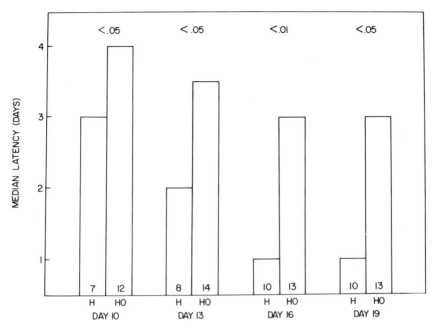

FIGURE 2. Median latencies to the onset of maternal behavior in hysterectomized (H) and hysterectomized-ovariectomized (HO) rats on days 10, 13, 16, and 19 of pregnancy. Testing began 48 hr postoperatively. Numbers within bars represent sample sizes. (Reprinted from Rosenblatt et al.[7] with permission of Academic Press, Inc.)

The pregnancy termination model offers several advantages. Pregnancy-terminated animals would be expected to be more similar in physiology and behavior to the parturient animal than the nulliparous animal. Experimental manipulations that interfere with maternal behavior indirectly through deleterious consequences on parturition can be used with pregnancy-terminated animals. For example, injections of progesterone during late pregnancy delay the onset of labor and can lead to fetal and maternal death; therefore the effects of progesterone on maternal behavior have been determined in pregnancy-terminated animals (see below). Finally, pregnancy termination allows for greater experimenter control. Animals hysterectomized (but not ovariectomized) on particular days of pregnancy are reliably maternal 2 days later during their first exposure to pups. This finding allowed researchers for the first time to define an interval during which to search for the relevant endocrine events leading to the onset of maternal behavior.

IDENTIFICATION OF THE MATERNAL HORMONES: PREGNANCY SIMULATION

Apart from the studies of Riddle et al.,[1,2] the first successful attempts at inducing maternal behavior used the radioimmunoassay results to simulate the salient endocrine characteristics of pregnancy by exogenous injections in ovariectomized virgin animals. The treatment used by Moltz et al.[13] employed all three hormones shown in FIGURE 1. The rats were injected with estradiol benzoate (EB) for 11 days, progesterone on days 6 through 9, and prolactin on days 9 and 10. Latencies measured from the time the pups were first introduced on day 10 were about 35 hr, a significant reduction in comparison to those of vehicle-treated controls. Groups receiving any of the combinations of two of the three hormones were not as responsive. The authors concluded that all three hormones were important and they were the first to place special emphasis on the behavioral effects of progesterone withdrawal.

Progesterone withdrawal refers to a period of high progesterone stimulation followed by its decline, a situation found during pregnancy and simulated in the Moltz et al. treatment. The withdrawal of progesterone is considered to increase the animal's responsiveness to other hormones and had previously been associated during pregnancy with lactogenesis and uterine contractions. Further evidence of the stimulating effects of progesterone withdrawal will be discussed below.

Short latencies to the onset of maternal behavior were also reported in a study designed to induce mammary gland growth and lactation in ovariectomized virgin rats.[14] The animals in this study were injected for 20 days with estrogen and progesterone and then daily with prolactin and cortisol until the end of testing. Pups were first introduced on day 25. The group treated with all four hormones had mean retrieval latencies of 1.4 days compared to 4-5 days for the controls. Two other groups were given three of the four hormones, either estrogen, progesterone, and prolactin or estrogen, progesterone, and cortisol. The latencies of these groups were 1.4 and 1.8 days, respectively.

SEARCH FOR SHORTER TREATMENTS

At about the same time that the pregnancy simulation studies were being carried out, Terkel and Rosenblatt[15,16] were using a different approach to induce maternal behavior. In the first of their studies, nulliparous rats given a single injection of blood plasma taken from recently parturient animals required about 2 days of pup exposure to become maternal. In the second more ambitious experiment, intact virgin rats received a 6-hr cross-transfusion of blood from either late-pregnant or parturient animals. The blood was most effective when the transfusion began as soon as the donor females had completed their deliveries. The virgin recipients were exposed to pups at the start of the transfusion and responded maternally with a mean latency of 14-15 hr. Transfusions from either late-pregnant (about 27 hr prepartum) or lactating (24 hr postpartum) animals were not as effective.

The injection and transfusion results modified both the interpretation of the pregnancy simulation experiments and the direction of the next series of studies. First, treatments involving a single injection of plasma or a 6-hr transfusion are much shorter than treatments employing 11 or 24+ days of the administration of three or four hormones.[13,14] Second, both pregnancy simulation studies involved progesterone withdrawal and it appears unlikely that such an effect played a role in the plasma injection and blood transfusion experiments. In both cases, progesterone would have declined to quite low levels in the mothers at the time their blood was removed or transfused. While the Terkel and Rosenblatt studies shifted the focus away from extended hormone treatments, they did not identify the critical factor(s) involved.

PREGNANCY TERMINATION

The usefulness of studying the pregnant animal and the demonstration that only a short treatment was necessary to induce maternal behavior in virgin animals formed the background for many of the pregnancy termination experiments. As mentioned earlier, hysterectomy during pregnancy resulted in short-latency maternal care beginning 48 hr later when pups were first presented.[11,12] Combined hysterectomy-ovariectomy resulted in significantly longer latencies. At this time it was hypothesized that hysterectomy during pregnancy was followed by an alteration in endocrine patterns that is normally associated with natural pregnancy termination at parturition.[12] These hormonal changes include a decline in progesterone and an increase in estrogen (and prolactin). This hypothesis was based primarily on two lines of evidence. The first was the accelerated onset of maternal responsiveness in animals hysterectomized during pregnancy compared to animals that either were hysterectomized-ovariectomized or remained intact. The second was the studies showing that hysterectomy during gestation was followed by a change in vaginal cytology in the direction of the resumption of estrous cyclicity.[17] Because the ovaries were critical, the major research question concerned the nature of ovarian secretion following hysterectomy.

Further confirmation of the suggested model of pregnancy termination-induced hormonal events came about unexpectedly. At the time, all experimental animals in Rosenblatt's laboratory were tested in large Plexiglas cages and the number of these cages determined the rate at which adequate sample sizes were obtained. Pregnancy-terminated animals were first presented with pups 48 hr after surgery and as soon as the animals displayed maternal care on two consecutive days, they were removed from their test cages to make room for new animals. On one day, several day 16 hysterectomized animals had retrieved their pups for the second time and were removed from the test cages and hastily placed together in another cage. In response to the tactile stimulation, some of these animals, now 4 days post-hysterectomy, showed lordosis behavior. This observation fit with the hypothesis presented above. Not only did these animals display an earlier onset of maternal behavior but also sexual behavior, a situation comparable to that of the animal who becomes maternal at parturition and then later that day enters postpartum estrus.

The relatively simple discovery that hysterectomy during pregnancy results in an increase in maternal and sexual responsiveness formed the basis of a great many studies beginning in the early 1970s and continuing to the present. One set of experiments determined the timing of sexual receptivity, ovulation, and the decline of progesterone relative to that of hysterectomy. FIGURE 3 shows these data along with the increase in maternal responsiveness when hysterectomy was performed on days 13, 16, and 19. The rate at which these events occurred was determined by the particular day on which the surgery was done; for example, the latency from hysterectomy to lordosis and ovulation was 5 to 6 days, 4 days, and 3 days for animals hysterectomized on days 13, 16, and 19, respectively. It is worth noting that progesterone first declines and then rises again around the time of sexual receptivity. This rise in progesterone is also observed in animals on the first day of lactation (see Reference 7).

The treatment to induce lordosis behavior in ovariectomized animals had been known for some time to consist of estrogen followed 2 days later by progesterone. Our first attempt to exogenously stimulate in hysterectomized-ovariectomized animals the maternal and sexual behavior of the hysterectomized animals made use of this estrogen-progesterone treatment. Pregnant animals were hysterectomized-

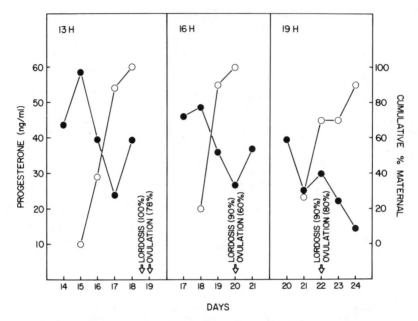

FIGURE 3. Circulating levels of progesterone (solid circles) and cumulative percentage of females showing maternal behavior (open circles) following hysterectomy (H) on days 13, 16, and 19 of pregnancy. Maternal behavior tests began 48 hr postoperatively. Time of lordosis and ovulation also shown. (Reprinted from Rosenblatt et al.[7] with permission of Academic Press, Inc.)

ovariectomized, given 100 μg/kg EB at surgery and 0.5 mg progesterone 44 hr later, and then tested with pups and sexually active males. As expected these rats displayed lordosis behavior, but more importantly, over 60% of the animals retrieved and crouched over the pups during their first hour of exposure to the young (see FIGURE 4). The progesterone was given to simulate the increase in this hormone that occurs postpartum after the parturient animal is already maternal and therefore, it was not thought to be necessary for maternal behavior. Consequently, additional groups were treated with one of two doses of EB without the progesterone. FIGURE 4 also shows that progesterone was not required for the maternal responsiveness.

We proceeded to apply the same treatment found effective in day 16 animals to rats on the other days of gestation that we had been studying. Essentially, the results were identical (see Reference 7) and now for the first time, using a more "natural" model, we identified a single hormone that appeared capable of stimulating short-latency maternal responsiveness. The importance of estrogen in our experimental preparation could be related to the endogenous increase in this hormone that occurs in the intact prepartum animal at a time when maternal responsiveness is also increasing (see Reference 12 and Mayer, this volume). We have assumed that estrogen increases following hysterectomy during pregnancy because the surgery is followed by the display of lordosis behavior and ovulation, two estrogen-dependent events.

The findings of three subsequent experiments helped confirm the role of estrogen. First, latencies to the onset of maternal behavior were also quite short in animals hysterectomized-ovariectomized on day 16 and given implants of estrogen in the medial preoptic area,[19] a site which when lesioned severely disrupts ongoing maternal care.[20] Second, because exogenous estrogen has been shown to release endogenous prolactin,[21] it is possible that the estrogen stimulated maternal behavior indirectly. However, estrogen had the same effects on the behavior of pregnancy-terminated animals when combined with prolactin release blockers.[19,22]

The third series of studies[23] focused on the nulliparous model. Virgin animals given a single injection of EB showed latencies to the onset of maternal behavior that were comparable to those of animals treated more extensively.[13,14] Two qualifications are necessary. In order to stimulate maternal behavior within 24 to 48 hr of pup exposure, the animals required large doses of EB and needed to be ovariectomized and hysterectomized. Single injections of 100 µg/kg EB were necessary in the initial studies; 50 µg/kg EB or lower doses were not effective.[24] The effect of hysterectomy was interpreted as removing a major estrogen target tissue thereby resulting in greater stimulation of brain areas by the hormone. This "uterine sponge" hypothesis generated additional studies demonstrating the existence of similar uterine-hormonal effects on other estrogen-dependent behaviors and associated molecular events (see Reference 25).

These studies also led to questions concerning the validity of the EB-treated virgin rat as an experimental model. The parturient animal is immediately responsive to pups at parturition but the virgin given EB still requires a period of pup exposure. It should be noted here that recent studies[26] using a more "naturalistic" approach have further reduced the latencies of the virgin animals. Ordinarily, rats deliver one pup at a time in the nest during the light phase of their photoperiod. More than 70% of EB-treated virgins became maternal within 1.5 hr by presenting them with one pup in the nest during the light followed by additional pups 30 min later.

Day 16 pregnancy-terminated animals given 20 µg/kg EB, or 5 µg/kg EB in later studies, were highly responsive to pups. The virgin animals required 100 µg/kg EB, a 20-fold-larger dose. Further, although no effect of endogenously released prolactin in pregnancy-terminated animals was found, similar studies in the virgin animals were not performed. Finally, there was the hysterectomy requirement, a finding that still needs to be reconciled with the role of the intact uterus in parturient rats. For these reasons, attention began to focus back on other hormonal conditions. In the next sections, I will discuss progesterone, prolactin, and oxytocin.

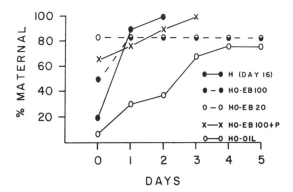

FIGURE 4. Cumulative percentage of day 16 pregnant rats showing maternal behavior after hysterectomy (H) or hysterectomy-ovariectomy (HO) plus 100 or 20 µg/kg estradiol benzoate (EB) or oil at surgery and either 0.5 mg progesterone or oil 44 hr later. Maternal behavior tests began 48 hr postoperatively. (Reprinted from Siegel and Rosenblatt[18] with permission of Academic Press, Inc.)

PROGESTERONE: INHIBITION AND FACILITATION

Ordinarily, progesterone levels are high during much of pregnancy and decline rapidly near term. The maintenance of high levels of progesterone delays or inhibits the onset of maternal behavior. This has been shown in pregnant animals[27] and in pregnancy-terminated and virgin animals given EB.[28,29] While these inhibitory effects have been confirmed,[30,31] the mechanism remains unknown.

A second more important effect of progesterone concerns its withdrawal-induced facilitation of maternal care. Moltz et al.[13] were the first to demonstrate and discuss the importance of progesterone withdrawal. Since then Bridges et al.[31] have argued that progesterone withdrawal is involved in the maternal behavior observed in pregnancy-terminated animals. One reason why virgin animals required large doses of EB may be that they were "missing" a prior period of progesterone stimulation followed by its withdrawal. We have since shown that the amount of EB required to induce

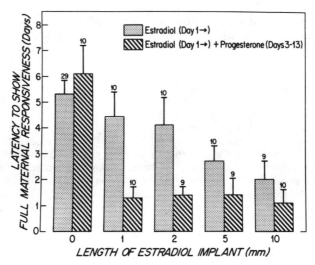

FIGURE 5. Effects of estradiol implants alone or combined with progesterone implants on latencies to the onset of maternal behavior in ovariectomized rats. Estradiol was implanted on day 1 and remained throughout testing. Progesterone was implanted on day 3 and removed on day 13. Testing began on day 14. (Reprinted from Bridges[33] with permission of The Endocrine Society.)

maternal care in nulliparous animals could be reduced to that found effective in pregnancy-terminated animals (5 μg/kg) if the virgins are given subcutaneous progesterone implants which are then removed prior to the administration of EB.[32] Bridges[33] has confirmed this effect and FIGURE 5 shows the results of his systematic studies on the relationship of the two hormones.

PROLACTIN

It is currently not possible to define a role for prolactin in maternal care. The early studies of Riddle et al.,[1,2] pointing to an effect of prolactin and still frequently cited, may have been confounded by sensitization and have not been replicated. In the Moltz *et al.* study,[13] animals receiving estrogen, progesterone, and prolactin had shorter and more uniform latencies than animals given estrogen and progesterone only.

Most of the remaining studies on prolactin have failed to demonstrate an effect of the hormone. In the similar but more extensive pregnancy simulation design,[14] animals treated with estrogen, progesterone, and cortisol with and without prolactin did not differ in their maternal latencies. Nulliparous rats given pituitary transplants to allow for high circulating levels of prolactin did not respond any earlier to pups than non-transplanted animals.[34]

Using an obvious approach, Obias[35] hypophysectomized animals after mid-pregnancy and observed their parturitional behavior. These results are difficult to interpret. The disruptive effects of hypophysectomy on the timing and duration of parturition and on the near-term decline in progesterone and increase in estrogen preclude an analysis of the role of prolactin in this case. Pregnant animals given prolactin release blockers were unable to lactate but showed adequate maternal care.[14,36] These drugs were also unable to interfere with the short-latency maternal behavior observed in EB-treated, pregnancy-terminated animals.[19,22]

Most recently, a role for prolactin has been suggested again.[37] Ovariectomized virgin rats were treated for 10 days with Silastic progesterone implants. On day 11 the progesterone implants were removed and estrogen implants inserted. Testing with pups began on day 12. These animals responded after a mean of about 2 days of pup exposure. However, the same treatment in hypophysectomized rats resulted in latencies of about 8 days. Latencies of steroid-treated hypophysectomized animals were reduced to 2 days if these animals were also given pituitary grafts or injected with prolactin on days 1 through 13.

OXYTOCIN

In the original studies published only several years ago,[38] oxytocin was shown to facilitate maternal behavior. Ovariectomized rats were treated with 100 μg/kg EB and 2 days later with a lateral ventricular infusion of oxytocin. The majority of these animals displayed maternal care within 2 hr of pup exposure which began immediately following the infusions. This effect of oxytocin has been replicated in nulliparous rats[39] and extended to pregnancy-terminated animals.[40] In the latter case, day 16 hysterectomized-ovariectomized, EB-treated animals were given either an antiserum to oxytocin or an analogue antagonist intracerebroventricularly. The rationale was that at least part of these animals' responsiveness to pups is due to their endogenous oxytocin. Both groups showed significant delays in the onset of maternal care.

At this time, some caution is warranted in the interpretation of these oxytocin findings. The treatment in virgins outlined above, a single injection of EB and a single infusion of oxytocin, has been found effective thus far in Sprague-Dawley rats from one particular supplier (Zivic Miller). Attempts to replicate this experiment using Sprague-Dawley rats from another supplier (Charles River) have failed (Reference 41 and unpublished results from our laboratory). There is a report that ovariectomized Charles River animals displayed short latencies but only after 10 days of estrogen priming and a doubling of the oxytocin dose.[42] However, the latencies of these animals were similar to those found in the experiment described earlier[26] in which virgin rats were given a single injection of EB and tested more naturalistically.

The inability to confirm the effect of oxytocin in other laboratories has identified additional issues worth noting. Higher percentages of the animals receiving EB and oxytocin became maternal if they were moved to a strange cage 2 hr prior to testing with pups. Animals tested in their home cages and those moved to a strange cage just 1 hr prior to testing were less responsive to the pups (S. E. Fahrbach, personal communication). While these findings are difficult to explain at present, they may provide clues regarding the nature of the oxytocin effect. Other factors that may be involved in the interlaboratory discrepancies include the age of the test pups and differences in the sources of the oxytocin. While several oxytocin preparations yield similar activity in a uterine bioassay, they vary considerably in their ability to facilitate maternal behavior.[39] It is expected that these points will be addressed in the near future.

CONCLUSIONS

Strong evidence exists for a role for estrogen and also progesterone in maternal behavior in the rat. Estrogen is effective in both the virgin and pregnancy termination models and it is necessary in treatments involving progesterone (inhibition and facilitation), prolactin, and oxytocin. The experimental evidence is consistent with the rising levels of the hormone prepartum when pregnant animals become responsive to pups. The two effects of progesterone are also quite clear. The maintenance of high levels of progesterone in the presence of estrogen delays the onset of maternal care while progesterone withdrawal followed by estrogen (the normal prepartum pattern) facilitates the display of the behavior.

The effects of prolactin and oxytocin still remain uncertain. Stimulatory effects of prolactin have only been shown in virgin animals and then only with additional steroid treatment. Although prolactin levels increase near term, it has not been possible to demonstrate a role for this hormone in pregnancy-terminated or parturient animals. There is evidence that oxytocin is related to maternal care. However, failures to replicate these findings and the associated methodological considerations require clarification.

Three types of preparations have been studied and each has advantages and disadvantages. While descriptive studies in pregnant/parturient animals have been valuable, the results of experimental manipulations including hypophysectomy, ovariectomy, and hormone administration have been difficult to interpret. Disruptive consequences of these manipulations on parturition and the animals' general health may indirectly and adversely affect maternal responsiveness. Also, one hormonal manipulation during pregnancy probably has multiple effects.

The use of the nulliparous animal has avoided many of the problems associated with studying the parturient female. Studies of virgin animals have been able to isolate specific endocrine conditions and separate them from the larger number of such events temporally correlated with the appearance of the behavior. These studies have confirmed the effectiveness of some manipulations identified in pregnancy-terminated animals (estrogen, progesterone inhibition), discovered unexpected findings (effect of hysterectomy), and suggested roles for other hormonal conditions (progesterone withdrawal, prolactin, and oxytocin). It is still necessary to relate the latter two types of results to the natural pregnant/parturient situation.

The pregnancy termination model provides perhaps the best of both worlds; parturition-related disturbances are avoided, endocrine factors after surgery can be studied, and the timing and display of the behavior are highly reliable. Here too, one must be careful in generalizing to the parturient condition. For example, preliminary evidence (see Mayer, this volume) suggests that part of the rapid induction of maternal care in pregnancy-terminated animals may be due to pelvic nerve excitation resulting from the surgery.

At this time, several different approaches to the further study of maternal behavior are evident. These include the approaches taken by the authors of the papers that follow in this section of this volume. In addition, work in our laboratory at the Institute of Animal Behavior currently consists of the following: detailed behavioral analysis of the prepartum period, descriptive and experimental studies of hypothalamic hormone receptor concentrations in all models, pharmacological determinants of the relevant hormonal and behavioral events, a continuing search for new and valid experimental models, and a comparative approach involving maternal behavior in the hamster.

REFERENCES

1. RIDDLE, O., E. L. LAHR & R. W. BATES. 1935. Maternal behavior induced in virgin rats by prolactin. Proc. Soc. Exp. Biol. Med. **32:** 730-734.
2. RIDDLE, O., E. L. LAHR & R. W. BATES. 1942. The role of hormones in the initiation of maternal behavior in rats. Am. J. Physiol. **137:** 299-317.
3. LOTT, D. & S. FUCHS. 1962. Failure to induce retrieving by sensitization or the injection of prolactin. J. Comp. Physiol. Psychol. **55:** 1111-1113.
4. BEACH, F. A. & J. WILSON. 1963. Effects of prolactin, progesterone, and estrogen on reactions of nonpregnant rats to foster young. Psychol. Rep. **13:** 231-239.
5. LOTT, D. 1962. The role of progesterone in the maternal behavior of rodents. J. Comp. Physiol. Psychol. **55:** 610-613.
6. ROSENBLATT, J. S. 1967. Nonhormonal basis of maternal behavior in the rat. Science **156:** 1512-1514.
7. ROSENBLATT, J. S., H. I. SIEGEL & A. D. MAYER. 1979. Progress in the study of maternal behavior in the rat: hormonal, nonhormonal, sensory, and developmental aspects. *In* Advances in the Study of Behavior. J. S. Rosenblatt, R. A. Hinde, C. Beer & M.-C. Busnel, Eds. **10:** 225-311. Academic Press, Inc. New York, N.Y.
8. ROSENBLATT, J. S. 1970. Views on the onset and maintenance of maternal behavior in the rat. *In* Development and Evolution of Behavior: Essays in Memory of T. C. Schneirla. L. R. Aronson, E. Tobach, D. S. Lehrman & J. S. Rosenblatt, Eds.: 489-515. Freeman. San Francisco, Calif.
9. WIESNER, B. P. & N. M. SHEARD. 1933. Maternal Behavior in the Rat. Oliver and Boyd. London, England.
10. MOLTZ, H., D. ROBBINS & M. PARKS. 1966. Caesarean delivery and maternal behavior of primiparous and multiparous rats. J. Comp. Physiol. Psychol. **61:** 455-460.
11. LOTT, D. F. & J. S. ROSENBLATT. 1966. Development of maternal responsiveness during pregnancy in the rat. *In* Determinants of Infant Behavior. B. M. Foss, Ed. **IV:** 61-67. Methuen. London, England.
12. ROSENBLATT, J. S. & H. I. SIEGEL. 1975. Hysterectomy-induced maternal behavior during pregnancy in the rat. J. Comp. Physiol. Psychol. **89:** 685-700.
13. MOLTZ, H., M. LUBIN, M. LEON & M. NUMAN. 1970. Hormonal induction of maternal behavior in the ovariectomized nulliparous rat. Physiol. Behav. **5:** 1373-1377.
14. ZARROW, M. X., R. GANDELMAN & V. H. DENENBERG. 1971. Prolactin: Is it an essential hormone for maternal behavior in mammals? Horm. Behav. **2:** 343-354.
15. TERKEL, J. & J. S. ROSENBLATT. 1968. Maternal behavior induced by maternal blood plasma injected into virgin rats. J. Comp. Physiol. Psychol. **65:** 479-482.

16. TERKEL, J. & J. S. ROSENBLATT. 1972. Humoral factors underlying maternal behavior at parturition: cross transfusion between freely moving rats. J. Comp. Physiol. Psychol. **80:** 365-371.
17. MORISHIGE, W. K. & I. ROTHCHILD. 1974. Temporal aspects of the regulation of corpus luteum function by luteinizing hormone, prolactin, and placental luteotrophin during the first half of pregnancy in the rat. Endocrinology **95:** 260-274.
18. SIEGEL, H. I. & J. S. ROSENBLATT. 1975. Hormonal basis of hysterectomy-induced maternal behavior during pregnancy in the rat. Horm. Behav. **6:** 211-222.
19. NUMAN, M., J. S. ROSENBLATT & B. R. KOMISARUK. 1977. The medial preoptic area and the onset of maternal behavior in the rat. J. Comp. Physiol. Psychol. **91:** 146-164.
20. NUMAN, M. 1974. Medial preoptic area and maternal behavior in the female rat. J. Comp. Physiol. Psychol. **87:** 746-759.
21. KALRA, P. S., C. P. FAWCETT, L. KRULICH & S. M. MCCANN. 1973. The effects of gonadal steroids on plasma gonadotropins and prolactin in the rat. Endocrinology **92:** 1256-1268.
22. RODRIGUEZ-SIERRA, J. & J. S. ROSENBLATT. 1977. Does prolactin play a role in estrogen-induced maternal behavior in rats: apromorphine reduction of prolactin release. Horm. Behav. **9:** 1-7.
23. SIEGEL, H. I. & J. S. ROSENBLATT. 1975. Estrogen-induced maternal behavior in hysterectomized-ovariectomized virgin rats. Physiol. Behav. **14:** 465-471.
24. SIEGEL, H. I., H. DOERR & J. S. ROSENBLATT. 1978. Further studies on estrogen-induced maternal behavior in hysterectomized-ovariectomized nulliparous rats. Physiol. Behav. **21:** 99-103.
25. AHDIEH, H. B. 1984. The role of the uterus in reproductive behavior: a theoretical review. Horm. Behav. **33:** 329-333.
26. MAYER, A. D. & J. S. ROSENBLATT. 1980. Hormonal interaction with stimulus and situational factors in the initiation of maternal behavior in nonpregnant rats. J. Comp. Physiol. Psychol. **94:** 1040-1059.
27. MOLTZ, H., R. LEVIN & M. LEON. 1969. Differential effects of progesterone on the maternal behavior of primiparous and multiparous rats. J. Comp. Physiol. Psychol. **67:** 36-40.
28. SIEGEL, H. I. & J. S. ROSENBLATT. 1978. Duration of estrogen stimulation and progesterone inhibition of maternal behavior in pregnancy-terminated rats. Horm. Behav. **11:** 12-19.
29. SIEGEL, H. I. & J. S. ROSENBLATT. 1975. Progesterone inhibition of estrogen-induced maternal behavior in hysterectomized-ovariectomized virgin rats. Horm. Behav. **6:** 223-230.
30. NUMAN, M. 1978. Progesterone inhibition of maternal behavior in the rat. Horm. Behav. **11:** 209-231.
31. BRIDGES, R. S., J. S. ROSENBLATT & H. H. FEDER. 1978. Serum progesterone levels and maternal behavior after pregnancy termination in rats: behavioral effects of progesterone maintenance and withdrawal. Endocrinology **102:** 258-267.
32. DOERR, H. K., H. I. SIEGEL & J. S. ROSENBLATT. 1981. Effects of progesterone withdrawal and estrogen on maternal behavior in nulliparous rats. Behav. Neural Biol. **32:** 35-44.
33. BRIDGES, R. S. 1984. A quantitative analysis of the roles of dosage, sequence, and duration of estradiol and progesterone exposure in the regulation of maternal behavior in the rat. Endocrinology **114:** 930-940.
34. BAUM, M. J. 1978. Failure of pituitary transplants to facilitate the onset of maternal behavior in ovariectomized virgin rats. Physiol. Behav. **20:** 87-89.
35. OBIAS, M. D. 1957. Maternal behavior of hypophysectomized gravid albino rats and the development and performance of their progeny. J. Comp. Physiol. Psychol. **50:** 120-124.
36. STERN, J. M. 1977. Effects of ergocryptine on postpartum maternal behavior, ovarian cyclicity and food intake in rats. Behav. Biol. **21:** 134-140.
37. BRIDGES, R. S., R. DIBIASE, D. D. LOUNDES & P. C. DOHERTY. 1985. Prolactin stimulation of maternal behavior in female rats. Science **227:** 782-784.
38. PEDERSEN, C. A. & A. J. PRANGE, JR. 1979. Induction of maternal behavior in virgin rats after intracerebroventricular administration of oxytocin. Proc. Natl. Acad. Sci. USA **76:** 6661-6665.
39. FAHRBACH, S. E., J. I. MORRELL & D. W. PFAFF. 1984. Oxytocin induction of short-

latency maternal behavior in nulliparous, estrogen-primed female rats. Horm. Behav. **18:** 267-286.
40. FAHRBACH, S. E., J. I. MORRELL & D. W. PFAFF. 1985. Possible role for endogenous oxytocin in estrogen-facilitated maternal behavior in rats. Neuroendocrinology. **40:** 526-532.
41. RUBIN, B. S., F. S. MENNITI & R. S. BRIDGES. 1983. Intracerebroventricular administration of oxytocin and maternal behavior in rats after prolonged and acute steroid pretreatment. Horm. Behav. **17:** 45-53.
42. ASCHER, J. A., C. A. PEDERSEN, D. E. HERNANDEZ & A. J. PRANGE, JR. 1982. Sources of variance in oxytocin-induced maternal behavior. Soc. Neurosci. Abstr. **8:** 368.

Maternal Responsiveness and Nest Defense during the Prepartum Period in Laboratory Rats[a]

ANNE D. MAYER

Institute of Animal Behavior
Rutgers-The State University of New Jersey
Newark, New Jersey 07102

Developmental analysis and tolerance of complexity are hallmarks of the approach to reproductive behavior practiced and taught by Daniel Lehrman and Jay Rosenblatt. Lehrman's analytic work on the parental behavior of ring doves[1] emphasized the seamlessness of their reproductive behavior as it progresses from courtship to selection of the nest site, construction of the nest, laying and incubation of the eggs, and feeding and finally weaning of the squabs. In their first studies of the maternal behavior of the rat, Rosenblatt and Lehrman[2] included extensive observations of females during gestation, as well as observations of their behavior during parturition, lactation, and weaning. Since that time Rosenblatt's research on maternal behavior has focused primarily on the later phases of the reproductive cycle, commencing with birth of the young. Studies of the prepartum period, however, have continued and indeed appear to possess a certain cyclicity of their own: the behavior of rats during gestation has been scrutinized and rescrutinized, each study yielding new insights but also raising new questions and again challenging our tolerance of complexity. This paper will review the published work on the prepartum period done at the Institute of Animal Behavior (IAB) under Rosenblatt's direction, and will also describe a number of studies in progress, illustrating the breadth of the questions which have been or are being addressed.

In their first work on the maternal behavior of the rat, published in 1963, Rosenblatt and Lehrman[2] observed pregnant rats behaving "freely" in their home cages, and also observed their responses to a standard stimulus consisting of several 5- to 10-day-old rat pups. Lactating females respond to this stimulus by licking and gathering the pups at a nest site, crouching over them, and in general behaving toward them as if toward their own offspring. Although there had been earlier reports that rats may display maternal behavior during the last days of pregnancy,[3,4] in this study the test pups did not elicit nursing or retrieving until after the females had given birth. A major contribution of this study was the observation that patterns of self-licking change during late pregnancy; licking of the nipple lines and anogenital areas greatly increases, while grooming of other body areas lessens. Subsequently, Roth and Rosenblatt[5] demonstrated that the tactile stimulation of self-licking augments nipple development, and so serves an important role in preparing for lactation.

Following these initial studies, Rosenblatt reintroduced the technique of sensitization, whereby nonlactating animals are exposed continuously to pups for up to 2

---

[a] Publication No. 431 of the Institute of Animal Behavior.

weeks, in time eliciting maternal behavior.[7] This procedure, described by Wiesner and Sheard[4] in 1933 but long ignored, made it possible to demonstrate that maternal behavior in the rat has a nonhormonal basis.[6,7] Moreover, it became possible to differentiate degrees of maternal responsiveness among rats, none of which immediately initiate maternal behavior, but which have shorter or longer latencies to become maternal when exposed continuously to standard test pups. By commencing the sensitization of pregnant females at different intervals after mating, it was found that sensitization latencies become shorter during the last third of pregnancy.[9] Cesarean-delivered and hysterectomized pregnant females also proved to have shorter latencies than nonpregnant females or sham-operated controls if these operations were performed on day 10 of pregnancy or later.[8,9] Indeed, pregnancy termination became and remains a major tool for analyzing the hormonal determinants of maternal behavior.

The work of Rosenblatt and Siegel[9,10] on the initiation of maternal behavior by pregnancy-terminated females led to the hypothesis, now confirmed, that increases in estrogen secretion, under the special conditions of late pregnancy, play a major role in the normal onset of maternal behavior. Since the prepartum rise in estrogen secretion begins well before parturition, this appeared to predict a prepartum onset of maternal responsiveness toward pups. Therefore Rosenblatt and his students again explored the maternal responsiveness of intact females during late pregnancy.[9] Beginning 40 hr prior to parturition, females were tested with standard pups at 2-hr intervals; in some cases the pups were removed between tests, while in others they remained with the females. Under both conditions females began retrieving the test pups between 14 and 18 hr prepartum, and by 2 hr prepartum 75% had retrieved pups at more than one test. This confirmed that under the proper circumstances most Sprague-Dawley females will initiate maternal behavior prior to parturition. A prepartum onset of maternal behavior has been observed in other strains as well, although lower percentages of females may show the behavior; in a Wistar strain, for example, about 50% commenced retrieving test pups prior to parturition.[11]

Many questions remained, however, concerning the precise timing of the prepartum appearance of maternal responsiveness. To answer them still another study of late-pregnant females was begun by the author under Rosenblatt's direction.[12] In contrast to earlier work, each experimental female was observed once only to avoid confounding an enhancement of maternal behavior caused by the physiological changes of the prepartum period with the effects of sensitization. In addition to briefly exposing females to groups of four 4- to 9-day-old pups as in previous experiments, a second standard test was employed which tends to elicit maternal behavior more readily. This second test reproduces some of the characteristics of the parturient female's first experience with pups; the test pups are newborn (delivered less than 2 hr previously), a single pup is placed in the female's nest 15 min before three additional pups are introduced some distance from the nest (to allow observations of the onset of retrieving), and tests are conducted during the light phase of the diurnal cycle when rats normally give birth. Whether testing with older or with newborn pups, observations for retrieving and gathering were terminated after 15 min. It has been found that 50% or more of hormonally primed, nonpregnant females retrieve and gather pups during newborn tests whereas less than 10% retrieve and gather older stimulus pups.[13] Pregnant females were tested on days 17, 20, 21, or on the morning of day 22 of gestation, the last being the expected day of parturition. To assess the possible effects of prior maternal experience, we observed both females in their first pregnancies and females that had given birth previously.

Tests with newborn pups revealed an unexpected pattern of changes in maternal responsiveness during the last third of pregnancy. A significant increase in maternal responsiveness was found beginning 1-2 days prior to parturition. This increase, however, proved to contain the reversal of a prior *decline* in readiness to behave maternally. Tested on days 17 and 20 of gestation, animals were less likely to respond maternally to newborn pups, and were more likely to cannibalize them, than females tested before becoming pregnant (FIGURE 1). It seems likely that this dip in maternal responsiveness during late pregnancy is related to a phenomenon noted by Rowland[14] among females that were simultaneously lactating and pregnant (having conceived at the postpartum estrus); such females show a marked, transient disruption of maternal care beginning 4-7 days before parturition, and ending in an abrupt resumption of maternal behavior about a day before birth of the second litter. Presumably hormonal factors are responsible, but they have not been identified.

While tests with newborn pups revealed that maternal responsiveness returns to nonpregnant levels and then rises above them about a day prior to parturition, *older* pups did not elicit maternal behavior from inexperienced females until shortly before the birth of their own pups. For these females maternal behavior toward older pups had a rather sharp onset about 3-4 hr prepartum (FIGURE 2). When tested on the day of parturition but 3.5 or more hours before their deliveries commenced, only 33% retrieved all four pups, whereas closer in time to parturition the percentage rose to over 80.

The sudden increase in maternal responsiveness occurring 3-4 hr prior to delivery focuses attention on the possible behavioral significance of the physiological events occurring in this brief period. Fuchs[15] has reported, and we also have observed, that in the rat regular, forceful uterine contractions commence at least 2 hr before the first pup is delivered. Kristal[16] has suggested that in the hormonally prepared, late-pregnant female, sensory feedback from uterine dilation and contractions, dilation of the cervix, and passage of pups through the birth canal acts centrally in some manner to enhance maternal responsiveness.

If sensory feedback from the early stages of labor mediates an additional increase in maternal responsiveness, the neuronal pathway is likely to include the pelvic nerves which innervate the cervix and vagina.[17,18] It has proved difficult to directly investigate the importance of the pelvic nerves in this regard, since pelvic neurectomy blocks parturition in the rat.[19,20] However, there are some indications that pelvic nerve stimulation also plays a role in enhancing maternal responsiveness after pregnancy termination by hysterectomy. In a recent experiment[21] we explored the hypothesis that traumatic excitation of the pelvic nerves contributes to the very rapid onset of maternal behavior 48 hr after pregnancy termination by hysterectomy-ovariectomy and injection of estradiol benzoate (HO-EB)[9,10]; 16-day pregnant females were subjected to either

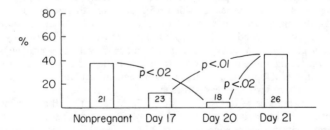

FIGURE 1. Percentages of nulliparous females that retrieved all of three newborn pups during 15-min tests. Tests were conducted on nonpregnant females or on day 17, 20, or 21 of gestation. Numbers per group are in the bars; probability values are based on χ^2 tests. (Adapted from Mayer and Rosenblatt.[12])

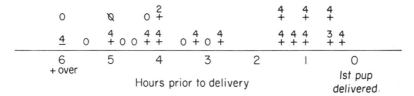

FIGURE 2. Responses of nulliparous females to four 4- to 9-day-old pups presented on the morning of the day of parturition. Numbers of pups retrieved and gathered are plotted against the hour prior to delivery of the first pup. A plus sign indicates that the female maintained contact with the retrieved pups over the last 5 min of the 15-min test. Pup killing is indicated by ɑ̷. (Adapted from Mayer and Rosenblatt.[12])

pelvic nerve section or a sham procedure immediately prior to HO-EB. When test pups were introduced 2 days later, median latencies to initiate maternal behavior were less than 15 min in the intact group, but 24 hr in the nerve-sectioned group, reflecting a significant delay ($p < 0.02$). These data tentatively suggest not only that pelvic nerve activity can augment hormonally mediated maternal responsiveness, but also that hysterectomy involves traumatic excitation of the pelvic nerves sufficient to have this effect, perhaps necessitating some reinterpretation of the pregnancy termination model to incorporate nonhormonal as well as hormonal factors.

It is possible that the central event linking stimulation of the pelvic nerves with an increase in maternal responsiveness is oxytocin release; pituitary release of oxytocin is known to follow stimulation of the cervix and vagina,[22,23] and centrally administered oxytocin evokes short-latency maternal behavior in estrogen-primed (nonpregnant) rats of at least one strain.[24,25] In further support of this idea, Fahrbach et al.[26] have reported that administration of oxytocin antagonist to 16-day, pregnancy-terminated HO-EB females temporarily delayed their initiation of maternal behavior.

If labor is (directly or indirectly) responsible for a final prepartum increment in maternal responsiveness, then the level of responsiveness attained *prior to* the beginning of labor would appear to be an appropriate end point for analyzing the contributions of the hormonal changes of late pregnancy to maternal behavior. A comparison of the responses to pups of females tested 4-6 hr prepartum and those tested 24 hr prepartum suggests that important changes occur over this 18-hr interval. Nearly all maternally experienced females, for example, will retrieve and gather older test pups 4-6 hr prior to parturition, whereas a day earlier only about 20% will show this behavior.[12] A more finely drawn chronology of prepartum developments in maternal responsiveness is a current goal of Rosenblatt's laboratory.

NEST DEFENSE

By virtue of studying rats in relatively small individual cages, well supplied with food and water, laboratories typically limit the maternal behavior displayed by females to interactions with the young (gathering, licking, nursing, etc.) and nest-building. It is known, however, that other kinds of behavior, such as food-hoarding[27] and aggression,[28,29] are altered during lactation and thus might properly be considered aspects of maternal behavior. Observations of wild Norway rats living under natural or seminatural conditions[27] indicate that the young typically are reared in burrows separate from those used by adults other than the mother (or occasionally mothers), suggesting that the establishment of isolated nests and presumably also their defense are regular components of maternal behavior in wild strains of this species. In our laboratory strains also, pregnant females living with nonpregnant animals tend to establish isolated nests some hours before giving birth if the housing conditions make this feasible; on occasion we have observed them to threaten or even attack cagemates that approach too closely.

To systematically investigate nest defense prepartum, late-pregnant Sprague-Dawley females were allowed to habituate to rather small individual cages; on the morning of the expected day of parturition an unfamiliar adult male was placed in each cage until attacked or for a maximum of 5 min. The females then were allowed to give birth without further disturbance, and the intervals between Intruder Tests and delivery of the first pups were determined retrospectively.[12]

Whereas nonpregnant females of this strain rarely attack intruding males, the pregnant females showed rather high levels of aggression. Moreover, among both maternally experienced and inexperienced females, attacks on the males became increasingly likely as the interval between tests and parturition grew shorter (FIGURE 3). When tested more than 3.5 hr before giving birth, 28% attacked the male whereas

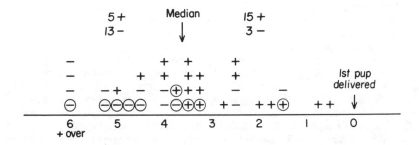

FIGURE 3. Responses of nulliparous females (uncircled) and of experienced breeders (circled) to 2-min Intruder Tests conducted on the morning of the day of parturition. Attack (+) and nonattack (−) are plotted against the number of hours prior to delivery of the first pup. Median test, $\chi^2 = 11.25$, $p < 0.001$. (Copyright 1984 by the American Psychological Association. Reprinted from Mayer and Rosenblatt[12] by permission of the publisher.)

after this point 83% attacked, indicating clearly that females increase in agonistic behavior during the immediate prepartum period. It remains to be determined whether this behavior is linked to the establishment of the isolated nest site, or simply reflects a heightened irritability, expressed regardless of surrounds. As with the increase in maternal responsiveness immediately prior to delivery, it is possible that the increase in aggressiveness is associated (temporally and perhaps causally) with the first stages of labor.

The observation that aggressive behavior increases prepartum has given rise to a number of studies currently in progress or recently completed. The chronology of maternal aggression in the Sprague-Dawley rat has been traced from implantation through lactation[30]; regarding the prepartum period, this work addressed the question of whether aggression toward intruders appears suddenly just before delivery, or has a more gradual development. Pregnant females were subjected to 10-min Intruder Tests (using unfamiliar males) at various points during gestation and lactation. The data revealed that the proportion of females attacking intruders increases by day 18 of gestation over the 10% rate characteristic of nonpregnant females. Latencies to attack decrease and the number of attacks launched within 10 min increases from day 18 to day 21, when over 50% of females attack at least once within a 10-min test. There appears to be some *decrease* in the probability of attacks when females are tested a few hours *following* rather than prior to delivery; observations suggest that this may in part reflect an avoidance of the parturitional nest by the intruder males, a hypothesis currently under investigation. From day 1 through day 9 of lactation, 65 to 85% of females attack one or more times, with no further significant changes in latencies or mean number of attacks. This pattern of postpartum maternal aggression is similar to that reported for both Sprague-Dawley and Long-Evans strains by Takushi et al.[31]

During the last days of pregnancy increases in aggressive nest defense appear to parallel increases in maternal responsiveness toward test pups. Are these changes mediated by the same hormonal factor? The hormonal determinants of maternal aggression are currently under study, using techniques which have proved useful in unraveling the hormonal determinants of maternal caretaking behavior. An important model has been the female whose pregnancy is terminated on day 16 or 17. As described elsewhere in this volume, at this point in gestation pregnancy termination by hysterectomy (H), or by hysterectomy-ovariectomy if exogenous estrogen is given (HO-EB), precipitates patterns of change in progesterone and estrogen levels which mimic those immediately preceding parturition: progesterone rapidly declines while estrogen rises (and then declines). If presented with older test pups (3- to 8- or 4- to 9-day-olds) 48 hr following pregnancy termination by H or by HO-EB, females initiate maternal behavior with median latencies of less than 15 min, whereas females whose pregnancies are terminated by HO *without* exogenous estrogen require 2 or more days of continuous cohabitation before displaying caretaking behavior.[9,10,32] The latencies to initiate maternal behavior of *nonpregnant* females also are reduced by HO combined with EB treatment, although they remain significantly longer than the latencies of pregnancy-terminated HO-EB females.[33,34] Both models have been used to investigate prepartum and maternal aggression.

The relative effects of these hormonal manipulations on levels of aggression in Intruder Tests conducted prior to introducing pups, thus before the females initiate maternal behavior, have been found to parallel their relative impact on maternal responsiveness (as determined either by subsequently exposing the females to pups, or by the sensitization latencies of independent groups). For example, in an experiment using day 16-17 pregnancy-terminated females,[35] 47% of H females and 53% of HO-EB (20 μg/kg) females attacked male intruders, whereas this behavior was shown by only 7% of Sham-H and 5% of HO-Oil groups. Subsequent to the Intruder Tests, all females were exposed to pups for 15 min. Maternal behavior was initiated within this time period by 63% of H females, 71% of HO-EB females, and only 7 and 24%, respectively, of Sham-H and HO-Oil groups.

More recently we studied groups of hormonally manipulated pregnant and nonpregnant females, observing their behavior toward intruders before and after sensitization.[21] The pregnant females were subjected on day 16 or 17 to HO-EB (20 μg/kg), HO-Oil, or a sham operation; the nonpregnant females were subjected to either HO-EB (100 μg/kg) or Sham-HO-Oil. Two days after surgery, the females were given the first of a series of two or three Intruder Tests. Ten minutes later continuous exposure to pups was begun. After 3 hr of pup exposure a second Intruder Test was administered; a third was given following the female's initiation of maternal behavior (if she had not become maternal by the second test).

Comparing groups for numbers of females attacking the intruder prior to initiating maternal behavior, it was found that all pregnancy-terminated, HO-EB females attacked at least once during the 10-min tests. Attacks also were launched by half of the pregnant, Sham-HO group, and by half of the nonpregnant, HO-EB females; however, only 10% of the females in the pregnancy-terminated, HO-Oil group and nonpregnant, sham group showed this level of aggressive behavior.

Until females had initiated maternal behavior, the presence of pups during testing did not appear to alter their responses to intruders. After commencing to behave maternally, however, nearly all pregnancy-terminated females, whether HO-EB (100%) or HO-Oil (90%), and nearly all (sham-operated) pregnant females (85%) attacked the intruders, often with extremely short latencies. Nonpregnant (sham-operated) females, on the other hand, seldom attacked (20%). The latter observation is in accord with previous reports that sensitized virgin females display only low levels of maternal aggression.[36,37] Whether the HO-EB manipulation increases the aggressiveness of nonpregnant females after sensitization has not yet been determined. Preliminary data suggest that those HO-EB females having relatively short latencies to become maternal tend to be aggressive, while those requiring longer pup exposure are not.

The enhanced aggressiveness prior to becoming maternal of pregnancy-terminated H and HO-EB females, and to a lesser extent of nonpregnant HO-EB females, points to a link between increased estrogen titers and prepartum aggression. Similarly, Bowden et al.[38] have recently reported that estrogen activates social aggression in (nonpregnant) female rats, although in their experiments aggression became manifest only when tendencies to show sexual behavior were suppressed (by administration of dihydrotestosterone). In evaluating the association between estrogen levels and prepartum aggression, however, the possibility must be considered that estrogen acts indirectly by stimulating prolactin release[39]; the latter hormone has been reported to contribute to maternal aggression in the hamster.[40–42]

Among females not showing maternal behavior there may be a reasonably good correlation between estrogen levels and aggression toward intruders. The mediation of intruder attack *following* the initiation of maternal behavior, however, is clearly more complex. For example, pregnancy-terminated HO-Oil females after becoming maternal displayed short-latency attacks on intruders although they then had been deprived of ovarian estrogen for 2-4 days. Work is continuing on this problem.

Finally, still another series of studies has been stimulated in part by the observation that aggression and maternal responsiveness increase in parallel during the prepartum period; in this case we have looked for a relationship between maternal aggression and olfaction. In 1974, Fleming and Rosenblatt[43,44] reported that olfactory deafferentation greatly reduces the latencies of virgin females to initiate maternal behavior. Indeed, many females with impaired olfaction appear to be "spontaneously" maternal, and begin retrieving and gathering almost as soon as test pups are presented. In recent years there have been hints that the rapid rise in maternal responsiveness shortly before parturition is mediated in part by changes within the olfactory system(s), perhaps conditioned by hormonal factors, or linked to the terminal stages of pregnancy in other ways. For example, lactating rats prefer the nest odors of lactating females to odors from clean bedding or the bedding of virgin rats; this preference has been shown to develop during the last days of pregnancy,[45] suggesting that prepartum hormonal or other events are capable of altering responsiveness to olfactory stimuli. In sheep, there is a sudden change shortly before parturition in the female's response to amniotic fluid, which becomes attractive rather than aversive.[46] A similar shift in response to amniotic fluid has been noted in the dog[47]; since taste and olfaction are closely linked, these studies also may implicate prepartum changes within the olfactory systems. Again in the sheep, a species in which olfaction plays a key role in the establishment of an exclusive mother-young bond, the brief period immediately after delivery during which the ewe will accept an alien lamb can be extended by estrogen treatment,[48] suggesting that this hormone may have its effect partially through the olfactory systems.

Recently we asked whether olfactory deafferentation also facilitates maternal aggression.[21] Nonpregnant females were exposed to pups until they became maternal, after which the pups were withdrawn. Then experimental groups were formed and subjected to intranasal infusion of zinc sulfate (which temporarily disrupts olfactory acuity[44]) or a sham procedure, and 1 day later pups again were introduced. Because of their prior sensitization, all females rapidly reinstated maternal behavior. Shortly thereafter they were given Intruder Tests with pups present. In line with previous reports that nonpregnant, sensitized females seldom show maternal aggression, only 20% of the females with intact olfaction attacked the intruder; however, 80% of the zinc sulfate-treated females showed this behavior. Among females that had *not* been sensitized, the zinc sulfate treatment did not increase aggression during Intruder Tests. Much more work needs to be done before this finding can be integrated into our understanding of the role(s) of olfaction in maternal behavior.

In their first publication on maternal behavior of the rat, Rosenblatt and Lehrman wrote the following:

> It is not to be expected that we will be able to discover and define "the" hormonal basis of maternal behavior, "the" role of experience in the origin of maternal behavior, "the" relative importance of autonomous and of stimulus-induced changes in physiological and behavioral condition, or "the" neural basis of maternal behavior. Rather, continuous and simultaneous attention to as many behavioral and physiological aspects of this complex pattern as possible may be rewarded by a gradual and continual increase in our understanding of the problems which we have posed and in our ability to pose further problems. (Reference 2, p. 53)

It is to be hoped that we are being so rewarded.

REFERENCES

1. LEHRMAN, D. S. 1965. Interaction between internal and external environments in the regulation of the reproductive cycle of the ring dove. *In* Sex and Behavior. F. A. Beach, Ed.: 355-380. John Wiley. New York, N.Y.
2. ROSENBLATT, J. S. & D. S. LEHRMAN. 1963. Maternal behavior of the laboratory rat. *In* Maternal Behavior in Mammals. H. L. Rheingold, Ed.: 36-56. John Wiley. New York, N.Y.
3. KARLI, P. 1956. The Norway rat's killing response to the white mouse: an experimental analysis. Behaviour **10**: 81-103.
4. WIESNER, B. P. & N. M. SHEARD. 1933. Maternal Behaviour in the Rat. Oliver & Boyd. London, England.
5. ROTH, L. L. & J. S. ROSENBLATT. 1967. Changes in self-licking during pregnancy in the rat. J. Comp. Physiol. Psychol. **63**: 397-400.
6. COSNIER, J. & C. COUTURIER. 1966. Comportement maternal provoqué chez les rattes adultes castrées. C. R. Soc. Biol. **160**: 789-791.
7. ROSENBLATT, J. S. 1967. Nonhormonal basis of maternal behavior in the rat. Science **156**: 1512-1514.
8. LOTT, D. & J. S. ROSENBLATT. 1969. Development of maternal responsiveness during pregnancy in the rat. *In* Determinants of Infant Behaviour. B. M. Foss, Ed. **4**: 61-67. Methuen. London, England.

9. ROSENBLATT, J. S. & H. I. SIEGEL. 1975. Hysterectomy-induced maternal behavior during pregnancy in the rat. J. Comp. Physiol. Psychol. **89:** 685-700.
10. SIEGEL, H. I. & J. S. ROSENBLATT. 1975. Hormonal basis of hysterectomy-induced maternal behavior during pregnancy in the rat. Horm. Behav. **6:** 211-222.
11. SLOTNICK, B. M., M. L. CARPENTER & R. FUSCO. 1973. Initiation of maternal behavior in pregnant nulliparous rats. Horm. Behav. **4:** 53-59.
12. MAYER, A. D. & J. S. ROSENBLATT. 1984. Prepartum changes in maternal responsiveness and nest defense in *Rattus norvegicus*. J. Comp. Psychol. **98:** 177-188.
13. MAYER, A. D. & J. S. ROSENBLATT. 1980. Hormonal interaction with stimulus and situational factors in the initiation of maternal behavior in nonpregnant rats. J. Comp. Physiol. Psychol. **94:** 1040-1059.
14. ROWLAND, D. L. 1981. Effects of pregnancy on the maintenance of maternal behavior in the rat. Behav. Neural Biol. **31:** 225-235.
15. FUCHS, A.-R. 1969. Uterine activity in late pregnancy and during parturition in the rat. Biol. Reprod. **1:** 344-353.
16. KRISTAL, M. B. 1980. Placentophagia: a biobehavioral enigma (or *De qustibus non disputandum est*). Neurosci. Biobehav. Rev. **4:** 141-150.
17. PETERS, L. C., M. B. KRISTAL & B. K. KOMISARUK. Sensory innervation of the external and internal genitalia of the female rat. Submitted for publication.
18. REINER, P., J. WOOLSEY, N. ADLER & A. MORRISON. 1981. A gross anatomical study of the peripheral nerves associated with reproductive function in the female albino rat. In Neuroendocrinology of Reproduction. N. T. Adler, Ed.: 545-548. Plenum. New York, N.Y.
19. BURDEN, H. W., R. C. GOREWIT, T. M. LOUIS, P. D. MUSE & I. E. LAWRENCE, JR. 1982. Plasma oxytocin and estradiol in pelvic neurectomized rats with blocked parturition. Endokrinologie **79:** 379-384.
20. LOUIS, T. M., I. E. LAWRENCE, JR., R. F. BECKER & H. W. BURDEN. 1978. Prostaglandin $F_{2\alpha}$, prostaglandin E_2, progesterone, 20α-dihydroprogesterone and ovarian 20α-hydroxysteroid dehydrogenase activity in preparturient pelvic neurectomized rats. Proc. Soc. Exp. Biol. Med. **158:** 631-636.
21. MAYER, A. D. & J. S. ROSENBLATT. Unpublished data.
22. FERGUSON, J. K. W. 1941. A study of the motility of the intact uterus at term. Surg. Gynecol. Obstet. **73:** 359-366.
23. MOOS, F. & P. RICHARD. 1975. Level of oxytocin release induced by vaginal dilatation (Ferguson reflex) and vagal stimulation (vago-pituitary reflex) in lactating rats. J. Physiol. (Paris) **70:** 307-314.
24. FAHRBACH, S. E., J. I. MORRELL & D. W. PFAFF. 1984. Oxytocin induction of short-latency maternal behavior in nulliparous, estrogen-primed rats. Horm. Behav. **18:** 267-286.
25. PETERSEN, C. A., J. A. ASCHER, Y. L. MONROE & A. J. PRANGE. 1982. Oxytocin induces maternal behavior in virgin female rats. Science **216:** 648-650.
26. FAHRBACH, S. E., J. I. MORRELL & D. W. PFAFF. 1985. Possible role for endogenous oxytocin in estrogen facilitated maternal behavior in rats. Neuroendocrinology **40:** 526-532.
27. CALHOUN, J. B. 1963. The Ecology and Sociology of the Norway Rat. U.S. Public Health Service Publication No. 1008.
28. ERSKINE, M. S., R. J. BARFIELD & B. D. GOLDMAN. 1978. Intraspecific fighting during late pregnancy and lactation in rats and effects of litter removal. Behav. Biol. **23:** 206-218.
29. OSTERMEYER, M. C. 1983. Maternal aggression. In Parental Behaviour of Rodents. R. W. Elwood, Ed.: 151-179. John Wiley. New York, N.Y.
30. REISBICK, S., B. GUARIGLIA, E. SHARPE, H. I. SIEGEL & J. S. ROSENBLATT. Unpublished data.
31. TAKUSHI, R. Y., K. J. FLANNELLY, D. C. BLANCHARD & R. J. BLANCHARD. 1983. Maternal aggression in two strains of laboratory rats. Aggress. Behav. **9:** 120.
32. BRIDGES, R. S., J. S. ROSENBLATT & H. H. FEDER. 1978. Stimulation of maternal responsiveness after pregnancy termination in rats: effect of time of onset of behavioral testing. Horm. Behav. **10:** 235-245.
33. SIEGEL, H. I. & J. S. ROSENBLATT. 1975. Latency and duration of estrogen induction of

maternal behavior in hysterectomized-ovariectomized virgin rats: effects of pup stimulation. Physiol. Behav. **14:** 473-476.
34. SIEGEL, H. I. & J. S. ROSENBLATT. 1975. Estrogen-induced maternal behavior in hysterectomized-ovariectomized virgin rats. Physiol. Behav. **14:** 465-471.
35. MAYER, A. D., C. MIZ & J. S. ROSENBLATT. Unpublished data.
36. ERSKINE, M. S., R. J. BARFIELD & B. D. GOLDMAN. 1980. Postpartum aggression in rats. II. Dependence on maternal sensitivity to young and effects of experience with pregnancy and parturition. J. Comp. Physiol. Psychol. **94:** 495-505.
37. GANDELMAN, R. & N. G. SIMON. 1980. Postpartum fighting in the rat: nipple development and the presence of young. Behav. Neural Biol. **28:** 350-360.
38. BOWDEN, N., N. E. VAN DE POLL, J. G. VAN OYEN, P. F. BRAIN & H. H. SWANSON. 1982. Gonadal steroids and aggressive behaviour in male and female rats. Aggress. Behav. **8:** 182-184.
39. BLAKE, C. A., R. L. NORMAN & C. H. SAWYER. 1972. Effects of estrogen and/or progesterone on serum and pituitary gonadotropin levels in ovariectomized rats. Proc. Soc. Exp. Biol. Med. **141:** 1100-1103.
40. BUNTIN, J. D., C. CATANZARO & R. D. LISK. 1981. Facilitatory effects of pituitary transplants on intraspecific aggression in female hamsters. Horm. Behav. **15:** 214-225.
41. WISE, D. A. 1974. Aggression in the female golden hamster: effects of reproductive state and social isolation. Horm. Behav. **5:** 235-250.
42. WISE, D. A. & T. L. PRYOR. 1977. Effects of ergocornine and prolactin on aggression in the postpartum golden hamster. Horm. Behav. **8:** 30-39.
43. FLEMING, A. & J. S. ROSENBLATT. 1974. Olfactory regulation of maternal behavior in rats. I. Effects of olfactory bulb removal in experienced and inexperienced lactating and cycling females. J. Comp. Physiol. Psychol. **86:** 221-232.
44. FLEMING, A. & J. S. ROSENBLATT. 1974. Olfactory regulation of maternal behavior in rats. II. Effects of peripherally induced anosmia and lesions of the lateral olfactory tract in pup-induced virgins. J. Comp. Physiol. Psychol. **86:** 233-246.
45. BAUER, J. H. 1983. Effects of maternal state on the responsiveness to nest odors of hooded rats. Physiol. Behav. **30:** 229-232.
46. LEVY, F., P. POINDRON & P. LE NEINDRE. 1983. Attraction and repulsion by amniotic fluids and their olfactory control in the ewe around parturition. Physiol. Behav. **31:** 687-692.
47. DUNBAR, I., E. RANSON & M. BUEHLER. 1981. Pup retrieval and maternal attraction to canine amniotic fluids. Behav. Proc. **6:** 249-260.
48. LE NEINDRE, P., P. POINDRON & C. DELOUIS. 1979. Hormonal induction of maternal behavior in non-pregnant ewes. Physiol. Behav. **22:** 731-734.

The Role of the Medial Preoptic Area in the Regulation of Maternal Behavior in the Rat

MICHAEL NUMAN

Department of Psychology
Boston College
Chestnut Hill, Massachusetts 02167

INTRODUCTION

A true understanding of the role of a particular hypothalamic nucleus in the control of a particular behavior would appear to require a two-stage process. The first stage entails showing that the nucleus contains neurons which are directly involved in regulating the behavior in question and the second stage involves determining the function of those neurons with respect to the behavior.

Concerning the first stage of analysis, the method most often used by physiological psychologists in order to determine whether a particular nuclear group is directly involved in regulating a particular behavior is the lesion technique. Valuable information can be gained from such a technique if two important problems are overcome:

(i) *The problem of lesioning fibers of passage.* Traditional lesioning methods not only destroy cell bodies, but also destroy axons that pass through the lesion site but have their origins elsewhere. Any change in behavior could be due to destruction of these axons, rather than the nuclear group.

(ii) *The problem of direct vs. indirect effects.* Any change in a behavior after a lesion to a particular area could be a secondary or indirect effect of the lesion. For example, there has been a recent trend to explain some of the deficits in somatic-motor responses observed after hypothalamic lesions as being secondary to lesion-produced disturbances in autonomic function.[1,2]

If a change in behavior can be shown to be a direct effect of damage to a nuclear group then we can conclude that elements within that nucleus are part of the neural machinery underlying the behavior being investigated. This, however, tells us nothing about the function these neurons play in the regulation of the behavior. To understand this we must uncover the mechanism by which these neurons influence the behavior. One of the ways this can be achieved is by uncovering the neural circuitry within which these neurons operate, given, of course, that we know something about the functions of the remaining parts of the circuit.

Physiological psychology is just beginning to ask questions concerning the mechanisms by which hypothalamic nuclei exert their influence on behavior, and this is, in large part, due to recent advances in our understanding of the neural connections of the various hypothalamic nuclei. Two recent reviews provide excellent examples of how an understanding of neural circuitry aids one in developing testable hypotheses about how particular hypothalamic nuclei influence particular behaviors.[1,3]

This brief introduction is related to my own research development in that many of my investigations have utilized the lesion technique in an attempt to show that the medial preoptic nucleus (MPOA) of the hypothalamus is directly involved in the control of maternal behavior in the rat. I am currently studying the neural circuitry of maternal behavior in order to gain insight into the mechanism by which the MPOA influences maternal behavior.

THE MPOA AND MATERNAL BEHAVIOR

Several findings suggest that the MPOA is directly involved in the control of maternal behavior in the rat. Lesions to this region depress nursing and nest-building behavior and abolish retrieving behavior.[4-8] There is also evidence that this effect is due to damage to the nucleus rather than to fibers of passage.[6] The question of whether the observed disruption is a direct effect of the lesion, rather than a secondary effect, has also been investigated and several pieces of evidence suggest we are dealing with a primary effect:

(i) The disruption of maternal behavior after MPOA lesions cannot be attributed to a neuroendocrine imbalance.[6,7]

(ii) After MPOA damage, nonmaternal rats show normal body weight regulation, locomotor activity, and female sexual receptivity.[6,7] This finding shows that a general physical debilitation is an unlikely cause of the lack of maternal responsivity.

(iii) After MPOA damage, postpartum female rats no longer retrieve their young, yet they are still capable of hoarding small pieces of candy approximating the size and weight of baby pups.[9] Therefore, the lack of retrieval behavior cannot be attributed to a general oral motor deficit.

All of the neurons of the MPOA, of course, are not considered to be related just to maternal behavior, but the evidence cited above does suggest that there are neurons within the MPOA that are specifically related to maternal behavior. There is an additional "secondary effect hypothesis," however, which deserves scrutiny. The MPOA has long been considered a hypothalamic region important in the regulation of body temperature.[10] It has been estimated that about 50% of MPOA neurons are thermosensitive.[10] Significantly, there are important relationships between body temperature and maternal behavior. Leon et al.[11] have reported that engaging in nursing raises a lactating rat's body temperature and that nursing bouts are terminated by the mother once her temperature rises above some critical level. One interpretation of this is that high body temperatures depress maternal responsiveness. Since preoptic damage which disrupts maternal behavior also results in hyperthermia,[5,7] it might be that the absence of maternal behavior after preoptic damage is secondary to a lesion-produced hyperthermia. As reviewed by Numan and Callahan,[7] the available evidence suggests that such a hypothesis is incorrect (also see Reference 5). In particular, animals with preoptic damage which are no longer hyperthermic still show an absence of maternal behavior. It should be pointed out that the fact that 50% of MPOA neurons are thermosensitive does not mean that 50% of MPOA neurons are involved in temperature regulation.[10] Some of the temperature-sensitive MPOA neurons may be "maternal" neurons, and their thermosensitivity may be the avenue by which body temperature affects maternal responsiveness.

One final study provides powerful evidence that the MPOA is directly involved in the control of maternal behavior. Numan et al.[12] have shown that estrogen implants into the MPOA can facilitate maternal behavior, and the estrogen is presumably acting on MPOA cell bodies which concentrate estrogen.[13] The fact that we can arouse maternal behavior through estrogen stimulation of the MPOA in addition to disrupting the behavior by destroying the area supports the direct involvement of this region in the behavior.

THE NEURAL CIRCUITRY OF MATERNAL BEHAVIOR

Given that the MPOA is involved in the control of maternal behavior, we can now ask what the MPOA does in order to influence maternal behavior. One way to approach this problem is to ask where MPOA efferents go in order to affect maternal responsiveness. Several studies have indicated that the lateral efferents of the MPOA are critical for maternal behavior, the most important finding being that knife cuts which sever the lateral connections of the MPOA selectively disrupt maternal behavior, while knife cuts which sever the anterior, posterior, or dorsal connections of this region do not.[7] Anatomical studies have shown that the following structures are among those innervated by the lateral efferents of the MPOA: septum, amygdala, other hypothalamic regions, ventral tegmental area (VTA) of the midbrain, and midbrain central gray.[14,15] Which, and how many, of these projections might be crucial for maternal behavior? FIGURE 1 shows a schematic diagram of the brain and indicates the location of the lateral MPOA cuts which have been found to disrupt maternal

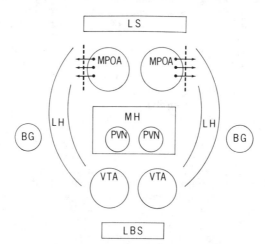

FIGURE 1. Schematic diagram of the brain, indicating the location of knife cuts (shown as broken lines) lateral to the medial preoptic area which have been found to disrupt maternal behavior. The lateral efferents of the medial preoptic area are shown as axons with arrows at their tips. The actual destinations of these projections are not shown. Abbreviations: BG = basal ganglia; LBS = lower brainstem; LH = lateral hypothalamus; LS = limbic system; MH = medial hypothalamus; MPOA = medial preoptic area; PVN = paraventricular hypothalamus; VTA = ventral tegmental area.

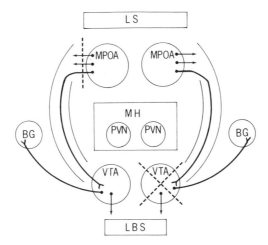

FIGURE 2. Schematic diagram of the brain, showing how a unilateral knife cut lateral to the medial preoptic area (shown as a broken line) paired with a contralateral electrolytic lesion of the ventral tegmental area (shown as a broken ×) would disrupt a pathway from the medial preoptic area to the ventral tegmental area and from the ventral tegmental area to the basal ganglia. The fact that other efferent projections from the medial preoptic area and ventral tegmental area exist is indicated by axons leaving these regions with arrows at their tips, their actual destinations not being shown. See FIGURE 1 caption for abbreviations.

behavior. The lateral cuts sever all of the lateral efferents of the MPOA, the diverse destinations of which are not shown. Concerning the destination of the efferents critical for maternal behavior, Numan and Smith[16] have provided evidence for the importance of MPOA input to the VTA. They found that if they paired a unilateral lateral cut of the MPOA with a contralateral electrolytic lesion of the VTA maternal behavior was disrupted as severely as in females which receive bilateral lateral cuts of the MPOA. The VTA projects to several brain regions,[17] which includes a projection to the basal ganglia (BG). Numan and Smith favor an MPOA-to-VTA-to-BG circuit as being crucial for maternal behavior because the BG is a major component of the extrapyramidal motor system. Such a pathway might be the mechanism by which MPOA neurons relevant to maternal behavior promote the somatic-motor processes underlying maternal responsiveness. FIGURE 2 shows that the asymmetrical lesions of Numan and Smith would bilaterally interrupt such a circuit.

Although an MPOA-to-VTA-to-BG pathway may be critical for maternal behavior, the problems associated with the nonselective nature of traditional lesioning techniques (knife cuts, electrolytic lesions, radiofrequency lesions) must be overcome before the importance of such a pathway can be established. Although some of these problems can be attacked with traditional lesioning methods, future progress will also depend upon the use of more selective lesioning and stimulation techniques coupled with advanced neuroanatomical methods (see Reference 18). Some of the problems associated with the findings of Numan and Smith, and how they might be resolved, are enumerated below:

(i) A unilateral cut of the MPOA paired with a contralateral lesion of the VTA not only disrupts MPOA output to the VTA but also interferes with other MPOA efferent projections, and it might be these latter projections which are important for maternal behavior. As one example, anatomical evidence indicates that the asymmetrical lesions of Numan and Smith would not only destroy an MPOA-to-VTA-to-BG circuit but would also destroy a pathway connecting the preoptic region with the paraventricular hypothalamic nucleus (PVN) and the PVN with the lower brainstem (LBS).[14,19] This is shown schematically in FIGURE 3. An MPOA-to-PVN-to-LBS pathway might be important for maternal behavior because the PVN is the major source of oxytocinergic fibers which descend to the lower brainstem[20,21] and because oxytocinergic pathways might be important for maternal behavior.[22] Numan and Corodimas[9] have recently investigated the importance of this pathway for maternal behavior in lactating rats. They found that PVN lesions did not disrupt maternal behavior, indicating that an MPOA-to-PVN-to-LBS pathway is not critical for maternal behavior. This provides indirect support for the importance of an MPOA-to-VTA-to-BG pathway, but much more work must be done to test this hypothesis more

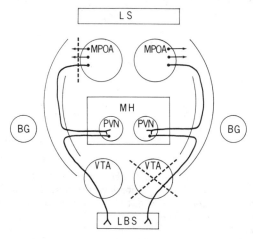

FIGURE 3. Schematic diagram of the brain, showing how a unilateral knife cut lateral to the medial preoptic area (shown as a broken line) and a contralateral lesion of the ventral tegmental area (shown as a broken ×) would not only destroy the pathway shown in FIGURE 2, but would also destroy a pathway from the medial preoptic area to the paraventricular nucleus and from the paraventricular nucleus to the lower brainstem. See FIGURE 1 caption for abbreviations.

directly. For example, VTA lesions may disrupt maternal behavior because they damage MPOA efferents passing through, rather than terminating in, the VTA.[14] The injection of drugs like ibotenic acid, which selectively destroy cell bodies,[23] into the VTA will help resolve the issue of whether VTA cell bodies are important for maternal behavior.

(ii) Although evidence strongly suggests that the lateral efferents of the MPOA are critical for maternal behavior, the possibility also exists that afferent lateral input to the MPOA may be essential for maternal behavior. Bilateral lateral cuts may sever both efferents and afferents, but a unilateral lateral cut of the MPOA paired with a contralateral VTA lesion may disrupt maternal behavior not by bilaterally interfering with the output of the MPOA, but by interfering with its input. Therefore, future research should focus on sources of afferent input to the MPOA which may be critical for maternal behavior, and examine whether such input travels between the MPOA and VTA. Alternatively, a possible approach to examine whether MPOA output to the VTA is critical for maternal behavior would be to first specify, with the aid of immunocytochemical techniques, for example, the neurotransmitters utilized by preoptic neurons which project to the VTA and then to use this information to neuropharmacologically interfere with the function of the VTA. If maternal behavior were affected by such neuropharmacological manipulation, support would be provided for a role of preoptic efferents to the VTA in the control of maternal behavior.

(iii) The efferent projections of the MPOA reach diverse brain regions and more than one circuit may be important for maternal behavior. Therefore, even if it becomes established that an MPOA-to-VTA circuit is critical for maternal behavior, this does not rule out the possibility that other circuits are equally essential.

THE FUNCTION OF THE MPOA WITH RESPECT TO MATERNAL BEHAVIOR

Research into the neural circuitry underlying maternal behavior is clearly underdeveloped and, therefore, an understanding of how the MPOA fits into this circuitry in order to influence maternal behavior is at best vague. The idea that an MPOA-to-VTA-to-BG pathway underlies maternal behavior is, in a sense, a hypothalamocentric point of view, suggesting that maternal behavior is in some way initiated and organized by the MPOA. The fact that hormonal stimulation at the time of parturition appears to be the trigger mechanism for maternal behavior,[24] and that estrogen action on the MPOA stimulates maternal behavior, tends to reinforce the view that maternal behavior is initiated by the MPOA. The MPOA may also be the site at which olfactory input[25,26] and temperature factors[27] influence maternal behavior. The attractive hypothesis that the MPOA influences maternal behavior by affecting the output of the extrapyramidal motor system via the VTA also conforms with a view that the MPOA organizes maternal responses. However, it is difficult to conceive how a complex behavior like maternal behavior can be completely organized by the output of the MPOA. One possibility is that the MPOA receives input relevant to maternal behavior from diverse sources and then projects not only to the VTA but to other brain regions as well, in this way ensuring the coordinated occurrence of adaptive maternal responses. An alternative view, however, is that the MPOA plays a simpler, yet still essential, role in the regulation of maternal behavior. Research certainly has indicated that the MPOA does not influence all maternal responses equally; the output of the MPOA appears more essential for the occurrence of retrieval behavior than for the occurrence of nursing.[7,8] Since VTA projections to the BG have been implicated in the control of oral motor responses,[28] perhaps the role of the MPOA is to facilitate activity in those neural circuits which regulate the performance of oral motor responses specifically related to retrieving.

These views that the MPOA serves either a complex or simple function with respect to maternal behavior both suggest that there are neurons in the MPOA specifically related to maternal behavior. However, research has indicated that the MPOA is not only involved in maternal behavior, but is also concerned with male sexual behavior, drinking behavior, and aggressive behavior.[29-31] Although different MPOA neurons with different projections may be involved in these different behaviors, it is also possible that the MPOA is involved in a general function which influences a certain limited class of behaviors. With respect to this issue, on the one hand there is recent evidence that different neurons within the MPOA underlie maternal behavior and male sexual behavior.[32] On the other hand, however, evidence exists which suggests that a neural pathway which travels between the preoptic area and the VTA is not only important for maternal behavior, but is also important for angiotensin-induced drinking in rats and quiet biting attack in cats.[31,33]

In conclusion, then, when the neural circuitry underlying maternal behavior, and other behaviors, is more exactly specified we should be in a better position to understand the function the MPOA performs for maternal behavior.

REFERENCES

1. PFAFF, D. W. 1980. Estrogens and Brain Function. Springer-Verlag. New York, N.Y.
2. SAWCHENKO, P. E., R. M. GOLD & S. F. LEIBOWITZ. 1981. Evidence for vagal involvement in the eating elicited by adrenergic stimulation of the paraventricular nucleus. Brain Res. **225:** 249-269.
3. SWANSON, L. W. & G. J. MOGENSON. 1981. Neural mechanisms for the functional coupling of autonomic, endocrine, and somatomotor responses in adaptive behavior. Brain Res. Rev. **3:** 1-34.
4. JACOBSON, C. D., J. TERKEL, R. A. GORSKI & C. H. SAWYER. 1980. Effects of small medial preoptic area lesions on maternal behavior: retrieving and nest building in the rat. Brain Res. **194:** 471-478.
5. MICELI, M., A. FLEMING & C. W. MALSBURY. 1983. Disruption of maternal behaviour in virgin and postparturient rats following sagittal plane knife cuts in the preoptic area-hypothalamus. Behav. Brain Res. **9:** 337-360.
6. NUMAN, M. 1974. Medial preoptic area and maternal behavior in the female rat. J. Comp. Physiol. Psychol. **87:** 746-759.
7. NUMAN, M. & E. C. CALLAHAN. 1980. The connections of the medial preoptic region and maternal behavior in the rat. Physiol. Behav. **25:** 653-665.
8. TERKEL, J., R. S. BRIDGES & C. H. SAWYER. 1979. Effects of transecting lateral neural connections of the medial preoptic area on maternal behavior in the rat: nest building, pup retrieval and prolactin secretion. Brain Res. **169:** 369-380.
9. NUMAN, M. & K. P. CORODIMAS. 1985. The effects of paraventricular lesions on maternal behavior in rats. Physiol. Behav. **35:** 417-425.
10. BOULANT, J. A. 1980. Hypothalamic control of thermoregulation. In Handbook of the Hypothalamus. Vol. 3, Part A: Behavioral Studies of the Hypothalamus. P. J. Morgane & J. Panksepp, Eds.: 1-82. Marcel Dekker. New York, N.Y.
11. LEON, M., P. G. CROSKERRY & G. K. SMITH. 1978. Thermal control of mother-young contact in rats. Physiol. Behav. **21:** 793-811.
12. NUMAN, M., J. S. ROSENBLATT & B. R. KOMISARUK. 1977. Medial preoptic area and onset of maternal behavior in the rat. J. Comp. Physiol. Psychol. **91:** 146-164.
13. PFAFF, D. W. & M. KEINER. 1973. Atlas of estradiol-concentrating cells in the central nervous system of the female rat. J. Comp. Neurol. **151:** 121-158.
14. CONRAD, L. D. & D. W. PFAFF. 1976. Efferents from the medial basal forebrain and hypothalamus in the rat. I. An autoradiographic study of the medial preoptic area. J. Comp. Neurol. **169:** 185-220.
15. SWANSON, L. W. 1976. An autoradiographic study of the efferent connections of the preoptic region in the rat. J. Comp. Neurol. **167:** 227-256.
16. NUMAN, M. & H. G. SMITH. 1984. Maternal behavior in rats: evidence for the involvement of preoptic projections to the ventral tegmental area. Behav. Neurosci. **98:** 712-727.
17. BECKSTEAD, R. M., V. B. DOMESICK & W. J. H. NAUTA. 1979. Efferent connections of the substantia nigra and ventral tegmental area in the rat. Brain Res. **175:** 191-217.
18. NUMAN, M., J. I. MORRELL & D. W. PFAFF. 1985. Anatomical identification of neurons in selected brain regions associated with maternal behavior deficits induced by knife cuts of the lateral hypothalamus in rats. J. Comp. Neurol. **237:** 552-564.
19. SWANSON, L. W. 1977. Immunohistochemical evidence for a neurophysin-containing auto-

nomic pathway arising in the paraventricular nucleus of the hypothalamus. Brain Res. **128:** 346-353.
20. LANG, R. E., J. HEIL, D. GANTEN, K. HERMANN, W. RASCHER & T. UNGER. 1983. Effects of lesions in the paraventricular nucleus of the hypothalamus on vasopressin and oxytocin contents in brainstem and spinal cord of rat. Brain Res. **260:** 326-329.
21. SWANSON, L. W. & P. E. SAWCHENKO. 1983. Hypothalamic integration: organization of the paraventricular and supraoptic nuclei. Annu. Rev. Neurosci. **6:** 269-324.
22. PEDERSEN, C. A. & A. J. PRANGE. 1979. Induction of maternal behavior in virgin rats after intracerebroventricular administration of oxytocin. Proc. Natl. Acad. Sci. USA **76:** 6661-6665.
23. KÖHLER, C. & R. SCHWARCZ. 1983. Comparison of ibotenate and kainate neurotoxicity in rat brain: a histological study. Neuroscience **8:** 819-835.
24. ROSENBLATT, J. S., H. I. SIEGEL & A. D. MAYER. 1979. Progress in the study of maternal behavior in the rat: hormonal, nonhormonal, sensory, and developmental aspects. *In* Advances in the Study of Behavior, Vol. 10. J. S. Rosenblatt, R. A. Hinde, E. Shaw & C. Beer, Eds.: 225-311. Academic Press. New York, N.Y.
25. FLEMING, A. S., F. VACCARINO, L. TAMBOSSO & P. CHEE. 1979. Vomeronasal and olfactory system modulation of maternal behavior in the rat. Science **203:** 372-374.
26. NUMAN, M. 1983. Brain mechanisms of maternal behaviour in the rat. *In* Hormones and Behaviour in Higher Vertebrates. J. Balthazart, E. Pröve & R. Gilles, Eds.: 69-85. Springer-Verlag. West Berlin, FRG.
27. WOODSIDE, B., R. PELCHAT & M. LEON. 1980. Acute elevation of the heart load of mother rats curtails maternal nest bouts. J. Comp. Physiol. Psychol. **94:** 61-68.
28. JONES, D. L. & G. J. MOGENSON. 1979. Oral motor performance following central dopamine receptor blockade. Eur. J. Pharmacol. **59:** 11-21.
29. BANDLER, R. 1979. Predatory attack behavior in the cat elicited by preoptic region stimulation. Aggressive Behav. **5:** 269-282.
30. MALSBURY, C. & D. W. PFAFF. 1974. Neural and hormonal determinants of mating behavior in adult male rats: a review. *In* Limbic and Autonomic Nervous Systems Research. L. DiCara, Ed.: 85-136. Plenum Press. New York, N.Y.
31. SWANSON, L. W., J. KUCHARCZYK & G. J. MOGENSON. 1978. Autoradiographic evidence for pathways from the medial preoptic area to the midbrain involved in the drinking response to angiotensin II. J. Comp. Neurol. **178:** 645-660.
32. GRAY, P. & P. J. BROOKS. 1984. Effect of lesion placement within the medial preoptic-anterior hypothalamic continuum on maternal behavior and male sexual behavior in female rats. Behav. Neurosci. **98:** 703-711.
33. HALLIDAY, R. & R. BANDLER. 1981. Anterior hypothalamic knife cut eliminates a specific component of the predatory behavior elicited by electrical stimulation of the posterior hypothalamus or ventral midbrain in the cat. Neurosci. Lett. **21:** 231-236.

Psychobiology of Rat Maternal Behavior: How and Where Hormones Act to Promote Maternal Behavior at Parturition

ALISON FLEMING

Department of Psychology
Erindale College
University of Toronto
Mississauga, Ontario, Canada L5L 1C6

The present paper provides a brief overview of the work that has been done by us and others concerning the psychoneuroendocrinology of maternal behavior. The studies discussed deal exclusively with work done on rats, as these laboratory creatures have provided us with the most extensive body of knowledge on this topic. However, some of these findings have been generalized to other species and may, in fact, constitute reasonable models for similar sorts of analyses in human mothers.[1] This work emanates quite directly and obviously from the work of Jay Rosenblatt and work done earlier by myself and others at the Institute of Animal Behavior of Rutgers University.

I will discuss, first, what maternal behavior looks like; second, what hormones influence the behavior; third, how the hormones may be acting to alter behavior; and finally, where in the brain these hormones exert their effects. I hope to convey the idea that hormones do not act in a unitary way, by reflexively eliciting a behavior which is controlled by a single center or pathway in the brain. Instead, I would like to argue that the appearance of a species-characteristic behavior like maternal behavior necessitates changes in a number of other behaviors, which are also influenced by the "maternal" hormones and which are controlled by separate but interconnected neural structures or systems.

MATERNAL BEHAVIOR IN THE NEW MOTHER

When a female rat gives birth to her first litter (consisting of some 10-15 pups) she shows a pattern of behavior which is quite complicated, made up of many different responses and performed with amazing efficiency considering her inexperience at pup rearing.[2-4] Prior to the birth, the mother builds a nest into which the young will be placed. At parturition the mother helps in the expulsion of the young by pulling them out with her mouth. She proceeds to pull off the amniotic sac, eat the attached umbilicus and placenta, and lick off the pup. Possibly between successive births, and certainly at the end, the mother retrieves pups to the nest site. She licks them and once they are piled together she assumes a crouch posture over them, and they attach to the teats and begin to suckle. This pattern—nest-building, retrieval, crouch posture, and pup-licking together—is what we mean when we say that an animal is maternal or shows maternal behavior.

For the purposes of the studies described below, we operationally define maternal behavior as the occurrence of retrieval of the entire litter to a nest site on two successive days and the adoption of a nursing or crouch posture on at least one of these days. Invariably licking will also occur and nest-building usually begins a few days later. Normally, maternal behaviors are seen for the first time at parturition and persist, although with decreasing intensity, throughout the postpartum period until the young are weaned at 20-25 days of age.

RESPONSIVENESS TO PUPS IN NULLIPAROUS FEMALES

If one contrasts the behavior of the new mother with the behavior of a nonmated virgin animal, one sees striking differences in their responses to neonatal rat pups. As indicated earlier the new mother is very responsive to newborn young—even if they are foster pups and not her own.[5]

If one places a litter of foster pups into the cage of a virgin animal one sees a different pattern of behavior. Some virgins approach the pups, sniff them, and then withdraw, showing a pattern of alternating approaches and withdrawals; some approach, become somewhat frantic, and cover them over; some attack and cannibalize.[2,5-7] However, if a new litter of foster pups is placed into the female's cage each day and left in for a 24-hr period such that the female is continuously exposed to pups, eventually the female will begin to show behavior which is very similar to the behavior of the new mother.[5,6] She will retrieve, lick, and adopt a nursing posture over the young, although she does not lactate.

This process, called pup induction or sensitization, can be divided into two phases: (1) an initial phase of variable length during which the female remains at a distance from the pups and, in fact, as we shall see, often actively avoids them, and (2) a final phase of a few hours to a day during which the female is in close proximity to the pups. It is during this latter phase that the female becomes sensitized to pups and shows the full repertoire of maternal behaviors (see Reference 8). How long this process takes depends on the strain of animal, the age of the pups being used, the age of the female, the size of the cage, etc.[7,9] In our laboratory, virgins take anywhere from 3 to >16 days to show maternal behavior.

Much of the work discussed below uses latencies to become maternal in days as the primary measure of the animals' underlying maternal responsiveness. Thus postpartum animals who are very responsive to young have a 0 latency, while virgin animals who are considerably less responsive have an average latency of approximately 8 days. As we shall see, some manipulations augment responsiveness in the virgin, thus shortening the onset latency, and others, like certain lesions, prevent or decrease responsiveness in the postpartum animal, thus increasing the onset latency.

HORMONAL CONTROL OF MATERNAL BEHAVIOR

The first question that intrigued investigators was, what accounts for the differences in level of maternal responsiveness between the new mother and the virgin? The early work of Terkel and Rosenblatt[10,11] and Moltz and co-workers[12,13] and the more recent studies of Siegel and Rosenblatt,[14-18] Bridges,[19-21] and others[22-26] have shown that the high level of maternal behavior seen at birth is due to the pattern of hormonal changes which occur during the latter half of pregnancy and the periparturitional period.

If one injects virgin females with the sequence of progesterone and estradiol that simulates these hormonal changes one finds virgin animals will respond maternally within a day of exposure to pups. In a recent study by Bridges,[21] virgin animals whose ovaries were removed were given Silastic implants containing high levels of progesterone and low levels of estradiol (all within the physiological range) for a 14-day period. To reproduce the periparturitional pattern of a rise in estrogen against a background of declining progesterone, the progesterone was removed on day 14, the estrogen was left in, and the animal was tested on day 15. Bridges found that animals receiving the hormone treatment became maternal more rapidly than did control animals. It seems, then, that 14 days of combined estradiol and progesterone followed by a shift in the progesterone/estrogen ratio is sufficient to induce almost immediate responsiveness.

Although animals show immediate responsiveness at birth, in the absence of stimulation from pups, their responsiveness declines over the first postpartum week.[3] This was demonstrated in a recent series of studies in which we cesarean-sectioned female rats at the end of their pregnancies (before they could give birth naturally) to prevent them from having any contact with pups, and tested them for maternal behavior with foster pups after varying periods following surgery. We found that responsiveness stays relatively high (and therefore latencies remain low) for the first 5 to 7 days, returning to virgin baseline levels by day 10[27] (see FIGURE 1). Although we assume that hormones are mediating this continued responding we have yet to establish which. We do not think postpartum hormonal changes are involved: removal of the ovaries at the time of pregnancy termination does not prevent the elevated responding 7 days later. Moreover, radioimmunoassay measurements of estradiol and progesterone in the blood on days 7 and 10 show these hormones to be at very low levels at both time points.[27]

It also seems that the hormones, estrogen and progesterone, which produce immediate responsiveness in tests undertaken immediately are *not* adequate to maintain responsiveness for the 7-day period. We have found that if virgin animals are given estrogen and progesterone in Silastic capsules for 14 days, using the Bridges[21] regimen, and are tested either immediately or 7 days later, the groups tested immediately show short-latency responding whereas the group tested 7 days later has long latencies, no different from controls[28] (see FIGURE 2). Extending the time of hormone exposure to 22 days and making it more comparable to the situation of a 22-day pregnancy similarly produces immediate responsiveness but no responsiveness at 7 days.[28] We are now exploring the possible role of prolactin in the 7-day retention of responsiveness seen in the normal postpartum animal. Preliminary results indicate prolactin may be involved in the retention of some components of maternal behavior (e.g., nest-building and retrieving) but not others.

HOW HORMONES INFLUENCE MATERNAL BEHAVIOR

What I would like to do now is discuss the variety of ways I believe the hormones may be acting to promote elevated responsiveness at birth and to ensure responding will occur beyond the period of elevated responsiveness, when we know that pregnancy influences are no longer in effect.

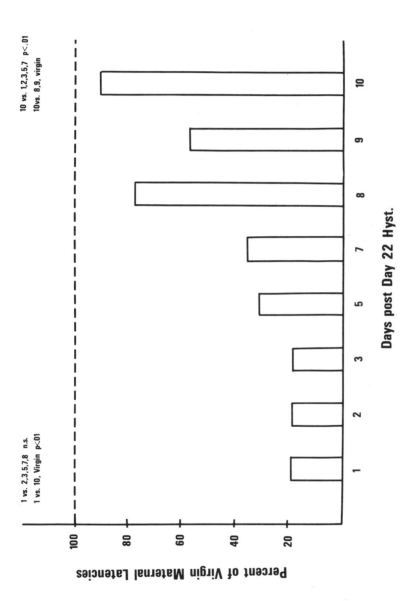

FIGURE 1. Maintenance of maternal responsiveness up to 10 days following pup removal at pregnancy termination (by cesarean delivery). Hyst 22 refers to the procedure of removing the pups and uteri on day 22 of gestation.

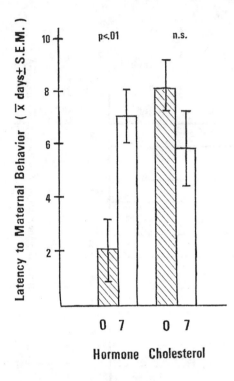

FIGURE 2. Effects of estradiol and progesterone on maternal behavior tested 0 and 7 days after hormone termination.

The argument I will be making is that all females possess an underlying readiness to express maternal behavior. Whether the behavior occurs or not will depend on whether or not there are competing responses which prevent or inhibit the appearance of the behavior. Thus one of the ways hormones act is by altering those other behavioral tendencies, thereby removing their interfering influences on maternal behavior.

What are these other behavioral tendencies which one assumes to be present in the parturient female but absent in the virgin, and do the maternal hormones alter these? Among the ones we will consider are (1) presence of any attraction to pup cues, and (2) absence of a tendency to actively avoid pups. We will also consider (3) differences between postpartum animals and virgins in their ability to consolidate and/ or retain maternal experiences.

Pup Preferences in Nulliparous and Primiparous Females

As indicated earlier, in contrast to the postpartum animal the virgin animal is not immediately maternal when presented with foster pups. On initial exposures to pups she does not spend time with them; if pups are placed opposite the virgin's nest site, she will not choose to lie near them. She does not apparently find them automatically attractive. However, the new mother clearly does. What features of the pups are important in this regard? Although I believe the full complement of stimuli—odors, visual cues, vocalizations, warmth—are probably important, in our work we have concentrated on chemosensory characteristics. This decision was based on the observation we made a number of years ago that if a virgin female is rendered anosmic through *olfactory bulb removal,* through nasal irrigation with zinc sulfate which destroys peripheral nasal epithelia, or through combined main and accessory bulb transections, the animal will show none of the avoidance responses described earlier and will become maternal very quickly.[29,30]

If the new mother is specifically attracted to the odor of the pups and the virgin is not, then one might expect differences between the two kinds of animals in preference for the odor of nesting material taken from the nest of a lactating female and her young. Moreover, if this preference is hormonally induced, exogenous hormones should produce it in the virgin.

The next three studies show both conditions to be true. (1) When the two groups are given a Y-maze preference test with lactating nest odor at one goal box and diestrous virgin nest odor at the other, we see a significant preference for the lactating nest odor by postpartum animals but not by virgins (A. Fleming, unpublished observation). These results support those recently reported by Bauer.[31] (2) Moreover, hormones that turn on maternal behavior can also induce that preference in ovariectomized (OVX) virgins (A. Fleming, unpublished observation) (see TABLE 1). (3) In a recent study comparing responses to nest material taken from the nest of a lactating female as opposed to clean nesting material, we found hormonal capsules induced a strong attraction to the lactating nest odor, while control capsules did not.[32] This is shown in FIGURE 3. These data suggest to us that in order for the postpartum animal to be willing to respond maternally to pups, it must first find the odor of pups attractive.

Differences between Nulliparous and Primiparous Females in Timidity

In addition to differences in their preferences for pup odors, virgins and postpartum animals also differ in fearfulness or timidity. The virgin does not simply refrain from approaching and remaining close to newborn pups placed at a distance from her, she also seems to actively avoid the pups; the postpartum animal does not.[8] This avoidance

TABLE 1. Hormonal Modulation of Olfactory Preferences in Nulliparous Females

		Time ($\bar{x}$ ± SEM, in sec) in Arm of Cage Containing:		Percentage of Animals Showing Preference for:	
Group	N	Lactating Nest Material	Nulliparous Nest Material	Lactating Nest Odor	Nulliparous Nest Odor
Nulliparous-OVX	33	127.6 ± 13.2	103.2 ± 12.6	65.6	34.4
Nulliparous-OVX + E-7, P + LTH	15	151.7 ± 16.8	65.6 ± 8.8	86.6	13.3[a]
Nulliparous-OVX + oil	15	107.9 ± 17.5	104.3 ± 15.1	53.3	46.7

[a] $p < 0.005$.

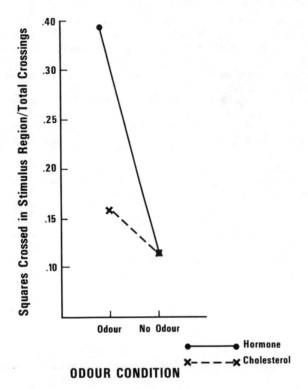

FIGURE 3. Effects of estradiol and progesterone on lactating nest odor preferences in virgin females.

means that the pups are not only not particularly attractive to the female, but that they may be aversive and, in response to this stimulus complex, the female actively moves away. To demonstrate that this active avoidance reflects a more general difference in overall timidity between the two types of animal, in a series of studies we compared virgin and postpartum animals on a variety of measures of timidity and found large differences in many of them.[8]

Postpartum animals (1) emerge more rapidly into an open, unfamiliar arena, (2) ambulate more in an open field, and (3) are more willing to ambulate into the center of the arena away from the walls. They are also less likely to be chased by intruders into their cages. These studies show these animals are, in general, more willing to enter into strange places, approach novel objects, etc.—features which are important if the new mother is to be willing to stay with her pups. We believe these differences in overall timidity are due to the parturitional hormones, estradiol (E) and progesterone (P).

If we implant virgin animals with the same Silastic regimen that turns on maternal behavior and test them in the open field, we find that these same hormones also reduce timidity to levels found in the postpartum animal. As shown in FIGURE 4, comparisons between hormone-treated and control animals show that hormone-treated animals ambulate more and emerge more quickly.[32] This is supported by data by other investigators which implicate estrogen in the reduction of fear, although the interpretation of these results is in some dispute.[33–36] These data suggest that among other things the hormones may change an animal from one which is more cautious, timid, and neophobic as a virgin to one that is less so, and increase an animal's attraction to pup odors.

Early Experience with Pup Cues Augments Maternal Responsiveness in Adults

If these propositions are true, then one would predict that manipulations which reduce an animal's fearfulness and which familiarize the animal with pup cues should augment a virgin animal's maternal responsiveness. A number of attempts by our laboratory to reduce fearfulness by extensive handling or by administration of tranquilizers to adult virgin animals have been without effect. However, Herrenkohl and Lisk[37] found that exposure of adult virgin animals to the odors, sight, and sound of pups between periods of induction testing resulted in a reduction in latency to show a number of maternal behaviors.

We have also found that if these exposure manipulations are applied during early development in animals tested in adulthood, clear experience effects can be demonstrated.[38] As shown in FIGURE 5, animals who are exposed to the odor (and vocali-

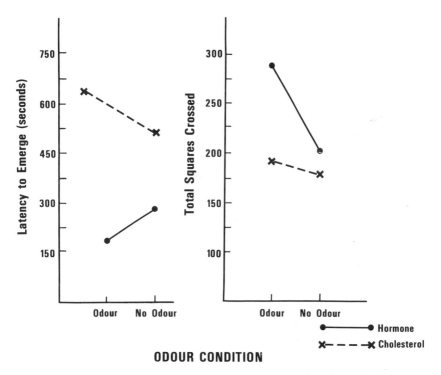

FIGURE 4. Effects of estradiol and progesterone on open-field behavior in virgins.

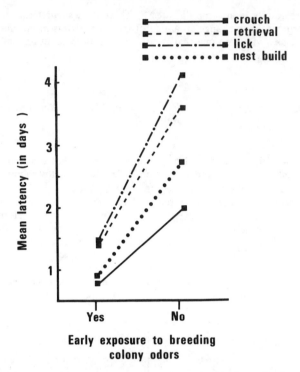

FIGURE 5. Effects of early exposure to breeding colony odors (and vocalizations) on maternal behavior in adulthood.

zations) of pups in a breeding room situation during the first 20 days of life have shorter retrieval, crouch, lick, and nest-building latencies in adulthood in comparison to animals not permitted this experience.

I believe under normal circumstances in the postpartum animal a reduction in timidity and an attraction to pups are due to hormonal changes of pregnancy and parturition, and that if both are fulfilled the animal will not be inhibited from responding maternally. However, if both are not fulfilled, maternal behavior will not occur. In support of this notion, we found that following a 14-day Silastic regime, virgin animals are maternal immediately but not 7 days later. If one looks at overall emotionality and attraction to lactating nest odors in the same animals, on the day after the termination of hormones one finds low timidity (high ambulation) and high attraction to the odors. Tests on day 7, however, show that animals still show low timidity, but that they no longer find pup odors to be as attractive.

Differences between Nulliparous and Primiparous Females in Retention of Maternal Experience

In addition to these influences of hormones on the onset of maternal responsiveness, I believe hormones may also influence the maintenance of responsiveness past the time when the relevant hormones are present and that they do so by ensuring that experiences acquired during that time will be retained for a long time. We have found, for instance, that as little as ½ hr of interaction with pups 1 day after birth will produce elevated responsiveness on day 10 when it has normally declined (see TABLE 2).[27] More impressively, Bridges and his colleagues found long-term effects of 1 hr of experience up to 25 days later.[19,23,39,40] A similar experience obtained by the virgin is not adequate to maintain the same level of responsiveness 10 or 25 days later.[28,41] On the basis of these results, it seems the early postpartum hormonal influences not only augment responsiveness but somehow produce better learning or greater retention of the behavior at a later point in time if animals interact or are exposed to the pups during the period of hormonally mediated elevated responsiveness.

NEURAL CONTROL OF MATERNAL BEHAVIOR

Where in the brain are hormones acting to produce these effects? We are only beginning to answer these questions. As shown in FIGURE 6, we have been working on a neural circuit which goes from the olfactory bulbs (including the main and accessory olfactory bulbs) via the lateral olfactory tract (LOT) to the corticomedial region of the amygdala, from the amygdala via the stria terminalis to the bed nucleus of the stria terminalis (BNST), and thence to the medial preoptic area (MPOA). The bed nucleus itself receives direct input from the accessory olfactory system and also projects down into the MPOA.[42-48]

As outlined in TABLE 3, this circuit is of interest in terms of the expression of maternal behavior because: (1) It constitutes the primary projection route of the olfactory fibers, a modality which we have seen is integrally involved in the expression of maternal behavior. (2) The amygdala and stria terminalis have been implicated by others as well as by ourselves in the regulation of emotional behavior.[49-55] We have

TABLE 2. Effects of Limited Maternal Experience during the First Postpartum Day on Maternal Behavior 10 Days Later

Group	N	Latency to Maternal Behavior [Median (IQR)]	Percentage Maternal in 24 hr	Percentage Not Maternal in 10 Days
No exposure	25	8 (3.25-10)[a]	6	35
Exposure but no retrieval	11	7 (1.75-9.25)[b]	18	36
Retrieval only	9	6 (1.25-10)[c]	22	33
Retrieval + 15 min	9	4 (0.25-8.25)[d]	44	33
Retrieval + 30 min	11	1 (0-2)[a,b,c,d]	64	0
Full exposure	17	1 (0-1)[a,b,c,d]	76	6

Note: IQR = inter-quartile range.
[a,b,c,d] $p \leq 0.05$ [meaning No exposure (group) different from Retrieval + 30 min (a), Exposure but no retrieval different from Retrieval + 30 min (b), etc.]

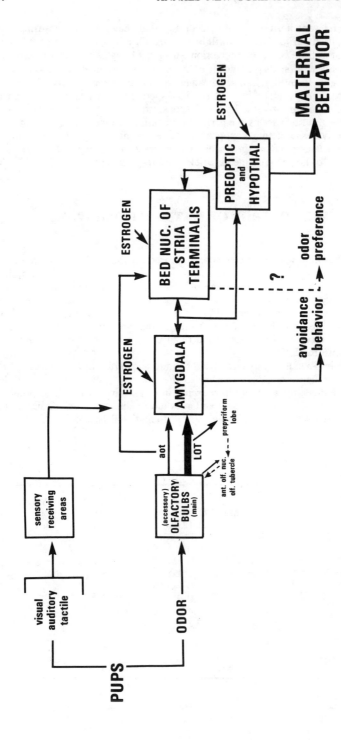

FIGURE 6. Schematic of olfactory-limbic-preoptic pathways mediating maternal behavior in rats. (LOT refers to lateral olfactory tract; aot refers to accessory olfactory tract).

found that lesions to the amygdala produce reduced neophobia, avoidance in response to novel objects, and increased ambulation in the open field, etc. (3) The MPOA, into which amygdala and BNST fibers project, has been shown by Numan and colleagues[56,57] and by others to be essential for the expression of maternal behavior—lesions of the MPOA or knife cuts transecting its connection with the medial forebrain bundle (MFB) as well as other laterally and caudally projecting fiber systems eliminate maternal behavior in postpartum animals and in virgins. (4) Estradiol is taken up and bound at sites within the circuit, almost exclusively—receptors being found in the olfactory tubercle, the BNST, the medial and cortical nuclei of the amygdala, and the MPOA.[58] This hormone is clearly involved in the expression of maternal behavior, reduced timidity, and preference for pup-related odors. Although its effectiveness is increased by a prior decline in progesterone, at higher concentrations it can activate behavior without prior progesterone priming.[14,15]

TABLE 3

(1) AMYGDALA AND BED NUCLEUS OF STRIA TERMINALIS ARE PRIMARY PROJECTION SITES OF OLFACTORY FIBERS:
  pup odors inhibit maternal behavior

(2) AMYGDALA MEDIATES AFFECTIVE/AVOIDANCE RESPONSES AND EVALUATES STIMULUS SALIENCE:
  lesions of the amygdala reduce timidity, as measured by conventional "emotionality" tests

(3) MEDIAL PREOPTIC AREA (MPOA) IS NECESSARY FOR MATERNAL BEHAVIOR:
  lesions of MPOA disrupt ongoing maternal behavior

(4) ALL STRUCTURES WITHIN THE CIRCUIT CONTAIN ESTRADIOL RECEPTORS:
  estrogen is the primary hormone in the hormonal induction of maternal behavior

To summarize a number of years of work, we have found that there is a disinhibition of maternal behavior (meaning a shorter latency to become maternal) in virgins given foster pups following lesions to the olfactory bulbs and their major efferents, the lateral olfactory tracts,[29,30] when compared to those given control/sham lesions. These animals are unable to smell the pups. A similar disinhibition of the behavior occurs with lesions to the two primary projection sites for the main and accessory olfactory bulbs.[51] These are the corticomedial nuclear groups within the amygdala which receive main and accessory olfactory input and the bed nucleus of the stria terminalis which receives accessory olfactory information. Animals with lesions to these sites are *not* anosmic and therefore we assume can smell the pups, and they respond maternally. Finally, knife cuts of the primary efferent pathway, the stria terminalis, projecting from the amygdala to both the BNST and the MPOA, also produce a short-latency maternal behavior.[59] However, facilitation of maternal behavior following lesions to many of these sites requires that the MPOA be intact. Lesions of the MPOA prevent the expression of maternal behavior in animals with amygdala lesions.[60] Observations indicate that, unlike animals with MPOA lesions only, animals with additional lesions to the amygdala show no avoidance of pups, and thus remain in close proximity to them although they will not respond maternally (see FIGURE 7).

246 ANNALS NEW YORK ACADEMY OF SCIENCES

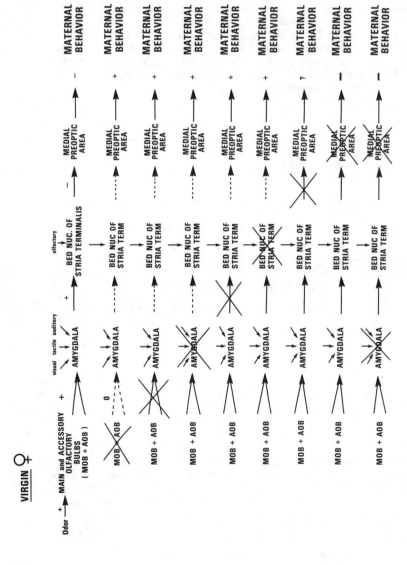

FIGURE 7. Schematic of effects of lesions to the different sites in the olfactory-limbic-preoptic pathways on maternal behavior.

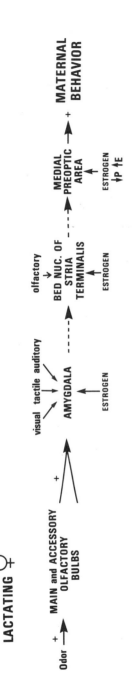
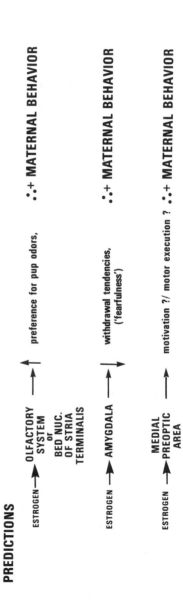

FIGURE 8. Schematic of action of estrogen in parturient (lactating) females.

Taken together, these data suggest to us that, in terms of this context, the amygdala's primary function is to interpret olfactory input from the pups in terms of its novelty and salience and to make the appropriate affective response. When the external pup stimulus is new, strange, or aversive, the amygdala becomes activated and causes withdrawal and avoidance. With repeated exposure to pups, as during inductions, the pups become less novel and more familiar, the amygdala activity declines, and the animals show less avoidance. Amygdala activity, and thus avoidance, is also prevented if the amygdala receives no olfactory input, which is the situation following olfactory bulb removal or if the amygdala is itself lesioned. Activity in the amygdala not only influences avoidance behavior directly but we assume also exerts inhibition on the MPOA. When the inhibition is removed animals show heightened maternal behavior.

What is the role of the BNST in this system? We do not know. However, one possibility is that it integrates information coming both from the olfactory bulbs and from the amygdala, and after integrating these two sources of information, may mediate preference behavior or attraction to biologically relevant odorants. If the odorants are interpreted as desirable, to be approached, the BNST conveys this information to the MPOA. Whether the BNST-MPOA relationship is an inhibitory one or an excitatory one is not known.

How does all this relate to hormonal effects on behavior, which normally occur in the postpartum animal? As indicated earlier, all these neural sites bind estradiol. Moreover, Numan found that estrogen implants in the MPOA facilitate short-latency maternal responding.[57] We are presently doing a large-scale implant study in which implants are being placed in the BNST, the amygdala, and the MPOA, and animals are being tested for odor preference, avoidance responsiveness, and maternal behavior.

Our predictions are that implants in the amygdala or the BNST will produce some reduction in maternal onset latencies, that hormonal stimulation of both sites together will produce still greater reductions, and that the effect of the hormone on the two sites will be somewhat different, producing a reduction in avoidance responding in the amygdala and an elevation of attraction to pup odors in the BNST.

On the basis of these diverse studies, we believe that in the parturient female the hormones of the periparturitional period act not only to prime maternal behavior but also to increase attraction to pups and their odors, decrease the tendency to avoid them, and consolidate experiences obtained while with the pups. As indicated in FIGURE 8, we think hormones act on a number of different limbic and hypothalamic sites to accomplish these ends.

REFERENCES

1. FLEMING, A. S. & B. G. ORPEN. 1986. Psychobiology of maternal behavior in rats, selected other species and humans. *In* Origins of Nurturance. A. Fogel & G. F. Melson, Eds.: 141-208. Lawrence Erlbaum Assoc. London, England.
2. WEISNER, B. P. & N. M. SHEARD. 1933. Maternal Behavior in the Rat. Oliver & Boyd. London, England.
3. ROSENBLATT, J. S. & D. S. LEHRMAN. 1963. Maternal behavior of the laboratory rat. *In* Maternal Behavior in Mammals. H. L. Rheingold, Ed. Wiley. New York, N.Y.
4. ROSENBLATT, J. S. 1965. The basis of synchrony in the behavioral interaction between the mother and her offspring in the laboratory nest. *In* Determinants of Infant Behavior, Vol. III. B. Foss, Ed. Methuen. London, England.
5. FLEMING, A. S. & J. S. ROSENBLATT. 1974. Maternal behavior in the virgin and lactating rat. J. Comp. Physiol. Psychol. **86:** 957-972.

6. ROSENBLATT, J. S. 1967. Nonhormonal basis of maternal behavior in the rat. Science 156: 1512-1514.
7. TERKEL, J. & J. S. ROSENBLATT. 1971. Aspects of nonhormonal maternal behavior in the rat. Horm. Behav. 2: 161-171.
8. FLEMING, A. S. & C. LEUBKE. 1981. Timidity prevents the virgin female rat from being a good mother: emotionality differences between nulliparous and parturient females. Physiol. Behav. 27: 863-868.
9. MAYER, A. D. & J. S. ROSENBLATT. 1979. Hormonal influences during the ontogeny of maternal behavior in female rats. J. Comp. Physiol. Psychol. 93: 879-898.
10. TERKEL, J. & J. S. ROSENBLATT. 1968. Maternal behavior induced by maternal blood plasma injected into virgin rats. J. Comp. Physiol. Psychol. 65: 479-482.
11. TERKEL, J. & J. S. ROSENBLATT. 1972. Humoral factors underlying maternal behavior at parturition: cross transfusion between freely moving rats. J. Comp. Physiol. Psychol. 80: 365-371.
12. MOLTZ, H., R. LEVIN & M. LEON. 1969. Differential effects of progesterone on the maternal behavior of primiparous and multiparous rats. J. Comp. Physiol. Psychol. 67: 36-40.
13. MOLTZ, H., M. LUBIN, M. LEON & M. NUMAN. 1970. Hormonal induction of maternal behavior in the ovariectomized nulliparous rat. Physiol. Behav. 5: 1373-1377.
14. SIEGEL, H. I. & J. S. ROSENBLATT. 1975. Hormonal basis of hysterectomy-induced maternal behavior during pregnancy in the rat. Horm. Behav. 6: 211-222.
15. SIEGEL, H. I. & J. S. ROSENBLATT. 1975. Estrogen induced maternal behavior in hysterectomized-ovariectomized virgin rats. Physiol. Behav. 14: 465-471.
16. SIEGEL, H. I. & J. S. ROSENBLATT. 1975. Progesterone inhibition of estrogen-induced maternal behavior in hysterectomized-ovariectomized virgin rats. Horm. Behav. 6: 223-230.
17. SIEGEL, H. I. & J. S. ROSENBLATT. Effects of pregnancy termination on maternal behavior, lordosis, ovulation and progesterone levels in the rat. Paper presented at the Eastern Conference on Reproductive Behavior, Storrs, Connecticut, June 1977.
18. SIEGEL, H. I. & J. S. ROSENBLATT. 1978. Duration of estrogen stimulation and progesterone inhibition of maternal behavior in pregnancy-terminated rats. Horm. Behav. 11: 12-19.
19. BRIDGES, R. S. 1978. Retention of rapid onset of maternal behavior during pregnancy in primiparous rats. Behav. Biol. 24: 113-117.
20. BRIDGES, R. S. A new and reliable preparation for examination of the hormonal regulation of maternal behavior in the rat. Abstract presented at the Conference on Reproductive Behavior, Nashville, Tennessee, June 1981.
21. BRIDGES, R. S. 1984. A quantitative analysis of the roles of estradiol and progesterone in the regulation of maternal behavior in the rat. Endocrinology 114: 930-940.
22. BRIDGES, R. S., J. S. ROSENBLATT & H. H. FEDER. 1978. Serum progesterone concentrations and maternal behavior in rats after pregnancy termination: behavioral stimulation following progesterone withdrawal and inhibition by progesterone maintenance. Endocrinology 102: 258-267.
23. BRIDGES, R. S., J. S. ROSENBLATT & H. H. FEDER. 1978. Stimulation of maternal responsiveness after pregnancy termination in rats: effect of time of onset of behavioral testing. Horm. Behav. 10: 235-245.
24. DOERR, H. K., H. I. SIEGEL & J. S. ROSENBLATT. 1981. Effects of progesterone withdrawal and estrogen on maternal behavior in nulliparous rats. Behav. Neural Biol. 32: 35-44.
25. KREHBIEL, D. A. & L. M. LEROY. 1979. The quality of hormonally stimulated maternal behavior in ovariectomized rats. Horm. Behav. 12: 243-252.
26. ZARROW, M. X., R. GANDELMAN & V. H. DENENBERG. 1971. Maternal behavior in the rabbit: critical period for nest building following castration during pregnancy. Proc. Soc. Exp. Biol. Med. 111: 537-538.
27. ORPEN, G. & A. S. FLEMING. Sensory and hormonal mediation of the postpartum transition in rats. Abstract presented at the Conference on Reproductive Behavior, Pittsburgh, Pennsylvania, June 1984.
28. ORPEN, G. & A. S. FLEMING. Hormonal control of the maintenance of maternal behavior in the rat. In preparation.

29. FLEMING, A. S. & J. S. ROSENBLATT. 1974. Olfactory regulation of maternal behavior in rats. I. Effects of olfactory bulb removal in experienced and inexperienced lactating and cycling females. J. Comp. Physiol. Psychol. **86:** 221-232.
30. FLEMING, A. S., F. VACCARINO, L. TAMBOSSO & P. CHEE. 1979. Vomeronasal and olfactory system modulation of maternal behavior in the rat. Science **203:** 372-374.
31. BAUER, J. H. 1983. Effects of maternal state on the responsiveness to nest odors of hooded rats. Physiol. Behav. **30:** 229-232.
32. KESSLER, Z., D. MORETTO & A. S. FLEMING. Effects of estradiol and progesterone on open field behavior and response to lactating nest odors. In preparation.
33. ARCHER, J. 1973. Tests for emotionality in rats and mice: a review. Anim. Behav. **21:** 205-235.
34. GRAY, J. A. 1971. Sex differences in emotional behavior in mammals including man: endocrine bases. Acta Psychol. **34:** 29-46.
35. IKARD, W. L., W. C. BENNETT, R. W. LUNDINO & R. C. TROST. 1972. Acquisition and extinction of the conditioned avoidance response. A comparison between male rats and estrous and nonestrous female rats. Psychol. Rec. **22:** 249-254.
36. QUADAGNO, D. M., J. SHRYNE, A. ANDERSON & R. A. GORSKI. 1972. Influences of gonadal hormones on social, sexual, emergence and open field behavior in the rat (*Rattus norvegicus*). Anim. Behav. **20:** 732-740.
37. HERRENKOHL, L. R. & R. D. LISK. 1973. The effects of sensitization and social isolation on maternal behavior in the virgin rat. Physiol. Behav. **11:** 619-624.
38. MORETTO, D., L. PACLIK & A. S. FLEMING. 1986. Effects of early rearing environment on maternal responsiveness in adult rats. Dev. Psychobiol. In press.
39. BRIDGES, R. S. 1975. Long-term effects of pregnancy and parturition upon maternal responsiveness in the rat. Physiol. Behav. **14:** 245-249.
40. BRIDGES, R. S. 1977. Parturition: its role in the long term retention of maternal behavior in the rat. Physiol. Behav. **18:** 487-490.
41. COHEN, J. & R. S. BRIDGES. 1981. Retention of maternal behavior in nulliparous and primiparous rats: effects of duration of previous maternal experience. J. Comp. Physiol. Psychol. **95:** 450-459.
42. BERK, M. L. & J. A. FINKELSTEIN. 1981. Afferent projections to the preoptic area and hypothalamic regions in the rat brain. Neuroscience **6:** 1601-1624.
43. KRETTEK, J. E. & J. L. PRICE. 1978. Amygdaloid projections to subcortical structures within the basal forebrain and brainstem in the rat and cat. J. Comp. Neurol. **178:** 225-254.
44. TURNER, B. H. & M. E. KNAPP. 1976. Projections of the nucleus and tracts of the stria terminalis following lesions at the level of the anterior commissure. Exp. Neurol. **51:** 468-479.
45. CONRAD, L. C. A. & D. W. PFAFF. 1976. Efferents from medial basal forebrain and hypothalamus in the rat. I. An autoradiographic study of the medial preoptic area. J. Comp. Neurol. **169:** 185-220.
46. CONRAD, L. C. A. & D. W. PFAFF. 1976. Efferents from medial basal forebrain and hypothalamus in the rat. II. An autoradiographic study of the anterior hypothalamus. J. Comp. Neurol. **169:** 221-262.
47. BARAS, S. & S. PAY. 1980. Neurosci. Lett. **17:** 265-269.
48. WINANS, S. S. & F. SCALIA. 1970. Amygdaloid nucleus: new afferent input from the vomeronasal organ. Science **170:** 330-332.
49. BLANCHARD, D. C. & R. J. BLANCHARD. 1972. Innate and conditioned reactions to threat in rats with amygdaloid lesions. J. Comp. Physiol. Psychol. **81:** 281-290.
50. COOVER, G., H. URSIN & S. LEVINE. 1973. Corticosterone and avoidance in rats with basolateral amygdala lesions. J. Comp. Physiol. Psychol. **85:** 111-122.
51. FLEMING, A. S., F. VACCARINO & C. LUEBKE. 1980. Amygdaloid inhibition of maternal behavior in the nulliparous female rat. Physiol. Behav. **25:** 731-743.
52. GALEF, B. G., JR. 1970. Aggression and timidity responses to novelty in feral Norway rats. J. Comp. Physiol. Psychol. **70:** 370-381.
53. GROSSMAN, S. P., L. GROSSMAN & L. WALSH. 1975. Functional organization of the rat amygdala with respect to avoidance behavior. J. Comp. Physiol. Psychol. **88:** 829-850.

54. KIMBLE, E. D. & J. A. NAGEL. 1973. Failure to form a learned taste aversion in rats with amygdaloid lesions. Bull. Psychon. Soc. **2:** 155-156.
55. NACHMAN, M. & J. ASHE. 1974. Effects of basolateral amygdala lesions on neophobia, learned taste aversions and sodium appetite in rats. J. Comp. Physiol. Psychol. **87:** 622-643.
56. NUMAN, M. 1974. Medial preoptic area and maternal behavior in the female rat. J. Comp. Physiol. Psychol. **87:** 746-759.
57. NUMAN, M., J. S. ROSENBLATT & B. R. KOMISARUK. 1977. Medial preoptic area and onset of maternal behavior in the rat. J. Comp. Physiol. Psychol. **91:** 146-164.
58. PFAFF, D. & M. KEINER. 1973. Atlas of estradiol-concentrating cells in the central nervous system of the female rat. J. Comp. Neurol. **151:** 121-158.
59. FLEMING, A. S. & M. MICELI. Effects of knife cuts to the dorsal stria terminalis on maternal behavior in rats. In preparation.
60. FLEMING, A. S., M. MICELI & D. MORETTO. 1983. Lesions of the medial preoptic area prevent the facilitation of maternal behavior produced by amygdaloid lesions. Physiol. Behav. **31:** 502-510.

Role of Prolactin in Avian Incubation Behavior and Care of Young: Is There a Causal Relationship?[a]

JOHN D. BUNTIN

Department of Biological Sciences
University of Wisconsin-Milwaukee
Milwaukee, Wisconsin 53201

The importance of humoral factors in the expression of parental activities in birds was first demonstrated over 50 years ago when Lienhart[1] reported that nonincubating fowl could be induced to sit on eggs by injecting them with serum from incubating hens. The most likely humoral candidate to mediate the behavioral changes that take place during the incubation and post-hatching periods has been widely assumed to be the pituitary hormone prolactin (PRL) since it also promotes many of the physiological changes that take place at these breeding stages, such as gonadal regression, brood patch formation, and, in columbiform species, crop "milk" formation (see References 2 and 3 for review). However, the question of whether prolactin is capable of "changing the carefree laying hen into a fowl tied to the nest and later defense of the young"[4] has yet to be resolved, owing largely to recent failures to corroborate earlier reports that PRL injections facilitate incubation onset, problems of data interpretation stemming from methodological inconsistencies, and a general lack of systematic investigation of the effects of exogenous PRL on reproductive behavior. It is also becoming apparent that participation in parental activity itself may directly or indirectly stimulate changes in PRL secretion, thereby adding further complexity to this relationship.

This review will focus specifically on the available evidence for a role of PRL in the expression of incubation and parental behavior in various avian species, with less emphasis being placed on the social and environmental control of these behaviors and the stimulus control of PRL secretion. The principal objectives of the review will be to identify and discuss some of the conceptual and methodological problems that have clouded the interpretation of the available evidence, to point out areas where additional corroboration of data is needed, and to suggest productive areas for future research.

---

[a] Work cited from my laboratory was supported by National Institute of Mental Health Grant MH 38160, National Science Foundation Grant DCB8303026, and a Faculty Research Grant from the Graduate School, University of Wisconsin-Milwaukee.

INDUCTION OF INCUBATION

Experimental Studies

Although PRL has long been assumed to play some role in the onset of incubation behavior in birds, direct evidence for such a relationship is surprisingly limited. In early studies, Riddle *et al.*[5] and Saeki and Tanabe[6] reported that daily injections of PRL (20-100 I.U.) were effective in inducing sitting behavior in "broody" strains of laying chickens, but recent attempts to confirm these findings have been unsuccessful.[7,8] In an extensive study in which hormone dosage, timing of daily hormone administration, and testing conditions were systematically varied, Opel and Proudman[7] found no evidence that ovine PRL injections induced the onset of incubation behavior in domestic turkey hens or in laying chickens with previous "broody" experience. PRL treatment also reportedly failed to induce incubation behavior in laying ducks[9] and in Japanese quail given injections of PRL and estradiol simultaneously.[10] Because the PRL preparations used in recent studies were more highly purified than those available to earlier investigators, Opel and Proudman suggested that the positive results obtained from these initial PRL injection studies may be due to the direct or indirect actions of other pituitary hormone contaminants present in the preparation. While the identity of these putative incubation-promoting agents is unknown, it is noteworthy that Riddle *et al.*[5] observed no effects of combined luteinizing hormone (LH) and follicle stimulating hormone (FSH) administration or of thyroid stimulating hormone (TSH) or progesterone alone on incubation onset, thereby suggesting that if these particular hormones are involved in incubation induction at all, they most likely are acting in concert with PRL.

The relationship between PRL and the initiation of sitting activity is further complicated by evidence suggesting that PRL acts directly in the brain to promote onset of incubation behavior in fowl. Opel[11] reported that sitting behavior was induced in laying hens following placement of bilateral PRL implants in the preoptic area, anterior hypothalamus, or lateral forebrain bundle. Implants placed in other forebrain loci failed to stimulate incubation, but some elicited related behavioral components such as clucking and aggressive defense. Gonadal regression accompanied the display of incubation in all cases, but induction of gonadal regression alone did not constitute a sufficient condition for the display of sitting behavior. These data suggest that a central site of PRL action should be seriously considered when interpreting the effects of systemic PRL injections. Moreover, it is possible that differences in the rate of PRL entry to the brain and/or the degree to which PRL is chemically altered as it traverses the blood-brain barrier[12] could account in part for the conflicting data obtained from these systemic PRL injection studies.

In view of the conflicting evidence for PRL-induced incubation behavior in galliform species, it is entirely possible that other physiological or environmental factors may normally participate in incubation onset, perhaps by increasing the sensitivity or responsiveness of the brain or other target tissues to PRL. The fact that hens must be in laying condition in order for PRL injections or intracranial implants to effectively induce incubation behavior suggests that the high titer of gonadal steroids present around the time of ovulation may set the stage for PRL-induced incubation onset.[5,6,11] In support of this concept, Wood-Gush and Gilbert[13] have demonstrated that estrogen and progesterone synergize to stimulate the initial stages of nesting activity in ovariectomized chickens. The possibility that steroids enhance PRL-induced incubation behavior is also supported by studies in the budgerigar, which indicate that estrogen-plus-PRL treatment is more effective than either hormone alone in promoting extended nest box occupation behavior and onset of incubation in ovariectomized females.[14] Significantly, estrogen-plus-PRL treatment was also found to be more effective than progesterone alone or estrogen-plus-progesterone administration in this regard.

The hypothesis that steroid hormones enhance the incubation-promoting effects of PRL in birds is parsimonious since it is consistent with the role of steroids in promoting other "parental" effects of PRL such as brood patch development[15] and phasing the period of responsiveness of various target tissues to PRL.[16] Nevertheless, it is not likely to be universally applicable. In ring doves, the relative importance of steroid hormones and PRL in stimulating sitting onset has been extensively debated,[17,18] but a convincing case can be made from hormone injection studies that steroid hormones play a more important role in the induction of incubation than does PRL. In this species, the transition from nesting activity to incubation is induced in both sexes through a complex interplay between hormonal stimuli and external stimuli from the mate and nest (see References 19 and 20 for review). While neither hormone alone is effective, combined estrogen-plus-progesterone treatment promotes intensive nest-building and sitting activity in ovariectomized females paired with courting males.[21] Moreover, progesterone is more effective than PRL in inducing sitting behavior in unpaired birds tested in isolation cages[22-24] and can act directly on the hypothalamus to facilitate incubation.[25] In addition, other gonadal steroids may act to enhance progesterone's action by increasing the concentration of cytoplasmic progestin receptors at this site.[26] Progesterone thus appears to be the principal hormonal stimulus for the induction of incubation in this species, but it would be premature to entirely rule out a role for PRL in promoting incubation onset since the effects of combined steroid-plus-PRL treatment have never been experimentally tested.

Correlational Studies

Several investigators have adopted a more indirect approach to the question of PRL involvement in incubation behavior by determining if changes in circulating PRL levels coincide with sitting activity during the reproductive cycle. Although there are reports to the contrary,[27,28] most radioimmunoassay data suggest that the display of incubation behavior is associated with elevated PRL levels in several galliform species,[29-34] in waterfowl,[35-38] in several passerines,[39-41] and in the ring dove.[42,43] However, more refined assay studies and corroborating experimental investigations are required to determine whether these elevated PRL levels induce, or are induced by, the onset of incubation behavior.

A rise in circulating PRL levels has been reported in several galliform species during egg-laying with peak levels coinciding with the onset of incubation.[29,30,31,33,34] This is consistent with the hypothesis that PRL promotes the onset of incubation behavior in galliform birds, but it may misrepresent or oversimplify the relationship between changes in PRL concentration and changes in responsiveness to the nest and eggs. A marked daily rhythm in plasma PRL concentrations has been reported in bantam hens[44] and White Rock hens[45] with the patterning and/or the amplitude of these fluctuations changing with reproductive stage. In bantams, Lea *et al.*[44] reported no marked changes in average PRL levels prior to the onset of incubation. However, significant changes in the periodicity of daily PRL fluctuations were observed, with a nocturnal rise in PRL levels occurring in females about 6 days before egg-laying ceased and incubation began. Significantly, this nocturnal rise was not observed in females which continued to lay eggs (see FIGURE 1). As the onset of incubation approached, the nocturnal period of elevated PRL levels began to lengthen until, by the first day of incubation, plasma PRL levels were elevated throughout the day and night. While they support the view that an increase in PRL concentration precedes the onset of incubation, these findings also imply that changes in the daily periodicity of PRL fluctuations may be at least as important as changes in the average titer of circulating PRL in determining PRL's effectiveness in promoting incubation onset. Such changes, in any event, cannot be adequately characterized by measuring PRL levels in blood samples collected at only one time of the day.

Among the non-galliform species investigated, few if any generalizations can be made regarding the presence of a pre-incubatory rise in circulating PRL. A rise in PRL has been reported during nest-building or egg-laying in some species (domestic duck,[38] pied flycatcher,[41] starling[40]) but not in others (canary,[39] ring dove[42]). The picture is particularly confused among the columbiform species, where PRL levels in females have been reported to undergo a sustained increase during the nest-building and egg-laying periods (band-tailed pigeon[46]), to exhibit a more modest and transitory post-ovulatory surge (ring dove[43]), or to remain unchanged from the nest-building stage to mid-incubation (ring dove[42]). Such discrepancies are probably due in large part to methodological differences associated with the type of radioimmunoassay system used for PRL measurement, blood collection procedures, and timing of blood sampling in relationship to underlying patterns of changes in PRL levels.

MAINTENANCE OF INCUBATION

Circulating PRL levels tend to be elevated during incubation in females of several species (see above) and in males of species in which incubation is shared between the breeding partners (ring dove,[42,43] Australian black swan[37]). More species diversity in incubation-phase PRL patterns is observed in males of species in which the female assumes all the incubation duties (canary,[39] starling,[40] goose,[36] duck[35]). However, sex differences in PRL levels during incubation are, as a rule, more pronounced in these species. This has prompted speculation that, apart from its hypothesized role in the induction of incubation, PRL may also help to maintain incubation in birds in which the behavior is already established. However, stimuli associated with or generated by sitting activity itself may facilitate PRL release[38,47,48] and accordingly, it is difficult from correlational data alone to establish the precise causal relationship between PRL secretion and sitting activity.

Relatively few studies have dealt specifically with the question of whether PRL maintains ongoing incubation behavior. However, several investigators have examined the issue of whether PRL maintains readiness to sit during a period of separation from the nest and eggs. While these problems are related, it is important to emphasize that the factors responsible for maintenance of incubation readiness in nest-deprived birds may not be the same as those involved in maintenance of ongoing incubation during a normal breeding cycle. Persistence of ongoing sitting behavior has been reported over a 72-hr period in ring doves[49] and over a 9-hr period in ducks[38] following experimental manipulations which induced a substantial decline in PRL levels. In bantam hens, moreover, injections of chicken PRL antiserum resulted in increased LH secretion but did not disrupt established incubation behavior, at least during the 24-hr period following antiserum administration.[33] Studies on short-term maintenance of sitting readiness in nest-deprived birds have yielded similar results. Most female canaries will resume sitting after a 48-hr period of separation from the nest despite

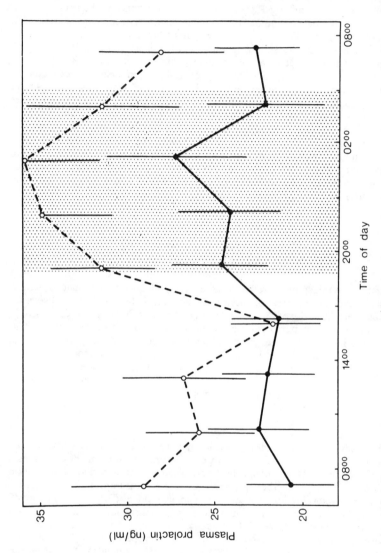

FIGURE 1. Mean (± SEM) daily variations in plasma PRL concentrations in 16 laying hens during a period in which incubation behavior was not subsequently displayed (solid circles) and in 13 of the same birds between 1 and 6 days before the onset of incubation (open circles). The stippled bar indicates the period of darkness. (Data from Lea et al.[44])

the fact that PRL levels decline markedly during this period.[50] Incubating turkey hens also exhibit a rapid decline in circulating PRL concentration when deprived of nesting opportunities, with a subsequent rise occurring after they are returned to their nests.[47] However, the resumption of incubation does not appear to depend upon this increase in PRL secretion since sitting is reestablished prior to any observable increase in PRL titer. Furthermore, this rise in PRL can be prevented pharmacologically without disrupting the resumption of sitting behavior.[48] These results clearly indicate that readiness to incubate and actual ongoing sitting activity can persist for 24-72 hr after serum PRL levels have declined. However, it would be premature to conclude that PRL plays no role in maintenance of these activities since a detectable change in PRL-dependent behavior could lag behind a change in circulating PRL concentration by several days. Additional studies involving long-term monitoring of the behavioral consequences of reduced PRL secretion may therefore be required in order to adequately evaluate PRL's role in incubation maintenance.

Data gathered in the ring dove provide the strongest and most direct evidence for a role of PRL in maintaining long-term responsiveness to a nest and eggs. Lehrman and Brody[51] reported that daily PRL injections maintained readiness to incubate in 60% of the females and 80% of the males tested after a 12-day period of isolation from mates, nests, and eggs. This facilitatory effect of PRL was confirmed in a more recent study[52] in which doves were tested following a 10-day isolation and hormone treatment period using a more stringent incubation criterion and testing procedure than that used by Lehrman and Brody. The incidence of incubation behavior was significantly higher in the PRL-treated groups than in the saline-injected control group in this study. However, in contrast to the pattern observed with PRL-induced crop sac, liver, and body weight changes, incubation responses were not closely related to PRL dosage (see FIGURE 2). Furthermore, the incidence of incubation behavior in PRL-treated groups was considerably below that observed in untreated normally incubating birds tested under the same conditions after a 30-min period of isolation from their mates and nests. On the basis of this evidence, PRL would appear to be important in maintaining normal responsiveness to the nest and eggs, but sitting readiness may also be influenced by other physiological or situational factors.

As it is presently characterized, the maintenance of sitting readiness represents a type of hormone-dependent change in behavioral state which is more complex and less well defined than most other hormone-behavior relationships. Whether PRL acts directly on the central nervous system (CNS) to promote this change remains to be determined but available data point to a possible peripheral site of action.[53] When given over a 10-day period of isolation from mates, nests, and eggs, twice-daily intracerebroventricular injections of ovine PRL (1 μg/2 μl) induced marked gonadal regression and a pronounced increase in food intake in both sexes but did not mimic the effect of subcutaneous injections of PRL in maintaining readiness to incubate (see TABLE 1). Although these data would argue against a central site of PRL action in maintaining sitting readiness, a conclusive answer to this question will require further tests with different PRL dosages, different PRL injection schedules, and different methods of intracranial PRL administration.

CARE OF THE YOUNG

In addition to its hypothesized role as an incubation-promoting hormone, PRL has also been implicated in the control of behavior patterns associated with care of

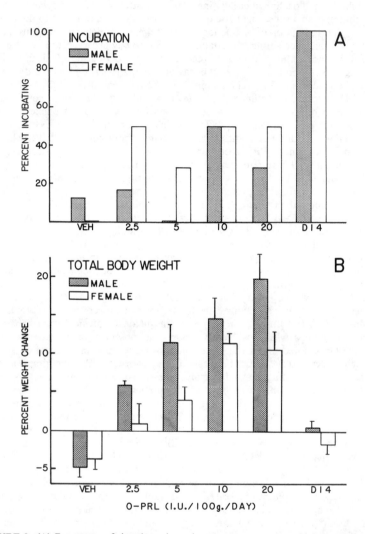

FIGURE 2. (A) Percentage of ring doves in various treatment groups which incubated during a 3-hr test scheduled at the end of a 10-day period of isolation from mates, nests, and eggs. In order to be classified as an incubator, the bird had to sit on the nest with ventral feathers erected and completely covering the eggs during 50% or more of the spot-checks conducted every 15 min during the 3-hr test. With the exception of group D14, birds were isolated on day 4 of incubation and received twice daily subcutaneous injections of saline (VEH) or one of four doses of ovine PRL (2.5, 5, 10, or 20 I.U./100 g body weight/day). D14 birds were allowed to incubate normally during the 10-day period and were tested after a 30-min isolation period. All birds were tested for incubation behavior while still separated from their mates. N's for each

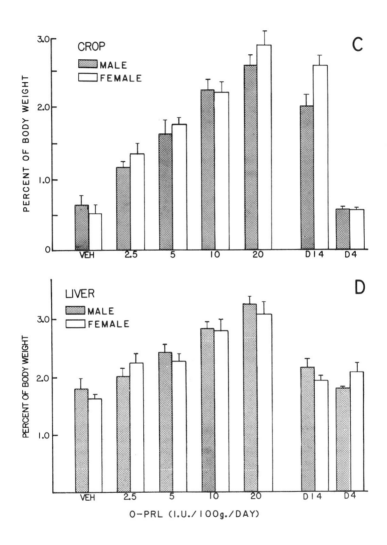

FIGURE 2. (*Continued.*)
group are as follows: VEH, 8 pairs; 2.5 I.U., 6 pairs; 5 I.U., 7 pairs; 10 I.U., 6 pairs; 20 I.U., 7 males and 6 females; D14, 8 pairs. (B) Mean (± SEM) percentage body weight change over the 10-day isolation and treatment period in various groups. (C) Mean (± SEM) crop sac weight measured as a percentage of body weight in various treatment groups at the end of the 10-day isolation and treatment period. (D) Mean (± SEM) liver weight measured as a percentage of body weight in various treatment groups at the end of the 10-day isolation and treatment period. (Data from Janik and Buntin.[52])

TABLE 1. Effects of Twice-Daily Intracerebroventricular Injections of Ovine PRL (1 μg/2 μl) or Saline Vehicle on Changes in Food Intake, Body Weight, Gonad Weight, and Crop Sac Weight and Maintenance of Incubation Readiness over a 10-Day Period in Ring Doves Separated from Mates, Nests, and Eggs on Day 4 of Incubation

Sex	Treatment Group	N	Percentage Incubating[a]	$\bar{x} \pm$ SEM Daily Food Intake (g)	$\bar{x} \pm$ SEM Body Weight Changes[b] (%)	$\bar{x} \pm$ SEM Gonad Weight[c] (g)	$\bar{x} \pm$ SEM Crop Sac Weight (g)
Male	PRL	8	0	21.0 ± 0.9[d]	+19.6 ± 1.1[d]	0.34 ± 0.07[d]	0.69 ± 0.08
	Saline	9	11	10.0 ± 0.7	− 6.4 ± 0.9	1.00 ± 0.06	0.68 ± 0.05
Female	PRL	8	0	17.6 ± 1.2[d]	+10.7 ± 1.5[d]	0.34 ± 0.05[d]	0.74 ± 0.04
	Saline	7	0	10.0 ± 1.0	− 4.9 ± 1.8	0.69 ± 0.12	0.64 ± 0.05

Note. Data from Buntin and Tesch.[53]
[a]See FIGURE 2 for incubation criterion.
[b]Percentage of body weight change over the 10-day isolation and treatment period.
[c]Values shown for females reflect combined weights of ovary and oviduct.
[d]Significantly different from control value ($p < 0.05$, Mann-Whitney U test).

the young after hatching. In several galliform species, PRL injections facilitate the induction of "broody" behavior, which consists of sheltering chicks under the wing, leading the chicks to food or away from danger, and in some species, calling to the young.[54-57] In chickens, the amount of PRL needed and/or the latency required to induce broodiness reportedly varies with the strain tested, with broody strains responding more quickly than nonbroody strains.[54] In a more recent study, Opel and Proudman[7] reported that injections of highly purified ovine PRL failed to induce incubation behavior in laying chickens of a broody strain but did induce the display of the maternal covering stance normally adopted during brooding of chicks. Although only 30-40% of the PRL-injected hens responded in this manner, situational conditions were not conducive for the display of broody behavior since hens were not provided with chicks during behavioral tests. Virtually all the PRL-injected females also showed "clucking" but the degree to which this behavior is functionally related to other broody behaviors is unclear since it is also exhibited by incubating hens (see Reference 7).

Although these PRL injection studies indicate that PRL plays some role in the expression of broody behavior, the significance of this relationship has been questioned by recent reports in galliform birds and in other species with precocial young that blood PRL levels decline at hatching and remain low during the brooding period.[30,35,36,37] However, it is difficult to determine from the available data whether periods of elevated blood PRL levels are entirely absent at this stage or whether they are present but either are shorter in duration or occur at different times of day than those which take place during incubation. In support of the latter view, the pattern of daily fluctuations in circulating PRL levels has recently been reported to change markedly after hatching in bantam hens.[44] Whereas levels remain elevated throughout the day and night during incubation, PRL levels in brooding hens are low during the day but increase substantially at night (see FIGURE 3). The issue of whether this nocturnal rise in PRL is causally related to the display of broodiness remains to be tested, but the point to be emphasized is that such a change could easily go undetected in hormone assay studies in which blood was collected at only one time of day. Because this blood sampling procedure was adopted in most of the published studies, the relationship between changes in circulating PRL levels after hatching and the display of brooding behavior needs to be reevaluated in other galliform and anseriform species.

With the exception of the pied flycatcher,[41] high circulating PRL levels tend to persist for some time after hatching in the few species with altricial young that have been studied (canary,[39] ring dove[42,43]). However, the ring dove represents the only altricial species for which data are available on the effects of PRL injections on parental responsiveness toward young. The secretion of PRL during the incubation and early post-hatching periods promotes extensive crop sac growth and formation of crop "milk" which is regurgitated to the young by both parents. In early studies by Lehrman,[58] ovine PRL was found to induce regurgitation feeding behavior in isolated doves tested with squabs after 7 days of hormone injections, provided that the birds had received prior breeding experience. Similar effects could be elicited in inexperienced birds provided that progesterone injections were also given.[59] Because Xylocaine diminished PRL's effectiveness in stimulating parental feeding when injected into the skin over the crop sac, Lehrman hypothesized that PRL acted peripherally to induce crop growth and engorgement with milk, thereby increasing the sensitivity of the overlying skin to emetic stimuli provided by movements of the squab's head against the parent's chest. However, subsequent studies have demonstrated that incubating birds show regurgitation feeding behavior when provided with squabs during early stages of incubation prior to detectable crop growth,[60] thus leaving open the possibility of a central site of PRL action.

CONCLUSIONS

Although there is now ample evidence to suggest that PRL secretion is influenced by stimuli associated with incubation behavior and care of the young in birds, the question of whether these behaviors are in turn influenced by PRL still remains largely unresolved despite 50 years of investigation. As discussed above, data bearing directly on the question of whether PRL stimulates incubation onset, long-term maintenance of sitting readiness, and care of the young are extremely limited. In recent years, research emphasis has been placed on correlational studies involving measurement of circulating PRL levels under varying physiological states. These studies have established the potential significance of PRL in regulating the expression of incubation

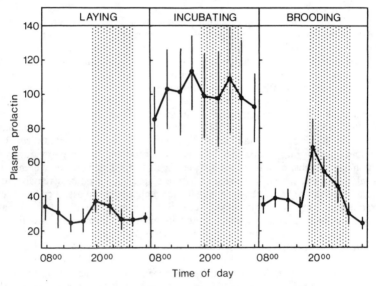

FIGURE 3. Mean (± SEM) daily variations in concentrations of plasma prolactin in seven bantam hens during the egg-laying period, on the first or second day of incubation, and on day 9 or 10 of the post-hatching period while hens were brooding young. Stippled bars indicate the periods of darkness. (Data from Lea et al.[44])

behavior during the normal breeding cycle and have provided insights into the temporal relationship between changes in PRL secretion and changes in response to eggs and young. However, they cannot substitute for PRL injection studies in conclusively establishing a causal relationship between these events. Lack of experimental data is particularly acute in the case of the effects of PRL on care of young in non-galliform birds, where only two studies on a single species (the ring dove) have been published. Clearly, more work will be required to resolve the conflicting results that have been obtained in the few PRL injection studies that have been published, but even in instances where a role for PRL appears reasonably well established (e.g., budgerigar incubation, maintenance of incubation readiness in ring doves) the effects themselves are still poorly characterized. In particular, more information is needed concerning dose-response relationships, hormonal specificity of the response, site of PRL action, the degree to which social and environmental stimuli influence the effectiveness of hormone treatment, and the effects of varying the timing or method of PRL administration on the quality and intensity of the behavioral response. Tests to determine whether the timing of PRL injections influences incubation behavior or parental care may prove to be particularly interesting in light of indications in bantam hens that changes in the patterning of circulating PRL fluctuations throughout the day seem to correlate with incubation onset and behavioral changes occurring during the posthatching period.[44] The hypothesis that PRL's behavioral effectiveness is influenced by these temporal factors is further supported by several experimental studies indicating that the nature and magnitude of certain PRL-dependent physiological and behavioral responses vary with the timing of daily PRL injections in relationship to daily rhythms of adrenal corticoid secretion.[16,61]

Although PRL's specific role in the regulation of incubation is still obscure, it is becoming clear that PRL is not the only factor responsible for onset of sitting behavior. Gonadal steroids may play a particularly important role as incubation-promoting agents but the degree to which these hormones interact with PRL to promote this activity needs to be tested directly by administering various combinations of hormones to gonadectomized birds in the presence of appropriate nesting facilities. In most species, incubation appears to arise as a natural outgrowth of the nest-related activities which precede it, the occurrence of which is dependent upon complex and subtle interactions among gonadal hormones, behavior, and environmental stimuli. Thus, it may be necessary to relate short-term fluctuations in steroid hormone levels and circulating PRL concentrations to changes in nesting behavior, alterations in brood patch sensitivity and other nesting-related stimuli, and changes in responsiveness to eggs introduced several days prior to spontaneous onset of incubation behavior in order to fully understand the physiological basis for induction of incubation. Valuable insights may also be gained through direct comparisons of hormone-behavior-environment interrelationships in broody and nonbroody strains of domestic fowl.

Apart from the fact that PRL levels are generally elevated in incubating birds and in birds brooding altricial young, few if any cross-species generalizations have emerged from more detailed studies of the relationship between changes in PRL levels and sitting activity or parental behavior during the breeding cycle. To a large extent these difficulties can be traced to methodological problems or to procedural differences. As discussed above, for example, blood sampling in most studies has not been frequent enough to adequately interpret the pre-incubatory rise in circulating PRL that has been reported in several galliform species. In addition, results may also be influenced by differences in behavioral testing procedures. In some studies with galliform species, for example, eggs were removed as they were laid[28,47,62] whereas in other studies eggs were allowed to accumulate in the nest.[30,31,33] Finally, the choice of the specific radioimmunoassay (RIA) system used for PRL measurement may influence the pattern of results obtained. In recent years, heterologous RIA systems have been used rather extensively to measure circulating PRL levels in various avian species. In such systems, the species which supplies the sample to be assayed differs from that which supplies the hormone preparation for antibody generation and radiolabeling. As Nicoll[63] has pointed out, while this approach offers a number of theoretical advantages over the use of homologous assays for measuring PRL in a variety of species, the potential for obtaining spurious results is present, since, in the absence of extensive validation tests, the cross-reacting material being detected in the RIA cannot be assumed a priori to be the PRL of the foreign species. While physiologically meaningful results appear to have been obtained in several species using this approach (see Reference 50 for review), the validity of heterologous RIA data obtained from other species can be seriously questioned. In the turkey, for example, an increase in circulating PRL levels during incubation is detectable by homologous RIA[31,34,47] but not by heterologous RIA.[27,28,62] Similar discrepancies were encountered when the same turkey plasma samples were assayed concurrently in a turkey PRL RIA and a chicken PRL RIA, despite the close taxonomic relationship between the two species.[64] These differences dramatically illustrate the potential danger of using heterologous RIA systems for measuring PRL in even closely related species, even though validation tests were usually conducted with these assays prior to use with turkey plasma samples.[27] In view of these problems, it would be prudent to cautiously interpret data obtained from heterologous RIAs until corroborating data are available from other assay systems.

Finally, the crucial question of whether PRL acts directly on the brain to influence incubation and parental care needs to be directly addressed through studies involving administration of PRL via intracerebroventricular injection or microinjection at specific neural loci. Because only minute quantities of hormone are needed for intracranial application, these methods also afford the opportunity to determine whether purified avian PRL preparations are more effective than mammalian PRL preparations in promoting these behavior changes. This is an important issue to resolve, but because of the limited quantities of purified avian PRL available, it would be difficult to conduct such a study using systemic hormone injection procedures. If central effects can be demonstrated, additional work will be required to determine the degree to which circulating prolactin is taken up and retained in the CNS, the characteristics and distribution of specific binding sites for PRL in the brain, and, if demonstrable, the possible role of locally produced "brain PRL"[65] in regulating the expression of incubation behavior. Such work should contribute significantly to our emerging understanding of the mechanisms by which peptide and protein hormones influence brain function and behavior.

ACKNOWLEDGMENTS

I wish to thank Dan Janik and Don Tesch for their research contributions and Mei-Fang Cheng and Rae Silver for their helpful comments on earlier drafts of the manuscript. I am also indebted to Elaine Ruzycki for technical assistance and Vicki Bennett for typing the manuscript.

REFERENCES

1. LIENHART, R. 1927. Contribution a l'etude de l'incubation. C. R. Soc. Biol. (Paris) **97:** 1296-1297.
2. ENSOR, D. M. 1978. Prolactin in birds. *In* Comparative Endocrinology of Prolactin. D. M. Ensor, Ed.: 93-188. Chapman & Hall. London, England.
3. DEVLAMING, V. 1979. Actions of prolactin among the vertebrates. *In* Hormones and Evolution. E. J. W. Barrington, Ed. II: 561-642. Academic Press. New York, N.Y.
4. RIDDLE, O. 1963. Prolactin in vertebrate function and organization. J. Natl. Cancer Inst. **31:** 1039-1110.

5. RIDDLE, O., R. W. BATES & E. L. LAHR. 1935. Prolactin induces broodiness in fowl. Am. J. Physiol. **111:** 352-360.
6. SAEKI, Y. & Y. TANABE. 1955. Changes in prolactin content of fowl pituitary during broody periods and some experiments on the induction of broodiness. Poult. Sci. **34:** 909-919.
7. OPEL, H. & J. A. PROUDMAN. 1980. Failure of a mammalian prolactin to induce incubation behavior in chickens and turkeys. Poult. Sci. **59:** 2550-2558.
8. SHARP, P. J. 1980. Female reproduction. *In* Avian Endocrinology. A. Epple & M. H. Stetson, Eds.: 435-454. Academic Press. New York, N.Y.
9. HOHN, E. O. 1971. Attempted hormonal induction of brood patches and broodiness in ducks. Auk **88:** 674-676.
10. HOHN, E. O. 1981. Failure to induce incubation behavior with estradiol and prolactin and hormonal induction of brood patches in Japanese quail (*Coturnix coturnix japonica*). Gen. Comp. Endocrinol. **44:** 396-399.
11. OPEL, H. 1971. Induction of incubation behavior in the hen by brain implants of prolactin. Poult. Sci. **50:** 1613.
12. MEISENBERG, G. & W. H. SIMMONS. 1983. Minireview: peptides and the blood-brain barrier. Life Sci. **32:** 2611-2623.
13. WOOD-GUSH, D. G. M. & A. B. GILBERT. 1973. Some hormones involved in the nesting behavior of hens. Anim. Behav. **21:** 98-103.
14. HUTCHISON, R. E. 1975. Effects of ovarian steroids and prolactin on the sequential development of nesting behavior in female budgerigars. J. Endocrinol. **67:** 29-39.
15. JONES, R. E. 1971. The incubation patch of birds. Biol. Rev. **46:** 315-339.
16. MEIER, A. H. 1975. Chronophysiology of prolactin in the lower vertebrates. Am. Zool. **15:** 905-916.
17. LEHRMAN, D. S. 1963. On the initiation of incubation behavior in doves. Anim. Behav. **11:** 433-438.
18. RIDDLE, O. 1963. Prolactin or progesterone as key to parental behavior: a review. Anim. Behav. **11:** 419-432.
19. SILVER, R. 1978. The parental behavior of ring doves. Am. Sci. **66:** 209-215.
20. CHENG, M.-F. 1979. Progress and prospects in ring dove research: a personal view. *In* Advances in the Study of Behavior, Vol. 9. J. S. Rosenblatt, R. A. Hinde, C. Beer & M.-C. Busnel, Eds.: 97-129. Academic Press. New York, N.Y.
21. CHENG, M.-F. & R. SILVER. 1975. Estrogen-progesterone regulation of nest building and incubation behavior in ovariectomized ring doves (*Streptopelia risoria*). J. Comp. Physiol. Psychol. **88:** 256-263.
22. LEHRMAN, D. S. 1958. Effect of female sex hormones on incubation behavior in the ring dove (*Streptopelia risoria*). J. Comp. Physiol. Psychol. **51:** 142-145.
23. LEHRMAN, D. S. & P. BRODY. 1961. Does prolactin induce incubation behaviour in the ring dove? J. Endocrinol. **22:** 269-275.
24. CHENG, M.-F. 1975. Induction of incubation behaviour in male ring doves (*Streptopelia risoria*): a behavioural analysis. J. Reprod. Fertil. **42:** 267-276.
25. KOMISARUK, B. R. 1967. Effects of local brain implants of progesterone on reproductive behavior in ring doves. J. Comp. Physiol. Psychol. **64:** 219-224.
26. BALTHAZART, J., J. D. BLAUSTEIN, M.-F. CHENG & H. H. FEDER. 1980. Hormones modulate the concentration of cytoplasmic progestin receptors in the brain of male ring doves (*Streptopelia risoria*). J. Endocrinol. **86:** 251-261.
27. MCNEILLY, A. S., R. J. ETCHES & H. G. FRIESEN. 1978. A heterologous radioimmunoassay for avian prolactin: application to the measurement of prolactin in the turkey. Acta Endocrinol. (Kbh) **89:** 60-69.
28. ETCHES, R. J., A. S. MCNEILLY & C. E. DUKE. 1979. Plasma concentrations of prolactin during the reproductive cycle of the domestic turkey (*Meleagris gallopavo*). Poult. Sci. **58:** 963-970.
29. ETCHES, R. J., A. GARBUTT & A. L. MIDDLETON. 1979. Plasma concentrations of prolactin during egglaying and incubation in the ruffed grouse (*Bonasa umbellus*). Can. J. Zool. **57:** 1624-1627.

30. SHARP, P. J., C. G. SCANES, J. B. WILLIAMS, S. HARVEY & A. CHADWICK. 1979. Variations in the concentrations of prolactin, luteinizing hormone, growth hormone and progesterone in the plasma of brooding bantams (*Gallus domesticus*). J. Endocrinol. **80:** 51-57.
31. BURKE, W. H. & P. T. DENNISON. 1980. Prolactin and luteinizing hormone levels in female turkeys (*Meleagris gallopavo*) during a photoinduced reproductive cycle and broodiness. Gen. Comp. Endocrinol. **41:** 92-100.
32. BURKE, W. H. & H. PAPKOFF. 1980. Purification of turkey prolactin and the development of a homologous radioimmunoassay for its measurement. Gen. Comp. Endocrinol. **40:** 297-307.
33. LEA, R. W., A. S. M. DODS, P. J. SHARP & A. CHADWICK. 1981. The possible role of prolactin in the regulation of nesting behavior and the secretion of luteinizing hormone in brooding bantams. J. Endocrinol. **91:** 88-92.
34. PROUDMAN, J. A. & H. OPEL. 1981. Turkey prolactin: validation of a radioimmunoassay and measurement of changes associated with broodiness. Biol. Reprod. **25:** 573-580.
35. GOLDSMITH, A. R. & D. M. WILLIAMS. 1980. Incubation in mallards (*Anser platyrhynchos*): changes in plasma levels of prolactin and luteinizing hormone. J. Endocrinol. **86:** 371-379.
36. DITTAMI, J. P. 1981. Seasonal changes in the behavior and plasma titres of various hormones in barheaded geese, *Anser indicus*. Z. Tierpsychol. **55:** 289-324.
37. GOLDSMITH, A. R. 1982. The Australian black swan (*Cygnus atratus*): prolactin and gonadotrophin secretion during breeding including incubation. Gen. Comp. Endocrinol. **46:** 458-462.
38. HALL, M. R. & A. R. GOLDSMITH. 1983. Factors affecting prolactin secretion during breeding and incubation in the domestic duck (*Anas platyrhynchos*). Gen. Comp. Endocrinol. **49:** 270-276.
39. GOLDSMITH, A. R. 1982. Plasma concentrations of prolactin during incubation and parental feeding throughout repeated breeding cycles in canaries (*Serinus canarius*). J. Endocrinol. **94:** 51-59.
40. DAWSON, A. & A. R. GOLDSMITH. 1982. Prolactin and gonadotrophin secretion in wild starlings (*Sturnus vulgaris*) during the annual cycle and in relation to nesting, incubation and rearing young. Gen. Comp. Endocrinol. **48:** 213-221.
41. SILVERIN, B. & A. R. GOLDSMITH. 1983. Reproductive endocrinology of free living pied flycatchers (*Ficedula hypoleuca*): prolactin and FSH secretion in relation to incubation and clutch size. J. Zool. (London) **200:** 119-130.
42. GOLDSMITH, A. R., C. EDWARDS, M. KOPRUCU & R. SILVER. 1981. Concentrations of prolactin and luteinizing hormone in plasma of doves in relation to incubation and development of the crop gland. J. Endocrinol. **90:** 437-443.
43. CHENG, M.-F. & W. H. BURKE. 1983. Serum prolactin levels and crop sac development in ring doves during a breeding cycle. Horm. Behav. **17:** 54-65.
44. LEA, R. W., P. J. SHARP & A. CHADWICK. 1982. Daily variations in the concentrations of plasma prolactin in broody bantams. Gen. Comp. Endocrinol. **48:** 213-221.
45. BEDRAK, E., S. HARVEY & A. CHADWICK. 1981. Concentrations of pituitary, gonadal, and adrenal hormones in serum of laying and broody white rock hens (*Gallus domesticus*). J. Endocrinol. **89:** 197-204.
46. MARCH, G. L. & B. A. MCKEOWN. 1973. Serum and pituitary prolactin changes in the band tailed pigeon (*Columba fasciata*) in relation to the reproductive cycle. Can. J. Physiol. **51:** 583-589.
47. EL HALAWANI, M. E., W. H. BURKE & P. T. DENNISON. 1980. Effect of nest deprivation on serum prolactin level in nesting female turkeys. Biol. Reprod. **23:** 113-123.
48. EL HALAWANI, M. E., W. H. BURKE & P. T. DENNISON. 1980. Effects of p-chlorophenylalanine on the rise in serum prolactin associated with nesting in broody turkeys. Biol. Reprod. **23:** 815-819.
49. RAMSEY, S. M., A. R. GOLDSMITH & R. SILVER. 1985. Stimulus requirements for prolactin and LH secretion in incubating ring doves. Gen. Comp. Endocrinol. **59:** 246-256.
50. GOLDSMITH, A. R. 1983. Prolactin in avian reproductive cycles. *In* Hormones and Behavior in Higher Vertebrates. J. Balthazart, E. Prove & R. Gilles, Eds.: 375-387. Springer-Verlag. New York, N.Y.

51. LEHRMAN, D. S. & P. BRODY. 1964. Effect of prolactin on established incubation behavior in the ring dove. J. Comp. Physiol. Psychol. **57:** 161-165.
52. JANIK, D. S. & J. D. BUNTIN. 1985. Behavioral and physiological effects of prolactin in incubating ring doves. J. Endocrinol. **105:** 201-209.
53. BUNTIN, J. D. & D. TESCH. 1985. Effects of intracranial prolactin administration on maintenance of incubation readiness, ingestive behavior, and gonadal condition in ring doves. Horm. Behav. **19:** 188-203.
54. NALBANDOV, A. V. & L. E. CARD. 1945. Endocrine identification of the broody genotype of cocks. J. Hered. **36:** 251-258.
55. CRISPENS, C. G. 1956. Prolactin: an evaluation of its use in ring neck pheasant propagation. J. Wildl. Manage. **20:** 453-455.
56. CRISPENS, C. G. 1957. Use of prolactin to induce broodiness in two wild turkeys. J. Wildl. Manage. **21:** 462.
57. CAIN, J. R., J. D. SNODGRASS & H. G. GORE. 1978. Induced broodiness and imprinting in wild turkeys. Poult. Sci. **57:** 1122-1123.
58. LEHRMAN, D. S. 1955. The physiological basis of parental feeding behavior in the ring dove (*Streptopelia risoria*). Behaviour **7:** 241-286.
59. LOTT, D. F. & S. COMERFORD. 1968. Hormonal initiation of parental behavior in inexperienced ring doves. Z. Tierpsychol. **25:** 71-75.
60. KLINGHAMMER, E. & E. H. HESS. 1964. Parental feeding in ring doves (*Streptopelia roseogrisea*): innate or learned? Z. Tierpsychol. **21:** 338-347.
61. MARTIN, D. D. & A. H. MEIER. 1971. Temporal synergism of corticosterone and prolactin in regulating orientation in the migratory white throated sparrow (*Zonotrichia albicollis*). Condor **75:** 369-374.
62. HARVEY, S., E. BEDRAK & A. CHADWICK. 1981. Serum concentrations of prolactin, luteinizing hormone, growth hormone, corticosterone, progesterone, testosterone, and oestradiol in relation to broodiness in domestic turkeys (*Meleagris gallopavo*). J. Endocrinol. **89:** 187-195.
63. NICOLL, C. S. 1975. Radioimmunoassay and radioreceptor assays for prolactin and growth hormone: a critical appraisal. Am. Zool. **15:** 881-903.
64. LEA, R. W. & P. J. SHARP. 1982. Plasma prolactin concentrations in broody turkeys: lack of agreement between homologous chicken and turkey radioimmunoassays. Br. Poult. Sci. **23:** 451-459.
65. KRIEGER, D. T. 1980. Pituitary hormones in the brain: what is their function? Fed. Proc. **29:** 2937-2941.

PART II. HORMONES AS REGULATORS OF REPRODUCTIVE BEHAVIOR
B. Sexual Behavior

Introduction

BARRY R. KOMISARUK

Institute of Animal Behavior
Rutgers-The State University of New Jersey
Newark, New Jersey 07102

CARLOS BEYER

Centro de Investigación en Reproducción Animal
CINVESTAV-Universidad Autónoma de Tlaxcala
Tlaxcala 90 000, Mexico

The nature of hormone action on activation and organization of sexual behavior is addressed in this section.

C. Beyer and *G. González-Mariscal* critically review current concepts of steroid hormone action on sexual receptivity, emphasizing evidence for an extragenomic site of action on the neuronal membrane. They discuss brain sites (particularly the ventromedial nucleus of the hypothalamus) of action of hormones, neurotransmitters, and neuromodulators in the control of this behavior. They present a unifying concept that all of these neuroactive agents may ultimately act via activation of intraneuronal cyclic AMP-dependent kinases, leading to phosphorylation of estrogen-induced proteins, which in turn mediates lordosis responding.

C. L. Williams presents and reviews evidence that organizational and activational effects of estradiol can occur at the same time in neonatal female rats. By administering estradiol systemically or directly to the anterior ventromedial hypothalamic area in 4-day-old female rats, she finds that lordosis responding is increased when the rats are tested 40 hr later. Hence the activational effect of estradiol occurs at a time when estrogen has also been shown to produce sexual differentiation of the brain. This demonstrates that the activational effect of estradiol starts as early as in the neonatal period.

J. F. Rodriguez-Sierra extends the concept of the organizational period of the brain by demonstrating that within 2 days, estradiol benzoate administered to 25-day-old female rats doubles the number of synaptic contacts in the arcuate nucleus of the hypothalamus but not in the cortex, septal area, or preoptic area. This effect does not occur in comparably treated males. Thus morphological alterations in neuroendocrine tissue can be produced by steroid hormones in juvenile female rats, thereby extending the concept of "organizational" effects of estrogen into this stage of development.

J. Balthazart and *M. Schumacher* contrast birds (especially quail) with rodents with regard to the organizing and activating actions of steroid hormones in the differentiation of sexual behavior. They emphasize that in contrast to rodents, postnatal ovarian hormones in quail are responsible for behavioral demasculinization during a critical period in post-hatch development. They critically review the comparative literature pointing out that organizing and activating effects of steroids can occur contemporaneously, and raising several caveats regarding interpretation of the literature.

C. Diakow advocates a closer analysis of individual differences in hormone-behavior relationships, emphasizing the evolutionary selective advantage provided by individual variability. A better understanding of individual variability in "neuroendocrine phenotypes" could provide insight into basic mechanisms of reproductive success and hormone-behavior interactions.

L. P. Morin discusses the effects that steroid hormones exert on biological rhythms. Steroid hormones modulate the frequency of oscillatory systems that control rhythmical singing, locomotion, EEG, heartbeat, uterine contractions, meal frequency, and circadian periodicity. Furthermore, the efficacy of the negative feedback of steroid hormones is modulated by an ultradian rhythm-generating process. This chronobiological perspective is an important aspect of hormonal influence on behavior and reproductive physiology that deserves increased attention.

Elevation in Hypothalamic Cyclic AMP as a Common Factor in the Facilitation of Lordosis in Rodents: A Working Hypothesis

CARLOS BEYER AND GABRIELA GONZÁLEZ-MARISCAL

Centro de Investigación en Reproducción Animal
CINVESTAV-Universidad Autónoma de Tlaxcala
Apdo. Postal 62 Tlaxcala
Tlaxcala 90 000, Mexico

INTRODUCTION

The hormonal regulation of sexual behavior in female mammals has been the subject of a large number of studies in the last 20 years. Although these reports have added significantly to our knowledge of the hormonal factors and the neural structures involved in sexual behavior, few explanations on the mode of action of sex steroids for stimulating sexual behavior have been explicitly proposed on the basis of these studies. The still-prevailing model of the hormonal regulation of sexual behavior is based on the biochemical model of steroid action proposed in 1968 by Jensen *et al.*[1] and Gorski *et al.*[2] According to the so-called two-step model of steroid action, these hormones enter the effector cells, bind to specific cytoplasmic receptors, and are translocated to the nucleus, where they act on the genome to stimulate protein synthesis. Several biochemical studies, however, have thrown some doubt on the validity of this model by showing that steroid receptors, at least for estrogen, may reside exclusively in target cell nuclei.[3,4] Moreover, a large number of reports, recently reviewed by Duval *et al.*,[5] have shown that many actions of steroids, including those of sex steroids (estrogens and progestins), are exerted through extragenomic mechanisms. Furthermore, since the discovery of the facilitatory action of some nonsteroidal hormones, e.g., luteinizing hormone-releasing hormone (LHRH), on lordosis behavior,[6-8] it has been clear that other cellular mechanisms, most likely not involving protein synthesis, can be activated to facilitate lordosis in estrogen-primed rodents.

In the present paper, we will discuss recent data from several laboratories which we believe support an extragenomic, possibly membrane, site of action of those hormones or neuromodulators that can stimulate lordosis in estrogen-primed rodents.

VARIOUS HORMONAL STIMULI CAN TRIGGER LORDOSIS IN ESTROGEN-PRIMED RATS

Information about the hormonal factors regulating lordosis comes from two experimental approaches: (a) endocrinectomies and subsequent replacement therapies and (b) studies correlating fluctuations of hormone levels in plasma to lordosis behavior. These experimental approaches have shown that estrogen is an indispensable hormone for the facilitation of estrous behavior in mammals.[9,10] Although some androgens can substitute for estrogen in facilitating sexual behavior, there is evidence that their effect is achieved through transformation to estrogen.[11-13]

In several species with prolonged periods of sexual receptivity, such as the cat and the rabbit, estrogen alone gradually stimulates sexual behavior. On the other hand, in species with short periods of estrus, such as rodents, a second hormone is required to trigger and to regulate the duration of sexual behavior.[11] Therefore, in these species, normal estrous behavior results from the sequential interaction of a priming hormone (estrogen) and a "triggering" hormone. It is generally believed that progesterone (P) is the triggering hormone in rodents. This belief is supported by the following findings: (a) the administration of P to estrogen-primed rodents consistently facilitates the display of sexual behavior,[11] and (b) a rise in P plasma levels occurs immediately prior to or coincident with the onset of sexual behavior.[14,15] However, several other hormones or neuromodulators also induce lordosis in estrogen-primed rats. For example, LHRH,[6-8] α-melanotropin (α-MSH),[16] prostaglandin E_2 (PGE_2),[17,18] desoxycorticosterone,[19] and 17β-estradiol[20] elicit lordosis in estrogen-primed rats with a latency similar to P (FIGURE 1). Interestingly, significant elevations in the concentration of some of these hormones, besides P,[14,15] namely, LHRH,[21] 17β-estradiol,[14,15] and PGE_2,[22] occur shortly before or coincident with the display of lordosis in normal cycling rats, thus suggesting that they play a role in the normal production of lordosis behavior.

The number and the heterogeneous chemical nature of hormones that facilitate lordosis in estrogen-primed rats apparently add complexity to the problem of the hormonal regulation of lordosis. However, such conditions may also aid in the understanding of this problem by pointing to common mechanisms and sites of action of the various triggering hormones.

TRIGGERING HORMONES PROBABLY SHARE A COMMON NEURAL TARGET IN THE VENTROMEDIAL HYPOTHALAMUS

There is evidence that the priming effect of estrogen is exerted on the ventromedial hypothalamus (VMH). Thus, by using small implants of diluted estrogen, Rubin and Barfield have clearly established that "estrogen stimulation of the region of the ventromedial hypothalamic nucleus alone is sufficient to prime the activation of estrous behavior in the ovariectomized rat."[23] This finding is consistent with the observation that VMH neurons selectively bind radioactive estradiol.[24] Similarly, it seems that the triggering hormones facilitate lordosis in estrogen-primed rats also by acting on the VMH. Thus, minute amounts of P implanted in the ventromedial hypothalamic nucleus elicit lordosis in estrogen-primed rats.[25-27] Implants only 0.5 mm away from this area usually fail to produce the response.[25]

It appears that the other triggering hormones also act on the VMH since recently, Rodriguez-Sierra and Komisaruk[27] found that a common site for the activation of lordosis by PGE_2 and LHRH was the VMH. These results agree with those reported by other investigators for PGE_2[28] and LHRH.[29,30] All these data point to the VMH area as the site of action and possibly interaction of hormones and neuromodulators to facilitate lordosis.

THE VENTROMEDIAL HYPOTHALAMUS EXERTS A FACILITATORY EFFECT ON THE LORDOSIS REFLEX ARC

Pfaff and his group[31] have systematically studied the role of the VMH in the regulation of lordosis behavior in the rat. They found that electrical stimulation of the VMH facilitated lordosis,[32] while its destruction[33] interfered with this response. Most interestingly, the facilitatory effect of VMH electrical stimulation was not immediate but required a prolonged period of excitation to become apparent. Moreover, the lesion[33] or the inactivation[34] of this area did not abruptly interrupt lordosis in estrogen-primed rats. From these and other studies, Pfaff concluded that the VMH exerts a facilitatory effect on the reflex arc for lordosis, but that this area itself is not a part of this reflex arc which is located in the lower brainstem and spinal cord (see Reference 31). The temporal characteristics of the effects on lordosis induced by either stimulation or lesion of the VMH are of interest since they may explain the relatively long latency with which most triggering hormones induce lordosis in estrogen-primed rats (see FIGURE 1).

As previously mentioned, triggering hormones facilitate lordosis when placed within or around the VMH. This effect must be correlated with an increased activity of VMH neurons. Such activation in turn may result either from alterations in the metabolism of VMH neurons or from alterations in the functioning of VMH synapses. Since in the latter case, hormones would act as neuromodulators regulating the afferent input to VMH neurons, it is important to briefly review the nature of the afferent input to this structure.

SOME HORMONES MAY ACT BY SELECTIVELY FACILITATING AFFERENT TRANSMISSION TO THE VENTROMEDIAL HYPOTHALAMUS

The VMH receives a rich afferent supply from the brainstem, as evidenced by the observation that degenerated nerve terminals are present in the VMH following lesion of various parts of the rat's brainstem.[35,36] The nature of the neurotransmitter present in these fibers has not been definitively established, but there is good evidence for the existence of noradrenergic, dopaminergic,[37,38] and serotonergic[38] fibers impinging on the VMH area. The fact, however, that a particular neurotransmitter is not found within the boundaries of the VMH does not rule out its possible participation in the regulation of VMH activity, since it is known that dendrites of VMH neurons are not confined to the nucleus but radiate in all directions from it.[39]

Anatomical studies in rodents have shown noradrenergic fibers from the ventral noradrenergic bundle[38,40] ending around the VMH. Millhouse,[39] in his Golgi study of the VMH, showed that "completely encircling the nucleus is a capsule formed partly by VMH afferents. Buried in this capsule are neurons whose dendrites curve along the periphery of the nucleus to form a dendritic grid." This capsule, according to Palkovits et al.,[41] is very rich in noradrenaline (NA), a finding consistent with the previous observation by Fuxe[42] of intense fluorescence of monoamine-containing fibers in the area around the VMH but not in the nucleus itself. Recently, similar findings have been reported by Chan-Palay et al.[43] using immunocytochemical methods to stain catecholaminergic fibers.

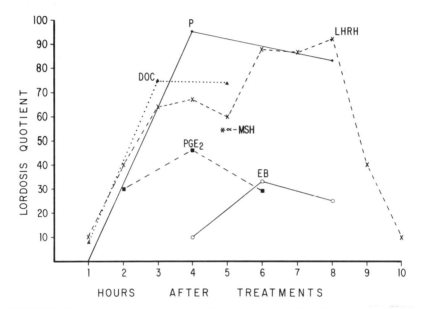

FIGURE 1. Temporal course and intensity of lordosis behavior induced by triggering hormones in ovariectomized, estrogen-primed rats. All drugs were administered subcutaneously. P = progesterone (2 mg)[115]; EB = 17β-estradiol benzoate (400 μg)[20]; DOC = desoxycorticosterone (300 μg)[19]; LHRH = luteinizing hormone-releasing hormone (0.5 μg)[7]; α-MSH = α-melanotropin (10 μg)[16]; PGE$_2$ = prostaglandin E$_2$ (500 μg).[18]

An alternative approach to establish the chemical nature of the synapses of a brain region is to study the responses of its neurons to the microiontophoretic administration of the putative neurotransmitters.[44] The effects of the microiontophoresis of several neurotransmitters upon the discharge frequency of VMH units have been studied by Moss et al.[45] They made the important observation that the application of NA provoked excitatory responses in most VMH neurons.[45] They also found that serotonin and dopamine induced variable and inconsistent responses while acetylcholine (ACh) mainly facilitated the discharge of VMH neurons.[45] These results suggest that ACh and NA may facilitate lordosis behavior by activating VMH neurons. This is indeed the case since Dohanich and Clemens[46] found that carbachol (a muscarinic agonist) readily facilitates lordosis in estrogen-primed rats when infused into the VMH and

Fernández-Guasti et al.[47,48] and Foreman and Moss[30] observed rapid facilitation of lordosis in similar preparations after the infusion of NA into this region.

More direct evidence for the participation of a noradrenergic link in the hormonal facilitation of lordosis comes from studies showing that lesions of the ventral noradrenergic bundle[49-51] or chemical depletion of brain NA[50,52] interferes with the display of lordosis in estrogen-progesterone-treated rats. Similarly, the administration of noradrenergic blockers antagonizes the facilitatory effects of both P[47] and LHRH[30] in estrogen-primed rats. In contrast, according to Clemens et al.,[53] the facilitation of lordosis by ACh seems not to involve an interaction with P since the combination of carbachol and P did not increase lordosis to a level higher than that induced by carbachol alone.

The noradrenergic system facilitating lordosis was suggested to be a component of the afferent pathway of the lordosis reflex arc.[50,51] However, the fact that adrenergic blockers antagonize lordosis only when it is induced by estrogen and P[47] but not when induced by chronic estrogen administration,[54] contradicts that suggestion. Rather, these findings support the idea that a noradrenergic system impinging on the VMH facilitates lordosis by favoring the activation of this structure.

CYCLIC AMP-DEPENDENT PROTEIN PHOSPHORYLATION: POSSIBLE COMMON CELLULAR PROCESS MEDIATING THE ACTION OF TRIGGERING HORMONES

There is evidence that the priming effect of estrogen on the hypothalamus involves protein synthesis. Findings in favor of this idea are: (a) stimulation of RNA polymerase II in the hypothalamus of ovariectomized rats by estrogen[55,56]; (b) identification of some estrogen-induced proteins in the hypothalamus, namely, P receptors,[57] estrogen receptors,[58] muscarinic receptors,[59] and creatine kinase[60]; and (c) interference of the behavioral action of estrogen by the intracerebral application of antibiotics.[61-65]

On the other hand, little is known about the cellular events underlying the facilitation of lordosis by P or by the other triggering hormones. Explanations in biochemical or molecular terms must consider that these hormones act on VMH neurons primed with estrogen since none of these agents elicits lordosis per se in ovariectomized females. Therefore, valid models should try to explain possible mechanisms of interaction between estrogen-induced processes and the triggering hormones. It has been proposed that the triggering effect of P on lordosis is mediated by genome activation and subsequent stimulation of protein synthesis.[26,66-68] However, some reservations regarding this idea arise from the finding that actinomycin D, a transcription blocker, fails to suppress the facilitation of lordosis induced by P in estrogen-primed rats.[69,70] Similarly, it has been shown that lordosis is induced in estrogen-primed rats within approximately 15 min by the intravenous administration of P or some of its derivatives,[71,72] a finding unlikely to support a mechanism involving de novo protein synthesis. On the other hand, some studies have shown that implants of protein synthesis inhibitors (translation inhibitors) into the VMH can interfere with the induction of lordosis provoked by P in estrogen-primed rats.[73,74] However, several data reviewed by Meisel and Pfaff,[75] indicate that antibiotics can antagonize lordosis through other mechanisms such as interference with the production of cyclic AMP (cAMP) in effector cells[76] and blockade of the synthesis[77,78] and release[79] of NA. Moreover, it must be recalled that even if P facilitates lordosis by acting on the genome, it is unlikely that the other hormones share this mechanism since amines and polypeptides are known to act mainly at the cell membrane.[80-83]

From these considerations, we recently proposed[84] a unitary model to explain the facilitation of lordosis by P and the other triggering hormones. A revised version of this model is shown in FIGURE 2. As previously mentioned, an essential step in the priming effect of estrogen is the production of specific proteins[1,2] (estrogen-induced proteins, EIP) related to neuronal excitability. The fact that estrogen alone, unless given in large amounts or in repeated dosages, does not facilitate lordosis in ovariectomized rats[11,85] suggests that most of these proteins are in an "inactive" state due to their conformation, their chemical composition, or their location. The induction of lordosis by large amounts or by repeated dosages of estrogen may result from an abnormally high synthesis of EIP which may yield a "critical" number of active proteins (active EIP) "sufficient" to initiate lordosis. However, under normal conditions, an additional triggering hormone is required for the activation of inactive EIP.

Protein activation is a common regulatory mechanism in cell function. This process involves changes in the conformation of a protein (allosteric processes) or changes in its chemical composition (e.g., methylation, acetylation, glycosylation, and phosphorylation). Thus, enzymes involved in a variety of biochemical pathways are regulated by these mechanisms.[86-88] However, as recently emphasized by Cohen[89] and by Nestler and Greengard,[90] protein phosphorylation is *the* major mechanism through which a variety of hormones regulate intracellular events in mammalian cells. Thus, several aspects of neurotransmission have been shown to involve protein phosphorylation, namely, changes in ionic conductance,[91,92] neurotransmitter release,[93] and activation of tyrosine hydroxylase (the rate-limiting enzyme in catecholamine biosynthesis).[94,95] (For review, see Reference 90.)

Protein kinases, the enzymes involved in the phosphorylation of proteins, are specifically activated by one or several second messengers, among them: cAMP, cyclic GMP, peptides, calcium, and phospholipids.[90,96,97] According to our model, the triggering hormones directly or indirectly stimulate protein kinases, thereby promoting the phosphorylation and consequent activation of inactive EIP (FIGURE 2). Indeed, it has been shown that some of the triggering hormones modify protein phosphorylation in peripheral tissues.[98-101]

Among the number of agents that can stimulate protein kinases, cAMP is a major mediator for a large number of hormones[90,102] and therefore a logical candidate to analyze for its possible mediation of the proposed phosphorylation of EIP by the triggering hormones. In order to consider that a chemical exerts its effect through an adenylate cyclase-cAMP system, Robison *et al*[102] proposed that the following three criteria should be met: (a) the drug should increase the concentration of cAMP in the target cell; (b) the effect produced by the administration of cAMP or its derivatives should mimic in the target cell the effect of the hormone, and (c) the administration of phosphodiesterase inhibitors should potentiate the effect of the drug.

These criteria have been met by P and some of the hormones that trigger lordosis in estrogen-primed rats. Thus, estradiol,[103,104] LHRH,[83,102] PGE_2,[106] and α-MSH[82,107] have been reported to raise cAMP levels in brain tissue. Moreover, it is particularly interesting that during proestrus, when levels of P,[14,15] LHRH,[21] and PGE_2[22] rise, a concomitant increase in hypothalamic cAMP occurs.[108,109]

The second criterion of Robison *et al*.[102] was explored in a series of studies made in our laboratory. The effect of several derivatives of cAMP and of various drugs that activate adenylate cyclase, namely, guanosine nucleotides,[110,111] forskolin,[112,113] and cholera toxin,[114] on the lordosis behavior of ovariectomized estrogen-primed rats was examined. Most of these drugs were found to elicit significant lordosis behavior in both ovariectomized and ovariectomized-adrenalectomized rats that were estrogen

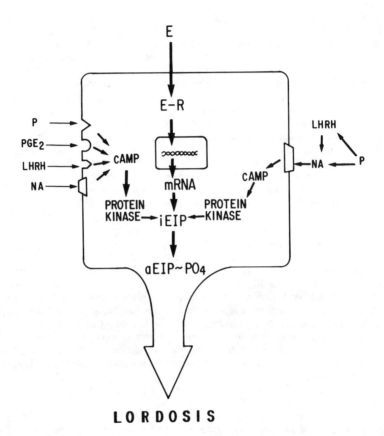

FIGURE 2. A proposed model for the activation of ventral hypothalamic neurons related to the facilitation of lordosis in ovariectomized rats. Estrogen (E) is bound to a cytoplasmic receptor (R) and the complex (E-R) is translocated to the nucleus where it stimulates the production of messenger RNA (mRNA) which, in turn, guides the synthesis of estrogen-induced proteins (EIP) most of which are in an inactive condition (iEIP). Activation of iEIP occurs by their phosphorylation through cyclic AMP (cAMP)-dependent protein kinases. cAMP may be produced by the binding of various triggering hormones to membrane receptors linked to adenylate cyclase, as shown on the left. A more likely possibility, at least for progesterone (P) and luteinizing hormone-releasing hormone (LHRH), is that cAMP is produced by the activation of noradrenaline (NA) receptors linked to adenylate cyclase, as shown on the right. aEIP=active EIP; $\sim PO_4$ = high-energy phosphate group transferred during phosphorylation reaction; PGE_2 = prostaglandin E_2.

primed.[115-118] Lordosis was induced in most cases by the infusion of these agents into the VMH, but GTP and some derivatives of cAMP and cyclic GMP were, surprisingly, also effective when administered systemically.

Finally, the third criterion of Robison et al.[102] has been met for two triggering hormones by the observation that phosphodiesterase inhibitors (theophylline, caffeine, and methyl-isobutylxanthine), that are known to increase cAMP levels by interfering with the breakdown of cAMP,[102] potentiated the effect of subthreshold doses of LHRH[119] and P.[120]

The fulfillment of the three criteria of Robison et al.[102] supports our proposition that a critical step in the facilitation of lordosis is the phosphorylation of EIP, mediated by the activation of cAMP-dependent protein kinases in neurons of the ventral hypothalamus. This proposition does not exclude the possibility that EIP may also be phosphorylated by kinases that are activated by agents other than cAMP (e.g., calcium, calmodulin, phospholipids, cyclic GMP). However, no experimental evidence exists in this regard.

TRIGGERING HORMONES: DIRECT OR INDIRECT ACTIONS ON THE VMH?

As shown on the left side of FIGURE 2, the stimulatory effect of triggering hormones on VMH neurons may result from their direct interaction with membrane receptors linked to adenylate cyclase. Evidence suggesting the existence of such receptors in brain tissue has been provided for LHRH[83,105] and prostaglandins.[106] Similarly, membrane binding sites for P and other steroids have been shown in hypothalamic synaptosomes,[121] though it is not known if they are coupled to adenylate cyclase.

On the other hand, there is evidence that the rise in hypothalamic cAMP caused by estrogen[103,104] and that occurring during proestrus[108,109] are mediated by NA since they are prevented by the administration of either α- or β-adrenergic antagonists.[103,104,108,109] Therefore, it is probable that P indirectly activates VMH neurons by stimulating or modulating the afferent noradrenergic component to this nucleus (FIGURE 2, right). Support for this idea comes from the observation that P administration facilitates the release of brain NA in vivo.[122]

There is good evidence that NA acts on the rat brain by raising cAMP levels.[80,123] Interestingly, Daly et al. showed that "the hypothalamus is unique in being the only brain region in which potentiative interactions between α-adrenergic and β-adrenergic mechanisms were found to be significant."[80] Thus, it appears that NA raises cAMP levels in VMH neurons by acting on both α- and β-postsynaptic receptors. This idea is strongly supported by the observation of Fernández-Guasti et al.[48] that lordosis is optimally induced in estrogen-primed rats by the combined administration of clonidine (α-agonist) and isoproterenol (β-agonist) while neither of these agonists alone facilitates lordosis.

The idea that P stimulates lordosis through the release or modulation of the action of hypothalamic NA (FIGURE 2, right) has been tested in our laboratory by studying the effect of α- and β-adrenergic blockers on the facilitatory effect of this steroid. Since, according to Daly et al.[80] and Palmer,[123] treatment with either α- or β-antagonists interferes with the stimulatory effect of NA on hypothalamic cAMP levels, it was predicted that a similar treatment would block the lordosis behavior induced by P in ovariectomized estrogen-primed rats. This prediction was correct since the administration of either phenoxybenzamine (α-antagonist) or propranolol (β-antagonist) blocked the facilitatory effect of P in estrogen-primed rats.[47]

These experimental data, therefore, support the idea that P triggers lordosis either by facilitating NA release or by favoring the effect of this neurotransmitter at the postsynaptic site. Evidence for a modulatory action of steroids—including P—on several peripheral adrenergic effector systems has been provided and recently reviewed by Davies and Lefkowitz.[124]

FIGURE 2 also shows the possibility that the facilitatory action of P on VMH activity, and therefore on lordosis, is reinforced through the release of or interaction with LHRH. This possibility is supported by the following data: (a) P facilitates the release of LHRH in estrogen-primed rats[125–129] and (b) the infusion of an LHRH antagonist into the third ventricle reduces the facilitation of lordosis induced by P in estrogen-primed rats.[130] LHRH could act directly on adenylate cyclase-linked receptors located on the membrane of VMH neurons (FIGURE 2, left) or indirectly through the release or modulation of NA (FIGURE 2, right). The latter suggestion is supported by the finding that either α- or β-adrenergic antagonists interfere with the action of LHRH on lordosis.[30]

In summary, we have proposed in this paper, as a working hypothesis, that one of the final cellular events leading to an increased excitability of the neurons related to the facilitation of lordosis is an elevation in hypothalamic cAMP. It seems that some of the peptides and prostaglandins reported to facilitate lordosis may act directly on membrane receptors linked to adenylate cyclase-cAMP systems to achieve this effect. On the other hand, P appears to facilitate lordosis by modulating, directly or indirectly, a noradrenergic system which would, in turn, raise hypothalamic cAMP and trigger the behavior.

REFERENCES

1. JENSEN, E. V., T. SUZUKI, T. KAWASHIMA, W. E. STUMPF, P. W. JUNGBLUT & E. R. DE SOMBRE. 1968. Proc. Natl. Acad. Sci. USA **59**: 632-638.
2. GORSKI, J., D. O. TOFT, G. SHYAMALA, D. SMITH & A. NOTIDES. 1968. Recent Prog. Horm. Res. **24**: 45-80.
3. CLARK, C. R. 1984. Trends Biochem. Sci. **9**: 207-208.
4. JORDAN, V. C., A. C. TATE, S. D. LYMAN, B. GOSDEN, M. F. WOLF, R. R. BAIN & W. V. WELSHONS. 1985. Endocrinology **116**: 1845-1857.
5. DUVAL, D., S. DURANT & F. HOMO-DELARCHE. 1983. Biochim. Biophys. Acta **737**: 409-442.
6. MOSS, R. L. & S. M. MCCANN. 1973. Science **181**: 177-179.
7. MOSS, R. L. & S. M. MCCANN. 1973. Neuroendocrinology **17**: 309-318.
8. PFAFF, D. W. 1973. Science **182**: 1148-1149.
9. BEACH, F. 1948. Hormones and Behavior. P. Hoeber. New York, N.Y.
10. FEDER, H., I. T. LANDAU & W. A. WALKER. 1979. *In* Endocrine Control of Sexual Behavior. C. Beyer, Ed.: 317-340. Raven Press. New York, N.Y.
11. MORALI, G. & C. BEYER. 1979. *In* Endocrine Control of Sexual Behavior. C. Beyer, Ed.: 33-75. Raven Press. New York, N.Y.
12. BEYER, C., N. VIDAL & A. MIJARES. 1970. Endocrinology **87**: 317-340.
13. BEYER, C. & B. KOMISARUK. 1971. Horm. Behav. **2**: 217-225.
14. NEQUIN, L. G., J. ALVAREZ & N. B. SCHWARTZ. 1975. J. Steroid Biochem. **6**: 1007-1012.
15. SMITH, M. S., M. E. FREEMAN & J. D. NEILL. 1975. Endocrinology **96**: 219-226.
16. THODY, A. J., C. A. WILSON & D. EVERARD. 1979. Psychopharmacologia **74**: 153-156.
17. DUDLEY, C. A. & R. L. MOSS. 1976. J. Endocrinol. **7**: 457-458.
18. RODRIGUEZ-SIERRA, J. F. & B. R. KOMISARUK. 1977. Horm. Behav. **9**: 281-289.

19. GORZALKA, B. B. & R. W. WHALEN. 1977. Horm. Behav. **8:** 94-99.
20. KOW, L. M. & D. W. PFAFF. 1975. Horm. Behav. **6:** 259-276.
21. SARKAR, D. K., S. A. CHIAPPA & G. FINK. 1976. Nature **264:** 461-463.
22. BROWN, C. G. & N. L. POYSER. 1984. J. Endocrinol. **103:** 155-164.
23. RUBIN, B. S. & R. J. BARFIELD. 1980. Endocrinology **106:** 504-509.
24. STUMPF, W. E. 1980. In The Endocrine Functions of the Brain. M. Motta, Ed.: 43-49. Raven Press. New York, N.Y.
25. RUBIN, B. S. & R. J. BARFIELD. 1983. Endocrinology **113:** 797-804.
26. PARSONS, E., T. C. RAINBOW, N. J. MACLUSKY & B. S. MCEWEN. 1982. J. Neurosci. **2:** 1446-1452.
27. RODRIGUEZ-SIERRA, J. F. & B. R. KOMISARUK. 1982. Neuroendocrinology **35:** 363-369.
28. HALL, N. R. & W. G. LUTTGE. 1977. Brain Res. Bull. **2:** 203-207.
29. MOSS, R. L. & M. M. FOREMAN. 1976. Neuroendocrinology **20:** 176-181.
30. FOREMAN, M. M. & R. L. MOSS. 1978. Pharmacol. Biochem. Behav. **9:** 235-241.
31. PFAFF, D. W. 1980. Estrogens and Brain Function. Springer-Verlag. New York, N.Y.
32. PFAFF, D. W. & Y. SAKUMA. 1979. J. Physiol. **288:** 189-202.
33. PFAFF, D. W. & Y. SAKUMA. 1979. J. Physiol. **288:** 203-210.
34. HARLAN, R. E., B. D. SHIVERS, L. M. DOW & D. W. PFAFF. 1983. Brain Res. **268:** 67-78.
35. ZABORSKY, L., C. S. ZERANTH, J. MARTON & M. PALKOVITS. 1973. In Hormones and Brain Function. K. Lissák, Ed.: 449-457. Akadémiai Kiadó. Budapest, Hungary.
36. ZABORSKY, L. & M. PALKOVITS. 1978. Acta Morphol. Acad. Sci. Hung. **26:** 49-71.
37. KIZER, J. S., M. PALKOVITS & J. M. SAAVEDRA. 1976. Brain Res. **108:** 363-370.
38. GRANT, L. D. & W. E. STUMPF. 1975. In Anatomical Neuroendocrinology. W. E. Stumpf & L. D. Grant, Eds.: 445-463. S. Karger. Basel, Switzerland.
39. MILLHOUSE, O. E. 1973. Brain Res. **55:** 71-87.
40. ROBBINS, T. W. & B. J. EVERITT. 1982. Int. Rev. Neurobiol. **23:** 303-365.
41. PALKOVITS, M., M. BROWNSTEIN, J. M. SAAVEDRA & J. AXELROD. 1971. Brain Res. **77:** 137-149.
42. FUXE, K. 1965. Acta Physiol. Scand. **64**(Suppl. 247): 38-85.
43. CHAN-PALAY, V., L. ZABORSKY, C. KÖHLER, M. GOLDSTEIN & S. L. PALAY. 1984. J. Comp. Neurol. **227:** 467-496.
44. HAYWARD, J. N. & T. A. REANES. 1980. In The Endocrine Functions of the Brain. M. Motta, Ed.: 21-41. Raven Press. New York, N.Y.
45. MOSS, R. L., M. J. KELLY & C. A. DUDLEY. 1978. Brain Res. **139:** 141-152.
46. DOHANICH, G. P. & L. G. CLEMENS. 1981. Horm. Behav. **15:** 157-167.
47. FERNÁNDEZ-GUASTI, A., K. LARSSON & C. BEYER. 1985. Pharmacol. Biochem. Behav. **22:** 279-282.
48. FERNÁNDEZ-GUASTI, A., K. LARSSON & C. BEYER. 1985. Pharmacol. Biochem. Behav. **22:** 613-618.
49. HERNDON, J. G., JR. 1976. Physiol. Behav. **17:** 143-148.
50. HANSEN, S., E. J. STANFIELD & B. J. EVERITT. 1980. Nature **286:** 152-154.
51. HANSEN, S. & S. B. ROSS. 1983. Brain Res. **268:** 285-290.
52. EVERITT, B. J., K. FUXE, T. HÖKFELT & G. JONSSON. 1975. J. Comp. Physiol. Psychol. **89:** 556-572.
53. CLEMENS, L. G., G. P. DOHANICH & J. A. WITCHER. 1981. J. Comp. Physiol. Psychol. **95:** 763-770.
54. LARSSON, K., A. FERNANDEZ-GUASTI & C. BEYER. Unpublished observations.
55. KELNER, K. L., A. L. MILLER & E. J. PECK. 1980. J. Rec. Res. **1:** 215-237.
56. MOHLA, S., E. R. DE SOMBRE & E. V. JENSEN. 1972. Biochem. Biophys. Res. Commun. **46:** 661-667.
57. MACLUSKY, N. J. & B. S. MCEWEN. 1980. Endocrinology **106:** 192-202.
58. CIDLOWSKY, J. A. & T. G. MULDOON. 1978. Biol. Reprod. **18:** 239-246.
59. RAINBOW, T. C., L. SNYDER, D. J. BERCK & B. S. MCEWEN. 1984. Neuroendocrinology **39:** 476-480.
60. KAYE, A. M. 1983. J. Steroid Biochem. **19:** 33-40.
61. HO, G. K. W., D. M. QUADAGNO, P. H. COOKE & R. A. GORSKI. 1973. Neuroendocrinology **13:** 47-55.

62. MEYERSON, B. J. 1973. Prog. Brain Res. **39:** 135-147.
63. QUADAGNO, D. M., J. SHRYNE & R. A. GORSKI. 1971. Horm. Behav. **2:** 1-10.
64. TERKEL, A. S., J. SHRYNE & R. A. GORSKI. 1973. Horm. Behav. **4:** 377-386.
65. WHALEN, R. E., B. B. GORZALKA, J. F. DE BOLD, D. M. QUADAGNO, G. HO & J. C. HOUGH. 1974. Horm. Behav. **5:** 337-343.
66. MCEWEN, B. S. 1981. Science **211:** 1303-1311.
67. PFAFF, D. W. & B. S. MCEWEN. 1983. Science **219:** 808-814.
68. NOCK, B. & H. H. FEDER. 1981. Neurosci. Biobehav. Rev. **5:** 437-447.
69. BEYER, C., E. CANCHOLA, F. CARACHEO & M. L. CRUZ. 1978. Resúmenes XXI Congreso Nacional Ciencias Fisiológicas, Chihuahua, México. P. 46.
70. SHIVERS, B. D., R. E. HARLAN, C. R. PARKER, JR. & R. L. MOSS. 1980. Biol. Reprod. **23:** 963-973.
71. KUBLI-GARFIAS, C. & R. E. WHALEN. 1977. Horm. Behav. **9:** 380-387.
72. LISK, R. D. 1960. Can. J. Biochem. Physiol. **38:** 1381-1383.
73. RAINBOW, T. C., M. Y. MCGINNIS, P. G. DAVIS & B. S. MCEWEN. 1982. Brain Res. **233:** 417-423.
74. GLASER, J. H. & R. J. BARFIELD. 1984. Neuroendocrinology **38:** 337-343.
75. MEISEL, R. L. & D. W. PFAFF. 1984. Brain Res. Bull. **12:** 187-193.
76. BROOKER, G., C. PEDONE & K. BAROVSKY. 1983. Science **220:** 1169-1170.
77. FLEXNER, L. B. & R. H. GOODMAN. 1975. Proc. Natl. Acad. Sci. USA **72:** 4460-4463.
78. SCHWERI, M. M. & L. A. CARR. 1982. J. Neural Transm. **54:** 41-50.
79. FREEDMAN, L. S., M. E. JUDGE & D. QUARTERMAIN. 1982. Pharmacol. Biochem. Behav. **17:** 187-191.
80. DALY, J. W., W. PADGETT, C. R. CREVELING, D. CANTACUZENE & K. L. KIRK. 1981. J. Neurosci. **1:** 49-59.
81. MCCANN, S. M. 1974. *In* Handbook of Physiology. R. O. Greep & E. B. Astwood, Eds.: 489-517. American Physiological Society. Washington, D.C.
82. CHRISTENSEN, C. W., C. T. HARSTON, A. J. KASTIN, R. M. KOSTRZEWA & M. A. SPIRTES. 1976. Pharmacol. Biochem. Behav. **5**(Suppl. 1): 117-120.
83. TIXIER-VIDAL, A. & D. GOURDJI. 1981. Physiol. Rev. **61:** 974-1011.
84. BEYER, C., E. CANCHOLA, M. L. CRUZ & K. LARSSON. 1980. *In* Endocrinology 1980. J. A. Cumming, J. W. Funder & F. A. O. Mendelsohn, Eds.: 615-618. Australia Academy of Science and Elsevier. Canberra, Australia and Amsterdam, the Netherlands.
85. BEYER, C., K. LARSSON & M. L. CRUZ. 1979. *In* Endocrine Control of Sexual Behavior. C. Beyer, Ed.: 365-387. Raven Press. New York, N.Y.
86. HOLZER, H. & W. DUNTZE. 1971. Annu. Rev. Biochem. **40:** 345-374.
87. KOSHLAND, D. E., JR. 1970. *In* The Enzymes. P. D. Boyer, Ed.: 341-396. Academic Press. New York, N.Y.
88. SHACTER-NOIMAN, E., P. B. CHOCK & E. R. STADTMAN. 1983. Philos. Trans. R. Soc. London Ser. B **302:** 157-166.
89. COHEN, P. 1982. Nature **296:** 613-620.
90. NESTLER, E. J. & P. GREENGARD. 1984. Protein Phosphorylation in the Nervous System. Wiley. New York, N.Y.
91. GREEN, D. J. & R. GILETTE. 1983. Nature **306:** 784-785.
92. HAIECH, J. & J. G. DEMAILLE. 1983. Philos. Trans. R. Soc. London Ser. B **302:** 91-98.
93. DE LORENZO, R. J. 1982. *In* Calcium and Cell Function. W. Y. Cheung, Ed.: 271-309. Academic Press. New York, N.Y.
94. EDELMAN, A. M., J. D. RAESE, M. A. LAZAR & J. D. BARCHAS. 1981. J. Pharmacol. Exp. Ther. **216:** 647-653.
95. YAMAUCHI, T. & H. FUJISAWA. 1981. Biochem. Biophys. Res. Commun. **100:** 807-813.
96. KREBS, E. G. 1983. Philos. Trans. R. Soc. London Ser. B **302:** 3-11.
97. NISHIZUKA, Y. 1983. Philos. Trans. R. Soc. London Ser. B **302:** 101-112.
98. REDDINGTON, M., R. RODNIGHT & M. WILLIAMS. 1973. Biochem. J. **132:** 475-482.
99. WILLIAMS, M. 1976. Brain Res. **109:** 190-195.
100. LIU, A. Y. C. & P. GREENGARD. 1976. Proc. Natl. Acad. Sci. USA **73:** 568-572.
101. LIU, A. Y. C., U. WALTER & P. GREENGARD. 1981. Eur. J. Biochem. **114:** 539-548.
102. ROBISON, G. A., R. W. BUTCHER & E. W. SUTHERLAND. 1971. Cyclic AMP. Academic Press. New York, N.Y.

103. GUNAGA, K. P., A. KAWANO & K. M. J. MENON. 1974. Neuroendocrinology **16:** 273-281.
104. WEISSMAN, B. A., J. W. DALY & P. SKOLNICK. 1975. Endocrinology **97:** 1559-1566.
105. LABRIE, F., P. BORGEAT, J. DROVIN, M. BEAULIEU, L. LAGACE, L. FERLAND & R. RAYMOND. 1979. Annu. Rev. Physiol. **41:** 555-569.
106. DISMUKES, K. & J. W. DALY. 1975. Life Sci. **17:** 199-210.
107. VAN CALKER, D., F. LÖFFLER & B. HAMPRECHT. 1983. J. Neurochem. **40:** 418-427.
108. KIMURA, F., M. KAWAKAMI, H. NAKANO & S. M. MCCANN. 1980. Endocrinology **106:** 631-635.
109. ZUBIN, P. & S. TALEISNIK. 1983. Brain Res. **271:** 273-277.
110. LIMBIRD, L. E. 1981. Biochem. J. **195:** 1-13.
111. SIEGEL, M. I. & P. CUATRECASAS. 1974. Mol. Cell. Endocrinol. **1:** 89-95.
112. SEAMON, K. B. & J. W. DALY. 1981. J. Cyclic Nucleotide Res. **7:** 201-224.
113. JOHNSON, G. L., H. R. KASLOW & H. R. BOURNE. 1978. J. Biol. Chem. **253:** 7120-7123.
114. NORTHUP, J. K., P. C. STERNWEIS, M. D. SMIGEL, L. S. SCHLEIFER, E. M. ROSS & A. G. GILMAN. 1980. Proc. Natl. Acad. Sci. USA **77:** 6516-6520.
115. BEYER, C., E. CANCHOLA & K. LARSSON. 1981. Physiol. Behav. **26:** 249-251.
116. BEYER, C., A. FERNANDEZ-GUASTI & G. RODRIGUEZ-MANZO. 1982. Physiol. Behav. **28:** 1073-1076.
117. FERNANDEZ-GUASTI, A., G. RODRIGUEZ-MANZO & C. BEYER. 1983. Physiol. Behav. **31:** 589-592.
118. FERNANDEZ-GUASTI, A., G. RODRIGUEZ-MANZO & C. BEYER. Unpublished observations.
119. BEYER, C., P. GOMORA, E. CANCHOLA & Y. SANDOVAL. 1982. Horm. Behav. **16:** 107-112.
120. BEYER, C. & E. CANCHOLA. 1981. Physiol. Behav. **27:** 731-733.
121. TOWLE, A. C. & P. Y. SZE. 1983. J. Steroid Biochem. **18:** 135-143.
122. NAGLE, C. A. & J. M. ROSNER. 1980. Neuroendocrinology **30:** 33-37.
123. PALMER, G. C. 1983. Prog. Neurobiol. **21:** 1-133.
124. DAVIES, A. O. & R. J. LEFKOWITZ. 1984. Annu. Rev. Physiol. **46:** 119-130.
125. JOHNSTON, P. G. & J. M. DAVIDSON. 1979. Neuroendocrinology **28:** 155-159.
126. KALRA, S. P. & P. S. KALRA. 1979. Acta Endocrinol. **92:** 1-7.
127. KALRA, S. P., P. S. KALRA, C. L. CHEN & J. A. CLEMENS. 1978. Acta Endocrinol. **89:** 1-9.
128. DROUVA, S. V., E. LAPLANTE & C. KORDON. 1985. Neuroendocrinology **40:** 325-331.
129. KIM, K. & V. D. RAMIREZ. 1985. Endocrinology **116:** 252-258.
130. DUDLEY, C. A., W. VALE, J. RIVIER & R. L. MOSS. 1981. Peptides **2:** 393-396.

A Reevaluation of the Concept of Separable Periods of Organizational and Activational Actions of Estrogens in Development of Brain and Behavior[a]

CHRISTINA L. WILLIAMS

Department of Psychology
Barnard College
Columbia University
New York, New York 10027

Steroid influences on brain and behavior have been characterized as being either "organizational" or "activational." The "organizational" actions occur when steroids interact with an immature neural substate and result in a relatively permanent blueprinting of the brain which determines its responses to steroids later in life. In contrast, "activational" actions occur when steroids trigger a mature neuroendocrine system and result in the initiation of transient behavioral or physiological responses (i.e., sexual behavior or gonadotropin release). Although the postpubertal activational actions of gonadal hormones have been recognized for many years (see reviews: References 1-4), the early organizing action of these hormones was not documented until the classic work of Phoenix and colleagues in 1959[5,6] (see reviews: References 4 and 7). Since this discovery, the conceptual framework which has directed research on the mechanisms of steroid action has focused on the differences between organizational and activational actions of steroids. The implication of this framework is that since organizational and activational actions of steroids are functionally different, their underlying mechanisms must also be different. Historically, this difference in steroid action has been explained in several ways: (1) The sites of steroid action may be different in infant and adult brains. (2) The mechanisms of steroid action, subsequent to the initial steroid-receptor interaction, may be different. (3) Steroids may exert different actions depending on the maturation state of the neural tissue at the time of exposure and not because of any difference in their site or mechanism of action.

The purpose of this paper is first, to assess the strength of these hypotheses in light of currently available research on the nature of hormone action in infant and adult brain, and second, to suggest an alternate conceptual framework of steroid action that focuses on the similarities rather than the differences between organizational and

---

[a]This work was supported by Grant MH 37523 from the National Institute of Mental Health and Grant NS 20671 from the National Institute of Neurological and Communicative Disorders and Stroke of the National Institutes of Health.

activational actions of steroids. In order to narrow the focus of this immense subject matter, I will limit the scope of this paper to the organizational action of estrogen on sexual differentiation and its activational effects on female sexual behavior, using research on the laboratory rat as a model. For an extensive review of hormone-behavior interactions in other laboratory animals and implications for human behavior I refer the reader to the excellent paper by Feder.[7] Beach[8] has also provided an extensive history of research on hormones and reproductive behavior.

LOCALIZATION OF ESTROGEN ACTION

Autoradiographic techniques which detect hormone-accumulating neurons in brain indicate that in the adult rat the medial preoptic area, the tuberal hypothalamus, a region of the midbrain, and regions of the limbic forebrain, such as medial amygdala and lateral septum, contain specific receptors for estradiol.[9] Early attempts to demonstrate that the neonatal brain contains estrogen receptors were confounded by the presence of α-fetoprotein. This estrogen-binding protein which is present in the blood of neonatal rats masked the estrogen receptor binding capacity of the neonatal brain. Thus, until 1974, when Sheridan et al. provided autoradiographic data demonstrating the existence of steroid receptors in neonatal brain,[10] estrogen receptors were not thought to be involved in the organizational actions of steroids. The fact that these neonatal estrogen receptors are occupied by estrogen in prenatal male rat brain[11] points to a role for these receptors in the organizational actions of estrogens.

Steroid implant and lesion studies have examined which steroid-sensitive brain areas are necessary and/or sufficient for estrogen activation of female sexual behavior in adults and for organizational actions of estrogen in infants. Komisaruk[12] and Pfaff[13] have reviewed the effects of neural lesions on steroid-activated lordosis behavior in adult female rats. They have concluded that bilateral lesions of the ventral medial hypothalamus (VMH) and large lesions of the anterior hypothalamus (AH) disrupt estrogen-facilitated sexual behavior. Preoptic area (POA) lesions in adult rats reduce the threshold for estrogen induction of lordosis behavior.[14] This effect may be due to an interruption of descending forebrain neurons, possibly from the septum, rather than through the destruction of neurons intrinsic to POA.[15-17] Thus, estrogen's normal action in the adult rat may be to inhibit neurons in or descending to the POA. This is supported by studies showing that electrical stimulation of the POA inhibits female sexual behavior.[18,19] To my knowledge, no lesion studies have attempted to examine the sites of estrogen's organizing actions in the infant.

Using small implants of estradiol (that do not diffuse far from the implant site), Davis et al.[20] and Rubin and Barfield[21] have shown that estradiol stimulation of the VMH alone is sufficient to elicit lordosis in adult rats. Larger implants (which may spread to VMH) facilitate lordosis when placed in POA, amygdala, and cortex (see review: Reference 22). The loci of estrogen's organizational action in the infant have also been examined. Christensen and Gorski[23] and Nordeen and Yahr[24] have shown that estradiol implants into POA of 1- to 2-day-old female rats cause masculinization (an increased tendency to display male mounting with testosterone treatment in adulthood); implants of estradiol into VMH cause defeminization (a decreased tendency to display lordosis with estradiol treatment in adulthood). As well, gonadal steroids present during the neonatal period cause structural changes in the POA region[25-27] (see review: Reference 28).

The results of these studies provide good evidence that estrogen activates female sexual behavior in adults and defeminizes the brains of infants through its actions on the AH-VMH region. The POA is also involved in estrogen's organizational and activational actions; however, the nature of its involvement is less clear. In the adult rat, estrogen may act to release the normal inhibition of the POA or descending inputs to the POA on female sexual behavior.[29] In the infant, estrogen acting on the POA seems to result in behavioral masculinization. Together, these data suggest that both the organizational and activational actions of estrogens are mediated through estrogen receptors in the AH-VMH region and on neurons in or descending to the POA. Thus it seems unlikely that the difference in action of estrogen in infant and adult brain can be accounted for by a change in estrogen's site of action.

MECHANISM OF ESTROGEN ACTION

Activation of sexual receptivity in adult females by estradiol has an 18- to 24-hr latency[30] and outlasts the removal of estradiol by 24 to 36 hr (see Reference 31). This time course suggests that estrogen is not working directly on the nerve cell membrane, but is causing longer-term changes. A wealth of information about the cellular mechanisms of estrogen activation of adult female sexual behavior has been published in the last few years. In order for estradiol to stimulate female sexual behavior, it must attach to specific receptor-like macromolecules in the soluble fraction of target cells (in the AH-VMH region, see above) which bind the steroid and carry it to the cell nucleus. The consequence of this translocation appears to be a change in the genomic action of the cell: an increase in protein synthesis, cell growth, and division which may be associated with changes in neurotransmitter synthesis, or synthesis of neurotransmitter and hormone receptors (see review: References 7 and 29). Recent studies have shown that the interruption of axonal transport with colchicine[32] or disruption of protein synthesis[33] prevents estrogen activation of lordosis. As well, the timing of these genomic events is critical for the activation of behavior. Two 1-hr exposures to estradiol are sufficient to induce lordosis, provided that they are at least 4 hr apart and not separated by more than 14 hr. Protein synthesis inhibition before the first, before the second, or between the first and the second estrogen implant will disrupt estrogen action.[34]

In the infant, considerably less research has examined estrogen's mechanism of action, but there is reasonably good evidence that the organizational actions of estrogens are also a consequence of estrogen binding to receptors and subsequent changes in cell genomic activity. Estrogen receptors are clearly present at the critical pre- and postnatal periods (see review: Reference 35). Studies using inhibitors of DNA, RNA, and protein synthesis to block organizational effects of estrogen do not always show positive results. However, studies have shown that some protein synthesis inhibitors disrupt estrogen's organizational actions, and certain RNA and DNA inhibitors protect against estrogen's masculinizing and defeminizing actions (see review: References 29, 36 and 37). One reason for some of the negative results using these techniques is that unlike our knowledge of estrogen's activational actions, we do not know very much about the temporal requirements of estrogen's organizational actions. It is not clear how long estrogen must be present, in what quantities it is effective, or at what time critical biochemical events occur. Until more information is obtained, unsuccessful inhibitor experiments should not be taken as evidence that the basic mechanisms of estrogen action are not similar in infant and adult brain.

It seems likely that at least the early steps of estrogen action subsequent to receptor binding are similar in infant and adult brains. The timing, or the products synthesized in later stages of estrogen action (i.e., which proteins are synthesized, or which neurotransmitter systems are modulated), may be different for organizational and activational actions.

CRITICAL PERIODS OF ESTROGEN ACTION

Since the site and physiochemical properties of estrogen receptors in neonatal and adult brain appear to be quite similar, it has been suggested that the change in estrogen's actions from organizational to activational must be due to the state of the neural tissue at the time of estrogen action.[36-38] This hypothesis is supported by data which demonstrate that there is a limited period for organizational actions of estrogen. And, only after this period has ended can estrogen have activational consequences. For example, male rats are normally masculinized and defeminized by a rise in endogenous androgens just before birth and within a few days after birth,[39] and female rats are particularly sensitive to the organizing actions of steroids within the first 5 days after birth.[40] Activational actions of steroids do not normally occur until around puberty, although precocious estrogen-facilitated lordosis[41,42] and gonadotropin release[43] have been demonstrated around postnatal day 15. These often-cited findings have strengthened the view that organizational and activational actions of steroids are separable in time.

There is other evidence, I believe, which suggests that organizational and activational actions of estrogen do not occur within clearly delineated periods. In fact, in some circumstances organizational and activational effects can occur at the same time. There is now a body of research which shows that steroids can cause organizational effects well after endogenous steroids would normally complete their action.[56] Feder and Goy,[44] for example, have found that neonatal administration of large doses of estrogen to female guinea pigs causes increased male mounting behavior in adulthood. More physiological doses of estrogen cause strong organizational actions only at midgestation and not after birth.[5,45] In the rat, there are a great number of studies that suggest that the process of sexual differentiation of female sexual behavior may extend into the peripubertal period (see References 46-50 and J. F. Rodriguez-Sierra, this volume). For example, Harlan and Gorski[51] have shown that if female rats are treated with a small dose of testosterone propionate (TP) at birth, ovarian steroids present around the time of puberty cause both organizational and activational effects. If these females are ovariectomized at puberty, they show an increased release of luteinizing hormone (LH) in response to progesterone treatment 60 days later. If the ovaries are left intact, or estrogen is administered to ovariectomized females, progesterone fails to induce an LH surge. Thus, in lightly androgenized females, the ovary can activate the mechanisms necessary for ovulation, but at the same time, ovarian secretions lead to an eventual loss of that function. Although Gorski (see review: Reference 36) has pointed out that this study may be conceptually important in our understanding of hormone action, I believe that its significance has been overlooked.

Not only can organizational actions of steroid hormones be demonstrated around the time of puberty, but recent research from my laboratory indicates that activational actions of estrogens can be demonstrated within the first 5 days of life in the rat.[52] I have found that as little as 10 μ/10 g body weight estradiol benzoate (EB), injected subcutaneously into 4-day-old rats, will facilitate lordosis and ear-wiggling 44 hr later.

This lordosis response resembles that of adult rats (FIGURE 1) and is induced by placing rats in isolation, in a warm (34°C), moist incubator, and stroking their flanks and lower back with a foam paintbrush. Although lordosis is easily elicited from oil-treated females, EB treatment facilitates the frequency ((number of lordosis responses displayed/four trials) × 100%), duration, and intensity (maximal curvature of the spine) of lordosis and frequency of ear-wiggling (FIGURE 2). Thus, estrogen can activate neural circuits for female sexual behavior during the period when estrogen has been shown to cause sexual differentiation of the brain. These data are particularly impressive because we have found that in the infant, as in the adult, the AH-VMH region is a site of estrogen's precocious activation of female sexual behavior (FIGURE 3). If 10 ng of estradiol is implanted in the AH-VMH region of 4-day-old rats, it significantly facilitates the frequency of lordosis compared to matched littermate control pups implanted with cholesterol. Interestingly, estradiol implants to the POA region significantly reduced lordosis frequency. In combination with the results of Christensen and Gorski[23] which showed organizational effects of postnatal estradiol implants in POA and VMH, these data suggest that both activational and organizational actions of estradiol can occur at the same time and in the same brain sites.

SOME SPECULATIONS ABOUT MECHANISMS OF ORGANIZATIONAL AND ACTIVATIONAL ACTIONS OF ESTROGENS

The evidence that I have presented thus far, suggests that both modes of estrogen action occur in the same brain sites, through similar mechanisms of action (at least

FIGURE 1. Photograph of a 6-day-old rat pup displaying lordosis in response to tactile stimulation of the flanks and lower back by a foam paintbrush. The pup was treated with 100 μg of estradiol benzoate 44 hr prior to testing in a warm (34°C), moist incubator.

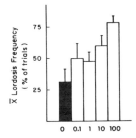
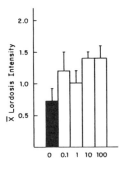
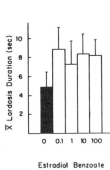
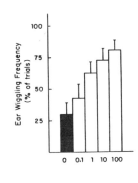

FIGURE 2. Frequency, intensity, and duration of lordosis and frequency of ear-wiggling were significantly increased by estradiol benzoate treatment 44 hr prior to testing compared to control, oil-treated littermates. Lordosis frequency: $F(4,36)=3.17$, $p < 0.05$; lordosis intensity: $F(4,36)=3.97$, $p < 0.05$; lordosis duration: $F(4,36)=2.69$, $p < 0.05$; ear-wiggling frequency: $F(4,36)=7.59$, $p < 0.05$.

initially), and may happen within the same time period. Instead of asking how estrogen target cells change from one mode of action to another, it is necessary to consider how organizational and activational actions of estrogen can occur simultaneously in the same brain sites. There are several possible mechanisms:

(1) Cells in the hypothalamus may have different rates of maturation; therefore, cells that mediate organizational effects may coexist with cells that have already matured to the adult, activational form. This explanation is a modification of the current notion that the maturation state of the neural tissue determines the response of the target cell; however, it accounts for the data indicating that both estrogen's modes of action can occur simultaneously. This idea is supported by the recent work of Nordeen and Yahr[24,53] which suggests that the different hemispheres of the brain may have different rates of maturation. They found that estrogen implants into the left hypothalamus of 2-day-old rats caused behavioral defeminization and implants into the right hypothalamus caused behavioral masculinization. Since there is ample reason to believe that the period for masculinization begins and ends earlier than the critical period for defeminization[54] these data suggest that the left hypothalamus matures faster than the right hypothalamus. If asymmetrical development continues, one might postulate that by 4 to 6 days of age cells in the left hypothalamus would have matured to their activational state, while cells in the right hypothalamus were

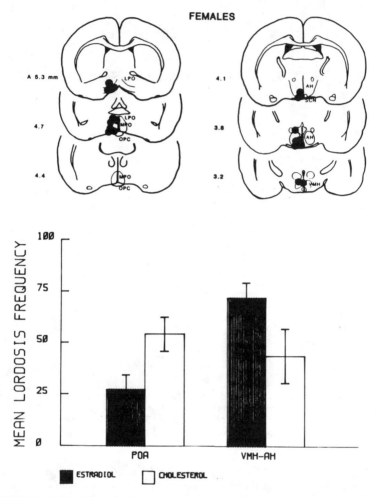

FIGURE 3. Depicted here is the frequency of lordosis of female 6-day-old rats given 10 ng unilateral estradiol implants in the preoptic region (POA) or the anterior and ventromedial hypothalamic regions (AH-VMH) and their matched littermate controls (pups implanted with cholesterol in the same sites). Implants were made 44 hr prior to testing. The exact location of each of the estradiol implants is shown in the drawings. Implants in the AH-VMH significantly increased lordosis frequency ($t=2.37$, $df=7$, $p < 0.05$). Implants in the POA significantly decreased lordosis frequency ($t=2.79$, $df=9$, $p < 0.05$). OPC, optic chiasm; MPO, medial preoptic area; LPO, lateral preoptic area; SCN, suprachiasmatic nucleus.

still being organized. Although this hypothesis might account for the data showing activational actions of estrogen at 4 days of age, it is a less likely explanation of organizational effects occurring peripubertally.

(2) Early activational actions of estrogens may not occur through classic receptor-mediated systems as activational actions of estrogens appear to be mediated in the adult. Thus far, I have no evidence that the mechanism by which estrogen activates

precocious lordosis and ear-wiggling is through estradiol's attachment to specific receptors and the subsequent changes in gene expression. In other words, although I do know that the AH-VMH region is a site of estrogen action in the infant as in the adult, I do not know much about the temporal requirements of, or the critical biochemical steps involved in, estrogen activation of behavior in the neonate. The effects of estrogen on pups' behavior might be due to nonspecific actions of the hormone on pups' metabolism or activity levels. I think this is an unlikely hypothesis for several reasons. First, it is not possible to facilitate lordosis in the infant by just activating the infant with amphetamine. This drug increases pups' overall activity level but does not increase lordosis frequency or duration. Thus, the effect of estrogen is not just one of increasing activity. Second, the fact that small estradiol implants into the AH-VMH region cause a specific facilitation of lordosis while similar implants into other hypothalamic sites either depress lordosis responding, or have little effect, argues against a nonspecific action of the hormone.

(3) Estrogen may cause two stages of action in estrogen target cells—an initial activational phase and a second, longer-term growth-promoting stage. Normally, the initial activational stage of estrogen action is not seen in infant rats because other necessary neural systems may not have matured, or the activation is masked by immature motor or sensory systems. If neonatal estrogen receptors are exposed to estrogen (as they are during normal development in the male) they may become activated, and then cause long-term growth-related events. Upon the next exposure to estrogen these estrogen target cells only show activational consequences. In the female, estrogen target cells are not normally exposed to large doses of estrogen until puberty and therefore they may retain their potential for "organizational actions" through the peripubertal period. At this time estrogen can alter central nervous system (CNS) morphology and cause long-term changes in how estrogen target cells respond to estrogen (see Rodriguez-Sierra, this volume).

This hypothesis appears to be an attractive one. It is parsimonious. Instead of postulating that estrogen target cells acquire a new function as development progresses (i.e., the ability to show reversible short-term effects), it is only necessary to suggest that as estrogen target cells mature they become more specialized and lose one of their functions. This is certainly in keeping with traditional theories of embryological development. Differentiating cells are born with their entire set of genetic instructions. As development proceeds, a gradual restriction of gene expression causes cells to become more specialized and lose some of their functions. For example, the work of LeDouarin and her colleagues in the early 1970s[55] showed that cells are born with the capacity to produce several different neurotransmitters, but the environment these cells encounter during development shapes their ultimate transmitter choice.

This hypothesis, that estrogen target cells are born with the ability to show both activational and organizational actions, may also be a testable one. If it were true, one would predict that a regimen of hormone treatment and administration of RNA or DNA inhibitors could be developed that would allow precocious activation of lordosis without organizational consequences. By administering an inhibitor after triggering estrogen's short-term activation of behavior, one should be able to block the permanent long-term programming of the brain.

In summary, I have attempted to provide a reexamination of the current concept of organizational and activational actions of estrogens as it applies to female sexual behavior. The current view of estrogen action emphasizes the differences between estrogen's two modes of action. This framework makes it difficult to see some of the interesting similarities between the sites, mechanisms, and timing of organizational and activational effects. I recognize that this dichotomy between neonatal and adult effects may be a convenient framework that has aided writers of review articles in this field and may, in fact, not be one that directs the research in this area. As Frank Beach says in his famous ramstergig paper:

> As is so often the case, it is not the initial, tentative, provisional interpretations proposed by the original investigators that cause the mischief. Rather, it is the subsequent second order writings formulated by others who read selectively only the positive sections of published experimental reports and who lack any firsthand knowledge of the qualifications and restrictions which formed essential components of the original conclusions. (Reference 2, p. 293)

My hope is that this paper will serve to focus attention on some of the exceptions to the current idea that organizational effects can always be contrasted with activational actions. Whatever the precise mechanisms of organizational and activational actions of estrogens on the brain, the interaction of these two processes, which appear to take place at the same time in the same brain areas, may be of considerable importance in our understanding of the nature of estrogen action on the brain.

ACKNOWLEDGMENTS

I thank Andrea Baron and Susan Yoo for their excellent research assistance and Rae Silver for helpful comments on an earlier version of the manuscript.

[Note added in proof: Since the completion of this paper an excellent review has been published[56] which describes evidence that organizational actions of steroids can occur in mature organisms.]

REFERENCES

1. BEACH, F. A. 1948. Hormones and Behavior. Hoeber. New York, N.Y.
2. BEACH, F. A. 1971. Hormonal factors controlling the differentiation, development, and display of copulatory behavior in the ramstergig and related species. *In* The Biopsychology of Development. E. Tobach, L. R. Aronson & E. Shaw, Eds.: 249-295. Academic Press. New York, N.Y.
3. YOUNG, W. C. 1961. The hormones and mating behavior. *In* Sex and Internal Secretions. 3rd edit. W. C. Young, Ed. **2**: 1173-1239. Williams & Wilkins. Baltimore, Md.
4. FEDER, H. H. 1981. Perinatal hormones and their role in the development of sexually dimorphic behaviors. *In* Neuroendocrinology of Reproduction. N. T. Adler, Ed.: 127-158. Plenum. New York, N.Y.
5. PHOENIX, C. H., R. W. GOY, A. A. GERALL & W. C. YOUNG. 1959. Organizing action of prenatally administered testosterone propionate on the tissues mediating mating behavior in the female guinea pig. Endocrinology **65**: 369-382.
6. GRADY, K. L., C. H. PHOENIX & W. C. YOUNG. 1965. Role of the developing rat testis in differentiation of the neural tissue mediating mating behavior. J. Comp. Physiol. Psychol. **59**: 176-182.
7. FEDER, H. H. 1984. Hormones and sexual behavior. Annu. Rev. Psychol. **35**: 165-200.
8. BEACH, F. A. 1981. Historical origins of modern research on hormones and behavior. Horm. Behav. **15**: 325-376.

9. PFAFF, D. W. 1976. The neuroanatomy of sex hormone receptors in the vertebrate brain. *In* Neuroendocrine Regulation of Fertility. J. C. Anand Kumar, Ed.: 30-45. S. Karger. Basel, Switzerland.
10. SHERIDAN, P. J., M. SAR & W. E. STUMPF. 1974. Autoradiographic localization of ^{3}H-estradiol or its metabolites in the central nervous system of the developing rat. Endocrinology **94**: 1386-1390.
11. MACLUSKY, N. J., C. CHAPTAL, I. LIEBERBURG & B. S. MCEWEN. 1976. Properties and subcellular inter-relationships of presumptive estrogen receptor macromolecules in the brains of neonatal and prepubertal female rats. Brain Res. **114**: 158-165.
12. KOMISARUK, B. K. 1978. The nature of the neural substrate of female sexual behavior in mammals and its hormone sensitivity: review and speculations. *In* Biological Determinants of Sexual Behavior. J. B. Hutchison, Ed.: 349-393. John Wiley. New York, N.Y.
13. PFAFF, D. W. 1980. Estrogens and Brain Function: Neural Analysis of a Hormone-Controlled Mammalian Reproductive Behavior. Springer-Verlag. New York, N.Y.
14. POWERS, B. & E. S. VALENSTEIN. 1972. Sexual receptivity: facilitation by medial preoptic lesions in female rats. Science **175**: 103-105.
15. NANCE, D. M., J. SHRYNE & R. A. GORSKI. 1975. Effects of septal lesions on behavioral sensitivity of female rats to gonadal hormones. Horm. Behav. **6**: 59-64.
16. NANCE, D. M., J. SHRYNE & R. A. GORSKI. 1975. Facilitation of female sexual behavior in male rats by septal lesions: an interaction with estrogen. Horm. Behav. **6**: 289-299.
17. YAMANOUCHI, K. & Y. ARAI. 1977. Possible inhibitory role of the dorsal inputs to the preoptic area and hypothalamus in regulating female sexual behavior in the female rat. Brain Res. **127**: 296-301.
18. MALSBURY, C. W., D. W. PFAFF & A. M. MALSBURY. 1980. Suppression of sexual receptivity in the female hamster: neuroanatomical projections from preoptic and anterior hypothalamic electrode sites. Brain Res. **181**: 267-284.
19. PFAFF, D. W. & Y. SAKUMA. 1979. Deficit in the lordosis reflex of female rats caused by lesions in the ventromedial nucleus of the hypothalamus. J. Physiol. **288**: 203-210.
20. DAVIS, P. G., B. S. MCEWEN & D. W. PFAFF. 1979. Localized behavioral effects of tritiated estradiol implants in the ventromedial hypothalamus of female rats. Endocrinology **104**: 898-903.
21. RUBIN, B. & R. BARFIELD. 1980. Priming of estrous responsiveness by implants of 17β-estradiol in the ventromedial hypothalamic nucleus of the female rat. Endocrinology **106**: 504-509.
22. GORSKI, R. A. 1976. The possible neural sites of hormonal facilitation of sexual behavior in the female rat. Psychoneuroendocrinology **1**: 371-387.
23. CHRISTENSEN, L. W. & R. A. GORSKI. 1978. Independent masculinization of neuroendocrine systems by intracerebral implants of testosterone or estradiol in the neonatal female rat. Brain Res. **146**: 325-340.
24. NORDEEN, E. J. & P. YAHR. 1982. Hemispheric asymmetries in the behavioral and hormonal effects of sexually differentiating mammalian brain. Science **218**: 391-394.
25. JACOBSON, C. D. & R. A. GORSKI. 1981. Neurogenesis of the sexually dimorphic nucleus of the preoptic area in the rat. J. Comp. Neurol. **196**: 519-529.
26. MATSUMONO, A. & Y. ARAI. 1980. Sexual dimorphism in "wiring pattern" in the hypothalamic arcuate nucleus and its modification by neonatal hormonal environment. Brain Res. **190**: 238-242.
27. RAISMAN, G. & P. M. FIELD. 1973. Sexual dimorphism in the neuropil of the preoptic area of the rat and its dependence on neonatal androgen. Brain Res. **54**: 1-29.
28. ARNOLD, A. P. & R. A. GORSKI. 1984. Gonadal steroid induction of structural sex differences in the central nervous system. Annu. Rev. Neurosci. **7**: 413-442.
29. PFAFF, D. W. & B. S. MCEWEN. 1983. Actions of estrogens and progestins on nerve cells. Science **219**: 808-813.
30. GREEN, R., W. G. LUTTGE & R. E. WHALEN. 1970. Induction of receptivity in ovariectomized female rats by a single intravenous injection of estradiol-17β. Physiol. Behav. **5**: 137-141.
31. PARSONS, B., N. J. MACLUSKY, L. C. KREY, D. W. PFAFF & B. S. MCEWEN. 1980. The temporal relationship between estrogen-inducible progestin receptors in the female rat

brain and the time course of estrogen activation of mating behavior. Endocrinology **107**: 774-779.
32. MEYERSON, B. J. 1982. Colchicine delays the estrogen-induced copulatory response in the ovariectomized female rat. Brain Res. **253**: 281-286.
33. RAINBOW, T. C., P. G. DAVIS & B. S. MCEWEN. 1980. Anisomycin inhibits the activation of sexual behavior by estradiol and progesterone. Brain Res. **194**: 548-555.
34. PARSONS, B., T. C. RAINBOW, D. W. PFAFF & B. S. MCEWEN. 1981. Oestradiol, sexual receptivity and cytosol progestin receptors in rat hypothalamus. Nature **292**: 58-59.
35. MCEWEN, B. S. 1978. Sexual maturation and differentiation: the role of the gonadal steroids. Prog. Brain Res. **48**: 291-307.
36. GORSKI, R. A. 1979. Nature of hormone action in the brain. *In* Ontogeny of Receptors and Reproductive Hormone Action. T. H. Hamilton, J. H. Clark & W. A. Sadler, Eds.: 371-392. Raven Press. New York, N.Y.
37. MCEWEN, B. S. 1981. Cellular biochemistry of hormone action in brain and pituitary. *In* Neuroendocrinology of Reproduction. N. T. Adler, Ed.: 485-518. Plenum. New York, N.Y.
38. MACLUSKY, N. & F. NAFTOLIN. 1981. Sexual differentiation of the central nervous system. Science **211**: 1294-1303.
39. GRADY, K. L., C. H. PHOENIX & W. C. YOUNG. 1965. Role of the developing rat testis in differentiation of the neural tissue mediating mating behavior. J. Comp. Physiol. Psychol. **59**: 176-182.
40. BARRACLOUGH, C. A. & R. A. GORSKI. 1962. Studies on mating behavior in the androgen-sterilized female rat in relation to the hypothalamic regulation of sexual behavior. J. Endocrinol. **25**: 175-182.
41. SODERSTEN, P. 1975. Receptive behavior in developing female rats. Horm. Behav. **6**: 307-317.
42. SODERSTEN, P. 1978. Lordosis behavior in immature male rats. J. Endocrinol. **76**: 233-240.
43. ANDREWS, W. W. & S. R. OJEDA. 1977. On the feedback actions of estrogen on gonadotropin and prolactin release in infantile female rats. Endocrinology **101**: 1517.
44. FEDER, H. H. & R. W. GOY. 1983. Effects of neonatal estrogen treatment of female guinea pigs on mounting behavior in adulthood. Horm. Behav. **17**: 284-291.
45. GOY, R. W., W. E. BRIDSON & W. C. YOUNG. 1964. Period of maximal susceptibility of the prenatal female guinea pig to masculinizing actions of testosterone propionate. J. Comp. Physiol. Psychol. **57**: 166-174.
46. BROWN-GRANT, K. 1975. A re-examination of the lordosis response in female rats given high doses of testosterone propionate or estradiol benzoate in the neonatal period. Horm. Behav. **6**: 351-378.
47. HENDRICKS, S. E. & J. A. DUFFY. 1974. Ovarian influences on the development of sexual behavior in neonatally androgenized rats. Dev. Psychobiol. **7**: 297-303.
48. KAWASHIMA, S. 1960. Influence of continued injections of sex steroids on the estrous cycle in the adult rat. Ann. Zool. Japan **33**: 226-229.
49. NIKELS, K. W. 1976. Ovarian modification of sexual behavior in neonatally androgenized female rats. Bull. Psychon. Soc. **7**: 59-64.
50. SODERSTEN, P. 1976. Lordosis behavior in male, female and androgenized female rats. J. Endocrinol. **70**: 409-420.
51. HARLAN, R. E. & R. A. GORSKI. 1978. Effects of postpubertal ovarian steroids on reproductive function and sexual differentiation of lightly androgenized rats. Endocrinology **102**: 1716-1724.
52. WILLIAMS, C. L. 1986. Estradiol benzoate facilitates lordosis and ear wiggling in 4- to 6-day old rats. Submitted for publication.
53. NORDEEN, E. J. & P. YAHR. 1983. A regional analysis of estrogen binding to hypothalamic cell nuclei in relation to masculinization and defeminization. J. Neurosci. **3**: 933-941.
54. GOY, R. W. & B. S. MCEWEN. 1980. Sexual Differentiation of the Brain. MIT Press. Cambridge, Mass.
55. LEDOUARIN, N. M. 1980. Migration and differentiation of neural crest cells. Curr. Top. Dev. Biol. **16**: 31-85.
56. ARNOLD, A. P. & S. M. BREEDLOVE. 1985. Organizational and activational effects of sex steroids on brain and behavior: a reanalysis. Horm. Behav. **19**: 469-498.

Extended Organizational Effects of Estrogen at Puberty[a]

JORGE F. RODRIGUEZ-SIERRA

Department of Anatomy
University of Nebraska Medical Center
Omaha, Nebraska 68105

It is well established that estrogen can exert two types of effect on reproductive function: an organizational and an activational effect. The organizational effect of estrogen is best exemplified during the sexual differentiation of the brain of the rat in the perinatal period (see review by Feder).[1] The activational effects of estrogen are most obvious during the induction of sexual behavior and/or gonadotropin secretion in the adult female rat.[2-3]

This conceptual framework was based on the classical works of Young[4] and his students[5] and borrows heavily from the literature on the effects of steroids on the development of the reproductive tract, particularly the classical work of Jost.[6] An analogy is made between the morphological alterations induced by steroids on the reproductive tract and the morphological changes induced in the brain by estrogen during the first few days before and after birth (see review by Arnold and Gorski).[7] The activational effects of estrogen are believed to occur due to *functional* changes in neuroendocrine systems established during the organizational period.

Implicit in this dichotomy of estrogen action is a temporal relationship: organizational effects occur exclusively during the perinatal period, while the activational effects occur during the pubertal and adult periods (see TABLES 1 and 2).

A series of experiments in my laboratory and work by other investigators have shown that estrogen can produce morphological alterations in the central nervous system (CNS) during the juvenile and adult periods. This evidence suggests that the "organizational" effects of estrogen can occur at critical stages of neuroendocrine function throughout the life of the organism.

We know that estrogen is concentrated and retained in neurons involved in neuroendocrine function[8] and that this steroid is transported to and accumulated within the nuclei of some neurons via a receptor-mediated mechanism.[9] It has been suggested that this nuclear estrogen-receptor complex initiates events at the genomic level to alter cellular function and morphology.[10]

Ramirez and Sawyer[11] demonstrated that physiological doses of estrogen administered to immature rats advanced the onset of vaginal opening (an external marker of puberty in the rat). Caligaris *et al.*[12] showed that administration of estradiol to female rats on days 22 and 23 of age induced a marked rise in serum luteinizing hormone (LH) and follicle-stimulating hormone (FSH) from the anterior pituitary gland by the second day of estrogen injection. Studies in our laboratory have shown that a single injection of estradiol benzoate (10 μg) to 25-day-old female rats induced

---

[a] This work was supported by funds from the University of Nebraska Medical Center and the National Institutes of Health (HD 13219).

TABLE 1. Classical View of Organizational Effects of Estrogen[a]

	Perinatal	Pubertal	Adult	Old Age
Gonads	***	**	*	—
Brain	***	—	—	—
2° sexual characteristics	*	***	*	—

[a]Symbols: —, no activity; *, some activity; **, active; ***, most active.

an acute rise of serum LH and prolactin on the afternoon of day 27 of age (see FIGURES 1 and 2). These surges of LH and prolactin are similar in magnitude to that in the preovulatory period in the adult female rat. The observations of the aforementioned studies could be interpreted (and were) as an activational effect of estrogen on neurocircuits laid out during neuroendocrine differentiation in the perinatal period. When we looked into the ultrastructure of the hypothalamus of these animals (see FIGURE 3) we found a doubling in the number of synaptic contacts in the arcuate nucleus[13] (see FIGURE 4). We found no effect of estrogen on the number of synapses in the cortex, septal area, or medial preoptic area.[13] Thus, it appears that estrogen exerts morphological alterations in some cells of the brain during the juvenile period and long after neuroendocrine sexual differentiation has taken effect.

In addition to the structural effects of estradiol in the brain, this steroid might accelerate the maturation of specific neurotransmitter systems involved in the control of gonadotropin and prolactin. We recently reported[14] that the density of nicotinic acetylcholine receptors (measured with the nicotinic agonist α-bungarotoxin) is significantly increased in the arcuate nucleus of the hypothalamus after estrogen administration (see FIGURE 5) in the same region where we find the most striking synaptogenic effect of estrogen. There is some evidence that estrogen can also increase the number of muscarinic receptors in the hypothalamus.[15]

We decided to test whether the synaptogenic effect of estradiol was generalizable to males and tested prepubertal male and female rats injected with estradiol (10 μg) on day 25 of life. Animals were sacrificed at 27 days of age and some females were sacrificed at 31 days of age. The older group of animals was tested to determine if the synaptogenic effect of estradiol was a spurious or transitory effect or whether it persisted for some time. We easily replicated our findings in the arcuate nucleus in the female rats at 27 and 31 days of age (that is, estrogen increased the number of synapses significantly), but the male rats showed no discernible effect of estrogen on the number of synapses (see FIGURE 6). The effects on the arcuate synapses correlated with the ability of estrogen to induce an LH surge in the females, but not the male rats (see FIGURE 7). The septal or preoptic areas (POA) showed no significant effect of estrogen in either sex. However, in both the POA and septal areas the number of synapses increased with age (see FIGURES 8 and 9) and in addition, analysis of the

TABLE 2. Classical View of Activational Effects of Estrogen[a]

	Perinatal	Pubertal	Adult	Old Age
Gonads	—	**	***	*
Brain	—	**	***	*
2° sexual characteristics	—	**	***	—

[a]Symbols: —, no activity; *, some activity; **, active; ***, most active.

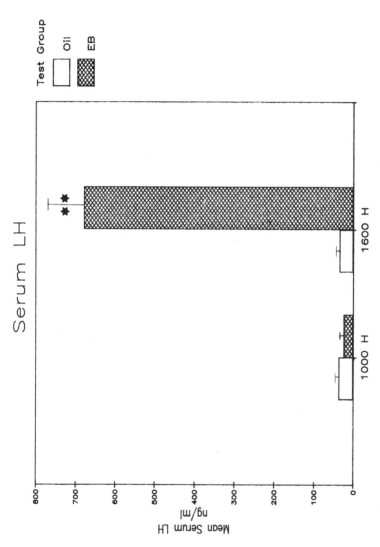

FIGURE 1. Mean (± SEM) serum concentration of luteinizing hormone (LH) in 27-day-old female rats treated with 10 μg of estradiol benzoate (EB) or oil 2 days earlier. Note that serum LH was at surge level at 1600 hr in the EB-treated animals.

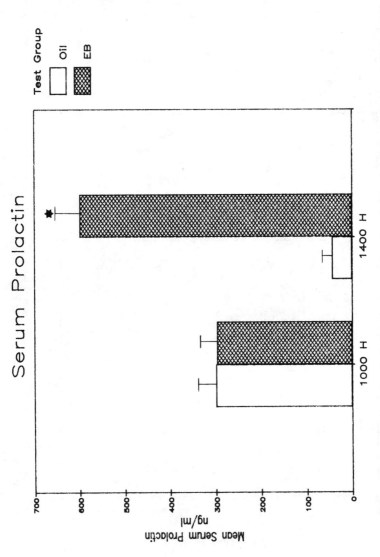

FIGURE 2. Mean (± SEM) serum concentration of prolactin in animals treated as described in the caption to FIGURE 1. Note that the EB-treated rats showed a significant surge in serum prolactin concentration at 1400 hr.

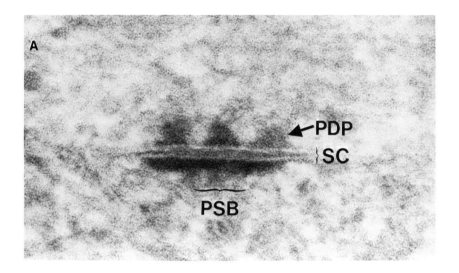
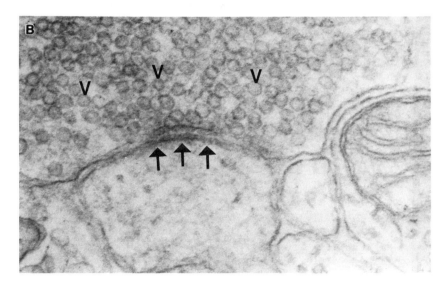

FIGURE 3. (A) Ethanol phosphotungstic (EPTA)-stained synapse in the arcuate nucleus of the hypothalamus (ARC). Note the presynaptic density projections (PDP), synaptic cleft (SC), and postsynaptic band (PSB). (B) Osmium-stained synapse in the ARC. Note the presynaptic ending with translucent vesicles (V). The postsynaptic band is marked by the arrows. The number of synapses was quantitated using stereological techniques described in detail elsewhere.[13]

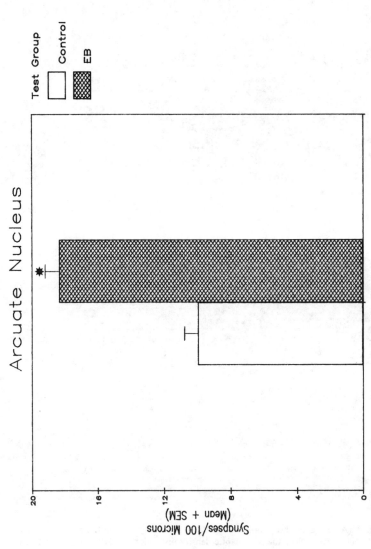

FIGURE 4. Mean (± SEM) number of synaptic densities in the ARC of female rats on day 27 of age, 2 days after EB or oil (control) treatment.

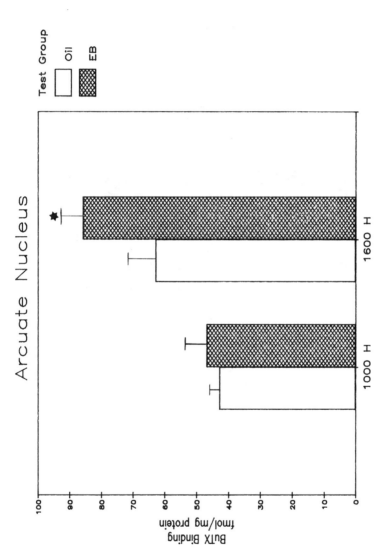

FIGURE 5. Mean (± SEM) tissue concentration of α-bungarotoxin binding sites in the ARC of female rats 2 days after EB or oil treatment. The 27-day-old female rats treated with EB showed a significant increase in binding sites when compared to the oil controls ($p < 0.01$).

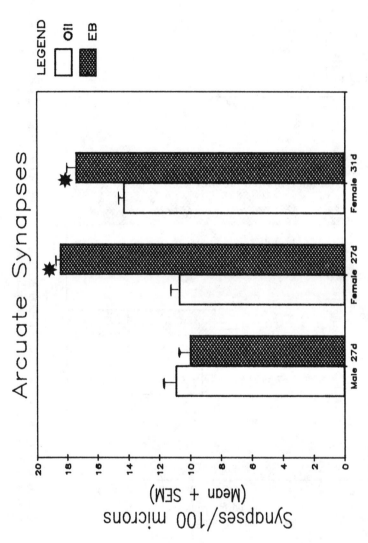

FIGURE 6. Mean (± SEM) synaptic area density of ARC in male and female rats 2 or 6 days after EB or oil treatment. The female rats treated with EB displayed a significant increase in the number of synapses over the oil controls at 27 or 31 days of age ($p < 0.01$). Male rats were not affected by the EB treatment.

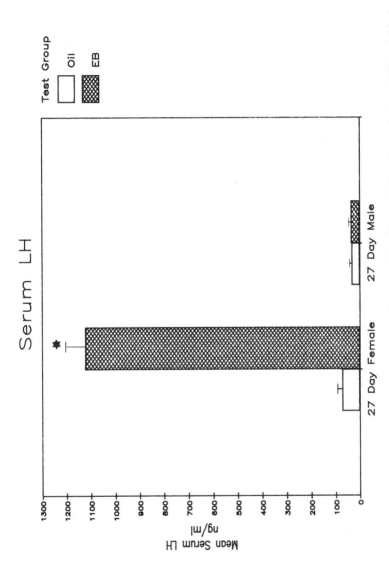

FIGURE 7. Mean (± SEM) serum LH concentration in the male and female rats from FIGURE 6. Female rats that were EB treated had a significant surge of serum LH concentration, but the oil control females did not ($p < 0.001$). Male rats showed no effect due to EB treatment.

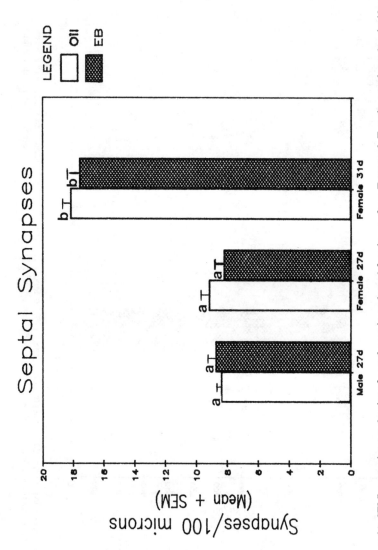

FIGURE 8. Mean (± SEM) synaptic area density of septal area in male and female rats from FIGURE 6. Female rats showed a significant increase in the number of synapses ($p < 0.01$) at 31 days of age when compared to 27-day-old females, regardless of the treatment. Groups with different superscripts are significantly different from each other ($p < 0.01$).

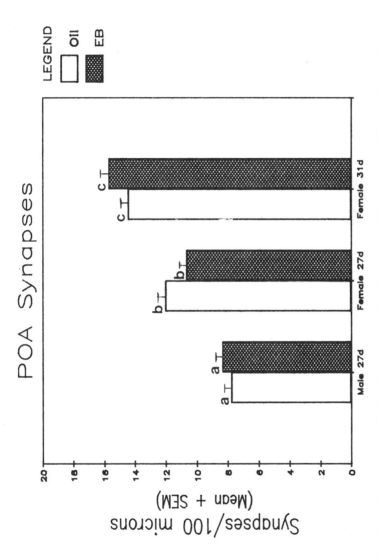

FIGURE 9. Mean (± SEM) synaptic area density of preoptic area (POA) in male and female rats administered EB at 25 days of age. Female rats showed a significantly greater number of synapses than male rats, regardless of treatment, at 27 days of age ($p < 0.01$). Female rats showed a significant increase in the number of synapses at 31 days of age, regardless of treatment ($p < 0.01$). Groups with different superscripts are significantly different from each other.

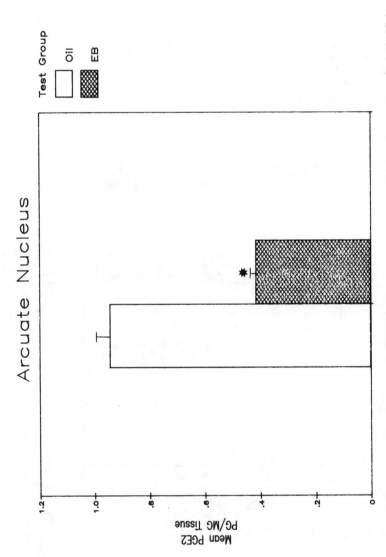

FIGURE 10. Mean (± SEM) concentration of prostaglandin E_2 (PGE_2) in the ARC of rats treated with EB at 25 days of age. At 27 days of age there is a significant reduction of PGE_2 just prior to the EB-induced LH surge.

POA revealed that female rats had a higher number of synaptic contacts than male rats at 27 days of age, but estrogen did not alter this sexual difference.

It appears that there is a profound maturational effect of estrogen in the arcuate nucleus during the juvenile period of development in female rats, and that this effect is sexually dimorphic: female rats respond to estrogen with accelerated synaptogenesis and LH surges, whereas males show neither of these responses. The synaptogenic effect of estrogen is not spurious since it was present for an additional 4 days.

These results show that estrogen can alter the morphology of the CNS in the prepubertal period in an area known to regulate the onset of puberty and gonadotropin secretion.

Estrogen can alter the morphology of another endocrine tissue, the uterus. In fact, administration of estradiol results in a dramatic increase in the number of myometrial gap junctions (MGJ).[16,17] The MGJ are responsible for cell communication and allow for the synchronization of muscle contractility during labor.[16] Administration of indomethacin, to block the cyclooxygenase pathway of prostaglandin synthesis, results in prevention of the facilitatory effects of estrogen on the number of MGJ in the uterus. The inhibitory effects of indomethacin can be reversed by administration of prostaglandin E_2 (PGE_2).[18] These results have been interpreted to mean that estrogens and prostaglandins are involved in the formation of MGJ in the uterus. The similarities between MGJ and neuronal synapses, both structurally and functionally, are striking.

TABLE 3. Revised View of the Organizational Effects of Estrogen[a]

	Perinatal	Pubertal	Adult	Old Age
Gonads	***	**	*	(?)
Brain	***	**	(?)	*
2° sexual characteristics	*	***	*	(?)

[a]Symbols: (?), possible activity; *, some activity; **, active; ***, most active.

Both appear as membranous structures that function to communicate information from one cell to another. We have found that estradiol can induce synaptogenesis, phasic release of LH, and deplete PGE_2 in the arcuate nucleus (see FIGURE 10).

Our work is now geared toward understanding how the new synapses are formed by estrogen and how prostaglandins are involved in this phenomenon. We are using the analogy of the MGJ and the synapse as a working hypothesis in which to test discrete experimental questions about synaptogenesis.

We hope that our investigations have provided evidence to modify the idea that "organizational" effects of estrogen are restricted to the perinatal period. In fact, there is further evidence that in adult animals estrogen can influence neuronal morphology. Nottebohm[19] showed that testosterone (which is aromatizable to estrogen) can induce a significant increase in the size of two telencephalic nuclei involved in song in the adult female canary. The ultrastructure of the supraoptic nucleus is markedly changed during pregnancy and lactation in the rat.[20]

In view of this new evidence, I would like to suggest a revision of the concept of "organizational" effects of estrogen (see TABLE 3). The major change will be in the extension of when estrogen could exert morphological alterations in neuroendocrine tissue. We have strong evidence that neuromorphological changes occur during the juvenile period. This change in synaptogenesis is probably comparable to what occurs during normal puberty onset. It would not be surprising if other neuromorphological alterations could be found during critical stages of neuroendocrine function: pregnancy, lactation, and cessation of ovarian cyclicity. Perhaps, the ability of neurons to grow and make new connections is intrinsically related to the neuroendocrine state of the organism, regardless of age. If this is so, a new age in developmental neurobiology is before us.

ACKNOWLEDGMENT

I would like to thank the faculty of the Institute of Animal Behavior of Rutgers University for providing me with, not only excellent scientific training, but the freedom to pursue and the criticism to improve my ideas.

REFERENCES

1. FEDER, H. H. 1981. Perinatal hormones and their role in the development of sexually dimorphic behaviors. *In* Neuroendocrinology of Reproduction. N. T. Adler, Ed.: 127-157. Plenum Press. New York, N.Y.
2. RODRIGUEZ-SIERRA, J. F. & B. R. KOMISARUK. 1982. Intrahypothalamic implants of LHRH, PGE_2, or progesterone into the mediobasal hypothalamus facilitates sexual receptivity in female rats. Neuroendocrinology **35**: 363-369.
3. RODRIGUEZ-SIERRA, J. F. & C. A. BLAKE. 1982. Catecholestrogens and release of anterior pituitary gland hormones. I. Luteinizing hormone. Endocrinology **110**: 318-324.
4. YOUNG, W. C. 1961. Hormones and mating behavior. *In* Sex and Internal Secretions. W. C. Young, Ed.: 1173-1239. Williams & Wilkins. Baltimore, Md.
5. PHOENIX, C. H., R. W. GOY, A. A. GERALL & W. C. YOUNG. 1959. Organizing action of prenatally administered testosterone propionate on the tissues mediating mating behavior in the female guinea pig. Endocrinology **65**: 369-382.
6. JOST, A. 1973. Studies on sex differentiation in mammals. Recent Prog. Horm. Res. **29**: 1-41.
7. ARNOLD, A. P. & R. A. GORSKI. 1984. Gonadal steroid induction of structural sex differences in the central nervous system. Annu. Rev. Neurosci. **7**: 413-442.
8. PFAFF, D. W. & M. KEINER. 1973. Atlas of estradiol concentrating cells in the central nervous system of the female rat. J. Comp. Neurol. **151**: 121-158.
9. MCEWEN, B. S. 1976. Steroid receptors in neuroendocrine tissues: topography, subcellular distribution and functional implications. *In* Subcellular Mechanisms in Reproductive Medicine. F. Naftolin, K. J. Ryan & J. Davis, Eds.: 277-304. Elsevier. Amsterdam, the Netherlands.
10. PFAFF, D. W. & B. S. MCEWEN. 1983. Action of estrogens and progestins on nerve cells. Science **219**: 808-814.
11. RAMIREZ, V. D. & C. H. SAWYER. 1965. Advancement of puberty in the female rat by estrogen. Endocrinology **76**: 1158-1168.
12. CALIGARIS, L., J. J. ASTRADA & S. TALEISNIK. 1972. Influence of age in the release of luteinizing hormone induced by oestrogen and progesterone in immature rats. J. Endocrinol. **55**: 97-103.
13. CLOUGH, R. W. & J. F. RODRIGUEZ-SIERRA. 1983. Synaptic changes in the hypothalamus of the prepuberal female rat administered estrogen. Am. J. Anat. **167**: 205-214.
14. MORLEY, B. J., J. F. RODRIGUEZ-SIERRA & R. W. CLOUGH. 1983. Increase in hypothalamic nicotine acetylcholine receptors in prepuberal female rats administered estrogen. Brain Res. **278**: 262-265.

15. DOHANICH, R. W., J. W. WITCHER, D. R. WEAVER & L. G. CLEMENS. 1982. Alteration of muscarinic binding in specific brain areas following estrogen treatment. Brain Res. **241:** 347-350.
16. DAHL, G. & W. BERGER. 1978. Nexus formation in the myometrium during parturition and induced by estrogen. Cell Biol. Int. Rep. **2:** 381-387.
17. GARFIELD, R. E., M. S. KANNAN & E. E. DANIEL. 1980. Gap junction formation in the myometrium: control by estrogens, progesterone and prostaglandins. Am. J. Physiol. **238:** C81-C89.
18. MCKENZIER, L. W., C. P. PURI & R. E. GARFIELD. 1983. Effect of estradiol-17β and prostaglandins on rat myometrial gap junctions. Prostaglandins **26:** 925-941.
19. NOTTEBOHM, F. 1980. Testosterone triggers growth of brain vocal control nuclei in adult female canaries. Brain Res. **189:** 429-436.
20. THEODOSIS, D. T. & D. A. POULAIN. 1984. Evidence for structural plasticity in the supraoptic nucleus of the rat hypothalamus in relation to gestation and lactation. Neuroscience **11:** 183-193.

A Two-Step Model for Sexual Differentiation[a]

JACQUES BALTHAZART[b] AND MICHAEL SCHUMACHER

Laboratoire de Biochimie Générale et Comparée
Université de Liège
B-4020 Liège, Belgium

During the last decade, the study of sexual differentiation in quail has been rewarding for two groups of reasons. First, the study of differentiation in an animal model quite different from rodents has for the first time progressed far enough to produce precise models and concepts. Results obtained in rodents can consequently be evaluated in a broader perspective and this should lead to redefinition of a number of notions. Quail research on differentiation also emphasizes that the rat is not a universal model for vertebrates and cautions against generalizations extrapolating rat data to primates or even humans (see Reference 1 for a detailed treatment).

Second, the quail has specific advantages which facilitate the study of sexual differentiation.[2] Embryonic development takes place in an isolated environment (the egg) which guards against external hormonal influences (from siblings *in utero*[3]; from the mother[4]) and at the same time facilitates treatment with exogenous hormones. Being precocial, young quail can be raised in isolation or at least in the absence of parental influence; parents could treat male and female (or hormone-treated and control) young differently and hence can be a source of confounding factors in an attempt to separate hormonal versus experiential effects (for a treatment of this problem in the rat, see the paper by C. L. Moore in this volume and Reference 5). Birds also show an extensive behavioral dimorphism, usually more pronounced than that in mammals. Finally, the lack of external genitalia in quail helps to avoid confusion between organizing effects of hormones on the brain and on peripheral tissues.

SEXUAL DIMORPHISM IN QUAIL

Sexual behavior is strongly differentiated in quail. The copulatory sequence (grab the neck feathers, mount, and cloacal contact movement) is seen only in males and never occurs in intact females.[6,7] This behavior pattern is thus homotypical-male in the sense of Cheng and Lehrman.[8] Squatting, which reflects sexual receptivity, is seen

---

[a] The research presented in this paper was supported by a Grant (2.4518.80) from the Fonds de la Recherche Fondamentale Collective (F.R.F.C.) to Professor E. Schoffeniels and Grants from the Belgian Fonds National de la Recherche Scientifique (F.N.R.S.) (credits aux chercheurs) to J. Balthazart. M. Schumacher is supported by a Grant from the F.N.R.S.

[b] Author to whom all correspondence should be addressed: J. Balthazart, Laboratoire de Biochimie Générale et Compareé, Université de Liège, 17 place Delcour, B-4020 Liège, Belgium.

mainly in females but is also displayed occasionally by submissive males[6] (Y. Delville and J. Balthazart, unpublished observations). Other behavior patterns are seen in both males and females (they are isomorphic; e.g., aggressive pecks) although with unequal frequencies (quantitative but not qualitative differences). Similarly, a number of morphological features are present in both sexes but show a quantitative dimorphism: males have a larger cloacal gland and larger sternotracheal (syringeal) muscles than females.[7]

These sex-related differences are not merely due to the different hormonal milieu to which each sex is exposed (i.e., ovarian versus testicular secretions). They are still observed in birds with regressed gonads (following exposure to short days) which are treated with adequate exogenous hormones: estradiol benzoate (EB) induces sexual receptivity in both males and females but testosterone propionate (TP) activates copulation in males only.[6,9,10] Residual secretions from the photoregressed ovary do not explain the differential responsiveness to TP: gonadectomized females still do not show male behavior in response to testosterone (T) treatment.[7,11] T also stimulates a larger growth of the cloacal gland and of the syringeal muscles in males than in females after gonadectomy or photoperiodic castration.[6,7,9,12]

In quail it is thus mainly the male homotypical behavior which is sexually differentiated. By contrast, in rats and rodents in general, the sexual dimorphism in the responsiveness to exogenous steroids essentially concerns the female behavior, that is, the induction of lordosis by sequential treatment with EB and progesterone (P). In rodents, females are bisexual and show masculine and feminine behavior in response to adequate hormone treatments (TP and EB+P, respectively) while it is extremely difficult to induce males to show lordosis. A quantitative sex difference is nevertheless also observed in the masculine behavior of rodents (see References 1 and 13 for reviews).

In this paper, we shall summarize the available data on sexual differentiation in quail by focusing on two separate aspects of this process: its hormonal specificity and its time course. New data will be presented which show that female demasculinization is not completed during a short period of embryonic life as was thought previously but rather takes place during the embryonic life and the first few weeks after hatching. Directions for future work will also be discussed.

THE PROCESS OF SEXUAL DIFFERENTIATION

Before 1975, the ontogeny of these sexual differences in birds was poorly understood. A few experiments had shown that treatment of embryos with exogenous hormones suppresses mating behavior in the adult birds[14-20] (see Reference 21 for review). With one exception,[20] these studies gave no evidence that the behavioral changes reflect neural differentiation rather than alterations in gonadal maturation.

In the last 10 years, Adkins and collaborators have carried out a very nice group of studies which now permit us to construct a coherent picture of the process of sexual differentiation in quail.[2,10,21,22]

They proved that differentiation of male behavior in quail is essentially a neural process. In all experiments, birds were tested for behavior after functional castration (exposure to short days[23]) and activation of behavior by exogenous steroids. This showed that the absence of copulation in females was not related to inadequate activation but reflected the androgen insensitivity of the neural mechanisms involved in behavioral activation. By treating eggs with steroid hormones and various drugs, it was shown that behavioral differentiation in quail mainly consists of the demasculinization of females by embryonic ovarian estrogens, i.e., the loss of the capacity to show male behavior even after adequate stimulation by testosterone. As a consequence, embryonic hormones were critical for the differentiation of females only and males could be considered as a "neutral" sex whose phenotype develops in the absence of hormonal control (the "anhormonal sex"[2]). These propositions were derived from the observation that EB-treated male embryos do not show male-typical sex behavior even if injected with TP in adulthood.[6,10] They are demasculinized by the exogenous hormone. It was then reasoned that this demasculinization was spontaneously taking place in untreated female embryos under the influence of ovarian estrogens. This was later supported by the demonstration that injection of an antiestrogen (CI-628) into eggs containing female embryos prevented their demasculinization.[24] This experiment also supported the concept of the male as a neutral sex; antiestrogens had no effect on the behavior of males.

DEMASCULINIZATION AND HORMONAL SPECIFICITY

An unexpected finding in these studies on bird sexual differentiation was the apparent interchangeability of exogenous testosterone and estradiol as agents inducing demasculinization of males.[6] This raised two critical questions:
(1) What is the specificity of this response and what is the hormone normally active in intact females?
(2) Why are males not demasculinized by their testicular secretions?

To answer the first question, Whitsett et al.[25] tested five different synthetic or natural estrogens (estradiol-17β, estradiol-17α, estrone, estriol, diethylstilbestrol) which were all capable of demasculinizing male quail at the dose of 2 μg/egg. By contrast, T at the same dose had no significant effect on male behavior. A detailed dose-response study later confirmed that demasculinization of male quail requires much higher doses of androgens compared to estrogens: threshold doses for statistically significant effects were approximately 500 μg for TP and 1 μg for EB.[10] These experiments provided a possible explanation as to why males were not demasculinized by their endogenous steroids: presumably circulating levels of testosterone were below threshold for demasculinization.

Studies on plasma levels of steroids in embryonic birds and on brain metabolism of steroids have confirmed this interpretation. Studies on the production of steroids by quail embryonic gonads cultured *in vitro* have demonstrated that testes do not produce more testosterone than ovaries while estrogen production (estradiol and estrone) is strongly dimorphic with much higher (5- to 10-fold) secretion in females.[26,27] Although plasma levels of androgens and estrogens in both male and female embryonic quail are still unknown (T only has been measured in males[28]), in the closely related chicken, sex differences in plasma testosterone (males > females[29]; this sex difference was not found by Gasc and Thibier[30]), estradiol (females > males[31]), and estrone (females > males[32]) have been demonstrated which support the proposed model of differentiation (see also Reference 33 for plasma levels of steroids in embryonic chickens).

Furthermore, in an extensive series of experiments, Adkins-Regan et al.[34] showed that aromatization mediates the testosterone-induced demasculinization. This conclusion is supported by the observations that aromatizable androgens (T, TP, and androstenedione) demasculinized copulation while nonaromatizable androgens (dihydrotestosterone propionate and androsterone) were inactive on this process, that the antiestrogen, tamoxifen, blocks the TP-induced demasculinization, and that the aromatization inhibitor, androstatrienedione (ATD), blocks the demasculinization induced by TP but not that induced by EB. It has also been established that brains of male and female chickens aromatize androstenedione[35] (nothing is known about embryonic brains in quail specifically) and the embryonic chicken brain contains estrogen receptors (autoradiographic studies).[36]

Taken together, these data satisfactorily explain why both TP and EB demasculinize behavior of quail embryos. Furthermore, in view of the very low enzymatic activity which is typical for the aromatase, they justify why high doses of T are necessary to demasculinize male embryos (only a small proportion of the molecules is transformed into active estrogens).

Another explanation for this requirement of unusually high doses of TP can be found in the brain metabolism of the hormone. In the bird brain, a very active 5β-reductase transforms T into 5β-dihydrotestosterone (5β-DHT) and the corresponding diols which are almost completely devoid of androgenic properties.[37-41] Only activational effects of 5β-androstanes have been tested and preliminary data suggest that they do not affect differentiation of sexual behavior; up to 2 mg of 5β-DHT injected on day 9 of incubation does not reduce male copulatory behavior (J. Balthazart, unpublished results). We showed recently that this enzymatic activity is extremely high (at least 10 times adult values) in the hypothalamus of male and female quail throughout the incubation period and only declines progressively during the first weeks after hatching to reach adult values at the time of sexual maturation, i.e., around 4-6 weeks of age.[42,43] Considering that 5β-reductase transforms T into behaviorally inactive compounds which are in addition not aromatizable, it can be stated that this enzymatic activity protects embryonic males from being demasculinized by their endogenous T through its aromatization. This also probably explains why such high doses of TP have to be injected into the eggs to demasculinize male embryos.

THE CRITICAL PERIOD

Organizing effects of steroids are characterized by periods of maximal sensitivity, the so-called critical periods. In an attempt to determine the end of the critical period for male demasculinization, Adkins[10] injected male embryos with 50 µg EB on day 10, 11, 12, 13, or 14 of incubation or with an oil vehicle on day 10 as the control. Birds that had been treated before day 12 were significantly demasculinized, a marginal nonsignificant decrease in copulation was still observed in day 12 birds, and no effect at all was observed in birds injected on day 13 or 14. We have recently confirmed this result in our strain of quail using two different doses of EB (5 and 25 µg) which were injected either on day 9 (active) or on day 14 (inactive) of incubation (M. Schumacher and J. Balthazart, unpublished data; see FIGURE 1). In this experiment, all birds were in addition gonadectomized 4 days after hatching to eliminate the confounding effects of postnatal steroids (see below). On the basis of this result (males cannot be demasculinized by exogenous EB after day 12), it was concluded that in quail the critical period for sexual differentiation ends on day 12 of incubation.[2,10] This implies that the EB-induced demasculinization in males is also a reliable model for the demasculinization process by ovarian secretions in females. That this is not necessarily so was first suggested by a study of Hutchison[44] who showed that if female quail were ovariectomized in the first 12 hr after hatching, they were not completely demasculinized and a large proportion showed male copulatory behavior when treated with testosterone. This experiment, however, did not include females castrated as

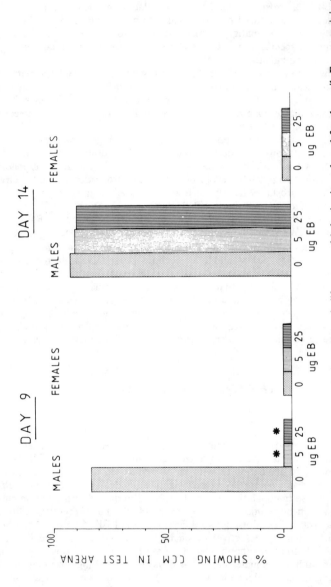

FIGURE 1. Effects of treatment with estradiol benzoate (EB) during embryonic life on sexual behavior in male and female quail. Eggs were injected with 0, 5, or 25 μg EB on day 9 or 14 of incubation; the young were gonadectomized 4 days after hatching and when adult received 60-mm Silastic implants of testosterone to activate behavior. CCM = cloacal contact movement. * = $2p < 0.05$, comparison with corresponding control (0 μg) by Fisher tests.

adults and receiving the same testosterone treatment as a control so that it was not possible to ascertain whether the presence of male behavior in these neonatally gonadectomized females was due to test conditions or strain differences. The following experiments rejected the notion that sexual differentiation ends on day 12 of incubation.

In a first pilot study we confirmed the presence of copulatory behavior in females gonadectomized after hatching (1 week after in this case).[45] This again suggested that female demasculinization was not completed at hatching. In the next experiment,[12] we gonadectomized male and female quail at different times after hatching (from 1 day to 6 weeks) to examine the time course of this postnatal demasculinization (see FIGURE 2). Homotypical male behavior was consistently seen, following stimulation by T in adulthood, in the females which had been gonadectomized within 2 weeks post-hatch. It almost never occurred in females ovariectomized at 4-6 weeks after

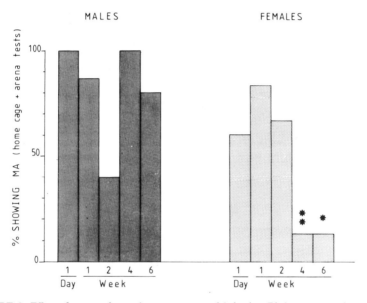

FIGURE 2. Effect of postnatal gonadectomy on sexual behavior. Birds were gonadectomized at different ages ranging from 1 day to 6 weeks post-hatch. When adult, sexual behavior was activated by 60-mm Silastic implants filled with T. Tests of sexual behavior were then performed both in the home cages and in a test arena. Pooled results are shown. MA = mount attempt. * = $p < 0.05$ and ** = $p < 0.01$, comparison with males of the same age by Fisher tests. (Modified from Schumacher and Balthazart.[12,46])

hatching. In contrast, copulatory behavior was not affected by the age of castration in males and most birds in all groups showed vigorous male behavior. The lack of copulatory behavior in females castrated after 4 weeks of age could still have been attributed to the stress of the recent surgical operation. Another experiment was thus performed to test this hypothesis.[46] Males and females were gonadectomized at 4 days or 4 weeks of age but this time, all birds were operated at both ages (sham or real gonadectomy). This confirmed that females gonadectomized early in life are not completely demasculinized but that demasculinization is nearly completed at the age of 4 weeks. Once again, the age of castration had no effect on the behavior of males (see FIGURE 3).

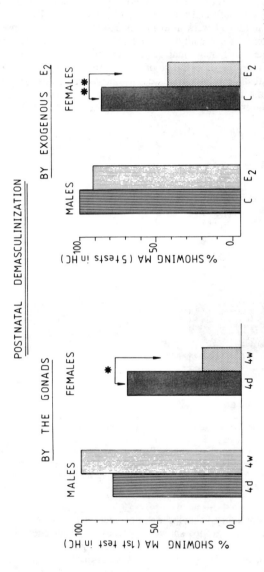

FIGURE 3. Effects of steroids after hatching on female demasculinization. Left: birds were gonadectomized at 4 days (4d) or 4 weeks (4w) post-hatch and when adult received a 60-mm T implant to activate behavior. Percentage of birds showing mount attempt (MA) during the first test performed in the home cages (HC) is shown. Right: all birds were gonadectomized at 4 days post-hatch and implanted with Silastic capsules filled with E2 or left empty (C). Implants were removed after 1 month and behavior was activated by 60-mm T implants. Behavior (mount attempt, MA) was tested 5 times in the home cages (HC). * = $2p < 0.05$ and ** = $2p < 0.01$ by Fisher tests. (Modified from Schumacher and Balthazart.[46,47])

This suggests that postnatal ovarian secretions are responsible for the complete demasculinization of female quail. We could then show that exogenous estradiol (E2) can mimic this effect of the ovary.[47] Male and female quail were gonadectomized at 4 days of age and immediately received E2-filled Silastic implants or empty implants as controls. These implants were removed at the age of 1 month and tests for copulatory behavior were then conducted with activation by exogenous T. Postnatal E2 treatment significantly reduced the sexual behavior in neonatally gonadectomized females but was without effect in males (see FIGURE 3, right). The postnatal sensitivity to the organizing action of E2 is thus sexually differentiated before hatching as females but not males (in agreement with previous data: see Reference 10 and FIGURE 1 in this paper) are demasculinized by exogenous E2.

A final series of experiments also established that while low doses of E2 can demasculinize females during their first month of life, even high amounts of hormone are without effect in adult birds (males and females). This suggested that the estrogen target tissues do not maintain their sensitivity to the organizing effect of E2 throughout life.

Although neonatally ovariectomized females consistently show homotypical male behavior if treated with T, it must be clearly stated that they are already partly demasculinized. This is mainly supported by the following facts: (1) their sexual behavior is less vigorous than that of males: they rarely show cloacal contact movements (full copulation) and their behavior is usually limited to neck grabs and mount attempts (see FIGURE 4); and (2) they show male behavior under optimal conditions: their sexual activity is, for example, very weak in the test arena that we normally use to test the behavior of males, but it becomes comparable to that of males if tests are performed in the home cages. The latter observations extend to birds a question recently raised by Feder[1] in the discussion of differentiation in rodents. Do steroids really organize the behavioral capacity of the animals or do they determine under what social or environmental conditions the behavior will be displayed (see also References 48-50)?

SEXUAL DIFFERENTIATION AS A BIPHASIC PROCESS

In quail, the information about the sensitive period during which E2 organizes male sexual behavior can be summarized as follows:

(1) Exogenous E2 demasculinizes male embryos up to day 12 of incubation and is not effective thereafter.

(2) Females gonadectomized soon after hatching retain partly the capacity to show male behavior.

(3) Demasculinization of females is completed during the first month of life under the influence of ovarian secretions. The sensitivity to the organizing effect of E2 is lost in adult females.

(4) Exogenous E2 demasculinizes neonatally ovariectomized females but is without effect in neonatally castrated males.

One set of data thus points to a brief sensitive period ending on day 12 of incubation and another set of data to a continuous process of female demasculinization which is only completed between 2 and 4 weeks post-hatch. To reconcile these data, we propose that sexual differentiation is a biphasic process. A limited "period of sensitization" corresponding to the classical critical period (ending on incubation day 12) would be followed by a continuous "period of organization" which would end only when birds approach puberty (see FIGURE 5). A short hormonal stimulus during the period of sensitization would change irreversibly the sensitivity of nervous tissues to the organizing effects of gonadal steroids. Once the sensitization effect takes place, the process of differentiation continues for an additional period of time, as in female quail. This model takes into account all the quail data available to date and also explains a number of paradoxical findings related to rodent sexual differentiation.

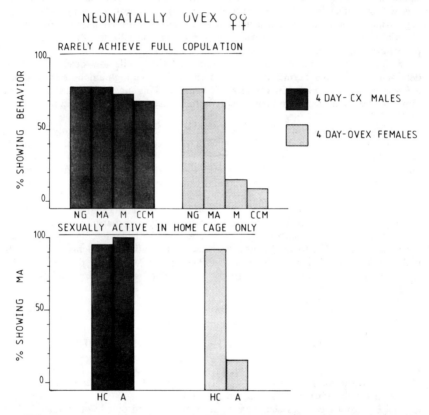

FIGURE 4. Neonatally ovariectomized (OVEX) females are already partly demasculinized. Top: their sexual behavior rarely includes mount (M) and cloacal contact movement (CCM) and is mainly limited to neck grab (NG) and mount attempt (MA). Results are relative to the first test performed in the home cages for a group of neonatally gonadectomized birds whose behavior was activated by 60-mm T implants. Bottom: they are active mainly in their home cage (HC) and show little activity in the test arena (A). Data concern the same birds as above tested six times in the home cages and three times in the arena.

In rats, sexual differentiation results from the exposure of males to T during the perinatal period (from a few days before birth to 5 days after). The lordosis response is lost (defeminization) and male behavior is reinforced (masculinization) if young rats are exposed to T during that period.[1] This model based on castration and replacement therapy experiments is, however, poorly supported by radioimmunoassays of plasma levels of T in young males and females.[51] Male rats have higher androgen

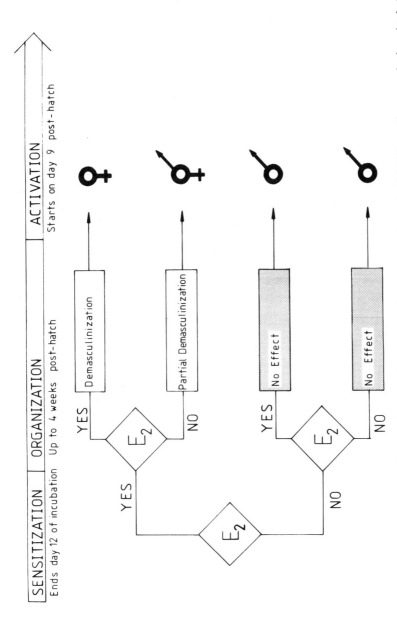

FIGURE 5. Model of the two-step sexual differentiation in quail. To obtain full demasculinization, estradiol (E2) must be present in the animals during a period preceding day 12 of incubation (period of sensitization) and during the post-hatching life (period of organization). If E2 is not present during the sensitization, it is no longer active later in life.

levels than females only during a short period of development (around day 18 postconception) and after that androgen levels almost completely overlap in both sexes (even though mean levels are always a little higher in males). The sexual differentiation thus cannot result from a prolonged exposure to a different hormonal milieu which suggested to Weisz and Ward[51] that differentiation of sexual behavior involves a sequence of two androgen-sensitive phases. High plasma testosterone on days 18-19 postconception sensitizes the fetal brain of males to the organizing action of androgens circulating in lower concentration at later stages of development. These data represent circumstantial correlative evidence suggesting that the model proposed here for quail development could have a more general applicability.

Experimental data also support this notion. Female rats show a cyclic pattern of luteinizing hormone (LH) secretion and in addition adequate treatment with E2 and progesterone (P) stimulates a dramatic LH surge. Males by contrast show a more tonic LH release and do not show a clear surge of LH in response to E2 and P.[52,53] This dimorphism is the result of androgen action on the brain during the first 5 days after birth. Androgen injections after the age of 6 days do not markedly alter the pattern of LH secretion in females. If females are lightly androgenized (with subthreshold doses of TP) on day 5 of life, they will display normal estrous cycles in adulthood (LH secretion is not defeminized). However, after a few cycles, they will lose this capacity and become anovulatory.[54-56] This delayed anovulatory syndrome (DAS) results from the exposure to postpubertal ovarian steroids as shown by ovariectomy experiments and is not observed if the females are not androgenized postnatally.[57] These experiments demonstrate that brief hormonal stimulation during the neonatal period in rats can alter for an extended period the sensitivity to steroids.[58] These data in the rat provide an example of the differentiation model that we propose here for quail. Future research should demonstrate if such a biphasic mechanism is a widespread phenomenon in vertebrate differentiation and what is its physiological basis. On the basis of current data, several possibilities can be considered:

(1) Steroid action during the sensitization phase causes survival of specific neurons which will differentiate during the organization period.[59]

(2) The sensitizing action of steroids induces in target cells steroid receptors or steroid metabolizing enzymes (e.g., aromatase) which may be required for the subsequent organizing action of steroids.[51]

(3) During the sensitization phase, E2 alters in a permanent way the conformation of specific genes or produces long-term accumulation of specific mRNA which would be translated later during the second exposure to the organizing hormone. Interestingly, it has already been suggested that the estrogen-induced vitellogenin synthesis in *Xenopus laevis* and in hens may involve such mechanisms: a primary hormonal stimulation enhances the effectiveness of a secondary exposure to estrogens.[60-62]

Experiments are now in progress with quail to collect additional data supporting this two-step differentiation model and to elucidate its physiological basis.

ISSUES AND DIRECTIONS FOR FUTURE WORK

Do Periods for Activation and Organization of Behavior Partly Overlap in Quail?

The above data clearly show that female demasculinization is not completed before 2 weeks post-hatch and that males cannot be demasculinized by exogenous E2 after day 12 of incubation. It is then interesting to note that male-type sexual behavior can be activated in young castrated males with 20-mm Silastic T implants during the first 2 weeks post-hatch (first positive tests on day 9 with implants placed on day 7 but no earlier age was tested; M. Schumacher and J. Balthazart, unpublished data). In this experiment, no sexual behavior was observed in T-treated females which supports the notion that they are partly demasculinized at hatching and thus probably have a higher threshold for behavior activation.

Separate experiments have, however, demonstrated that T or DHT Silastic implants can stimulate crowing in both male and female quail chicks (first 3 weeks of life) with equal efficiency (same frequencies and latencies of occurrence were noted and dose-response curves were similar in both sexes[63]) and it has been clearly established that the hormonal induction of crowing in adult quail is sexually dimorphic.[6,7,11] Thus in quail, organization and activation of male sexual behavior partly overlap in time as previously shown for female sexual behavior in the rat[64,65] (see also the paper by C. L. Williams in this volume). The concept of independent periods for these two processes should thus be reconsidered and experimental work should be undertaken to understand their precise specificity (e.g., do the same hormonal stimuli in young animals activate and at the same time differentiate the behavior?).

Do We Have an Adequate Terminology for Differentiation?

Comparison of quail and rat studies on differentiation also suggests a few reflections about the taxonomy of differentiation processes and the use of this terminology in the literature. Four independent processes can be distinguished on the basis of the presence-absence of the considered behavior in the "anhormonal" condition and of the differentiating effects of steroids. Homotypical male behavior can be present in the anhormonal condition and lost after exposure to steroids, a process called demasculinization, which apparently explains differentiation of copulatory behavior in quail. Alternatively, male behavior can be acquired (or increased) under the influence of steroids (it is absent in anhormonal conditions). Examples of this masculinization process can be found in the differentiation of copulation in rodents[66] and in the acquisition of song by zebra finches.[2,67,68]

Mirror images of these processes can be defined for differentiation of homotypical female behavior patterns. The lordosis response of rodents is differentiated by a defeminization of the males; lordosis is present in the anhormonal state and lost following exposure of males to testosterone. It is more difficult to find examples of the feminization process, that is, the acquisition following exposure to steroids of a female behavior pattern which is absent in the anhormonal condition (see, however, Reference 69 for a possible feminization of the lordosis response in female rats during sexual maturation).

On the basis of these logical definitions, several remarks can be made. First, confusion exists in the common use of terms. The loss of male behavior in male quail following TP or EB treatment is frequently referred to as feminization, and similarly the maintenance of male behavior in females treated in the egg with antiestrogens[24] has been called masculinization when it should be absence of the demasculinization. A similar confusion exists in the literature concerning rodents. This sloppiness in terminology not only complicates communication between scientists but it also conceals important theoretical problems.

Are We Asking All Pertinent Questions?

A complete description of sexual differentiation in a single species should include the study of the organization of both male and female homotypical behaviors, the identification of their status in anhormonal conditions, and description of the effects of steroids at different times in the ontogeny. This task has not yet been achieved in any species.

In quail, the best-known avian model, the only differentiation process which has received detailed attention is demasculinization. Female behavior is reputed to be isomorphic (it can be elicited in both males and females with EB injection) and no study has thus been made on the role of steroids on its organization. This is unfortunate for two reasons. First, it is not certain that because the behavior is present in both sexes, it does not require at some stages of development the organizing action of steroid hormones. Second, it is quite possible that female behavior, although present in both sexes, is quantitatively differentiated. All activation experiments performed so far used large doses of EB (50 μg; see References 6 and 9) which probably produce in males supraphysiological plasma levels of estrogens (in females 50 μg EB raises plasma E2 to 1215 pg/ml while levels in intact birds are around 100-200 pg/ml; see References 2 and 70). Sexual differences in the threshold for activation of squatting are thus quite possible. On the other hand, male behavior is supposed to be present in the anhormonal state and is differentiated only by demasculinization of the females. What is really the evidence for the absence of a hormonal requirement in the organization of homotypical male behavior? It is actually very limited. We know that the behavior is normally organized in birds castrated just after hatching (there is thus no postnatal masculinization[46]) and in embryos treated on day 9 of incubation with the antiestrogen, CI-628,[24] which presumably means that the presence of estrogens is not required for its organization. No experimental data can, however, be found to reject the idea that male behavior requires for its normal development the presence during the ontogeny of physiological levels of testosterone (it would be masculinized).

The same argument can be developed for the organization of lordosis in rodents. The lordosis response is supposedly present in the anhormonal state and is differentiated by a specific loss of the behavior (defeminization) in males following exposure to testosterone (acting via neural aromatization[66]). Once again what is the evidence that no hormonal stimulus (estrogens) is required for the normal development of the homotypical behavior? Some experiments suggest that low levels of androgens around the time of birth potentiate feminine as well as masculine behavior in female rodents (see references in MacLusky and Naftolin[66]). Many studies have demonstrated the morphogenetic effect of E2 on brain structures (organization of the sexually dimorphic nucleus of the preoptic area[71]; growth of dendritic processes[72]). The importance of estradiol to neural development clearly challenges the notion that the female neural and behavioral phenotype merely results from the absence of androgen exposure.[73,74] The effects of steroids on *in vitro* differentiation of neuronal cells clearly suggest that no pattern of differentiation is intrinsic to the nervous tissue, but rather that both male and female patterns require organization by steroids.

Should We Use the Concept of Anhormonal Development?

At the behavioral level, evidence for the anhormonal development of a given behavior is always limited with respect to the hormone concerned (there is no need for T in the normal development of lordosis in rats, no need for E2 in the development of copulation in quail) and with respect to the period of development studied (testicular secretions play no role in the postnatal differentiation of copulation in quails but what about before hatching; is the organization of lordosis really independent of estrogens during the embryonic stages[75] or even during juvenile life[69]?). The absence of a hormonal requirement is probably impossible to demonstrate because an embryo without hormone is impossible to obtain and would not be viable. The notion of anhormonal or neutral development should thus be used cautiously by making reference to the hormone and the period of development concerned (i.e., behavior develops normally without this hormone during that period). At the same time, experimental work should refine our knowledge on hormonal requirements in the behavioral organization in order to go beyond the simplistic view that sexual differentiation mainly proceeds by loss in one sex of behavioral characteristics which develop in the other sex without any hormonal control.

ACKNOWLEDGMENT

We are indebted to Professor E. Schoffeniels for his continued interest in our work.

REFERENCES

1. FEDER, H. H. 1984. Hormones and sexual behavior. Annu. Rev. Psychol. **35:** 165-200.
2. ADKINS-REGAN, E. 1983. Sex steroids and the differentiation and activation of avian reproductive behaviour. *In* Hormones and Behaviour in Higher Vertebrates. J. Balthazart, E. Prove & R. Gilles, Eds.: 218-228. Springer-Verlag. West Berlin, FRG.
3. VOM SAAL, F. S. 1983. The interaction of circulating oestrogens and androgens in regulating mammalian sexual differentiation. *In* Hormones and Behaviour in Higher Vertebrates. J. Balthazart, E. Prove & R. Gilles, Eds.: 159-177. Springer-Verlag. West Berlin, FRG.
4. WARD, I. L. & J. WEISZ. 1980. Maternal stress alters plasma testosterone in fetal males. Science **207:** 328-329.
5. MOORE, C. L. Developmental elaborations of hormonally based sex differences. Abstract presented at the 15th Conference on Reproductive Biology, Medford, Mass., 1983.
6. ADKINS, E. 1975. Hormonal control of sexual differentiation in the Japanese quail. J. Comp. Physiol. Psychol. **89:** 61-71.
7. BALTHAZART, J., M. SCHUMACHER & M. A. OTTINGER. 1983. Sexual differences in the Japanese quail: behavior, morphology, and intracellular metabolism of testosterone. Gen. Comp. Endocrinol. **51:** 191-207.
8. CHENG, M. F. & D. LEHRMAN. 1975. Gonadal hormone specificity in the sexual behavior of ring doves. Psychoneuroendocrinology **1:** 95-102.
9. ADKINS, E. K. & N. T. ADLER. 1972. Hormonal control of behavior in the Japanese quail. J. Comp. Physiol. Psychol. **81:** 27-36.
10. ADKINS, E. K. 1979. Effect of embryonic treatment with estradiol or testosterone on sexual differentiation of the quail brain. Neuroendocrinology **29:** 178-185.
11. SCHUMACHER, M. & J. BALTHAZART. 1983. The effects of testosterone and its metabolites on sexual behavior and morphology in male and female Japanese quail. Physiol. Behav. **30:** 335-339.
12. SCHUMACHER, M. & J. BALTHAZART. 1983. Effects of castration on postnatal differentiation in the Japanese quail (*Coturnix coturnix japonica*). IRCS Med. Sci. **11:** 102-103.

13. FEDER, H. H. 1981. Perinatal hormones and their role in the development of sexually dimorphic behaviors. *In* Neuroendocrinology of Reproduction Physiology and Behavior. N. T. Adler, Ed.: 127-157. Plenum Press. New York, N.Y.
14. DOMM, L. V. 1939. Intersexuality in adult brown leghorn males as a result of estrogenic treatment during early embryonic life. Proc. Soc. Exp. Biol. Med. **42:** 310-312.
15. DOMM, L. V. & D. E. DAVIS. 1948. The sexual behavior of intersexual domestic fowl. Physiol. Zool. **21:** 14-31.
16. KAUFMAN, L. 1956. Experiments on sex modification in cocks during their embryonal development. World's Poult. Sci. J. **12:** 41-42.
17. GLICK, B. 1961. The reproductive performance of birds hatched from eggs dipped in male hormone solutions. Poult. Sci. **40:** 1408. (Abstract.)
18. GLICK, B. 1965. Embryonic exposure to testosterone propionate will adversely influence future mating behavior in male chickens. Fed. Proc. **24:** 700. (Abstract.)
19. WENTWORTH, B. C., B. G. HENDRICKS & J. STURTEVANT. 1968. Sterility induced in Japanese quail by spray treatment of eggs with mestranol. J. Wildl. Manage. **32:** 879-887.
20. WILSON, J. A. & B. GLICK. 1970. Ontogeny of mating behavior in the chicken. Am. J. Physiol. **218:** 951-955.
21. ADKINS-REGAN, E. K. 1981. Early organizational effects of hormones. An evolutionary perspective. *In* Neuroendocrinology of Reproduction Physiology and Behavior. N. T. Adler, Ed.: 159-228. Plenum Press. New York, N.Y.
22. ADKINS, E. K. 1978. Sex steroids and the differentiation of avian reproductive behavior. Am. Zool. **18:** 501-509.
23. ADKINS, E. K. 1973. Functional castration of the female Japanese quail. Physiol. Behav. **10:** 619-621.
24. ADKINS, E. K. 1976. Embryonic exposure to an antiestrogen masculinizes behavior of female quail. Physiol. Behav. **17:** 357-359.
25. WHITSETT, J. M., E. W. IRVIN, F. W. EDENS & J. P. THAXTON. 1977. Demasculinization of male Japanese quail by prenatal estrogen treatment. Horm. Behav. **8:** 254-263.
26. GUICHARD, A., D. SCHEIB, K. HAFFEN, TH. M. MIGNOT & L. CEDARD. 1980. Comparative study in steroidogenesis by quail and chick embryonic gonads in organ culture. J. Steroid Biochem. **12:** 83-87.
27. SCHEIB, D., A. GUICHARD, TH. M. MIGNOT & L. CEDARD. 1981. Steroidogenesis by gonads of normal and of diethylstilbestrol-treated quail embryos: radioimmunoassays on organ cultures. Gen. Comp. Endocrinol. **43:** 519-526.
28. OTTINGER, M. A. & M. BAKST. 1981. Peripheral androgen concentrations and testicular morphology in embryonic and young male Japanese quail. Gen. Comp. Endocrinol. **43:** 170-177.
29. WOODS, J. E., R. M. SIMPSON & P. L. MOORE. 1975. Plasma testosterone levels in the chick embryo. Gen. Comp. Endocrinol. **27:** 543-547.
30. GASC, J. M. & M. THIBIER. 1979. Plasma testosterone concentration in control and testosterone-treated chick embryos. Experientia **35:** 1411-1412.
31. WOODS, J. E. & D. M. BRAZZIL. 1981. Plasma 17β-estradiol levels in the chick embryo. Gen. Comp. Endocrinol. **44:** 37-43.
32. WOODS, J. E., D. D. CONGORAN & R. C. THOMES. 1982. Plasma estrone levels in the chick embryo. Poult. Sci. **61:** 1729-1733.
33. TANABE, Y., T. NAKAMURA, K. FUJIOKA & O. DOI. 1979. Production and secretion of sex steroid hormones by the testes, the ovary and the adrenal glands of embryonic and young chicken (*Gallus domesticus*). Gen. Comp. Endocrinol. **39:** 26-33.
34. ADKINS-REGAN, E. K., P. PICKETT & D. KOUTNIK. 1982. Sexual differentiation in quail: conversion of androgen to estrogen mediates testosterone-induced demasculinization of copulation but not other male characteristics. Horm. Behav. **16:** 259-278.
35. CALLARD, G. V., Z. PETRO & K. J. RYAN. 1978. Conversion of androgen to estrogen and other steroids in the vertebrate brain. Am. Zool. **18:** 511-523.
36. MARTINEZ-VARGAS, M. C., D. B. GIBSON, M. SAR & W. E. STUMPF. 1975. Estrogen target sites in the brain of the chick embryo. Science **190:** 1307-1308.
37. MASSA, R., L. BOTTONI & V. LUCINI. 1983. Brain testosterone metabolism and sexual behaviour in birds. *In* Hormones and Behaviour in Higher Vertebrates. J. Balthazart, E. Prove & R. Gilles, Eds.: 230-236. Springer-Verlag. West Berlin, FRG.

38. STEIMER, TH. & J. B. HUTCHISON. 1981. Metabolic control of the behavioural action of androgens in the dove brain: testosterone inactivation by 5β-reduction. Brain Res. **107**: 9-16.
39. DEVICHE, P., L. BOTTONI & J. BALTHAZART. 1982. 5β-Dihydrotestosterone is weakly androgenic in the adult Japanese quail (*Coturnix coturnix japonica*). Gen. Comp. Endocrinol. **48**: 421-424.
40. BALTHAZART, J. & M. SCHUMACHER. 1983. Testosterone metabolism and sexual differentiation in quail. *In* Hormones and Behaviour in Higher Vertebrates. J. Balthazart, E. Prove & R. Gilles, Eds.: 237-260. Springer-Verlag. West Berlin, FRG.
41. BALTHAZART, J. 1983. Hormonal correlates of behavior. *In* Avian Biology. D. S. Farner, J. R. King & K. C. Parkes, Eds. **7**: 221-365. Academic Press. New York, N.Y.
42. BALTHAZART, J. & M. SCHUMACHER. 1984. Changes in testosterone metabolism by the brain and cloacal gland during sexual maturation in the Japanese quail (*Coturnix coturnix japonica*). J. Endocrinol. **100**: 13-18.
43. BALTHAZART, J. & M. A. OTTINGER. 1984. 5β-Reductase activity in the brain and cloacal gland of male and female embryos in the Japanese quail (*Coturnix coturnix japonica*). J. Endocrinol. **102**: 77-81.
44. HUTCHISON, R. E. 1978. Hormonal differentiation of sexual behavior in the Japanese quail. Horm. Behav. **11**: 363-387.
45. SCHUMACHER, M. & J. BALTHAZART. 1983. The postnatal differentiation of sexual behaviour in the Japanese quail (Coturnix coturnix japonica). Behav. Proc. **8**: 189-195.
46. SCHUMACHER, M. & J. BALTHAZART. 1984. The postnatal demasculinization of sexual behavior in the Japanese quail (*Coturnix coturnix japonica*). Horm. Behav. **18**: 298-312.
47. BALTHAZART, J. & M. SCHUMACHER. 1984. Estradiol contributes to the postnatal demasculinization of female Japanese quail (*Coturnix coturnix japonica*). Horm. Behav. **18**: 287-297.
48. THORNTON, J. E., J. V. CADWALLADER & R. W. GOY. An unusual organizational effect of prenatal testosterone on lordosis in guinea pigs. Abstract presented at the 14th Conference on Reproductive Behavior, East Lansing, Mich., 1982.
49. BROWN-GRANT, K. 1975. A re-examination of the lordosis response in female rats given high doses of testosterone propionate or estradiol benzoate in the neonatal period. Horm. Behav. **6**: 351-378.
50. CLEMENS, L. G., M. HIROI & R. A. GORSKI. 1969. Induction and facilitation of female mating behavior in rats treated neonatally with low doses of testosterone propionate. Endocrinology **84**: 1430-1438.
51. WEISZ, J. & I. L. WARD. 1980. Plasma testosterone and progesterone titers of pregnant rats, their male and female fetuses, and neonatal offspring. Endocrinology **106**: 306-316.
52. BROWN-GRANT, K. 1974. Steroid hormone administration and gonadotrophin secretion in the gonadectomized rat. J. Endocrinol. **62**: 319-332.
53. GORSKI, R. A. & J. W. WAGNER. 1965. Gonadal activity and sexual differentiation of the hypothalamus. Endocrinology **76**: 226-239.
54. PFEIFFER, C. A. 1936. Sexual differences of the hypophysis and their determination by the gonads. Am. J. Anat. **58**: 195-226.
55. GORSKI, R. A. 1968. Influence of age on the response to paranatal administration of a low dose of androgen. Endocrinology **82**: 1001-1004.
56. HARLAN, R. E. & R. A. GORSKI. 1977. Steroid regulation of luteinizing hormone secretion in normal and androgenized rats at different ages. Endocrinology **101**: 741-749.
57. HARLAN, R. E. & R. A. GORSKI. 1978. Effects of postpubertal ovarian steroids on reproductive function and sexual differentiation of lightly androgenized rats. Endocrinology **102**: 1716-1724.
58. HARLAN, R. E., J. H. GORDON & R. A. GORSKI. 1979. Sexual differentiation of the brain: implications for neuroscience. *In* Review of Neuroscience. D. M. Schneider, Ed. **4**: 31-71. Raven Press. New York, N.Y.
59. MCEWEN, B. S. 1982. Sexual differentiation of the brain: gonadal hormone action and the current concepts of neuronal differentiation. *In* Molecular Approaches to Neurobiology. I. R. Brown, Ed.: 195-219. Academic Press. New York, N.Y.
60. CHAN, L., A. R. MEANS & B. W. O'MALLEY. 1978. Steroid hormone regulation of specific gene expression. Vitam. Horm. **36**: 259-295.

61. Baker, H. J. & D. J. Shapiro. 1977. Kinetics of estrogen induction of *Xenopus laevis* vitellogenin messenger RNA as measured by hybridization to complementary DNA. J. Biol. Chem. **252:** 8428-8434.
62. Mullinx, K. P., W. Netekam, R. G. Deeley, J. I. Gordon, M. Meyers, K. A. Kent & R. F. Goldberger. 1976. Induction of vitellogenin synthesis by estrogen in the avian liver: relationship between level of vitellogenin mRNA and vitellogenin synthesis. Proc. Natl. Acad. Sci. USA **73:** 1442-1446.
63. Balthazart, J., M. Schumacher & G. Malacarne. 1984. Relative potencies of testosterone and 5α-dihydrotestosterone on crowing and cloacal gland growth in the Japanese quail (*Coturnix coturnix japonica*). J. Endocrinol. **100:** 19-23.
64. Williams, C. L. & A. Baron. The sexual enlightenment of the infant rats. Abstract presented at the 15th Conference on Reproductive Biology, Medford, Mass., 1983.
65. Williams, C. L. Steroids induce lordosis and ear wiggling in infant rats. Abstract presented at the 11th Conference on Reproductive Biology, New Orleans, La., 1979.
66. MacLusky, N. J. & F. Naftolin. 1981. Sexual differentiation of the central nervous system. Science **211:** 1294-1303.
67. Gurney, M. E. & M. Konishi. 1980. Hormone-induced sexual differentiation of the brain and behavior in zebra finches. Science **208:** 1380-1383.
68. Pohl-Apel, G. Induction of male sexual behavior in female zebra finches. Abstract presented at the 15th Conference on Reproductive Biology, Medford, Mass., 1983.
69. Södersten, P. 1976. Lordosis behaviour in male, female and androgenized female rats. J. Endocrinol. **70:** 409-420.
70. Doi, O., T. Takai, T. Nakamura & Y. Tanabe. 1980. Changes in the pituitary and plasma LH, plasma and follicular progesterone and estradiol, and plasma testosterone and estrone concentrations during the ovulatory cycle of the quail (*Coturnix coturnix japonica*). Gen. Comp. Endocrinol. **41:** 156-163.
71. Döhler, K. D., A. Coquelin, M. Hines, F. Davis, J. E. Shryne & R. A. Gorski. 1983. Hormonal influence on sexual differentiation of rat brain anatomy. *In* Hormones and Behaviour in Higher Vertebrates. J. Balthazart, E. Prove & R. Gilles, Eds.: 194-203. Springer-Verlag. West Berlin, FRG.
72. Toran-Allerand, C. D. 1980. Sex steroids and the development of the newborn mouse hypothalamus and preoptic area in vitro. II. Morphological correlates and hormonal specificity. Brain Res. **189:** 413-427.
73. Toran-Allerand, C. D. 1978. Gonadal hormones and brain development: cellular aspects of sexual differentiation. Am. Zool. **18:** 553-565.
74. Toran-Allerand, C. D. 1981. Gonadal steroids and brain development: *in vitro* veritas? Trends Neurosci. May 1981: 118-121.
75. Döhler, K. D., S. S. Sristava, J. E. Shryne, B. Jarzab, A. Sipos & R. A. Gorski. 1984. Differentiation of the sexually dimorphic nucleus in the preoptic area of the rat brain is inhibited by postnatal treatment with an estrogen antagonist. Neuroendocrinology **38:** 297-301.

Reciprocity in Neuroendocrine and Evolutionary Approaches to the Study of Reproductive Behavior

CAROL DIAKOW

Biology Department
Adelphi University
Garden City, New York 11530

Behavioral neuroendocrinologists must deal with a lot of complexity within the endocrine systems underlying reproductive behavior. Some reasons for this complexity are the existence of numerous hormones which affect reproductive behavior, hormonal interactions within individuals, and quantitative hormonal differences between individuals and species. The object of this paper is to point out that a full understanding of such physiological complexity can expand our appreciation of the evolutionary process, and, conversely, that an evolutionary perspective can provide an understanding of these complex physiological mechanisms.

Rosenblatt and Komisaruk[1] have written that "Daniel S. Lehrman's lifelong love and study of animal behavior gave us a wealth of new insights into reproductive behavior and evolution." In creating the Institute of Animal Behavior of Rutgers University, Dr. Lehrman provided an environment in which physiological and evolutionary perspectives on behavior could, to use his own words, "intersect" and "compliment" each other.

We behavioral neuroendocrinologists accept as given, and are now fairly adept at handling, complex mechanisms underlying behavior; this was not always so. In the 1960s our approach, a reasonable one given our knowledge of psychoneuroendocrinology at that time, was to look for a simple neural button that would generate mating postures if "pushed" and for the steroid hormone that "pushed" it. In fact, for rat lordotic behavior, it appeared likely that the neural button was in the hypothalamus and that the steroid hormone, estrogen, triggered it. Of course, many investigators, including Drs. Lehrman and Rosenblatt, recognized that hormone-behavior relationships were affected by social and other environmental factors, but within an organism, we looked for simple mechanisms. Nevertheless, a complicated picture quickly emerged even in the "simple system" underlying rat lordosis, and it became clear that it was incorrect to look for one hormone that serves as a "button pusher." Our work at the Institute[2] illustrated that lordosis is elicited by multisensory stimulation in the absence of a steroid hormone of ovarian or adrenal origin, that either estrogen or progesterone alone potentiates it, and that both hormones acting together have a synergistic effect. Since then, prostaglandins,[3,4] luteinizing hormone-releasing hormone,[5,6] and adrenocorticotrophic hormones[7] have also been implicated in the mechanism underlying lordosis.

Now that we recognize that many hormones may act to potentiate a single behavioral response, how do we deal with the multiplicity? We consider hormones one at a time and ask which ones act in the animal when it normally exhibits the behavior in question. We ask how these hormones interact. We give more depth to the analysis by analyzing molecular, cellular, and genetic mechanisms of the hormones. All of these approaches are illustrated by papers presented in this volume.

In this paper, I would like to emphasize the notion that we behavioral physiologists should welcome the complexity we reveal within the neuroendocrine mechanisms we study, welcome the multiple mechanisms and the inter- and intraspecies variation, because the more we understand these, the more we will be able to appreciate the relationships of physiology and evolution.

HOW UNDERSTANDING SIMILARITIES AND DIFFERENCES IN BEHAVIORAL PHYSIOLOGY CAN PROVIDE A TOOL FOR UNDERSTANDING ITS EVOLUTION

The comparative method can help us understand the evolution of the hormonal basis of reproductive behavior. Describing the evolution of behavior patterns is, of course, a classical use of the comparative approach. It has been used very successfully by behavioral evolutionists who have compared behaviors of closely related species to reveal the phylogeny of many behavior patterns. These include the evolution of the honeybee dance and fly courtship behavior.[8] For any comparison, similarities may represent evolution from a common ancestor or convergent evolution in accord with common selection pressures. Differences can relate to different environmental pressures.

The following are examples of successful application of the comparative method to understanding the evolution of behavioral physiology. The first is Scharrer's analysis,[9] which is at the phyletic level. In comparing invertebrates and vertebrates, she classified the hormones underlying behavior as first or second order. The first order consists of those hormones produced by the nervous system and which affect behavior (for example, luteinizing hormone-releasing hormone) and the second order consists of those that are produced by nonneural target organs (for example, estrogen). She proposed an evolutionary scheme within which the first-order systems evolved before the second, and while both exist in vertebrates and advanced invertebrates, the second order became more of an essential feature in vertebrates.

Another example of the successful use of comparative analysis of physiological mechanisms of behavior to reveal principles of evolution lies in the work of Adkins-Regan.[10] She derived principles for the hormonal basis of the developmental processes underlying reproductive behavior by comparing vertebrate classes. Her main conclusion, a "fundamental theme of vertebrate organization," is that the basic mechanism of behavioral sex differentiation involves control by embryonic gonadal secretions. This echoes, for behavior, the mechanism for differentiation of anatomical secondary sexual characteristics of vertebrates.

Crews (see Reference 11 and his paper in this volume) has also used the comparative method to propose a sequence for the evolution of the associations among gamete production, gonadal hormone secretion, and mating behavior. He has proposed that gonadal hormone action on the neural tissues which influence sex behavior evolved coincident with, or consequent to, the neural feedback action of gonadal hormones.

HOW AN EVOLUTIONARY PERSPECTIVE CAN HELP US UNDERSTAND INTERINDIVIDUAL VARIATION IN PHYSIOLOGICAL RELATIONSHIPS TO BEHAVIOR

Environments are dynamic; thus, there is a selective advantage to genetic and phenotypic variability. Mayr[12] has pointed out that the selective advantage of variability in genotype is so strong that there is a large number of genetic mechanisms to preserve it; these include recessiveness, modification of penetrance, heterosis, linkage, and epistasis. What does this view of evolution mean for those of us who analyze behavior at the physiological level? It means that there must be variability in the physiology of the individuals of any species we study.

Most behavioral neuroendocrinologists work conscientiously to develop consistent techniques of rearing and housing animals, drawing blood, measuring hormone levels, injecting, performing surgery, and observing animals. We do this to reduce variability due to confounding influences in our measurements of physiology and behavior. No matter what we do, however, we cannot eliminate variability per se. We call it experimental "error," and conceptualize it as occurring for peripheral and chance reasons. We capture it with statistics, and then ignore it in our analysis of "central tendencies." We should reorient ourselves to the notion that this variation is not "error"—this variation in phenotype represents variation in genotypes. This "error" represents the raw material of natural selection!

Some sources of evolutionarily relevant variation in behavior might be variation in hormone levels in individuals, in responsivity of individuals to hormones, or in fundamentally different neuroendocrine controlling mechanisms. This idea of different reproductive success of individuals with different neuroendocrine phenotypes leads to experimentally testable hypotheses about differential reproductive success of individuals with different hormone level profiles. There have been a few such studies. The first was that of Beach and Holz-Tucker[13] which related amounts of testosterone injected into male rats to sexual performance and later, Hardy and DeBold[14] studied lordosis. More recently, Harding and Feder[15] and Damassa et al.[16] measured endogenous levels of testosterone in relation to mating tendencies in individual rodents. Although they found no direct relationship, we can do more sophisticated analyses today to get at the general question of the relationship of phenotypic variation in physiology and reproductive success. This analysis might include measuring more than one hormone in the same blood sample, measuring the rate of steroid receptor induction, or analyzing postuptake cellular mechanisms of steroid action. If we were to use this approach to its fullest, we would be able to test hypotheses about hormone profiles or mechanisms that lead to maximum reproductive success.

OTHER SUGGESTIONS FOR VIEWING EVOLUTION OF THE ENDOCRINE BASIS OF BEHAVIOR

We should view the evolution of hormonal mechanisms underlying behavior from the point of view of the evolution of the endocrine target organ and not necessarily of the evolution of the hormones themselves.

Barrington[17] traced the evolution of the molecular structures of hormones and made it clear that these molecular structures and their fundamental relationships to molecular processes are evolutionarily conservative. Feder[18] also supported cross-species extrapolations of cellular mechanisms of hormone action on the brain. In Barrington's view, "flexibility in adaptation" is due to modification of the target cells. Based on this, we would expect to find similar hormones in different species, but their target organs will have different physiologies.

In 1958, Beach[19] recognized this and, citing Medawar,[20] wrote that "endocrine evolution is not an evolution of hormones but an evolution of the uses to which they are put." In a comparison of the endocrine bases of behavior in the different vertebrate classes, Beach contrasted the "apparent evolutionary consistency of testis function" and the tremendous behavioral variation in the effects of androgenic stimulation. He concluded that changes that gave rise to the differences in behavior that occurred during the evolution of vertebrates occurred in physiological systems other than the testis, and suggested the nervous system as the primary target of evolutionary change.

I would like to introduce the idea that a functional constellation of hormones is a target of natural selection. Mayr[12] reminds us that the "target of selection is always an entire individual, that is, an entire interacting system of genes" and that natural selection targets a genotype indirectly through the phenotype. So, natural selection will be acting on the functional constellation of endocrine mechanisms underlying the behavior of an individual. If an individual reproduces successfully, it will pass on the genetic material that allowed it to develop an endocrine apparatus leading to successful reproduction in the environment in which it developed and functioned. This view makes it understandable why the attempts (cited above) to correlate reproductive behavior and testosterone levels failed. It would probably be more fruitful to correlate with reproductive success a profile established by the simultaneous measurement of several hormones.

CONCLUSION

I began this paper by pointing out that there are a large number of interactions and complexities in the hormonal mechanisms of behavior and by showing that behavioral neuroendocrinologists study each element in turn, then look for the interactions among these elements. This paper proposes the idea that there is a reciprocity in understanding that can be achieved by relating evolutionary concepts and physiological complexity. First, these physiological complexities will help us understand evolution. Comparing similarities and differences in the physiological bases of behavior can provide an understanding of the path of evolution of the hormonal basis of behavior and of the relation of physiology to the environmental factors that influenced evolution. Second, understanding evolutionary processes will help us put physiological phenomena into perspective. Knowing that the adaptation of a species to a changing environment depends on the existence of variability in the genetic composition of the individuals in a population, we can accept individual variation as relevant, not as experimental error, and we can study individual variation in neuroendocrinology in terms of differential reproductive success.

I remember an extended discussion at an Eastern Conference on Reproductive Behavior concerning contradictory findings from two laboratories about where estrogen acted in the diencephalon to affect lordosis. The participants went on and on in discussing experimental differences in their procedures (estrogen doses? estrogen vehicle? route of administration?). Then someone pointed out that in one case, hamsters were the subjects, and in the other, rats. There was a second of silence while presumably the question of species differences was weighed, then conversation went back to techniques. That was the best we could do then; after all, we needed to reassure ourselves that differences really were not due to methodological errors before we could attribute them to species differences. Perhaps it is still too soon to ask the question of how the differences in the evolutionary history of two very different rodents can relate to differences in sites of action of estrogen. Nevertheless, there are ways to relate physiological and evolutionary perspectives even now. As Feder[18] has pointed out, "It is here that students of comparative neuroendocrinology can make their greatest impact on the further analysis of hormone-behavior interaction." We of the Institute, because of Dr. Lehrman's orientation, have associated with both physiological and evolutionary approaches to behavior rather intimately. We are in an excellent position to contribute to the development of the synthesis of these two perspectives.

ACKNOWLEDGMENTS

Many thanks to J. Wayne Lazar and David Crews for their comments on the manuscript, and to Donald Dewsbury, for discussion.

REFERENCES

1. ROSENBLATT, J. S. & B. R. KOMISARUK (Eds.). 1977. Reproductive Behavior and Evolution. Plenum Press. New York, N.Y.
2. KOMISARUK, B. R. & C. DIAKOW. 1973. Lordosis reflex intensity in rats in relation to the estrous cycle, ovariectomy, estrogen administration and mating behavior. Endocrinology **93:** 548-557.
3. RODRIGUEZ-SIERRA, J. F. & B. R. KOMISARUK. 1977. Effects of prostaglandin E and indomethacin on sexual behavior in the female rat. Horm. Behav. **9:** 281-289.
4. RODRIGUEZ-SIERRA, J. F. & B. R. KOMISARUK. 1978. Lordosis induction in the rat by prostaglandin E systemically or intracranially in the absence of ovarian hormones. Prostaglandins **15:** 513-524.
5. MOSS, R. L. & S. M. MCCANN. 1973. Induction of mating behavior in rats by luteinizing hormone releasing factor. Science **181:** 177-179.
6. PFAFF, D. W. 1973. Luteinizing hormone-releasing factor potentiates lordosis behavior in hypophysectomized ovariectomized female rats. Science **182:** 1148-1149.
7. WILSON, C. A., A. J. THODY & D. EVERARD. 1979. Effect of various ACTH analogs on lordosis behavior in the female rat. Horm. Behav. **13:** 293-300.
8. ALCOCK, J. 1984. Animal Behavior: An Evolutionary Approach. 3rd edit. Sinauer Associates, Inc. Sunderland, Mass.
9. SCHARRER, B. 1977. Evolutionary aspects of neuroendocrine control processes. *In* Reproductive Behavior and Evolution. J. S. Rosenblatt & B. R. Komisaruk, Eds. Plenum Press. New York, N.Y.
10. ADKINS-REGAN, E. 1981. Early organizational effects of hormones: an evolutionary perspective. *In* Neuroendocrinology of Reproduction. N. Adler, Ed. Plenum Press. New York, N.Y.
11. CREWS, D. Functional associations in behavioral endocrinology. *In* Masculinity/Femininity: Basic Perspectives. J. M. Reinisch, L. A. Rosenblum & S. A. Sanders, Eds. Oxford University Press. Oxford, England.
12. MAYR, E. 1970. Populations, Species, and Evolution. Belknap Press. Cambridge, Mass.

13. BEACH, F. A. & A. M. HOLZ-TUCKER. 1949. Effects of different concentrations of androgen upon sexual behavior in castrated male rats. J. Comp. Physiol. Psychol. **42:** 433-453.
14. HARDY, D. F. & J. F. DEBOLD. 1971. The relationship between levels of exogenous hormones and the display of lordosis by the female rat. Horm. Behav. **2:** 287-294.
15. HARDING, D. F. & H. H. FEDER. 1976. Relation between individual differences in sexual behavior and plasma testosterone levels in the guinea pig. Endocrinology **98:** 1198-1205.
16. DAMASSA, D., E. R. SMITH, B. TENNENT & J. DAVIDSON. 1977. The relationship between circulating testosterone levels and male sexual behavior in rats. Horm. Behav. **8:** 275-286.
17. BARRINGTON, E. J. W. 1964. Hormones and Evolution. D. Van Nostrand Co., Inc. Princeton, N.J.
18. FEDER, H. H. 1984. Hormones and sexual behavior. Annu. Rev. Psychol. **35:** 165-200.
19. BEACH, F. A. 1958. Evolutionary aspects of psychoneuroendocrinology. *In* Behavior and Evolution. A. Roe & G. G. Simpson, Eds.: 81-102. Yale University Press. New Haven, Conn.
20. MEDAWAR, P. B. 1953. Some immunological and endocrinological problems raised by the evolution of viviparity in vertebrates. *In* Evolution. Symposium of the Society for Experimental Biology, No. 7., pp. 320-328. Cambridge University Press. Cambridge, England. (Cited by Beach.[19])

A Concept of Physiological Time: Rhythms in Behavior and Reproductive Physiology[a]

L. P. MORIN

Department of Psychiatry and Behavioral Science
Health Sciences Center
State University of New York at Stony Brook
Stony Brook, New York 11794

INTRODUCTION

Coordination among parts of the body and their functions is achieved by two methods of physiological scaling. Spatial scaling was represented by Galileo as the cross-sectional area of a bone being proportional to the cube of its length. Newton's second principle of mechanical similarity has been invoked to describe the relationship among body mass, length, and gravitational acceleration. From that relationship, physiology can be reconstrued in the time domain. It can be shown that time is proportional to body mass raised to the 0.25 power, giving rise to the concept of temporal scaling among mammals. The classic observation is that basal metabolic rate = 70 Body Mass$^{0.75}$ leading to the discovery that the period of many biological rhythms can be described approximately by a constant, K, times the Body Mass$^{0.25}$. These include heart rate, respiration, meal cycle, generation time, longevity, and even population cycles.

Biological rhythms are critically important in mammalian physiology and their periodicities can be influenced by steroid hormones. In turn, the functional effectiveness of steroid hormones can be influenced by the periodicity of certain biological rhythms. Regulation of body function can be achieved via rhythm frequency modulation ("homeokinetics"). Concepts of "homeokinetics" are distinctly different from traditional ideas concerning regulation by negative feedback homeostasis. Examples are provided demonstrating the utility of frequency modulation in behavioral and neuroendocrine control. Physiological time scaling is described for small rodents and related to clock-regulated homeostasis of feeding and rhythmic dynamics in a prey-predator ecosystem.

---

[a] Preparation of the manuscript was supported by NIH funds to the State University of New York at Stony Brook (Grant RR05736), funds from the Department of Psychiatry, and NICHD Grant HD 16231.

TOWARD A CONCEPT OF PHYSIOLOGICAL TIME

In order to indulge in a unified field and laboratory approach to the study of behavior there must be unifying principles which transcend the differences so readily apparent between such fields as neuropharmacology and behavioral ecology. One variable which transcends the separate disciplines and can serve as a focus common to all is the concept of time in the domain of biology. A concept of physiological time may be derived from basic physical principles and has demonstrable utility across a wide variety of physiological measures.

Two historical scientific observations have joined to influence the emergence of a generalized concept of physiological time. The first concerns the study of mammalian metabolic rate. In the early part of this century, there were numerous demonstrations that basal metabolic rate could not be accurately predicted from the surface area of the mammal from which the measure was obtained.[1-3] This conclusion was contrary to the physical rules governing heat loss from inanimate objects and its wide acceptance has been a relatively recent phenomenon. Instead, basal metabolic rate (BMR) is best predicted[1] by the relationship

$$\text{BMR} \propto \text{Body Mass}^{0.75}$$

(FIGURE 1). Although the precise exponent which best describes the relationship is still under discussion,[4,5] the discovery that BMR deviates from a surface law prediction may be taken as historical fact. (Unfortunately, the exponent has generally been calculated from animal data which fail to consider biological time; i.e., BMR has been measured during the rest phase of small nocturnal rodents and the active phase of larger diurnal mammals.[5]) A second, much older set of observations from the field of comparative morphology has also had a significant impact on the concept of physiological time. FIGURE 2 is an illustration from Galileo in 1638. It demonstrates a characteristic of which we have been made amply aware by Beer,[6,7] namely, the homology between morphological characteristics of widely separate species.

Another characteristic which is particularly noteworthy in this figure, and which was described by Galileo, relates to the thickness of the bones shown. The major criterion used to determine which bone belongs to the larger mammal is the diameter of the particular bone in relationship to its length, with diameter reflecting the cross-sectional area. Indeed, Galileo[8] went on to state that the cross-sectional area of a bone is proportional to the cube of its length. In making this observation, Galileo introduced the principle of morphological scaling. To reiterate, FIGURE 2 illustrates two significant concepts: first, the concept of homology between morphological features and second, the concept of morphological scaling of those homologous features.[4] The body plan of an organism must be such that all components are spatially integrated. In general, there cannot be exaggeration of one component without simultaneous compensation for that exaggeration by all the other components. The concept of spatial coordination can now be extended to the time domain permitting development of the concept of physiological time.[4,9]

The following series of relationships have been described in detail by Lindstedt and Calder[9] and Economos.[4] From Newton we obtain the first principle,

$$F = MG, \qquad (1)$$

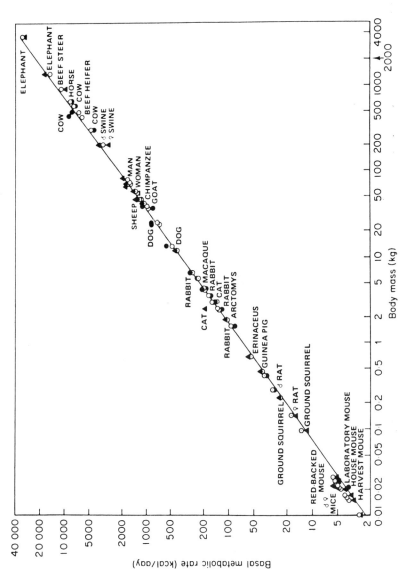

FIGURE 1. Basal metabolic rate (BMR) as a function of mammalian body mass. The solid line shows BMR = 70 Mass$^{0.75}$. (Reprinted with permission from Economos.[4] Copyright 1982 by Academic Press Inc. (London) Ltd.)

where F = force, M = mass, and G = acceleration due to gravity. For homologous tissues, this can be expanded to

$$F = M L/T^2, \qquad (2)$$

where L = tissue length and T = time.

Mechanical stress on a tissue is directly proportional to force and inversely proportional to the cross-sectional area, d^2, of the tissue. Thus, relationship (2) is further expanded such that for homologous tissues, mechanical stress becomes a constant whereby

$$\text{Stress} = Ld^2\, L/T^2\, 1/d^2 \qquad (3)$$

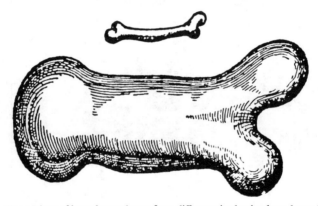

FIGURE 2. Comparison of homologous bones from different-sized animals as drawn by Galileo. (Reprinted with permission from Drake.[8])

(mass is proportional to Ld^2). Finally, by reducing relation (3) you see a most important, yet simple relationship:

$$T \propto L. \qquad (4)$$

As stated, Ld^2 is proportional to the mass of the given muscle. Further derivation from relationship (3) can be achieved to show that

$$\text{muscle power} \propto ML^2/T^3. \qquad (5)$$

This leads directly to the conclusion that

$$\text{ideal muscle power} \propto M^{3/4}. \quad (6)$$

That is precisely the relationship between basal metabolic rate and body mass with basal metabolic rate reflecting energy requirements of an organism simultaneously resisting gravity and losing heat through its surface area.[10] Certain organs, such as the brain[11] (mass) are similarly scaled to Body Mass$^{3/4}$.

Nevertheless, numerous tissues are not. In these instances, function, rather than size, is often scaled to Mass$^{3/4}$. For example, the mammalian heart is scaled directly to body mass, but its function, the rate of blood circulation, is scaled to Mass$^{3/4}$ (see References 9, 12, and 13). Further derivation reveals that

$$L \propto M^{1/4}. \quad (7)$$

This ideal relationship has been supported for large animals by the general empirical observation[4] that length is proportional to Mass$^{0.27}$. A major source of variation not appropriately accounted for in this relationship may be the significant tail or ear mass of small mammals not represented by the overall length measurement.

Because of relations (4) and (7) above, ideal physiological time may be described as

$$T \propto M^{1/4}. \quad (8)$$

Thus, in the foregoing material, we have gone from old and enduring concepts of spatial coordination and spatial scaling to new and evolving concepts of temporal coordination and temporal scaling.

FIGURE 3 shows the various measures of the time scale displayed for comparison with functionally analogous dimensions on the space scale.[14] These scales have been described as the "dynamic spectrum of man" and serve to guide our thinking about spatiotemporal organization. Given the relationships among time, space, and body mass, the question now arises whether the time and space scales are adjustable according to the particular mass of the species under study. The answer is affirmative and several of the following examples specifically address that issue.

FIGURE 4 shows a composite of many of these examples.[15] Both the ordinate and the abscissa are logarithmic. At the lower end of the ordinate the rhythm domain is very high frequency, while at the upper end of the ordinate the rhythm domain is rather low frequency. In a more understandable context, the ordinate units of time range from fractions of seconds on through centuries. Each line represents a large number of points as determined by the method of least squares. Observe first that pulse time (interpulse interval) increases as body mass increases. The rate of increase is nearly exactly the same as the increase in the interval between breaths as body size increases (FIGURE 4). Temporal correlation between pulse frequency and respiratory frequency permits maximal efficiency of oxygen utilization.[16]

The next cycle in this series of examples is the sleep cycle time, which is usually measured as the interval from the onset of one REM sleep episode to the next. Again, there is a striking similarity between the rate of increase in sleep cycle time as a

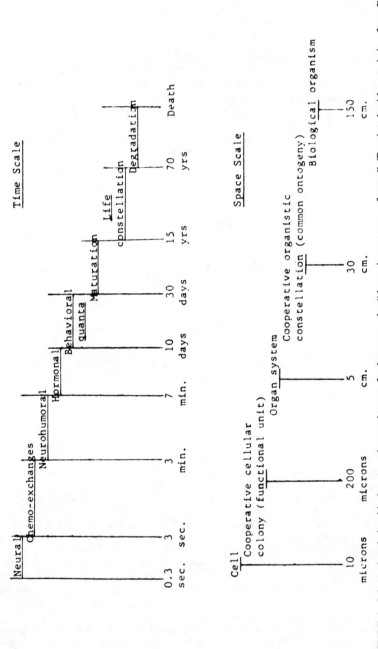

FIGURE 3. Schematic relationship between time and space for humans—the "dynamic spectrum of man." (Reprinted with permission from Iberall and McCulloch.[14])

function of brain mass and that for respiratory interval or pulse time. Further along the time dimension is the microtine rodent meal cycle or intermeal interval. That meal timing can be expressed as a rhythm is a particularly important observation to which we will return. At the moment, it is sufficient to say that the intermeal interval is predictable according to the mass of the organism which raises the question of whether meal timing is regulated by a biological clock or an hourglass process. Finally,

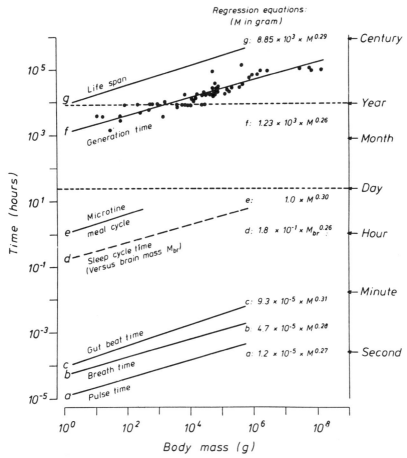

FIGURE 4. Relationships between biological rhythm period and body mass. (Reprinted with permission from Daan and Aschoff[15]; see that article for references.)

at the extreme end of the ordinate are two measures of cycles which take much longer periods of time. The first is generation time, defined as the duration of gestation plus time to sexual maturity. The second is life span. It is quite clear that the rate of increase of each measure is entirely consistent with the increase in interval length of microtine meal cycle, sleep cycle time, and much higher frequency rhythms. All of these observations support the contention that animals function in a temporally coordinated fashion, and further, that the time constants for a given species are scaled according to the body mass of that species.

Recently, the utilization of physiological time has been extended beyond the physiology of a single individual. In particular, it has been invoked to explain the waxing and waning in the size of certain animal populations.[17] FIGURE 5 shows a plot of population cycle data recently compiled from several sources (see Reference 17). The data are necessarily limited and certain characteristics of each of the data sets can be legitimately criticized. Nevertheless, the composite does support the idea that mammalian population cycles, at least of some species, are related to the body mass of the particular species. This has been recently emphasized as a general rule for herbivore populations. Peterson et al.[18] have provided evidence from 41 species of birds and mammals which shows population cycle period as being related to Body Mass$^{0.26}$.

This is not to say, however, that all species which show population cycles generate such cycles based upon a concept of time which is endogenous to the individuals of the population. For example, in many areas of Canada, the lynx has distinct 10-year population cycles. Ten years is a period of time shorter than would be expected according to the mass of this carnivore. Careful examination of lynx ecology[19] has revealed that the population cycles are, in fact, exogenously driven and are specifically related to the prevalence of snowshoe hares. Of all species studied to date, the 10-year population cycle of snowshoe hares is best described. Nevertheless, although the basic endocrinology and ecology of the 10-year cycle are known,[20] there is no direct information concerning its causation.

One attractive approach to the physiology of population cycles has been provided by the ecological "logistic equation" which describes the asymptotic approach of a population to the carrying capacity of the environment.[21] A principal parameter of that description is the birthrate of the population, with birthrate being dependent to a large extent upon life span (i.e., reproductive longevity) and generation time. These variables are, of course, related to Mass$^{1/4}$. We have also determined that the reciprocal of young per adult female per year graphs with a slope ($b = 0.243$, y intercept = -0.78, $r = 0.79$) virtually identical to that in FIGURE 5 (1/[young/litter $\times$ litters/year]; based on information from 41 North American mammal species[22]). It is therefore of significant interest that slight modifications of the birthrate parameter can yield a wide range of stable, albeit hypothetical, population rhythms from the same equation that predicts the gradual approach to stable population equilibrium.[21] Thus, there may be an intrinsic capacity of mammals to show long-term oscillations in population size with the period of such cycles governed by reproductive rate constants which are related to Mass$^{1/4}$.

Within the time domain of animal physiology, steroid hormones may play a particularly active and important role. The following example relates to the well-known coordination between a stimulus and its particular sensory organ. In weakly electric fish from the genus *Sternopygus*, there is a pacemaker system consisting of electrotonically coupled cells which display spontaneous, rhythmic depolarizations. This pacemaker system generates oscillations in weak electrical signals from highly specialized electric organs. Interestingly, the electric organ discharges demonstrate sex differences.[23] During development, the discharge frequency from the electric organ increases in females while the discharge frequency decreases in males. During the season of reproductive activity, the discharge frequency increases further in females, and it decreases further in males. Thus, there is a sex-related divergence between the frequencies of electric organ discharges that is developmental as well as seasonal.[23] The discharge frequencies of *Sternopygus* have proven to be dependent on the availability of gonadal hormones. In the female, the increase in discharge frequency is

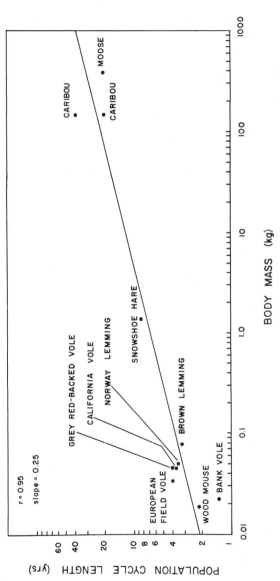

FIGURE 5. Relationship between reported population cycle for various species and the species body mass. See Calder[17] for references.

apparently due to the presence of estradiol, whereas in the male, the decrease in frequency is in response to dihydrotestosterone (DHT). Data of this type support the general observation that one effect of steroid hormones is to perform frequency modulation of oscillatory systems. In addition to the above example, this generalization appears to be true for zebra finch singing rate,[24] uterine contractile frequency,[25] locomotor rate,[26] optimal driving frequency of the hippocampal theta rhythm,[27,28] meal frequency,[29] heart rate,[30] and circadian periodicity,[31] among others.

The electric organ discharge frequency shift induced by DHT in males is approximately 30%.[23] This is a substantial shift. It might be expected that, as a result of this marked change in frequency, the receptors optimally tuned to the prior rate of discharge might be unable to correctly receive the new rate of discharge. That this is not the case has been clearly demonstrated by an experiment[32] in which electric organ discharge has been related to the optimal tuning frequency in the electroreceptors of the same individual.

FIGURE 6 shows the frequency of the electric organ discharge under the influence of DHT.[32] There is a substantial decrease in the rate at which the organ discharges over the course of 14 days. Moreover, it has also been shown that as the discharge of the electric organ decreases under the influence of DHT so does the optimal tuning frequency at which the electroreceptor is capable of responding. It is important to note that the electroreceptor optimal tuning frequency change is independent of the electric organ to the extent that the electric organ need not discharge in order to obtain a hormone-induced decrease in the optimal receptor frequency. Thus, in this particular species, not only does the hormone alter the frequency with which the electric organ discharges,[23] but it alters the optimal frequency to which the electroreceptor can respond to that discharge.[32] In doing so, temporal coordination between stimulus output and sensory input is maintained.

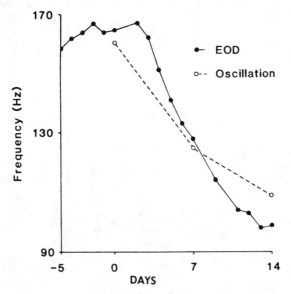

FIGURE 6. Frequency of electric organ discharge (EOD) and optimal oscillation frequency in the electroreceptor of an adult male *Sternopygus* treated for 14 days with dihydrotestosterone beginning at day 0. (Reprinted with permission from Meyer.[32] Copyright 1982 by the AAAS.)

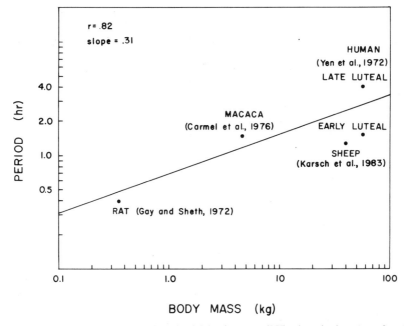

FIGURE 7. Period of pulsatile pituitary luteinizing hormone (LH) release is plotted as a function of body mass. Except for the humans, data are from ovariectomized animals. References: Gay and Sheth[33]; Carmel et al.[34]; Karsch et al.[35]; Yen et al.[36]

The pulsatile control of gonadotropin secretion has been the subject of much recent research. Despite differing pulse frequencies across species, we have speculated that temporal control mechanisms may be similar. FIGURE 7, showing data from ovariectomized animals and luteal-phase humans, provides an initial approach to the problem. The number of species for which data are available is small, but the observed temporal scaling is close to that predicted according to the power law. That is, there is a high correlation between the log period of gonadotropin release under the influence of releasing hormones and the log Mass$^{1/4}$ of the animal. This supports the idea that there is physiological scaling of luteinizing hormone release timing and is consistent with data showing that time required for a single cycle of blood circulation is similarly scaled.[13] The significance of this observation lies in the fact that hormonal information is retained intact precisely because of the temporal synchrony among the hypothalamus, pituitary, blood, and, presumably, target tissues.

FIGURE 8 provides an example of pulsatile gonadotropin release. The data are from ewes.[35] In addition to showing the clear periodicity of luteinizing hormone (LH) release, the figure indicates that exogenous estradiol caused increased LH pulse frequency thereby mimicking the action of endogenous estradiol which is released after luteal-phase progesterone withdrawal. As expected from most other studies (e.g., References 25, 29, and 31, but cf. Reference 30) where rhythmic events are concerned, estradiol increased the rhythm frequency.

Of what consequence is the timing of this pulsatile control of gonadotropin release? It is becoming increasingly obvious that frequency modulation is one of the primary methods for the brain to communicate with the pituitary thereby regulating the release of gonadotropins. For example, it is now clear in male rats that the ability of DHT to exert negative feedback is exquisitely modulated by the timing of luteinizing hormone-releasing hormone (LHRH) surges and it is the timing that matters rather than the absolute amount of hormone that is available.[37] If the interval between LHRH pulses is too long (e.g., 1 hr), then the ability of DHT to exert negative feedback on the release of LH is virtually nonexistent. If the LHRH pulses are given at an interval mimicking the expected pulsatile LH release interval ($\frac{1}{2}$ hr), then the negative feedback of DHT will reduce LH output by about 50%. Thus, there is an emergent, elegant, and novel phenomenon in the field of neuroendocrinology that utilizes the concept of ultradian clocks as modulators of steroid feedback efficacy.

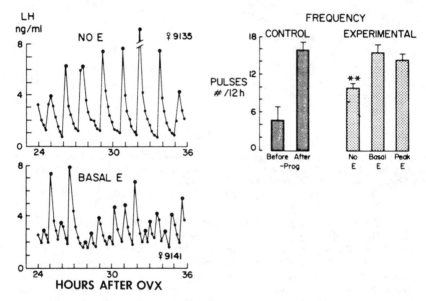

FIGURE 8. Pulsatile LH release in ewes. The left portion shows pulsatile LH in individual ovariectomized animals with sources of estradiol removed (No E) or maintained at 1-2 pg/ml serum (Basal E). On the right, the mean number of LH pulses per 12 hr is shown for intact animals in which progesterone implants suppressed endogenous estradiol release (Before) and for animals with elevated estradiol after withdrawal of the progesterone (After). Comparable data are also shown for the ovariectomized ewes given No, Basal, or Peak (5.3 pg/ml) estradiol treatments. (Reprinted with permission from Karsch et al.[35])

How might this change of pulsatile hormone periodicity be accomplished under the influence of gonadal steroids? One mechanism that has received modest experimental attention has been hormonal alteration of the rate at which neurons can discharge by affecting the refractory period. It has recently been demonstrated that estrogen is able to influence the absolute refractory period of at least some neurons.[38,39] Estradiol, but not DHT, can shorten the absolute refractory period of rat corticomedial amygdala neurons by about 33%.[38] This observation is also consistent with the general view that estradiol increases the frequency of cyclic events.

Having made this foray to the level of individual neurons and their temporal modification by hormones, the following discussion returns to a more physiological concept of time and addresses the issue of rhythmic homeostasis and its ecological significance. The classical concept of feedback homeostasis utilizes, as its primary analogy, the typical residential thermostat in which a temperature signal exceeds or drops below a fixed threshold, thereby turning the heat generating system off or on. An inevitable consequence of such negative feedback is uncontrolled fluctuation of temperature around the threshold. In an attempt to achieve stability, instability is automatically created. Thus, according to some engineers, it is a preferable procedure to design control systems with known rhythmic properties.[40] Because so many bodily functions and constituents are oscillatory[41] it would be most appropriate to have physiological control systems which are also oscillatory. The word "homeokinetics" has been coined to describe this form of physiological control.[42] A recent review of the thermoregulatory literature has specifically hypothesized that mammalian body temperature is regulated through a complex interaction of many oscillators of differing frequencies contributing to thermoregulatory stability.[43]

Using that general rhythmic model as a guide, let us now look at the complexity of temporal constraints on eating by a vole, *Microtus arvalis,* and the effect of time on the relationship between the vole and its predators.[44] *M. arvalis* can be captured and recaptured in the wild at rather discrete intervals throughout the 24-hr day. The surface activity, the period during which these voles are regularly caught, consists of foraging for food (FIGURE 9). The question is, of what significance are these peaks in foraging behavior, and how do they relate to timing mechanisms of this particular species? In an ecological context, of what importance are these cyclic fluctuations in foraging behavior to other species in this animal's ecosystem? This vole, first of all, eats a meal approximately every 120 min. It is a small animal, weighing approximately 15-18 g, but it generates an intermeal interval almost exactly predicted by the least squares regression line of microtine meal cycles.

FIGURE 10 shows that meals, as they are measured by this feeding apparatus, are temporally linked to the animal's subjective daytime and subjective nighttime. As the locomotor rhythm freeruns in constant darkness, the feeding rhythm remains synchronized to the locomotor rhythm, and likewise freeruns in constant darkness. This demonstration has an important implication. In conjunction with the prior assertion that there is such a concept as physiological time which is scaled to the mass of the individual species, the possibility is raised that intermeal interval is regulated by one or more biological clocks. The relative coordination between the feeding rhythms, which are high-frequency rhythms, and the running rhythm, which is a lower-frequency rhythm, further supports the contention that the feeding rhythms are based on a biological clock mechanism. Despite the fact that the feeding rhythm is not in the circadian domain, it seems clear that each particular meal is synchronized to a particular phase of a 24-hr rhythm.

Taking this logic of clock control over meal timing one step further, Daan and Slopsema[44] made the noteworthy observation that when food was withheld from the voles, the animals continued to go to the food hopper at times expected according to the prior meal pattern. The mealtimes of rats and hamsters appear to share this characteristic with those of voles. Unlike the vole, neither rats nor hamsters synchronize their feeding times to the 24-hr light/dark cycle[45] (manuscript in preparation). Nor do the feeding times freerun synchronous with the locomotor rhythm of the particular species. However, both rats and hamsters demonstrate ultradian rhythmicity in their feeding patterns, although the feeding patterns of both species are considerably more complex than those of the voles. When food is deprived, both rats and hamsters continue to enter the feeders at times that are generally predicted by the prior meal pattern and by the subsequent meal pattern.

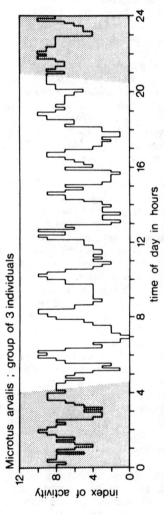

FIGURE 9. Locomotor activity between two housing compartments of three voles (*M. arvalis*) housed together. The activity peaks are coincident with feeding. Darkness is shown by the gray areas. (Reprinted with permission from Daan and Slopsema.[44])

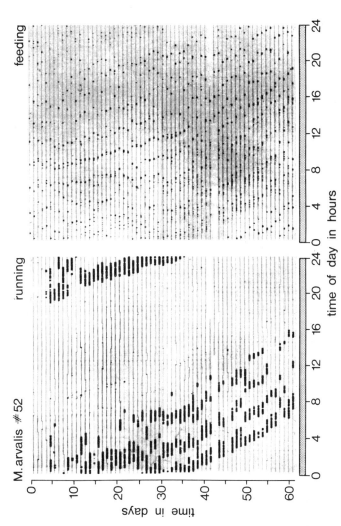

FIGURE 10. Circadian wheel-running activity of the vole, *M. arvalis*, recorded simultaneously with eating from a food hopper. Note that the timing of meals is temporally related to the period of the locomotor circadian rhythm. (Reprinted with permission from Daan and Slopsema.[44])

To reiterate then, a concept of feedback homeostasis for meal patterning states that hunger waxes and wanes according to some level of nutritional repletion or depletion.[46] Such a hypothesis predicts that if an animal is sated with a large amount of food, then the postprandial interval is likely to be large. If the animal eats a small number of calories, then the postprandial interval is likely to be small. But, it is possible to turn that concept around because of the fact that it is simply a correlation. If, for example, a species consults an internal clock at the onset of every meal and physiologically asks the question, "How much time until the next meal?", the amount of food taken at the current meal can then be directly related to the interval to the next expected meal. This is not an unusual proposition inasmuch as a wide variety of species including bees, birds, and mammals are able to use internal clocks for orientation of goal-directed behavior.[47-51]

Returning to *M. arvalis,* we have an individual species which is utilizing one or more clocks to regulate its locomotor activity. Its general locomotion is controlled by at least a circadian clock system.[44] Eating-related activity is controlled by several circadian clocks closely phased to one another with activity peaks approximately 120 min apart, or by a single 120-min ultradian rhythm. Either alternative could produce meal patterns synchronized with the locomotor rhythm. Although the vole, because of its clock-like regulation, is off and on inactive across the entire 24 hr, most of its predators are active only during the daylight hours. The kestrel (*Falco tinnunculus*) and the hen harrier (*Circus cyaneus*) are two such day-active predators. Daan and colleagues[44,52] have shown that when the hunting activity is examined closely (FIGURE 11), a mild oscillation in the percentage of birds hunting at any given time of day may be found. This oscillation is synchronized with the vole surface activity. At the moment it is not possible to rule out an internally generated ultradian rhythm in predator hunting activity.

A glance at the hunting yield immediately reveals a rhythm of greater importance. FIGURE 11 shows quite clearly that the hunting yield, in terms of voles caught per hour of hunting, goes up dramatically as the amount of vole surface activity goes up. This occurs despite the fact that there is not necessarily a large increase in the percentage of birds actually hunting, and in at least one case, there is an actual decrease in the percentage of birds hunting.

Of what ecological significance is this? An answer is only possible in a general sense. For the predators, the effect is relatively clear. When the voles are very active, the hen harriers need spend much less energy in order to capture prey. Thus, the voles, by synchronizing their population foraging activity, maximize the hunting benefits to their chief predators. At a first glance, that seems to be countereffective. However, from a vole's perspective, the probability of being captured by a kestrel or hen harrier goes down enormously during one of these foraging phases because the capture probability decreases in proportion to the number of voles foraging.

Thus, between the vole, *M. arvalis,* and its avian predators, there is an interesting relationship between internal and external rhythms. For the voles, an internal ultradian periodicity generates peaks of surface foraging activity throughout the day. This is synchronous with an entrained circadian rhythm of general locomotion. The kestrel is a diurnal species that generates virtually all of its locomotor activity during the daylight hours. This species at least partially synchronizes its hunting activity to an exogenous ultradian rhythm, namely, the vole surface activity. As a consequence, there is a clear ultradian pattern in the probability of the predator catching a vole.

Temporal constraints on the prey-predator relationship appear to be even more complex because of the predator's ability to learn temporal information and to address an internal clock.[15] Kestrels show a strong tendency to learn circadian rhythm times. Thus, when an individual kestrel makes a kill, there is an increased tendency for it to hunt at the same time the next day, suggesting that the predator has employed circadian time cues to acquire information about the habits of its prey. This phenomenon reiterates the point that a variety of species are able to use internal clocks to time the performance of learned events.[47-51]

It is unfortunate that the significance of time in an ecological setting relies so heavily on a few papers. Nevertheless, the limited data available demonstrate the need to consider time and periodicity as critical elements in day-to-day social interactions.

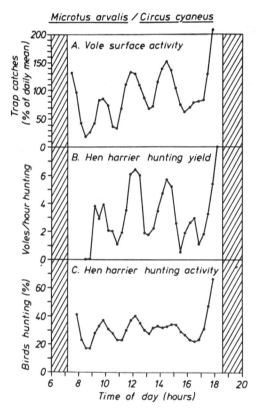

FIGURE 11. Temporal interactions between prey (*M. arvalis*) and an avian predator (*C. cyaneus*). (A) Vole surface activity; (B) voles caught per hour of hen harrier hunting; (C) percentage of hen harriers hunting. The hatched portion indicates darkness. (Reprinted with permission from Daan and Aschoff.[15])

The fascinating ecological oscillations observed by Daan and colleagues are expected to be modulated by still other rhythms. For example, the relationship between prey and/or predator reproductive cycles and the level of predation has not been studied. Reproductive cycles, at least of some species, are modulated according to circadian and annual rhythms.[53-56] These are able to alter responsiveness to steroid hormones,[57,58] possibly through frequency modulation of the neuroendocrine system.[59] Availability of gonadal steroids during yearly phases of reproductive activity as well as within reproductive cycles will alter general metabolic rate, fat deposition, and patterns of food intake.[60] Such phenomena will be demonstrable within individuals, but the availability of, and responsiveness to, gonadal steroids will also presumably alter oscillatory models of social interaction. The existence and impact of frequency-modulated social encounters on social bonds, group formation, and societal stability remain to be demonstrated (but see note 27 in Reference 18).

A substantial flaw in the proposed relationships between rhythmic variables and body mass exists in their reliance on data obtained from adults of differing species. Insufficient data currently limit valid estimation of the extent to which existing generalizations reflect the physiology and behavior of developing individuals of a given species. To be more explicit, there has been little investigation of the functional effects of rhythmic events within individuals and virtually none concerning functional relationships between rhythmic variables. Two recent exceptions have been the demonstration (1) of a correlative link between nocturnal sleep duration in the human sleep-wake cycle and body temperature rhythm[61] and (2) of the phase locking of respiratory and gait rhythms of bipedal or quadrapedal runners.[62] It is noteworthy that the former example involves modulation of one rhythm by another with a similar period. In the second case, however, a longer-period rhythm is synchronized with a rhythm having an entirely different fundamental frequency. There is a great need for understanding the temporal relationships between rhythms of different frequencies within the same individual.[63] Until more research is performed, the list of similar time-body mass relationships which are demonstrable across species (partly shown in FIGURE 4) exists simply as a fascinating set of coincidences with little explanatory value.

SUMMARY

1. Biological clocks are self-sustained and endogenous.
2. Animals have biological clocks in a variety of frequencies.
3. Biological clocks provide temporal coordination among physiological, behavioral, and environmental events.
4. One important function of steroid hormones may be to alter the temporal coordination among those events.
5. Biological clocks provide a referent mechanism for the timing of both current and future endogenous and exogenous events.
6. Physiological time is a unifying principle in biology.[64,65]

ACKNOWLEDGMENT

I would like to thank Dr. Serge Daan for his critical reading of the manuscript.

REFERENCES

1. KLEIBER, M. 1947. Body size and metabolic rate. Physiol. Rev. **27:** 511-541.
2. KLEIBER, M. 1961. The Fire of Life. Wiley. New York, N.Y.
3. HOFFMAN, M. A. 1983. Energy metabolism, brain size and longevity in mammals. Q. Rev. Biol. **58:** 495-512.
4. ECONOMOS, A. C. 1982. On the origin of biological similarity. J. Theor. Biol. **94:** 25-60.
5. PROTHERO, J. 1984. Scaling of standard energy metabolism in mammals. I. Neglect of circadian rhythms. J. Theor. Biol. **106:** 1-8.
6. BEER, C. G. 1974. Comparative ethology and the evolution of behavior. *In* Ethology and Psychiatry. N. F. White, Ed.: 173-181. University of Toronto Press. Toronto, Ontario, Canada.
7. BEER, C. G. 1977. What is a display? Am. Zool. **17:** 155-165.
8. DRAKE, S. 1974. Translation of *Two New Sciences* by Galileo (1638). Pp. 5-306. University of Wisconsin Press. Madison, Wisc.
9. LINDSTEDT, S. L. & W. A. CALDER III. 1981. Body size, physiological time, and longevity of homeothermic animals. Q. Rev. Biol. **56:** 1-16.
10. ECONOMOS, A. C. 1979. Gravity, metabolic rate and body size of mammals. Physiologist **22:** 571-572.
11. ARMSTRONG, E. 1983. Relative brain size and metabolism in mammals. Science **220:** 1302-1304.
12. ADOLPH, E. F. 1949. Quantitative relations in the physiological constitutions of mammals. Science **109:** 579-585.
13. STAHL, W. R. 1967. Scaling of respiratory variables. J. Appl. Physiol. **22:** 453-460.
14. IBERALL, A. S. & W. S. MCCULLOCH. 1969. The organizing principle of complex living things. J. Basic Eng., Trans. ASME Ser. D **91:** 290-294.
15. DAAN, S. & J. ASCHOFF. 1982. Circadian contributions to survival. *In* Vertebrate Circadian Rhythms. J. Aschoff, S. Daan & G. A. Groos, Eds.: 305-321. Springer-Verlag. New York, N.Y.
16. GEBBER, G. L. 1980. Central oscillators responsible for sympathetic nerve discharge. Am. J. Physiol. **239:** H143-H155.
17. CALDER, W. A., III. 1983. An allometric approach to population cycles of mammals. J. Theor. Biol. **100:** 275-282.
18. PETERSON, R. O., R. E. PAGE & K. M. DOGE. 1984. Wolves, moose, and the allometry of population cycles. Science **224:** 1350-1352.
19. BERGERUD, A. T. 1983. Prey switching in a simple ecosystem. Sci. Am. **249:** 130-140.
20. DAVIS, G. J. & R. K. MEYER. 1973. FSH and LH in the snowshoe hare during the increasing phase of the 10-year cycle. Gen. Comp. Endocrinol. **20:** 53-60.
21. MAY, R. M. 1974. Biological populations with nonoverlapping generations: stable points, stable cycles, and chaos. Science **186:** 645-647.
22. BURT, W. H. & R. P. GROSSENHEIDER. 1964. Field Guide to the Mammals. Houghton Mifflin. Boston, Mass.
23. MEYER, J. H. 1983. Steroid influences upon the discharge frequencies of a weakly electric fish. J. Comp. Physiol. **153:** 29-37.
24. ARNOLD, A. P. 1975. The effects of castration and androgen replacement on song, courtship, and aggression in zebra finches (Poephila guttata). J. Exp. Zool. **191:** 309-326.
25. REXROAD, C. E. 1980. Estradiol regulation of the frequency and site of origin of uterine contractions in ewes. J. Anim. Sci. **151:** 1139-1147.
26. ROY, E. J. & G. N. WADE. 1975. Role of estrogens in androgen-induced spontaneous activity in male rats. J. Comp. Physiol. Psychol. **89:** 573-579.
27. VALERO, I., J. STEWART, N. MCNAUGHTON & J. A. GRAY. 1977. Septal driving of the hippocampal theta rhythm as a function of frequency in the male rat: effects of adrenopituitary hormones. Neuroscience **2:** 1029-1032.
28. DREWETT, R. F., J. A. GRAY, D. T. D. JAMES, N. MCNAUGHTON, I. VALERO & I. J. DUDDERIDGE. 1977. Sex and strain differences in septal driving of the hippocampal theta rhythm as a function of frequency: effects of gonadectomy and gonadal hormones. Neuroscience **2:** 1033-1041.

29. BLAUSTEIN, J. D. & G. N. WADE. 1976. Ovarian influences on the meal patterns of female rats. Physiol. Behav. **17:** 201-208.
30. MCCABE, P. M., S. W. PORGES & C. S. CARTER. 1981. Heart period variability during estrogen exposure and withdrawal in female rats. Physiol. Behav. **26:** 535-538.
31. MORIN, L. P., K. M. FITZGERALD & I. ZUCKER. 1977. Estradiol shortens the period of hamster circadian rhythms. Science **196:** 305-307.
32. MEYER, J. H. 1982. Androgens alter the tuning of electroreceptors. Science **217:** 635-637.
33. GAY, V. L. & N. A. SHETH. 1972. Evidence for periodic release of LH in castrated male and female rats. Endocrinology **90:** 158-162.
34. CARMEL, F. W., S. ARAKI & M. FERIN. 1976. Pituitary stalk portal blood collection in rhesus monkeys: evidence for pulsatile release of gonadotropin-releasing hormone (GnRH). Endocrinology **99:** 243-248.
35. KARSCH, F. J., D. L. FOSTER, E. L. BITTMAN & R. L. GOODMAN. 1983. A role for estradiol in enhancing luteinizing hormone pulse frequency during the follicular phase of the estrous cycle of sheep. Endocrinology **113:** 1333-1339.
36. YEN, S. S. C., C. C. TSAI, F. NAFTOLIN, G. VANDENBERG & L. AJABOR. 1972. Pulsatile patterns of gonadotropin release in subjects with and without ovarian function. J. Clin. Endocrinol. Metab. **34:** 671-675.
37. NANSEL, D. D. & D. F. TRENT. 1979. Frequency modulation of pulsatile luteinizing hormone-releasing hormone stimulation can alter the effectiveness of direct androgen feedback on luteinizing hormone-releasing hormone-induced luteinizing hormone release. Endocrinology **104:** 532-535.
38. KENDRICK, K. M. & R. F. DREWETT. 1980. Testosterone-sensitive neurones respond to oestradiol but not to dihydrotestosterone. Nature **286:** 67-68.
39. KENDRICK, K. M. 1982. Inputs to testosterone-sensitive stria terminalis neurones in the rat brain and the effects of castration. J. Physiol. (London) **315:** 437-447.
40. GARFINKLE, A. 1983. A mathematics for physiology. Am. J. Physiol. **245:** R455-R466.
41. RAPP, P. E. 1979. Bifurcation theory, control theory and metabolic regulation. *In* Biological Systems, Modelling and Control. Peregrinus. New York, N.Y.
42. YATES, F. E. 1982. Outline of a physical theory of physiological systems. Can. J. Physiol. Pharmacol. **60:** 217-248.
43. GORDON, C. J. & J. E. HEATH. 1983. Reassessment of the neural control of body temperature: importance of oscillating neural and motor components. Comp. Biochem. Physiol. A **74:** 479-489.
44. DAAN, S. & S. SLOPSEMA. 1978. Short-term rhythms in foraging behaviour of the common vole, Microtus arvalis. J. Comp. Physiol. **127:** 215-227.
45. MORIN, L. P. Ultradian rhythms in hamster and rat eating. Abstract (#18.9) presented at the 11th Annual Meeting of the Society for Neuroscience, Los Angeles, Calif., 1981.
46. LEMAGNEN, J. 1981. The metabolic basis of dual periodicity of feeding in rats. Behav. Brain Sci. **4:** 561-575.
47. HOFFMAN, K. 1954. Versuche zu er in Richtungsfinden der Vogel enthaltenen Zeitschatzung. Z. Tierpsychol. **11:** 453-475.
48. RENNER, M. 1957. Ein Transozeanversuch zum Zeitsinn der Honigbiene. Naturwissenschaften **42:** 540-541.
49. HOLLOWAY, F. A. & R. A. WANSLEY. 1973. Multiple retention deficits at periodic intervals after passive avoidance learning. Science **180:** 208-210.
50. HOLLOWAY, F. A. & R. A. WANSLEY. 1973. Multiple deficits after active and passive avoidance learning. Behav. Biol. **9:** 1-14.
51. ROSENWASSER, A. M., R. J. PELCHAT & N. T. ADLER. 1984. Memory for feeding time: possible dependence on coupled circadian oscillators. Physiol. Behav. **32:** 25-30.
52. RIJNSDORP, A., S. DAAN & C. DIJKSTRA. 1981. Hunting in the kestrel, Falco tinnunculus, and the adaptive significance of daily habits. Oecologia (Berlin) **50:** 391-406.
53. EVERETT, J. W. & C. H. SAWYER. 1950. A 24-hour periodicity in the "LH-release apparatus" of female rats, disclosed by barbiturate sedation. Endocrinology **47:** 198-218.
54. ALLEVA, J. J., M. V. WALESKI & F. R. ALLEVA. 1971. A biological clock controlling the estrous cycles of the hamster. Endocrinology **88:** 1368-1379.
55. SADLEIR, R. M. F. S. 1969. The Ecology of Reproduction in Wild and Domestic Animals. Methuen. London, England.

56. ZUCKER, I. & P. LICHT. 1983. Seasonal variations in plasma luteinizing hormone levels of gonadectomized male ground squirrels (Spermophilus lateralis). Biol. Reprod. **29:** 278-285.
57. MORIN, L. P. & I. ZUCKER. 1978. Photoperiodic regulation of copulatory behaviour in the male hamster. J. Endocrinol. **77:** 249-258.
58. TUREK, F. W. & G. B. ELLIS. 1981. Steroid-dependent and steroid-independent aspects of the photoperiodic control of seasonal cycles in male hamsters. *In* Biological Clocks in Seasonal Reproductive Cycles. B. K. Follett & D. E. Follett, Eds.: 251-260. Wright. Bristol, England.
59. LINCOLN, G. A. & R. V. SHORT. 1980. Seasonal breeding—nature's contraceptive. Recent Prog. Horm. Res. **36:** 1-52.
60. BARTNESS, T. J. & G. N. WADE. 1984. Photoperiodic control of body weight and energy metabolism in Syrian hamsters (Mesocricetus auratus): role of pineal gland, melatonin, gonads and diet. Endocrinology **114:** 492-498.
61. CZEISLER, C. A., E. D. WEITZMAN, M. C. MOORE-EDE, J. C. ZIMMERMAN & R. S. KNAUER. 1980. Human sleep: its duration and organization depend on its circadian phase. Science **210:** 1264-1267.
62. BRAMBLE, D. M. & D. R. CARRIER. 1983. Running and breathing in mammals. Science **219:** 251-256.
63. PORGES, S. W., R. E. BOHRER, M. N. CHEUNG, F. DRASGOW, P. M. MCCABE & G. KEREN. 1980. New time-series statistic for detecting rhythmic co-occurrence in the frequency domain: the weighted coherence and its application to psychophysiology. Psychol. Bull. **88:** 580-587.
64. YATES, F. E., D. J. MARSH & A. IBERALL. 1972. Integration of the whole organism—a foundation for a theoretical biology. *In* Challenging Biological Problems. J. A. Behnke, Ed.: 118-132. Oxford University Press. New York, N.Y.
65. PRIGOGINE, I. & I. STENGERS. 1984. Order Out of Chaos. P. 349. Bantam. New York, N.Y.

PART III. BIOCHEMICAL BASES OF HORMONAL ACTION ON
REPRODUCTIVE BEHAVIOR

Introduction

HARVEY H. FEDER

Institute of Animal Behavior
Rutgers-The State University of New Jersey
Newark, New Jersey 07102

In this section, biochemical bases of steroid-dependent behaviors, especially behaviors involved in reproduction, are examined from several perspectives. The contribution by R. E. *Whalen* examines some currently held concepts of hormonal action at the cellular level, and points out that generalizations across genotypes generally fail when we begin to look closely at the details of hormone-behavior relationships. In a similarly critical vein, J. E. *Thornton* examines some widely held assumptions about heterotypical sexual behavior patterns and the ways in which hormones may influence the display of these patterns by actions on the differentiating nervous system. C. F. *Harding* deals with the role of metabolism of androgen in the activation of male behavior, and draws attention to the importance of interactions among 5α-reduced and aromatized products of testosterone metabolism. I. T. *Landau* discusses the uses and limitations of steroid antagonists in the study of reproductive behavior, and concludes on a pessimistic note. Again, generalization across species, tissues, and behaviors seems difficult, and initial hopes that steroid antagonists would provide a means of understanding cellular bases of hormone-behavior interactions appear to have dimmed. G. N. *Wade* contrasts the effects of steroids on reproductive behaviors and on energy balance. He argues that general principles of hormone-behavior interactions might emerge more readily from study of hormone actions on a wide range of behaviors rather than through more restricted focus on a narrow range of behaviors and functions. J. D. *Blaustein* reviews the evidence indicating a critical role for receptors in mediating the effects of steroids on reproductive behavior. While this is currently a powerful integrating theme in studies of hormone-behavior relationships, Blaustein is careful to point out alternative mechanisms, and he discusses future directions that can be taken in order to pinpoint locations at which steroids interact with neurotransmitter systems to modulate reproductive behaviors. B. *Nock* provides an overview of the ways in which neurotransmission (particularly noradrenergic) regulates the concentration and/or distribution of brain progestin receptors. From this work one might be able to develop models of how environmental and behavioral events could selectively alter the sensitivity of discrete brain areas to the actions of steroids. W. R. *Crowley* takes the argument a step further by developing the idea that catecholamine-regulated reproductive processes may be locally modulated by opioid neuropeptides. Another neurotransmitter system, the serotoninergic system, has also been studied extensively in relation to steroids and reproductive behavior. L. H. *O'Connor* and C. T. *Fischette* discuss some clinical implications of these interactions and draw contrasts between two syndromes (serotonin behavioral syndrome in rats and myoclonus in guinea pigs). Finally, C. P. *Reboulleau* and J. N. *Wilcox* discuss two approaches to the study of steroid action at the cellular level that have had an obvious and increasing impact. Reboulleau's contribution reviews the use of cell culture of clonal cell lines to study mechanisms of steroid action in the process of neural differentiation. Wilcox's contribution describes *in situ* hybridization techniques that can be used to localize pro-opiomelanocortin mRNA containing cells in the brain, and to study the effects of steroids on these cells.

Hormonal Control of Behavior—A Cautionary Note

RICHARD E. WHALEN

Department of Psychology
University of California
Riverside, California 92521

In 1849 Berthold published what is usually considered to be the first true experiment in endocrinology. He castrated roosters and in some, he reimplanted testes. For the latter group Berthold noted,

> So far as voice, sexual urge, belligerence and growth of comb and wattles are concerned, such birds remained true cockerels. Since, however, transplanted testes are no longer connected to their original innervation . . . it follows that the results in question are determined by the productive functions of the testes, i.e., by their actions on the blood stream. . . . (Reference 1, p. 328)

As Beach[1] has pointed out, the early endocrinologists frequently used behavioral end points in their studies. This practice disappeared as the field of endocrinology became more biochemically oriented. Only during the past 15-20 years have biochemical and behavioral studies been reunited as both biochemists and behaviorists have become involved in modern neurobiology. I believe that we are making great advances in our understanding of how and where hormones act at the cellular level to regulate behavior. But, I also believe that we have made some mistakes along the way. Thus, this cautionary note.

Our quest, as behavioral biologists, to define reality, is neverending. It must be so because the fundamental unit with which we work, a genotype, is not a stable entity. I mean this in the sense that our conclusions about how steroid hormones regulate behavior are always limited by the genetic material with which we work. Our relative failure to include genotype as a factor in our thinking has, I believe, led us to propose "principles" of hormone action prematurely and that this has limited our view of reality.

As an example, I offer our enthusiasm to establish the principle that testosterone (T) regulates male-typical mating behavior via its intracellular metabolism to estradiol (E2), the aromatization hypothesis. McDonald *et al.*[2] found that under their test conditions, castrated male rats administered dihydrotestosterone (DHT) failed to initiate mating. Since it was known that T, but not DHT, can be aromatized to E2, they postulated that such a conversion might be obligatory for T's ability to maintain or reinstate mating in castrated male rats.

There followed literally dozens of experiments attempting to confirm this hypothesis as a principle of hormone action rather than as one mechanism by which testicular testosterone could regulate mating in one genotype. This occurred in spite of the fact that in 1973 Luttge and Hall[3] demonstrated convincingly that DHT's actions are genotype specific. They showed that DHT can lead to intromission behavior in castrated male Swiss-Webster mice, but that this steroid is without effect in CD-1 mice. In both strains, DHT stimulated seminal vesicle growth.

In 1974, Alsum and Goy[4] showed that DHT was almost as effective as T in inducing mating in prepubertally castrated guinea pigs; we found that high doses of DHT would fully maintain mating behavior in male hamsters[5]; and Phoenix[6] reported that DHT would reinstate mating in castrated male rhesus monkeys. More recently it has been found that DHT can even stimulate male mating behavior in at least one strain of rats, the King-Holtzman strain.[7,8] Thus, aromatization is not a *principle* of hormone action, but rather one mechanism that is characteristic of certain genotypes by which testosterone can regulate male-typical mating responses.

It should be noted that the line of reasoning that led to a reification of the aromatization hypothesis has not been restricted to students of sexual behavior. Students of another type of testis-dependent behavior, intermale fighting in mice, have championed one or another hypothesis as the principle of testicular hormone action. Brain and Bowden[9] have favored the aromatization hypothesis; Gandelman[10,11] and colleagues have suggested direct androgenic control of fighting; and Finney and Erpino[12] have recommended a combined aromatization-5α-reduction hypothesis. Interestingly, these different hypotheses have been derived from studies of the effects of diverse steroids on fighting in different strains of mice.

Recently, we[13] compared the effectiveness of estrogens and androgens on the stimulation of fighting in three strains of mice. Methyltrienolone, the synthetic androgen, was quite potent in CF-1 males, moderately potent in CD-1 males, and weakly active in CFW males. In contrast, both CFW and CF-1 males responded to the synthetic estrogen, diethylstilbestrol, while the CD-1 males did not. Again, we are forced to conclude that the effectiveness of androgens and their metabolites upon testis-dependent behavior is determined by genotype.

Before leaving the topic of genotype and hormonal regulation of behavior, I will review briefly two sets of findings, ones which may prove to be as interesting as those stimulated by the aromatization hypothesis, but that have received less attention. These concern female sexual behavior. In rodents, female-typical sexual behaviors, such as the lordotic response, are thought to be controlled by the ovarian steroids E2 and progesterone (P). Even though E2 is "less metabolic" than T, one can detect estrone (E1) and even estriol (E3) in the brain of rodents after the administration of estradiol. Which of these estrogens are important for behavior?

In 1971, Beyer *et al.*[14] compared the behavioral effects of various estrogens in female rats and found that at low doses E2 was more potent than E1 and that E3 was ineffective. At higher doses, E2 and E1 were equally potent and both were more potent than E3. These data are consistent with the widely held notion that E2 is "the" active estrogen and that E1 and E3 are "weak" estrogens (presumably because of differences in their affinity for the estrogen receptor).

If E1 and E3 are indeed weak estrogens, this should hold across genotypes. It does not. Feder and Silver[15] found that E1, E2, and E3 will all induce lordosis in guinea pigs and that the effective dose is about the same for all three estrogens. Thus, the potency of steroids appears to be genotype specific for female as well as for male behaviors.

Again looking at female-typical behavior, we find that P itself seems to be the most potent natural progestin. Species and strains differ, however, in response to the metabolites of P such as dihydroprogesterone (DHP). In rats, under certain conditions of subcutaneous administration, P and DHP can be equipotent in facilitating lordosis in estrogen-primed animals[16]; however, when given intravenously, DHP is almost completely ineffective.[17] In guinea pigs, DHP stimulates receptive behavior, but is less potent than P.[18,19]

In mice, the nature of progestin action is more complex than in rats or guinea pigs. Following ovariectomy and estrogen treatment, female mice are apparently insensitive to P. We[20] found that five weekly pairings of estrogen and P were needed to induce maximal receptivity in Swiss-Webster and CD-1 mice. On the eighth week of treatment with estrogen, we substituted DHP for the usual P treatment. Lordotic behavior did not occur. We continued weekly treatments with estrogen and DHP. Females of the CD-1 strain became progressively more receptive; Swiss-Webster females did not. Thus, CD-1 females were capable of responding to both P and DHP, while the Swiss-Webster females were limited to a response to P. Luttge and Hall[21] subsequently confirmed the differential sensitivity of these strains to the actions of DHP.

Again we see that genotype plays a major role in determining the behavioral actions of steroid hormones. We should be cautioned, therefore, regarding overgeneralization of our findings and the postulation of principles where none exist.

The studies of mice described above should trigger another caution with respect to the establishment of principles of hormone action. There has been a great deal of research during the past few years indicating that in rats and guinea pigs estrogen stimulation leads to the induction of progestin receptors in the brain.[22,23] Since P facilitates lordotic behavior in estrogen-primed female rats and guinea pigs, our tendency is to conclude not only that receptor induction is critical for P to have an effect, but that this is a principle of hormone action. In the mouse strains studied, progestin priming is critical for the estrogen-progestin combination to become behaviorally active. Moreover, great specificity exists—in CD-1 females priming with P does not sensitize the animal to DHP and priming with DHP does not sensitize the animal to P. Clearly, the rules by which estrogens and progestins regulate sexual receptivity in at least some strains of mice are different from the rules of ovarian hormone action in rats and guinea pigs.

I am not suggesting that we avoid model building. It can be quite useful in guiding research. For example, Jensen's two-step model of steroid hormone action at the cellular level has been the impetus for thousands of experiments.[24] The notion of soluble cytoplasmic receptor proteins that bind steroids with high affinity followed by translocation of the hormone-receptor complex into the nucleus for binding to the chromatin and/or DNA has graced our textbooks for 20 years. This model was based in part on the oft-repeated observations that in the gonadectomized animal (1) unoccupied receptors are found in the cytosolic fraction following homogenization and high-speed centrifugation, (2) following hormone administration, unoccupied receptors disappear from the cytosol and (3) then appear in the nuclear fraction in a steroid-bound form.

This model has now been challenged in two studies using nonhomogenization techniques to locate the unoccupied receptor. King and Greene[25] used monoclonal antibodies to localize the receptor and reported that the unoccupied receptor is associated with the nucleus rather than with the cytoplasm. Welshons et al.[26] used cytochalasin B to enucleate receptor-containing cells and also reported that the unoccupied receptor is associated with the nucleus rather than the cytoplasm. These studies suggest that what we have been calling a cytoplasmic or cytosolic receptor must now be characterized as an "unoccupied receptor that is readily available for steroid binding following homogenization."

This conceptualization should remind us of the elegant study by Pietras and Szego[27] which showed that the location of estradiol binding sites in the uterus is very much a function of homogenization conditions. Our commonly used procedures guarantee a cytosolic location of the unoccupied receptor. With "gentle" procedures, estradiol binding was found primarily associated with the plasma membrane.

When equally compelling studies lead to different conclusions, we are forced to reexamine our models and the nature of the data on which they are based. It is fair to say that the location of the unoccupied receptor remains an open issue and merits further study. Nonetheless, this may be considered a "detail." As Schrader[28] has pointed out, the new findings do not alter the essence of Jensen's theory, namely, that occupied and unoccupied receptors are functionally different and that occupied receptors influence gene action. Our challenge as behavioral biologists is to determine how the interaction of hormone-receptor complexes with DNA alters neuronal function in ways that are relevant for our understanding of the hormonal regulation of behavior. Of course, even this goal must be questioned on occasion. Recall the fascinating report by Dryden and Anderson[29] showing that the sexual behavior, uterus, and vagina do not change when the Asian musk shrew (*Suncus murinus*) is ovariectomized. Ovariectomized shrews allow the male to intromit and ejaculate and ovariectomy causes no weight or histological changes in the uterus or vagina, even though nuclei from the uterus and vagina bind [^{3}H]estradiol at substantially higher levels than nuclei from the diaphragm or kidney. Indeed, there are no universals in biological systems.

The Jensen model of steroid hormone action has led us to believe that gonadal hormones act through their gene products, proteins. This belief led Quadagno and colleagues[30-33] through a series of experiments showing that the application of actinomycin-D (Act-D) directly to preoptic and ventromedial nuclei of rats inhibits the estrogen induction of lordosis. Act-D inhibits DNA-regulated RNA synthesis; thus, the conclusion that estrogen acts by activating the genome.

This conclusion has been relatively easy to accept given what we know about the time course of estrogen action. In our own work[34] we found that lordotic behavior did not appear before 16 hr after E2 treatment, many hours after the estrogen has been bound to hypothalamic nuclei.[35,36] Since studies of antiestrogen blockade of estrogen-induced lordosis show that there is a strong correlation between the inhibition of lordosis and the inhibition of nuclear binding of estrogen in the hypothalamus,[37-39] we are led to the conclusion that the activation of DNA by estrogen is *the* mechanism by which estrogen stimulates lordosis.

As noted earlier, one gene product of estrogen action is the progestin receptor. Estrogen stimulates the production of progestin receptors in several hypothalamic and limbic areas[40] and it has been argued that this receptor induction is important for the induction of lordotic behavior,[22,41] although Etgen[42] has reported that a similar induction of progestin receptor by estrogen occurs in males and neonatally androgenized females, animals that do not readily respond to estrogen and progesterone with the display of lordosis.

It is not easy to foresee how this controversy will be resolved. The fact that there is a good correlation between progestin receptor induction and lordotic behavior in female rats does not prove that receptor induction is necessary for lordosis (see Reference 43 for review). As shown recently by Rubin and Barfield,[44] cells localized in the ventromedial nucleus of the hypothalamus can respond sequentially to estrogen and progesterone to induce lordosis. Moreover, both estrogen and progesterone are accumulated by cells in the ventromedial nucleus. Thus, it might prove very difficult to dissociate a "pure" estrogen action from the estrogen induction of progestin receptor.

We must also be careful to dissociate pharmacological from physiological effects. We already know that we can induce lordosis in rats with high doses of estrogen alone[45] and we know that sequential treatment with estrogen followed by estrogen will induce lordosis,[46] but these effects may be simply pharmacological, providing little insight into the physiological mechanisms of lordosis control.

Returning to the Jensen model, it is possible to ask whether progesterone acts in the estrogen-primed animal via DNA activation and gene products. The issue was raised by Kent and Liberman's early finding[47] that intraventricular injection of P would lead to lordosis within 10 min and by the observations of Lisk[48] who reported the onset of lordotic behavior within 2.5-10 min of the intravenous administration of P. It was generally thought that this rapid onset of progesterone action more likely reflected a membrane effect of P than a protein synthetic action of the hormone. Since progesterone, like other gonadal steroids, can have short latency effects on brain electrical activity (see Reference 49 for review) such a mechanism seemed possible. With the recent demonstration by Towle and Sze[50] of specific binding of progesterone (and other steroids) to synaptic plasma membranes of rats, the possibility that P acts to facilitate lordosis by interaction with cell surface receptors will doubtless receive renewed attention.

Nonetheless, the hypothesis that P acts through protein synthesis cannot be discarded. Rainbow et al.[51] have reported that the protein synthesis inhibitor anisomycin blocks progesterone-induced lordosis when applied to the ventromedial nucleus. This finding has been replicated by Meisel and Pfaff.[52] These findings would suggest that P indeed acts by DNA activation of protein synthesis. However, this conclusion conflicts with the observation that Act-D does not inhibit progesterone-induced lordosis. Act-D appears to inhibit estrogen action, but not P action with respect to lordosis. Presumably, if P acts through protein synthesis, its action should be blocked by both Act-D and anisomycin. Meisel and Pfaff[52] have proposed an interesting resolution to this dilemma. They noted that "inhibitors of protein synthesis, but not actinomycin-D, can block cyclic AMP accumulation *in vitro*.... Thus, if progesterone acted on lordosis through specific membrane receptors, which stimulated cyclic AMP release, then one might expect a protein synthesis inhibitor, but not an RNA synthesis inhibitor, to be effective in this system" (p. 190). Thus, while we can correlate the inhibition of protein synthesis, whether produced by Act-D or by a protein synthesis inhibitor, with the inhibition of a behavioral response, such as lordosis, we must be cautious about establishing still another principle of hormone action, namely, that steroid hormones act solely via protein synthesis regulation.

What, then, are the principles by which hormones act on the brain to regulate behavior? I am suggesting that there may be none of great generality. The enormous genetic diversity, even amongst the limited number of species that we study in the laboratory, ensures that what is true of one genotype may not be true of another. This should not discourage us, for knowing the precise details of hormone action in one genotype would be a marvelous accomplishment. But we should never forget genetic diversity—it is what makes research in the biobehavioral sciences so exciting.

REFERENCES

1. BEACH, F. A. 1981. Historical origins of modern research on hormones and behavior. Horm. Behav. **15**: 325-376.
2. MCDONALD, P., C. BEYER, F. NEWTON, B. BRIEN, R. BAKER, H. S. TAN, S. SAMPSON,

P. KITCHING, R. GREENHILL & D. PRITCHARD. 1970. Failure of 5α-dihydrotestosterone to initiate sexual behavior in the castrated male rat. Nature **227:** 964-965.
3. LUTTGE, W. G. & N. R. HALL. 1973. Differential effectiveness of testosterone and its metabolites in the induction of male sexual behavior in two strains of albino mice. Horm. Behav. **4:** 31-44.
4. ALSUM, P. & R. W. GOY. 1974. Actions of esters of testosterone dihydrotestosterone or estradiol on sexual behavior in castrated male guinea pigs. Horm. Behav. **5:** 207-217.
5. WHALEN, R. E. & J. F. DEBOLD. 1974. Comparative effectiveness of testosterone, androstenedione and dihydrotestosterone in maintaining mating behavior in the castrated male hamster. Endocrinology **95:** 1674-1679.
6. PHOENIX, C. H. 1974. Effects of dihydrotestosterone on sexual behavior of castrated male rhesus monkeys. Physiol. Behav. **12:** 1045-1055.
7. OLSEN, K. L. 1979. Induction of mating behavior in androgen-insensitive (tfm) and normal (King-Holtzman) male rats: effect of testosterone propionate, estradiol benzoate and dihydrotestosterone. Horm. Behav. **13:** 66-84.
8. OLSEN, K. L. & R. E. WHALEN. 1984. Dihydrotestosterone activates male mating behavior in castrated King-Holtzman rats. Horm. Behav. **18:** 380-392.
9. BRAIN, P. F. & N. J. BOWDEN. 1979. Sex steroid control of intermale fighting in mice. In Current Developments in Psychopharmacology. W. B. Essman & L. Valzelli, Eds.: 403-475. SP Medical and Scientific Books. New York, N.Y.
10. GANDELMAN, R. 1980. Gonadal hormones and the induction of intraspecific fighting in mice. Neurosci. Biobehav. Rev. **4:** 133-140.
11. SCHECHTER, D., S. M. HOWARD & R. GANDELMAN. 1981. Dihydrotestosterone promotes fighting in female mice. Horm. Behav. **15:** 233-237.
12. FINNEY, H. C. & M. J. ERPINO. 1976. Synergistic effect of estradiol benzoate and dihydrotestosterone on aggression in mice. Horm. Behav. **7:** 391-400.
13. SIMON, N. G. & R. E. WHALEN. 1986. Hormonal regulation of aggression: evidence for a relationship among genotype, receptor binding and behavioral sensitivity to androgen and estrogen. Aggressive Behav. In press.
14. BEYER, C., G. MORALI & R. VARGAS. 1971. Effect of diverse estrogens on estrous behavior and genital tract development in ovariectomized rats. Horm. Behav. **2:** 273-277.
15. FEDER, H. H. & R. SILVER. 1974. Activation of lordosis in ovariectomized guinea pigs by free and esterified forms of estrone, estradiol-17β and estriol. Physiol. Behav. **13:** 251-255.
16. GORZALKA, B. B. & R. E. WHALEN. 1977. The effects of progestins, mineralocorticoids, glucocorticoids and steroid solubility on the induction of receptivity in mice. Horm. Behav. **8:** 94-99.
17. KUBLI-GARFIAS, C. & R. E. WHALEN. 1977. Induction of lordosis behavior in female rats by intravenous administration of progestins. Horm. Behav. **9:** 380-386.
18. CZAJA, J. A., D. A. GOLDFOOT & H. KARAVOLIS. 1974. Comparative facilitation and inhibition of lordosis in the guinea pig with progesterone, 5α-pregnane-3,20-dione or 3α-hydroxy-5α-pregnan-20-one. Horm. Behav. **5:** 261-274.
19. WADE, G. N. & H. H. FEDER. 1972. Effects of several pregnane and pregnene steroids on estrous behavior in guinea pigs. Physiol. Behav. **9:** 773-775.
20. GORZALKA, B. B. & R. E. WHALEN. 1974. Genetic regulation of hormone action: selective effects of progesterone and dihydroprogesterone (5α-pregnane-3,20-dione) on sexual receptivity in mice. Steroids **23:** 499-505.
21. LUTTGE, W. G. & N. R. HALL. 1976. Interactions of progesterone and dihydroprogesterone with dihydrotestosterone on estrogen activated sexual receptivity in female mice. Horm. Behav. **7:** 253-257.
22. MOGUILEWSKY, M. & J.-P. RAYNAUD. 1979. The relevance of hypothalamic and hypophyseal progestin receptor regulation in the induction and inhibition of sexual behavior in the female rat. Endocrinology **105:** 516-522.
23. BLAUSTEIN, J. D. & H. H. FEDER. 1979. Cytoplasmic progestin receptors in guinea pig brain: characteristics and relationship to the induction of sexual behavior. Brain Res. **169:** 481-497.
24. JENSEN, E. V. & E. R. DESOMBRE. 1973. Estrogen-receptor interaction. Science **182:** 126-134.

25. KING, W. J. & G. L. GREENE. 1984. Monoclonal antibodies localize oestrogen receptor in the nucleus of target cells. Nature **307:** 745-747.
26. WELSHONS, W. V., M. E. LIEBERMAN & J. GORSKI. 1984. Nuclear localization of unoccupied oestrogen receptors. Nature **307:** 747-749.
27. PIETRAS, R. J. & C. M. SZEGO. 1979. Estrogen receptors in uterine plasma membrane. J. Steroid Biochem. **11:** 1471-1483.
28. SCHRADER, W. T. 1984. New model for steroid hormone receptors? Nature **308:** 17-18.
29. DRYDEN, G. L. & J. N. ANDERSON. 1977. Ovarian hormone: lack of effect on reproductive structures of female Asian musk shrews. Science **197:** 782-784.
30. QUADAGNO, D. M., J. SHRYNE & R. A. GORSKI. 1971. The inhibition of steroid-induced sexual behavior by intrahypothalamic actinomycin-D. Horm. Behav. **2:** 1-10.
31. HO, G. K. W., D. M. QUADAGNO, P. H. COOKE & R. A. GORSKI. 1973/1974. Intracranial implants of actinomycin-D: effects on sexual behavior and nucleolar ultrastructure in the rat. Neuroendocrinology **13:** 47-55.
32. HOUGH, J. C., G. K. W. HO, P. H. COOKE & D. M. QUADAGNO. 1974. Actinomycin-D: reversible inhibition of lordosis behavior and correlated changes in nucleolar morphology. Horm. Behav. **5:** 367-375.
33. WHALEN, R. E., B. B. GORZALKA, J. F. DEBOLD, D. M. QUADAGNO, G. HO & J. C. HOUGH. 1974. Studies on the effects of intracerebral actinomycin-D implants on estrogen-induced receptivity in rats. Horm. Behav. **5:** 337-343.
34. GREEN, R., W. G. LUTTGE & R. E. WHALEN. 1970. Induction of receptivity in ovariectomized female rats by a single intravenous injection of estradiol-17β. Physiol. Behav. **5:** 137-141.
35. ZIGMOND, R. E. & B. S. MCEWEN. 1970. Selective retention of oestradiol by cell nuclei in specific brain regions of the ovariectomized rat. J. Neurochem. **17:** 889-899.
36. WHALEN, R. E. & J. MASSICCI. 1975. Subcellular analysis of the accumulation of estrogen by the brain of male and female rats. Brain Res. **89:** 255-264.
37. ROY, E. J. & G. N. WADE. 1977. Binding of ^{3}H-estradiol by brain cell nuclei and female rat sexual behavior: inhibition by antiestrogens. Brain Res. **126:** 73-87.
38. WADE, G. N. & J. D. BLAUSTEIN. 1978. Effects of an anti-estrogen on neural estradiol binding and on behavior in female rats. Endocrinology **102:** 245-251.
39. ETGEN, A. M. 1979. Antiestrogens: effects of tamoxifen, nafoxidine and CI-628 on sexual behavior, cytoplasmic receptors and nuclear binding of estrogen. Horm. Behav. **13:** 97-112.
40. PARSONS, B., T. C. RAINBOW, N. J. MACLUSKY & B. S. MCEWEN. 1982. Progestin receptor levels in rat hypothalamus and limbic nuclei. J. Neurosci. **2:** 1446-1452.
41. PARSONS, B., N. J. MACLUSKY, L. KREY, D. W. PFAFF & B. S. MCEWEN. 1980. The temporal relationship between estrogen-inducible progestin receptors in the female rat brain and the time course of estrogen activation of mating behavior. Endocrinology **107:** 774-779.
42. ETGEN, A. M. 1981. Estrogen induction of progestin receptors in the hypothalamus of male and female rats which differ in their ability to exhibit cyclic gonadotropin secretion and female sexual behavior. Biol. Reprod. **25:** 307-313.
43. ETGEN, A. M. 1984. Progestin receptors and the activation of female reproductive behavior: a critical review. Horm. Behav. **18:** 411-430.
44. RUBIN, B. S. & R. J. BARFIELD. 1983. Induction of estrous behavior in ovariectomized rats by sequential replacement of estrogen and progesterone to the ventromedial hypothalamus. Neuroendocrinology **37:** 218-224.
45. EDWARDS, D. A., R. E. WHALEN & R. D. NADLER. 1968. Induction of estrus: estrogen-progesterone interactions. Physiol. Behav. **3:** 29-33.
46. KOW, L-M. & D. W. PFAFF. 1975. Induction of lordosis in female rats: two modes of estrogen action and the effect of adrenalectomy. Horm. Behav. **6:** 259-276.
47. KENT, G. C. & M. J. LIBERMAN. 1949. Induction of psychic estrus in the hamster with progesterone administered via the lateral brain ventricle. Endocrinology **45:** 29-32.
48. LISK, R. D. 1960. A comparison of the effectiveness of intravenous as opposed to subcutaneous injection of progesterone for the induction of sexual behavior in the rat. Can. J. Biochem. Physiol. **38:** 1381-1383.

49. PFAFF, D. A. 1981. Electrophysiological effects of steroid hormones in brain tissue. *In* Neuroendocrinology of Reproduction. Physiology and Behavior. N.T. Adler, Ed.: 533-541. Plenum. New York, N.Y.
50. TOWLE, A. C. & P. Y. SZE. 1983. Steroid binding to synaptic plasma membrane: differential binding of glucocorticoids and gonadal steroids. J. Steroid Biochem. **18:** 135-143.
51. RAINBOW, T. C., M. Y. MCGINNIS, P. G. DAVIS & B. S. MCEWEN. 1982. Application of anisomycin to the lateral ventromedial nucleus of the hypothalamus inhibits the activation of sexual behavior by estradiol and progesterone. Brain Res. **233:** 417-423.
52. MEISEL, R. L. & D. W. PFAFF. 1984. RNA and protein synthesis inhibitors: effects on sexual behavior in female rats. Brain Res. Bull. **12:** 187-193.

Heterotypical Sexual Behavior: Implications from Variations[a]

JANICE E. THORNTON

Institute of Animal Behavior
Rutgers-The State University of New Jersey
Newark, New Jersey 07102

One generally thinks of sexual behavior by males and females of any given species as being sexually dimorphic. That is, certain behaviors are associated with one sex and not the other. In general, males show mounting, intromissive, and ejaculatory behaviors, whereas females display certain sexual solicitations and lordosis or presenting postures. It has been proposed that many, if not all, of these behavioral dimorphisms are dependent upon the action of gonadal hormones.[1-3] However, for many species complete sexual behavior dimorphism does not exist.[2,4] The display of heterotypical sexual behavior is quite common.[5,6] Heterotypical sexual behavior refers to instances where males show female-typical and/or females show male-typical copulatory behavior. Homotypical sexual behavior refers to the case of females showing female-typical or males showing male-typical copulatory behavior. In the present discussion I will first review the current model for the role of hormones in the sexual differentiation and activation of homotypical sexual behaviors. Secondly, I will review how this model has been applied to the heterotypical display of sexual behavior, describe some cases of heterotypical sexual behavior, and then see how well the model can account for them. Lastly, because some of these cases are not completely explained by the extant model, I will discuss the implications of this mismatch on the formulation of the model. Just as genetic anomalies can tell us something about the normal functioning of genes, variations of behavior sexual differentiation that do not fit the extant model may highlight the complexities involved and give us further clues about the underlying principles of sexual differentiation.

GONADAL HORMONES AND DIMORPHIC SEXUAL BEHAVIOR

The currently accepted model for the role of gonadal hormones on dimorphic behavior is that most, if not all, behavioral dimorphisms are dependent upon exposure to gonadal hormones at one or both of two periods of time: a "sensitive" period early in development (generally perinatally) and in adulthood.[2] Dimorphic sexual behaviors are generally dependent upon exposure to hormones at both of these times.

---

[a]Parts of this work were supported by National Institutes of Health Grant RR00167 and National Institute of Mental Health Grant MH21312.

It has long been known that gonadal hormones maintain sexual and other dimorphic characters in adulthood.[7] Generally, the full complement of sexual behavior normally shown by adult female and male mammals is dependent upon ovarian hormones (estrogens and progestins) or testicular hormones (androgens and/or estrogens), respectively. This "activational" effect is reversible in that gonadectomy of either sex leads to decreased sexual behavior which can then be restored with appropriate hormones (see Reference 8).

For most species it is apparent that adult hormone levels are not the sole determinants of sex-typical behavior. Generally one cannot cause a complete sex reversal of copulatory behavior by administration of ovarian hormones to adult gonadectomized males or testicular hormones to adult gonadectomized females.[5]

It has been suggested that this inability to cause a complete sex behavior reversal in adulthood is at least partly attributable to the presence or absence early in development of gonadal hormones which have permanent "organizational" effects on the differentiation of the brain mechanisms which later control adult dimorphic sexual behavior.[1,4] It appears that the neural systems mediating mammalian sexual behavior have an intrinsic tendency to develop according to the female pattern. For the most part, the capability to display female behavior in adulthood is not dependent upon the perinatal presence of ovaries and their secretory products (e.g., Reference 9; although see Reference 10 for contradictory evidence). On the other hand, if testicular hormones are present during the perinatal period, as in normal males, they act to permanently reorganize or differentiate the developing brain to respond in a male fashion.

Exposure of developing animals to testicular hormones has been conceptualized as having two major effects on adult dimorphic sexual behavior, masculinization and defeminization.[1,11,12] Masculinization refers to an increase in the occurrence of behaviors normally better developed in males than in females such as mounting, intromission, and/or ejaculation. Defeminization refers to a decrease in the occurrence of behaviors typically better developed in females than in males such as lordosis, presentation, and/or certain sexual solicitations. Masculinization and defeminization are independent processes which may be affected differentially depending on the dosage, timing, and type of hormone used. Thus one can induce masculinization without concomitant defeminization and vice versa.[11,13-16] Additionally, although earlier research suggested gonadal androgens mediate early masculinization and defeminization of adult mammalian sexual behavior, later work has indicated that in some species and/or strains it may actually be perinatal estrogens, derived from androgens through aromatization, that are the active substances.[17-21]

In sum, the general model is that adult gonadal hormones activate the brain mechanisms which control sexual behavior and these brain mechanisms have been masculinized and defeminized in males by perinatal exposure to the organizational effects of testicular androgens (and/or estrogens).

GONADAL HORMONES AND LORDOSIS IN MALES, MOUNTING IN FEMALES

Because males have testes and hence testicular hormones perinatally (and in adulthood) whereas females do not, the model just presented could be interpreted as propounding that mounting should be shown exclusively by normal males whereas only normal females should show lordosis. However, this condition of complete dimorphism generally does not exist. According to Beach,[6] mounting by females occurs in at least 13 species of five different orders. Although lordosis by males may be less common than mounting by females[5] it still occurs in more than a few species.

To account for the lack of complete sex behavior dimorphism, further elaborations have been made on the model. Beach[22] suggested that activational hormones from the animal's own gonads maintained spontaneous heterotypical sexual behavior. Although no theory has been promulgated regarding the particular type of activational hormone involved in the control of heterotypical sexual behavior, if heterotypical sexual behavior is analogous to the same behavior in the sex which usually displays it, then it may depend on the same activational hormones as in the sex which usually shows it. A role for organizational hormones has also been proposed.[2,23] It has been hypothesized that anomalies in endocrine stimulation during the early stages of development lead to the display of heterotypical sexual behavior. More specifically, it has been suggested that the lordosis or presenting seen in some males is due to a deficiency of or insensitivity to perinatal hormonal stimulation (i.e., there is a lack of defeminization), whereas the mounting seen in females of many species is due to supervening exposure to perinatal testicular hormones (i.e., they are masculinized). I will now examine some of the cases where females mount or males show lordosis. I will describe first the adult hormonal conditions which elicit the behavior and then see if the presence of these heterotypical behaviors is due to perinatal exposure to excess androgens (and/or estrogens) in females or to the lack of androgens (and/or estrogens) in males.

Mounting by Females

Although mounting has been reported for females of many species,[6] the role of adult and perinatal hormones on female mounting in most of these species is unknown. However, two species where the role of hormones on female mounting behavior has been examined extensively are the rat and guinea pig.

Mounting in adult female rats does not appear to depend upon adult ovarian hormones. Female rats mount at low levels independent of the presence of ovaries and mounting is not facilitated in ovariectomized (ovx) females by sequential estrogen (E) and progesterone (P) administration.[24] However, the mounting shown by gonadectomized female rats is similar to that shown by males as it is enhanced by testosterone administration in adulthood.[25,26] The mounting shown by adult female rats appears to be due, at least partially, to prenatal exposure to testicular androgens (and/or estrogens) from male siblings. Mounting behavior in adulthood is highest for females located *in utero* closest to, and lowest for those most distal to, male siblings.[27] Furthermore, antiandrogen treatment *in utero* diminishes the amount of adult mounting behavior shown by female rats.[28]

In contrast to rats, the mounting shown by normal adult female guinea pigs depends upon adult ovarian hormones. Female guinea pigs regularly mount around the time of heat onset.[29-31] When ovariectomized, estrogen and progesterone administration will also lead to mounting and the most effective regimen of E and P is that which also produces lordosis.[32] In contrast to males, testosterone administration in adulthood to females has much less effect on mounting than does estradiol and progesterone treatment.[32,33] It is unlikely that this mounting in response to E and P is due to prenatal exposure to hormones from male siblings since even females with no siblings will mount.[34] This mounting is probably not the result of exposure to prenatal androgens from some source other than male siblings because preliminary data indicate that prenatal administration of the antiandrogen flutamide does not significantly decrease E and P induced mounting in adult female guinea pigs.[35] Additionally, prenatal androgen exposure does not increase mounting in adulthood in response to estrogen and progesterone although it does increase mounting in response to testosterone.[1]

Lordosis in Males

Lordosis, or some equivalent of the female sexual receptivity posture, is shown by normal adult males from a number of species.

Similar to females, male hamsters will readily show lordosis if given estrogen and progesterone as adults.[36,37] If males are treated neonatally with testosterone, their adult lordosis to E and P is eliminated.[37]

Although adult gonadally intact male ferrets have not been tested, it is known that adult castrated males will readily show the female-typical receptivity posture. As in female ferrets, the response is shown after estradiol treatment in adulthood.[38] It has been suggested that male ferrets readily show the female receptivity posture because they are insensitive to the defeminizing actions of perinatal testicular hormones on this behavior.[39] Consistent with this interpretation, high doses of androgens or estrogens given perinatally do not block the display of the receptivity posture in adulthood in either males or females.[38,40] This is not a complete insensitivity to hormones since perinatal testosterone exposure will masculinize adult behavior.[38,40,41]

Lordosis is also shown by neonatal male guinea pigs. Both male and female guinea pigs show lordosis for a period of time beginning within 1 hr and lasting for approximately 3 hr after birth.[42] Although the lordosis shown by the neonatal animals appears similar to that shown by adult females in response to estradiol and progesterone, it is probably not dependent on gonadal hormones.[42,43] Prenatal androgen administration also has no effect on the display of this neonatal lordosis.[44] In contrast, prenatal androgens will block the lordosis normally shown by adult females to estrogen and progesterone.[1] Rather than an example of true female copulatory behavior, this neonatal lordosis may instead be part of the normal elimination pattern of the infant guinea pig.[43,45]

Lastly, male guinea pigs from inbred strain 2 will readily show lordosis independent of gonadal hormone stimulation.[46] That is, castration does not decrease the display of lordosis and castrated males who show lordosis do not have higher plasma estrone, estradiol, or androgen levels than ovx females (none of which show lordosis without exogeneous estrogen treatment). Additionally, housing in larger groups (e.g., six) suppresses the display of lordosis by strain 2 males (but not by strain 2 females). This display of lordosis by strain 2 males is not simply due to a lack of defeminization as a result of a deficiency of or insensitivity to prenatal gonadal hormones. When strain 2 males are given additional testosterone prenatally they continue to show lordosis in adulthood and, interestingly, prenatally testosterone treated strain 2 females display lordosis in adulthood under the conditions in which strain 2 males show the behavior rather than under those in which the female normally shows the behavior. That is, prenatally androgen treated strain 2 females show hormone-independent lordosis when they are isolated but not when they are housed in large groups.[47]

IMPLICATIONS

Although the model of activational and testicular organizational hormones may adequately explain the development of female-typical copulatory behavior in females and male-typical copulatory behavior in males for most of the species examined, the way the model has been applied to heterotypical sexual behavior does not always account for the presence of heterotypical behavior in either sex. Although in some cases it may be true that spontaneous display of heterotypical behavior depends upon the presence of gonads as suggested by Beach,[22] this is not true for others, e.g., mounting in female rats and lordosis in neonatal male guinea pigs or strain 2 male guinea pigs. It is also clear that heterotypical sex behavior is not always controlled by the same activational hormones as when that behavior is displayed by the sex which usually shows it. Thus mounting in adult male guinea pigs is dependent upon activational androgens but mounting in female guinea pigs is dependent on estradiol and progesterone. The role of organizational hormones in adult display of heterotypical sex behavior also varies. In some cases it does indeed appear that mounting in females is due to perinatal androgen (and/or estrogen) exposure (e.g., female rats) whereas lordosis in males is due to perinatal androgen insufficiency (e.g., male hamsters). In other instances, this is not the case (e.g., female guinea pig mounting and strain 2 male lordosis).

This examination of the various instances of heterotypical sexual behavior has led me to a reformulation of the concept of heterotypical sexual behavior. Previous investigators have considered all cases of male lordosis and female mounting automatically as instances of heterotypical sexual behavior, i.e., female-typical or male-typical behavior, respectively. It makes some sense to consider lordosis as female typical and mounting as male typical since to reproduce, females of all species must show lordosis or its equivalent and males must show mounting. Nevertheless, it is important to note that in some cases the lordosis shown by males is not controlled by the same mechanisms as lordosis shown by females. Analogous to this, female mounting is also sometimes controlled by different mechanisms than mounting in males. Therefore, it appears that there are really two different categories of heterotypical sexual behavior. One is true heterotypical behavior, the other is quasi-heterotypical behavior. The two terms are defined as follows: True heterotypical sexual behavior refers to those cases where females mount under the same eliciting stimulus conditions (this includes both environmental and concurrent hormonal conditions) as males of that same strain and/or species. The parallel case holds for male lordosis or presentation postures. In essence, under this schema the designation of female mounting or male lordosis as truly male typical or female typical, respectively, would be context and species/strain specific. Quasi-heterotypical sexual behavior then refers to cases of female mounting and male lordosis which are not shown under the same stimulus conditions as when the behavior is shown as a sexual behavior by males and females, respectively, of that same strain/species. Basically, quasi-heterotypical sexual behaviors are cases where males show lordosis under conditions not typical of females of that strain and mounting is shown by females under conditions not typical of males of that strain. Cases of true heterotypical sexual behavior include mounting in female rats and lordosis in male hamsters. The sexual receptivity posture in male ferrets also probably fits under this rubric. In contrast, female guinea pig mounting, neonatal guinea pig lordosis, and strain 2 male lordosis are really examples of quasi-heterotypical sexual behavior. The example that points out the need for the distinction between true and quasi-heterotypical sexual behavior perhaps most clearly is the strain 2 male guinea pig. Since strain 2 males show lordosis under different adult hormonal and social conditions than females and prenatal androgen exposure causes females to show lordosis under the conditions in which males, rather than females, show the behavior, it is most appropriate to call strain 2 male lordosis quasi-heterotypical, or perhaps even male-typical behavior.

The division of heterotypical sexual behavior into two distinct categories may resolve the lack of agreement between the organizational hypothesis and the occurrence of heterotypical sexual behavior. The basic premise of the organizational hypothesis is really not that perinatal hormones increase mounting and decrease lordosis but that they increase male-typical and decrease female-typical behavior. From the data available, it appears that indeed those cases of female mounting which are true male-typical behaviors in females are due to perinatal masculinzation and cases of male lordosis which are true female-typical behaviors in males appear to be due to a lack of defeminization by gonadal hormones. This holds for both female rat mounting behavior and male hamster lordosis. The sexual receptivity posture in male ferrets may also be explained in this manner. Further testing needs to be done with other species to determine if this postulate has wide generalizability.

The role of organizational hormones in the display of quasi-heterotypical sexual behavior cannot be readily categorized. Even if the argument is taken one step further and the definitions of male-typical and female-typical behaviors are made species and context specific, only some cases fit the organizational hypothesis. For example, if hormonally independent, socially suppressed lordosis is considered to be male typical for strain 2 then, just as the organizational hypothesis would predict, females are masculinized (i.e., show male-type lordosis) by prenatal androgens. On the other hand, even if mounting in response to E and P in female guinea pigs is considered female typical for that species, there is no indication that it is due to a lack of defeminization by prenatal androgens. Prenatally androgenized females still show mounting to E and P.[1] It may only be after more cases are accumulated that generalities will emerge.

In summary, there are two types of heterotypical sexual behaviors, true and quasi. One needs to be careful not to automatically assume that any display of lordosis is controlled by the same mechanisms as lordosis that is shown by sexually behaving adult females. The analogous case is true for mounting. Furthermore, it appears that adult mounting may be present without exposure whereas lordosis may be present even with exposure to perinatal androgens. This suggests that the existence of the neural circuits controlling mounting and lordosis is not dependent upon perinatal androgen exposure. As a corollary of this, it is important to note that the display of mounting by females is not necessarily indicative of masculinization, nor can one infer that the presence of lordosis or presenting postures is indicative of a lack of defeminization by perinatal hormones. For many species, the organizational effect of hormones may not be to determine the presence or absence of lordosis and mounting but rather to determine under what social and/or adult hormonal conditions the behavioral patterns will be displayed. Perinatal gonadal hormones may only be one of a number of substances which affect the probability of display of either behavior.

It is unknown if there is a final common neural pathway for all the cases of lordosis or all the cases of mounting. It is also unknown at what level variations in the control mechanisms occur. Nor is much known about the role of specific neurotransmitters or receptors (either perinatally or in adulthood) in determining what factors will or will not influence the display of heterotypical sexual behavior. These are a few of the many areas which remain to be explored. Perhaps through further exploration of questions such as these we can learn more about the mechanisms involved in the control of heterotypical sexual behavior. The study of variations in heterotypical sexual behavior may also offer us a different perspective with which to gain new insights into the underlying principles involved in the control of adult homotypical sexual behavior and the general process of sexual differentiation.

ACKNOWLEDGMENTS

The author wishes to thank R. W. Goy, D. A. Goldfoot, and K. Wallen for helpful discussions and H. Feder, M. Loose, H. Adieh, and B. Inman for critical comments on an earlier draft of the manuscript.

REFERENCES

1. PHOENIX, C. H., R. W. GOY, A. A. GERALL & W. C. YOUNG. 1959. Organizing actions of prenatally administered testosterone propionate on the tissues mediating mating behavior in the female guinea pig. Endocrinology 65: 369-382.
2. GOY, R. W. & B. S. MCEWEN. 1980. Sexual Differentiation of the Brain. MIT Press. Cambridge, Mass.
3. MCEWEN, B. S. 1983. Gonadal steroid influences on brain development and sexual differentiation. In Reproductive Physiology IV, International Review of Physiology. R. O. Greep, Ed. 27: 99-145. University Park Press. Baltimore, Md.
4. GOY, R. W. & D. A. GOLDFOOT. 1973. Hormonal influences on sexually dimorphic behavior. In Handbook of Physiology, Endocrinology II, Part 1. R. O. Greep & E. B. Astwood, Eds.: 169-186. American Physiological Society. Washington, D.C.
5. YOUNG, W. C. 1961. The hormones and mating behavior. In Sex and Internal Secretions. W. C. Young, Ed.: 1173-1239. Williams and Wilkins, Baltimore, Md.
6. BEACH, F. A. 1968. Factors involved in the control of mounting behavior by female mammals. In Perspectives in Reproduction and Sexual Behavior. M. Diamond, Ed.: 83-131. Indiana University Press. Bloomington, Ind.
7. BERTHOLD, A. 1849. Transplantation der hoden. Arch. f. Anat. u. Physiol. S: 42.
8. CLEMENS, L. G. & B. A. GLADUE. 1979. Neuroendocrine control of adult sexual behavior. Rev. Neurosci. 4: 73-103.
9. BAUM, M. J. 1979. Differentiation of coital behavior in mammals: a comparative analysis. Neurosci. Biobehav. Rev. 3: 265-284.
10. DOHLER, K. D., J. L. HANCKE, S. S. SRIVASTAVA, C. HOFMANN, J. E. SHRYNE & R. A. GORSKI. 1984. Participation of estrogens in female sexual differentiation of the brain: neuroanatomical, neuroendocrine and behavioral evidence. In Progress in Brain Research. G. J. DeVries, J. P. C. De Bruin, H. B. M. Uylings and M. A. Corner, Eds. 61: 99-117. Elsevier. Amsterdam, the Netherlands.
11. BEACH, F. A., T. E. KUEHN, R. H. SPRAGUE & J. J. ANISKO. 1972. Coital behavior in dogs. XI. Effects of androgenic stimulation during development on masculine mating responses in females. Horm. Behav. 3: 143-168.
12. BEACH, F. A. 1975. Hormonal modification of sexually dimorphic behavior. Psychoneuroendocrinology 1: 3-23.
13. GOLDFOOT, D. A., H. H. FEDER & R. W. GOY. 1969. Development of bisexuality in the male rat treated neonatally with androstenedione. J. Comp. Physiol. Psychol. 67: 41-45.
14. GOLDFOOT, D. A. & J. J. VAN DER WERFF TEN BOSCH. 1975. Mounting behavior of female guinea pigs after prenatal and adult administration of the propionates of testosterone, dihydrotestosterone and androstanediol. Horm. Behav. 6: 139-148.
15. DEBOLD, J. F. & R. E. WHALEN. 1975. Differential sensitivity of mounting and lordosis control systems to early androgen treatment in male and female hamsters. Horm. Behav. 6: 197-209.

16. WHITSETT, J. M. & J. G. VANDENBERG. 1975. Influence of testosterone propionate administered neonatally on puberty and bisexual behavior in female hamsters. J. Comp. Physiol. Psychol. **88**: 248-255.
17. WHALEN, R. E. & R. D. NADLER. 1963. Suppression of the development of female mating behavior by estrogen administered in infancy. Science **141**: 273-274.
18. LEVINE, S. & R. J. MULLINS, JR. 1964. Estrogen administered neonatally affects adult sexual behavior in male and female rats. Science **144**: 185-187.
19. PAUP, D. C., L. P. CONIGLIO & L. G. CLEMENS. 1972. Masculinization of the female golden hamster by neonatal treatment with androgen or estrogen. Horm. Behav. **3**: 123-131.
20. CONIGLIO, L., D. C. PAUP & L. G. CLEMENS. 1973. Hormonal factors controlling the development of sexual behavior in the male golden hamster. Physiol. Behav. **10**: 1087-1094.
21. MCEWEN, B. S., I. LIEBERBERG, C. CHAPTAL & L. C. KREY. 1977. Aromatization: important for sexual differentiation of the neonatal rat brain. Horm. Behav. **9**: 249-263.
22. BEACH, F. A. 1976. Cross species comparisons and the human heritage. Arch. Sex. Behav. **5**: 469-485.
23. GOY, R. W. & D. A. GOLDFOOT. 1975. Neuroendocrinology: animal models and problems of human sexuality. Arch. Sex. Behav. **4**: 405-420.
24. BEACH, F. A. & P. RASQUIN. 1942. Masculine copulatory behavior in intact and castrated female rats. Endocrinology **31**: 393-409.
25. BALL, J. 1937. The effect of male hormone on the sex behavior of female rats. Psychol. Bull. **34**: 725-732.
26. BEACH, F. A. 1942. Execution of the complete masculine copulatory pattern by sexually receptive female rats. J. Genet. Psychol. **60**: 137-142.
27. CLEMENS, L. G. & L. CONIGLIO. 1971. Influence of prenatal litter composition on mounting behavior of female rats. Am. Zool. **11**: 617.
28. WARD, I. L. & F. J. RENZ. 1972. Consequences of perinatal hormone manipulation on the adult sexual behavior of female rats. J. Comp. Physiol. Psychol. **78**: 349-355.
29. AVERY, G. T. 1925. Notes on reproduction in guinea pigs. J. Comp. Physiol. Psychol. **5**: 373-396.
30. YOUNG, W. C., E. C. DEMPSEY & H. I. MEYERS. 1935. Cyclic reproductive behavior in the female guinea pig. J. Comp. Physiol. Psychol. **19**: 313-335.
31. GOY, R. W. & W. C. YOUNG. 1957. Strain differences in the behavioral responses of female guinea pigs to alpha-estradiol benzoate and progesterone. Behaviour **10**: 340-354.
32. YOUNG, W. C. & B. RUNDLETT. 1939. The hormonal induction of homosexual behavior in the spayed female guinea pig. Psychosom. Med. **1**: 449-460.
33. DIAMOND, M. 1965. The antagonistic actions of testosterone propionate and estrogen and progesterone on copulatory patterns of the female guinea pig. Am. J. Physiol. **116**: 201-209.
34. THORNTON, J. E. & R. W. GOY. Unpublished.
35. THORNTON, J. E., S. IRVING & R. W. GOY. Unpublished.
36. TIEFER, L. 1970. Gonadal hormones and mating behavior in the adult golden hamster. Horm. Behav. **1**: 189-202.
37. EATON, G. 1970. Effect of a single prepubertal injection of testosterone propionate on adult bisexual behavior of male hamsters castrated at birth. Endocrinology **87**: 934-940.
38. BAUM, M. J., C. A. GALLAGHER, J. T. MARTIN & D. A. DAMASSA. 1982. Effects of testosterone, dihydrotestosterone or estradiol administered neonatally on sexual behavior of female ferrets. Endocrinology **111**: 773-780.
39. BAUM, M. J. & C. A. GALLAGHER. 1981. Increasing dosages of estradiol benzoate activate equivalent degrees of sexual receptivity in gonadectomized male and female ferrets. Physiol. Behav. **26**: 751-753.
40. BAUM, M. J. 1976. Effects of testosterone propionate administered perinatally on sexual behavior of female ferrets. J. Comp. Physiol. Psychol. **90**: 399-410.
41. BAUM, M. J. & M. S. ERSKINE. 1984. Effect of neonatal gonadectomy and administration of testosterone on coital masculinization in the ferret. Endocrinology **115**: 2440-2444.
42. BOLING, J. L., R. J. BLANDAU, J. G. WILSON & W. C. YOUNG. 1939. Post-parturitional heat responses of newborn and adult guinea pigs. Proc. Soc. Exp. Biol. Med. **42**: 128-132.

43. BEACH, F. A. 1966. Ontogeny of 'coitus-related' reflexes in the female guinea pig. Proc. Natl. Acad. Sci. USA **56:** 526-532.
44. GOY, R. W., C. H. PHOENIX & R. MEIDINGER. 1967. Postnatal development of sensitivity to estrogen and androgen in male, female and pseudohermaphroditic guinea pigs. Anat. Rec. **157:** 87-96.
45. HARPER, L. V. 1972. The transition from filial to reproductive function of 'coitus-related' responses in young guinea pigs. Dev. Psychobiol. **5:** 21-34.
46. THORNTON, J. E., K. WALLEN & R. W. GOY. Submitted. Lordosis behavior in males of two inbred strains of guinea pig.
47. THORNTON, J. E., J. V. CADWALLADER & R. W. GOY. 1982. An unusual organizational effect of prenatal testosterone on lordosis in guinea pigs. Conference on Reproductive Behavior, East Lansing, MI.

The Role of Androgen Metabolism in the Activation of Male Behavior[a]

CHERYL F. HARDING

Biopsychology Program
Hunter College, City University of New York
New York, New York 10021

Department of Ornithology
American Museum of Natural History
New York, New York 10024

Although we have begun to realize the importance of hormone metabolism in modulating an individual's response to particular gonadal hormones, the full import of normal metabolic influences still escapes us. For many years, scientists interested in endocrine function in the male focused primarily on the production of testosterone (T). T metabolism was viewed primarily as a catabolic process, resulting in the breakdown of the active hormone and its excretion. A large number of androgenic metabolites were identified, but most of them were ignored because they had minimal biological potency when their activity was assessed in various bioassay systems such as rat seminal vesicle or chick comb. One exception to this was 5α-dihydrotestosterone (DHT). First, DHT was shown to be more potent than testosterone in several bioassay systems, and then in the sixties, conclusive evidence was published which documented the local formation of DHT in male sexual accessory tissues. These findings sparked research on the importance of local metabolism in increasing, as well as decreasing, the biological activity of gonadal hormones.

Research on the formation and mechanism of action of DHT in the sexual accessory tissues proved invaluable in developing a general model of androgen metabolism and action. The higher concentrations and more uniform distribution of hormones and hormone receptors in peripheral tissues allowed researchers to carry out studies on these tissues which were not technically or practically feasible in brain tissue. Ironically, the focus on sexual accessory tissues led many to ignore metabolic pathways other than 5α-reduction, but when scientists began to study the role of androgen metabolism in the activation of behavior, it became obvious that other metabolic pathways were also important, since DHT showed limited effectiveness in activating normal levels of male social behavior. This pointed to the importance of another metabolic process, the aromatization of androgens to estrogens, for estrogenic metabolites play an important role in activating male behaviors in many species. This research has highlighted the possibility that many behaviors which have been described as androgen dependent will prove in fact to be dependent on estrogens formed from androgenic precursors.

Aromatization and 5α-reduction are opposing metabolic pathways. That is, once an androgen is 5α-reduced, most evidence suggests that it can no longer be converted

---

[a] This work was supported by City University of New York PSC-BHE Research Awards 13011, 13495 and 664182, and by NICHD Grant HD15191.

to an estrogen. Conversely, once an androgen is aromatized to an estrogen, all evidence suggests that it cannot be converted back to an androgen and 5α-reduced. Comparative studies indicate that aromatizing and 5α-reducing enzymes are present in the brains of representatives of every major vertebrate group, suggesting that the ability to metabolize androgens such as T is a primitive characteristic of the brain which has been widely conserved.[1] The coexistence of these two metabolic pathways may be a specialization of brain tissue; most tissues possess only one of these enzyme systems. Data from those species which have been investigated indicate that both of these metabolic pathways are involved in the activation of behavior. In many species, particular behaviors, most often copulation or high-intensity aggressive behaviors, appear to be activated by very low doses of estrogen (e.g., finch,[2] hamster,[3] Japanese quail,[4] mice,[5] red deer,[6] and guinea pig.[7] Treatment with DHT also reinstates some level of sexual and/or aggressive behavior in castrated males, though it is usually not as effective as T (e.g., hamsters,[8] rabbits,[9] some strains of mice,[10] guinea pigs,[11] and rhesus monkeys[12]).

Such data have led to heated controversy over the specificity of the hormonal activation of male social behaviors. This controversy has focused for the most part on the relative efficacy of androgens versus estrogens in activating behavior. This controversy is complicated by many issues, including differences in the relative efficacies of these two classes of steroids across species and even across closely related strains, the imperfect specificity of androgen and estrogen receptors, the inherent limitations of castration-hormone replacement experiments, the multiple sites of hormone action in activating any one behavior, and the multiple sites of hormone metabolism. In the midst of this controversy, researchers have often shown a parochial interest in demonstrating that a given male behavior is controlled by a particular androgenic or estrogenic metabolite, ignoring the complex issues outlined above. Unfortunately, ignoring these issues has often led to poor experimental design, the misinterpretation of data, and a high proportion of conflicting conclusions between studies.

The fact that the metabolites which are best able to activate behavior vary from species to species and even from strain to strain has made it difficult to develop a general model of androgen metabolism and the modulation of behavior. For example, in many species such as rats, rabbits, and mice, T appears to be the most effective treatment for restoring the behavioral repertoire of castrated males, but in other species including hamsters, zebra finches, and pigeons, another aromatizable androgen, androstenedione (AE), is more effective than T. The underlying cause of this variability in efficacy is unknown. Similarly, in some species such as rats, estradiol alone is fairly effective in restoring particular male behaviors, such as copulation, and DHT has little effect.[1] However, in the guinea pig, the opposite is true. DHT is more effective than estradiol in restoring the copulatory behavior of male guinea pigs.[11] Strain differences in the efficacy of testosterone metabolites are nicely illustrated by the work of Luttge and his colleagues.[13] They found that testosterone was able to elicit sexual behavior in castrated Swiss Webster (SW) or CD-1 mice, being slightly more effective in SW males. In contrast, DHT was slightly less effective than T in Swiss Webster males, but totally ineffective in CD-1 males. Testosterone activated intramale aggression in castrated males of both strains, being slightly more effective on this measure in CD-1 males. But once again, DHT was effective only in SW males. In both strains, DHT was an effective promoter of seminal vesicle growth. Thus, CD-1 males were not totally insensitive to DHT, but the neural mechanisms modulating behavior certainly appeared to be insensitive to this metabolite. No satisfactory explanation has been proposed for the inter- and intraspecies variation in the behavioral efficacy of various metabolites of testosterone.

The variation in behaviorally active metabolites is paralleled by interspecific variation on the predominant metabolic pathways. Variation in androgen metabolism is the rule rather than the exception. For example, DHT appears to be *the* important metabolite of T in stimulating the growth of sexual accessory structures, and in feedback regulation of gonadotropin levels in most species, but there are important exceptions to this generalization. Two exceptions occur in dogs, in which 5α-androstane-3α,17α-diol (3α-diol), not DHT, is the active metabolite in the prostate gland and androstenedione, not DHT, is the active metabolite in the submaxillary salivary gland.[14] In many species of birds, unlike mammals, AE rather than DHT appears to be the primary metabolite of T produced in the brain, pituitary, and sexual accessory structures.[1] The 5β-reduction pathway also appears to be very important in the metabolism of androgens in both brain and peripheral tissues of birds in contrast to mammals. Although this appears to be an inactivation pathway, its importance should not be underestimated. One investigator has found that on average 80% of T incubated with hypothalamic tissue is converted to 5β-reduced metabolites, diverting the bulk of the available androgen from conversion to active metabolites.[15]

While hormone metabolism and action in peripheral tissues have served as models for these processes in the central nervous system (CNS), we should not expect such processes to be identical in all tissues. Characteristics of the 5α-reductase system have been shown to vary somewhat from one peripheral tissue to another.[16] For example, it has been shown that the reaction kinetics, substrate specificity, and subcellular localization of the enzyme are different in liver and prostate. As the process of hormone metabolism in brain tissue receives further study, we should not be surprised to discover differences between peripheral organs and brain tissue or between different brain areas.

One problem with the division of behaviors into estrogen- and androgen-sensitive categories is the imperfect specificity of neural estrogen and androgen receptors.[14] While some authors have found sex-steroid receptors to be fairly specific, other laboratories have found that both steroidal and nonsteroidal estrogens may compete with androgens for androgen receptors. Similarly, both 5α- and 5β-DHT have been shown to compete with estrogen for cytosolic estrogen receptors, and in the former case, translocation of the steroid-receptor complex to the nucleus and the induction of specific proteins have been shown. Thus, one cannot assume a priori that an androgenic metabolite is acting through the androgen receptor or an estrogenic metabolite through an estrogen receptor, particularly when animals are treated with supraphysiological doses, as is often the case.

Another level of complexity is added to the issue of hormonal specificity by the fact that behaviors are not simply switched on and off by the action of one hormone at one specific site in the brain. Instead, hormonal activation of behavior is often determined by the actions of multiple metabolites at multiple sites. The control of singing behavior in passerine birds is a good example. In my laboratory, we have shown that singing behavior in male zebra finches requires a combination of androgenic and estrogenic stimulation.[2] Eight discrete brain nuclei are known to be involved in the control of vocal behavior in this species, and five of these nuclei are steroid sensitive. Four of them contain both androgen and estrogen receptors. The fifth nucleus is known to contain androgen receptors, but estrogen receptors have not yet been assayed. Thus, androgens and estrogens appear to act at multiple sites in the brain from the telencephalon to the motor nucleus innervating the vocal organ to modulate singing behavior. The precise role of the different nuclei in controlling singing behavior and the role androgens and estrogens play in modulating their function remain to be elucidated. In addition to the multiple sites of steroid action in the brain, steroids also affect the functioning of the syrinx, the avian vocal organ. In contrast to neural target tissues, the syrinx appears to respond only to androgenic metabolites.[17] Thus, the hormone specificity of different tissues involved in the same behavior differs. This appears to be a fairly common finding. Hormone-dependent peripheral tissues in males of other species, such as scent glands, antlers, or combs and wattles, appear to be purely androgen dependent, while the neural tissues which control the behavioral use of these structures, in scent marking or fighting for example, are often dependent on the actions of estrogen alone or in combination with androgenic metabolites.[14]

Investigation of hormonal specificity is also restricted to some extent by limitations in the castration-hormone replacement paradigm that is used for much of this research. First of all, castration does not remove all sources of sex steriods. Although the testes are the primary source of androgens in the male, significant quantities may be produced by the adrenal glands and additional androgens may be provided by metabolism of glucocorticoids outside the adrenals.[14] Thus, castration does not guarantee that all sources of behaviorally active androgenic or estrogenic metabolites are removed. In some bird species, castration actually results in a temporary increase in circulating estrogen levels, and it takes several weeks following castration for estrogen levels to reach baseline.[18,19] In men, the administration of diethylstilbestrol has been shown to be more effective than castration in lowering circulating androgen levels.[14] Much of the interstudy variation in the behavioral efficacies of various metabolites in hormone replacement studies is perhaps due to the interaction of exogenous hormone treatments with varying levels of endogenous adrenal steroids. Very few studies have used castrated, adrenalectomized males. And in some species, adrenalectomy is not a practical option, because of the difficulty of the operation and/or the ill health of adrenalectomized animals.

Variation in hormone dosage, the form of hormone used, and method and route of administration have also contributed to interstudy variation in the efficacy of hormone replacement therapy and have at times led to misinterpretation of results. For example, some of the first studies examining the efficacy of DHT compared injection of free DHT to the same dose of testosterone propionate, and concluded that DHT was without effect. Other workers quickly pointed out that this was an unfair comparison, since esterified hormones are more potent because they are metabolized more slowly. Studies using DHT propionate (DHTP) found that it was more effective than the free hormone. However, rapid metabolism has proved to be a continuing problem in studying the efficacy of DHT. For example, while DHT administered in propylene glycol was more effective than T in supporting seminal vesicle growth and feedback suppression of LH levels, it proved incapable of maintaining the sexual behavior of castrated male mice. However, when administered in an oily vehicle which slowed down its release and metabolism, DHT was nearly as potent as T in maintaining sexual behavior, illustrating the importance of mode of administration.[13] Perhaps the most reasonable paradigm now available to compare the efficacies of various gonadal steroids is administration in Silastic capsules. Radioimmunoassay can then be used to monitor plasma hormone levels so that capsule sizes may be adjusted to provide comparable circulating levels of the various steroids. Although time-consuming, this offers a degree of control not available from the techniques typically employed in such studies.

Greater attention to the issues outlined above could go a long way to resolving the many conflicting results published in the last 20 years. One recently published study[20] provides a case in point. One of the most agreed upon findings in the literature is that estrogenic metabolites are necessary for the restoration of normal levels of sexual behavior in castrated rats. Treatment with nonaromatizable androgens alone, such as DHT, results in very modest increases in behavior, if any. However, recently a study was published showing that a synthetic nonaromatizable androgen, 17β-hydroxy-17α-methyl-estra-4,9,11-triene-3-one (R1881), was just as potent as estradiol or testosterone benzoate in restoring the behavior of castrated males. As expected,

DHT benzoate did not reinstate behavior, though it was quite effective in stimulating prostate growth. Superficially, these data suggest that, contrary to most of what has previously been published, estrogenic mechanisms are not necessary to stimulate normal levels of sexual behavior in the male rat. However, closer examination calls this conclusion into question. Even if R1881 proves to be nonaromatizable in brain tissue as it has in sexual accessory organs, its ability to elicit sexual behavior does not indicate that estrogen-sensitive mechanisms do not also play a role. First of all, the possibility that R1881 may interact with estrogen receptors in rat brain has not been ruled out. Second, the rats used in this experiment were not adrenalectomized, so adrenal estrogens could have synergized with the R1881 to stimulate behavior. The inability of DHT benzoate to elicit behavior under the same conditions may be attributable to a high rate of metabolism as the authors suggest. Thus, data which on initial examination appeared to conflict with previous results on the specificity of the hormonal control of behavior, may not conflict at all. But additional data addressing these issues are needed before a proper interpretation can be made.

These data highlight a growing realization that it is not a question of whether androgens or estrogens activate behavior, but a question of how all of the metabolites of T interact to modulate behavior. The best data now available suggest that androgenic and estrogenic metabolites interact at the neural level to induce the full pattern of male behavior in most species.[1,21] Treatment with aromatizable androgens such as T or androstenedione which can be converted to both estrogenic and 5α-reduced androgenic metabolites leads to consistent restoration of behavior in castrated males of all species, and concurrent treatment with a combination of 5α-reduced androgen and estrogen is more effective than treatment with either steroid alone.[2,9,15,22-24] Research must now begin to examine not only the ability of various androgenic and estrogenic metabolites to activate behavior, but how these two classes of hormones normally interact in eliciting male behaviors. We must also seek to understand why such interactive hormonal control of behavior evolved. What advantages does such multiple-metabolite (interactive) control of behavior provide?

In the past, most studies on hormone specificity regarded patterns of hormone metabolism as a rather fixed characteristic of a given tissue. It is now becoming clear that hormone metabolism is in fact quite labile, and a variety of internal and environmental factors have been shown to effect T metabolism in the male. One factor is the level of androgens available. Castration has a significant impact on T metabolism in many different tissues. Castration severely reduces the level of 5α-reductase in rat prostate[14] but causes a dramatic, fivefold increase in this enzyme in the pituitary gland.[25] No changes in 5α-reduction in the hypothalamus have been noted following castration, but significant increases in aromatase levels in several hypothalamic nuclei have been documented.[25] In birds, hypothalmic 5β-reduction increases following castration.[26] Hormone replacement therapy with T can reverse the effects of castration on enzyme levels and hormone metabolism. The ultimate effect of these changes in metabolism on behavior is unclear, since some appear to be increasing the availability of active metabolites in response to a reduction in circulating androgen levels, while others would seem to cause a further reduction in active metabolites. Clearly, more research is needed to clarify this issue.

Available levels of other hormones can also effect T metabolism. For example, progesterone, 11-deoxycorticosterone, 11-deoxycortisol, and corticosterone are all physiological substrates for 5α-reductase and 3-hydroxysteroid dehydrogenase. Thus, increasing levels of these hormones may competitively inhibit in a dose-dependent fashion the conversion of T to DHT.[1] This is one way in which the increased corticoid secretion caused by stressful situations might result in decreases in behaviors dependent on DHT formation. Prolactin has also been shown to decrease 5α-reduction in the hypothalamus, though the mechanism involved in this interaction is not known. The 5α-reduction and aromatization pathways have also been shown to be mutually inhibitory, the end products of one pathway inhibiting hormone metabolism by the other enzyme system.[1] Examination of the enzyme activity in various hypothalamic nuclei using the Palkovits punch technique indicates that, as one might expect, aromatase and 5α-reductase tend to be concentrated in different hypothalamic nuclei.[27]

Age has also been shown to affect hormone metabolism. Most of the age-related changes have been documented in reference to physiological rather than behavioral measures. However, in birds there is a developmental switchover from the prepubertal pattern of inactivation of androgens by 5β-reduction to the adult pattern of forming behaviorally active metabolites through 5α-reduction.[28] Recently, some very exciting data became available demonstrating that declining sexual behavior in aging Japanese quail was correlated with decreased production of DHT by the anterior hypothalamus.[29] Thus, both the onset and decline of sexual behavior in male quail are correlated with changes in androgen metabolism.

Two environmental factors that affect androgen metabolism are photoperiod and social interactions. Photoperiod affects 5α-reductase activity. For example, exposure of adult male rats to constant dark decreases 5α-reduction in the hypothalamus, while administering the pineal hormone melatonin increases the activity of this enzyme.[1] This suggests that there may be important seasonal changes in hormone metabolism correlated with changes in reproductive behavior. Massa and his colleagues[30] have found seasonal changes in 5β-reduction in avian brain tissue independent of seasonal changes in circulating androgen levels. Similarly, the formation of active metabolites is affected by social interactions. Aromatization rates in the hypothalamus of male rats are affected by social interactions; depriving male rats of both contact with females and their odor caused significant increases in hypothalamic rates of aromatizataion. Changes in aromatization rates were inversely correlated with plasma testosterone levels, but changes in aromatization rate were found to be independent of changes in circulating androgens.[31]

Perhaps the most disturbing fact about research on androgen metabolism and the control of behavior is how little we have learned in the past 20 years. We do not even know whether testosterone itself is capable of mediating any changes in behavior or if, as many assume, it can only act indirectly through its metabolites. At best, we have come to appreciate the complexity of the issues involved. Research has suffered from our tendency to dichotomize, to assume that a behavior had to be activated by androgen or by estrogen, by aromatizable or by nonaromatizable androgens, by a central or by a peripheral mechanism. A more integrative approach is needed.

First, we need to examine the role of a greater number of metabolites before focusing on one or two. Studies examining the behavioral efficacies of several metabolites from each hormone class have demonstrated differences in efficacy between metabolites in the same class (e.g., nonaromatizable androgens such as DHT, androsterone, R1881) which suggests our classification scheme is inadequate. Even if DHT and estradiol prove to be the most important androgen metabolities in terms of eliciting behavior, other metabolites may serve as important intermediates, providing precursor pools which can be metabolized to active hormones as needed. Such precursor pools may be extremely important in the maintenance of adquate hormone levels since T and particularly DHT have very short half-lives in plasma.

Second, we need to develop an overall picture of androgen metabolism in different species, including the major sites of androgen metabolism and the major hormones formed. We need to know what proportion of plasma hormones are actually available to various target tissues, not bound to plasma proteins, and whether the active metabolites are taken from plasma, formed in the target tissues, or some combination of the two. While local formation can be advantageous in that the metabolite is readily available and less likely to be inactivated by carrier proteins or rapid degradation, local formation is not always the rule. There are data available from dogs showing that only one-third of the DHT found in the prostate is formed locally; surprisingly, the majority comes from general circulation.[14] Many species of birds are known to have significant quantities of behaviorally active metabolites in plasma; male zebra finches, for example, have significant quantities of DHT and estradiol available in plasma, and like most birds, no evidence of sex-steroid binding globulin.[19,32] But it is not known whether the metabolites which act on the CNS in these species are formed locally or concentrated from the bloodstream.

Third, we need to document the mechanism of steroid action in the brain. It is unclear whether there is one androgen receptor with differing affinities for different metabolites or multiple androgen receptors relatively specific for different metabolites.[33] We need to determine if the binding of androgens to estrogen receptors and vice versa is a normal phenomenon of behavioral significance *in vivo* or merely an artifact of the available techniques. Although we can document the binding of various metabolites to receptors in the brain, our knowledge of the mechanism involved in the activation of behavior begins to weaken there. There have been relatively few studies documenting changes in protein synthesis as a result of hormone stimulation, and none which would elucidate how changes in protein synthesis might affect behavior. Presumably, neurotransmission is eventually being affected, but such effects have yet to be clearly demonstrated.

Fourth, we need to extend the studies examining how various internal and external factors can affect patterns of androgen metabolism. Additional work is needed to determine what other internal and environmental stimuli may affect patterns of hormone metabolism and to document the resulting metabolic changes. Data are also needed demonstrating that these changes in hormone metabolism actually alter behavior and determining how significant such effects may be. Finally, research is needed to uncover the underlying mechanisms that mediate these changes in androgen metabolism. Although there has been some work illustrating that manipulating neurotransmitter levels in particular brain regions can alter hormone metabolism in these areas, this work is just suggestive of how internal or environmental stimuli may act to affect androgen metabolism.

Hopefully, by being more open-minded, yet rigorous in experimental design, and by investigating the question of androgen metabolism and the activation of behavior from many different perspectives, we will gain the knowledge we need to develop a general integrative model of this complex phenomenon.

ACKNOWLEDGMENT

I thank Dr. Harvey Feder for his suggestions on the manuscript.

REFERENCES

1. MARTINI, L. 1982. Endocr. Rev. **3**: 1-25.
2. HARDING, C. F., K. SHERIDAN & M. J. WALTERS. 1983. Horm. Behav. **17**: 111-133.

3. VANDENBERGH, J. G. 1971. Anim. Behav. **19:** 589-594.
4. ADKINS, E. K., J. J. BOOP, D. L. KOUTNICK, J. B. MORRIS & E. E. PNIEWSKI. 1981. Physiol. Behav. **24:** 441-446.
5. EDWARDS, D. A. & K. G. BURGE. 1971. Horm. Behav. **2:** 239-245.
6. FLETCHER, T. J. & R. V. SHORT. 1974. Nature **248:** 616-618.
7. ANTLIFF, H. R. & W. C. YOUNG. 1956. Endocrinology **59:** 74-82.
8. WHALEN, R. E. & J. F. DEBOLD. 1974. Endocrinology **95:** 1674-1679.
9. BEYER, C., L. DE LA TORRE, K. LARSSON & G. PEREZ-PALACIOS. 1975. Horm. Behav. **6:** 301-306.
10. LUTTGE, W. G. & N. R. HALL. 1973. Behav. Biol. **8:** 725-732.
11. ALSUM, P. & R. W. GOY. 1974. Horm. Behav. **7:** 417-429.
12. PHOENIX, C. H. 1974. *In* Reproductive Behavior. W. Montagna & W. A. Sadler, Eds.: 249-258. Plenum Press. New York, N.Y.
13. LUTTGE, W. G., N. R. HALL & C. J. WALLIS. 1974. Physiol. Behav. **13:** 553-561.
14. MAINWARING, W. I. P. 1977. The Mechanism of Action of Androgens. Monographs of Endocrinology, Vol. 10. Springer-Verlag. New York, N.Y.
15. BALTHAZART, J. & M. SCHUMACHER. 1983. *In* Hormones and Behaviour in Higher Vertebrates. J. Balthazart, E. Pröve & R. Gilles, Eds.: 237-260. Springer-Verlag. New York, N.Y.
16. VERHOEVEN, G., G. LAMBERIGTS & P. DE MOOR. 1974. J. Steroid Biochem. **5:** 93-100.
17. LUINE, V. N., C. F. HARDING & W. V. BLEISCH. 1983. Brain Res. **279:** 339-342.
18. WINGFIELD, J. C. Personal communication.
19. HARDING, C. F. Unpublished data.
20. SODERSTEN, P. & J. A. GUSTAFSSON. 1980. J. Endocrinol. **87:** 279-283.
21. PEREZ-PALACIOS, G., K. LARSSON & C. BEYER. 1975. J. Steroid Biochem. **6:** 999-1006.
22. DENISSON, W. E., E. W. KLOSTERMAN & M. L. BUCHANON. 1951. J. Anim. Sci. **10:** 885-888.
23. FOOTE, R. E., P. J. DRADDY, M. BREITE & E. A. B. OLTENACU. 1977. Horm. Behav. **9:** 57-68.
24. JOSHI, H. S. & J. I. RAESIDE. 1973. J. Reprod. Fertil. **33:** 411-423.
25. JOUAN, P. & S. SAMPEREZ. 1980. *In* The Endocrine Functions of the Brain. M. Motta, Ed.: 95-115. Raven Press. New York, N.Y.
26. STEIMER, T. & J. B. HUTCHISON. 1981. Nature **292:** 345-347.
27. SELMANOFF, M. K., L. D. BRODKIN, R. I. WEINER & P. SIITERI. 1977. Endocrinology **101:** 841-848.
28. MINGUELL, J. J. & W. D. SIERRALTA. 1975. J. Endocrinol. **65:** 287-315.
29. BALTHAZART, J., R. TUREK & M. A. OTTINGER. 1984. Horm. Behav. **18:** 330-345.
30. MASSA, R., L. BOTTONI & V. LUCINI. 1983. *In* Hormones and Behaviour in Higher Vertebrates. J. Balthazart, E. Pröve & R. Gilles, Eds.: 230-236. Springer-Verlag. New York, N.Y.
31. DESSI-FULGHERI, F., C. LUPO, G. M. CIAMPI, M. CANONACO & K. LARSSON. 1983. *In* Hormones and Behaviour in Higher Vertebrates. J. Balthazart, E. Pröve & R. Gilles, Eds.: 305-313. Springer-Verlag. New York, N.Y.
32. WINGFIELD, J. C., K. S. MATT & D. S. FARNER. 1984. Gen. Comp. Endocrinol. **53:** 281-292.
33. SHERIDAN, P. J. 1983. Endocr. Rev. **4:** 171-178.

Steroid Hormone Antagonists and Behavior

I. THEODORE LANDAU

Department of Psychology
Oakland University
Rochester, Michigan 48063

Gonadal steroid hormones influence the behavior of a wide variety of animals. An underlying assumption is that these hormonal effects are mediated by their action in the brain and one of the primary goals of recent research in this area has been to elucidate the neurochemical mechanisms that account for these behavior effects (see Reference 1).

Steroid hormone antagonists (generally compounds developed by pharmaceutical companies) are so named because they appear to inhibit the action of naturally occurring steroid hormones in at least some situations.[2] These synthetic compounds have been used to study the mechanisms of action of steroid hormones in nonneural as well as neural target tissues.

It is beyond the scope of this paper to review all the relevant literature relating steroid hormone antagonists to behavior. This paper will limit itself to research on rodents, particularly laboratory strains of rats. It will further focus on activational and organizational effects[3] of steroid antagonists on sexual behavior, though these antagonists have also been used in such areas as the hormonal regulations of aggression[4,5] and food intake.[6] Finally, this paper will address primarily the research relating antiestrogens to female sexual behavior, though again there is a considerable amount of literature on the effects of antiandrogens and other antagonists on male sexual behavior (see references below).

ANTIESTROGENS

Antagonism of Estrogen Effect on Lordosis

In 1968, Meyerson and Lindstrom[7] first reported an inhibitory effect of an antiestrogen (MER-25) on estradiol-stimulated lordosis in ovariectomized rats. In a similar study, administration of another antiestrogen, CI-628, successfully antagonized estradiol-stimulated lordosis.[8] These basic effects were documented in numerous other studies over the next several years, which employed a growing variety of antiestrogens including nafoxidine and clomiphene.[9-13]

The basic design of these studies was to give a systemic injection of antiestrogen around the same time that a single injection of estrogen (either estradiol or estradiol benzoate) was administered. Behavioral testing occurred around 48 hr later, subsequent to an injection of progesterone several hours prior to behavioral testing.

Mechanisms of Action in Nonneural Target Tissues

Initial hypotheses concerning the mechanisms of action of antiestrogen inhibition of lordosis were derived from research studying nonneural target tissues (e.g. estrogen-stimulated uterine growth). This research established that antiestrogens could competitively block radioactive estradiol from binding to presumed cytoplasmic estrogen receptors (ER).[2,14] The earliest proposed model of steroid hormone action stated that estrogen-ER complexes were translocated from the cytoplasm to the nucleus where they then bound to nuclear acceptor sites on the chromatin, initiating a chain of events leading to the hormone's ultimate physiological effects. Only estrogen bound to ER was presumed to be translocatable (see Reference 15). Antiestrogens significantly interfered with this process by binding to ER, preventing the translocation of estrogen-ER complexes to the nucleus.[14] Antiestrogen-ER complexes were themselves translocated to the nucleus where they were presumed to be without biological effect, thereby accounting for their competitive inhibition of estrogen-stimulated uterine growth.

More recent research has challenged this direct competitive inhibition model of antiestrogen action in several ways. Clark and his colleagues have demonstrated that antiestrogens initially mimic at least some of the early effects of estrogen, making the term "antiestrogen" somewhat of a misnomer.[16,17] The inhibitory effect became apparent only with subsequent injections (24 hr later or longer), due to the fact that the initial antiestrogen prevented the replenishment of cytoplasmic estrogen receptors for a far greater period of time than did estrogen itself (receptor depletion model).[17]

There is now considerable evidence that neither model offers a comprehensive explanation of antiestrogen action (see Reference 18). For example, some antiestrogens have antagonistic effects in the absence of correlated receptor depletion[19] while others do not show the agonistic effects reported by Clark's laboratory.[20] Other ideas on antiestrogen actions have emphasized the long biological half-life of the antiestrogen-ER complex in the nucleus,[16,21,63] which may somehow disrupt normal nuclear "processing."[22] Also, antiestrogen-ER complexes may bind in the nucleus to a different (salt-extractable) binding site than estrogen-ER complexes.[23,24] Markaverich et al.[25] have reported a correlation between an antiestrogen's antagonistic effects and its inability to stimulate what they have described as nuclear Type II binding sites. It has even been suggested that there is a separate cytoplasmic antiestrogenic receptor,[26] though whether this can account for antiestrogens' inhibitory effects is not clear.[27] Others have demonstrated that antiestrogens are not a uniform category and different ones may have separate mechanisms of action,[28] while the same antiestrogen may be an antagonist in one tissue yet have an agonistic effect in another.[25]

Finally, Katzenellenbogen and Ferguson[29] reported that estrogen could get into the nucleus even after prior administration of antiestrogen had depleted receptor levels. This finding, which is counter to the "classic" model of steroid hormone action described earlier, may be consistent with the current hypothesis that cytoplasmic receptors are an artifact of isolation procedures and that there are actually only nuclear steroid receptors.[30,31]

Mechanisms of Action in Neural Tissues

Luttge,[65] using a cannula containing antiestrogen implanted in the hypothalamus of rats, demonstrated an inhibition of estradiol-stimulated lordosis. This supported the notion of a neural localization of antiestrogenic effects on behavior. However, Whalen and Gorzalka[33] reported a lack of competitive inhibition of radioactive estradiol uptake by CI-628, using a whole-cell hypothalamic preparation. This absence of inhibition was probably due to the difficulty in assessing such effects at the whole-cell level in a heterogeneous tissue such as the hypothalamus. When measurements of retention of radioactivity were limited to the nuclear fraction of hypothalamic-preoptic area tissue (in rats), clear inhibition of estradiol uptake by both CI-628 and nafoxidine was observed.[34-36]

In an initial report, Whalen et al.[37] reported a hypothalamic estrogen receptor depletion as a result of CI-628 treatment, similar to what had been reported for uterine tissue (though the duration of the depletion in hypothalamic tissue was shorter). However, subsequent studies questioned the validity of this model of neurobehavioral systems.[13,35,38] For example, nafoxidine caused a reduction in cytoplasmic ER that was considerably longer than its effect on behavior.[13] Also, unlike for uterine growth, no clear agonistic effects of antiestrogens have been reported for lordosis in rats (see Reference 35). However, some antiestrogens may have some facilitative effects on lordosis in guinea pigs[39] and have clear estrogenic effects on food intake and body weight in rats.[13]

As in uterine tissues, antiestrogens have a prolonged period of retention in neural target tissues, remaining at significant levels for at least as long as 15 days.[13] Roy[40] suggested that the prolonged presence of antiestrogen in plasma may cause a continual translocation of receptors to the nucleus, thereby obscuring any receptor repletion, as well as resulting in the reported prolonged period of nuclear retention. Also, as was reported for uterine tissue, antiestrogen binds in the nucleus of neural target tissue in a form that is more easily salt extractable than does estrogen[40] though this apparently does not correlate with the antiestrogenic inhibition of lordosis.[41]

Another lack of correlation occurs between CI-628's effects on lordosis and its effects on estrogen retention in neural tissues. For example, Landau[35] reported a greater inhibition of nuclear estradiol benzoate (EB) uptake in the hypothalamus of rats if CI-628 was given 24 hr prior to EB rather than concurrently with EB, even though concurrent administration of antiestrogen and estrogen generally resulted in the strongest inhibition of lordosis.[8,35]

Antiestrogens such as CI-628[35] and MER-25[39,42] can inhibit estradiol facilitation of lordosis, even if the antiestrogen is not given until relatively later (24 hr or later) in the "priming" process. Even if CI-628 is given concurrently with EB, its maximal level of inhibition of hypothalamic estrogen retention does not occur until more than 12 hr later.[35] These data have been taken as evidence for the importance of a continued presence of estradiol throughout the priming period, though this remains a subject of debate.[43,44]

Hormonal facilitation of lordosis in rats and guinea pigs is usually the result of an injection of estrogen followed by an injection of progesterone 24-48 hr later. The progesterone greatly facilitates estrogen's facilitatory actions on lordosis (see Reference 45). Recent research has reported the presence of progestin receptors in the hypothalamus and that the levels of these receptors are increased as a result of estrogen priming.[46,47] Roy et al.[48] reported that CI-628 could inhibit this induction of progesterone receptors by estrogen and may thus account for its inhibitory effects on lordosis. However, CI-628 is an equally effective antagonist of estrogen-facilitated lordosis whether as a result of estrogen plus progesterone or estrogen alone.[49] This suggests that the inhibitory effects of antiestrogens on lordosis are not simply a result of their inhibition of the induction of progestin receptors. More recently, Wilcox and Feder[50] reported that the antiestrogen enclomiphene may actually increase nuclear progestin receptor levels in brain.

RELATED STEROID ANTAGONIST EFFECTS

Antiestrogens are also able to inhibit lordosis in female or male rats stimulated by androgens.[51,52] Brown and Blaustein[53] have reported an inhibitory effect of a progestin antagonist or hormone-stimulated lordosis in guinea pigs.

Consistent with the hypothesis that testosterone's effects on male sexual behavior are dependent upon its conversion to estrogen (aromatization), the antiestrogen CI-628 was reported to antagonize testosterone's effects on male sexual behavior in castrated male rats.[32] However, other studies failed to confirm this effect.[54] In general, the literature on steroid antagonism of androgen-induced male sexual behavior by antiestrogens, antiandrogens,[55,56,64] or aromatization inhibitors[57-59,62] has yielded inconsistent results. There is also conflicting evidence as to the ability of antiestrogens to inhibit androgen-induced aggression in castrated male mice.[4,60] A particular problem is that many of these antagonists may themselves be weak but behaviorally active and androgens.[55,58,61]

ANTAGONISTS AND SEXUAL DIFFERENTIATION

In addition to the activational effects of steroid hormones discussed previously, these hormones also have well documented effects on sexual differentiation during critical prenatal and perinatal periods. Further, for many mammalian species, this process can be divided into two separate processes: masculinization (an increased sensitivity to the stimulation of male sexual behavior as an adult) and defeminization (a decreased sensitivity to the stimulation of female sexual behavior as an adult). For some mammalian species, only masculinization may normally occur (see Reference 66 for review). Androgens, secreted by the male sex during these critical periods, are hypothesized as the cause of these effects.

Antiandrogens, such as cyproterone acetate and flutamide, have been given perinatally to male and female rats, often in combination with other steroid hormone treatments and/or neonatal castration, in an effort to observe their effect on subsequent sexual differentiation of behavior.[67,68,69] In general, these antiandrogens had their greatest effect when given prenatally, preventing the defeminization typically caused by prenatal exposure to androgens. This is in agreement with other recent research that suggests a direct role of androgens in the process of sexual differentiation.[70,71]

However, the predominant emphasis over the past decade has been on the role of estrogens in the process of sexual differentiation. This is based on the presumed importance of the aromatization of androgens to estrogens in the mediation of androgenic effects.[72,73] As a result, much research in this area has focused on the effect of antiestrogens and aromatization inhibitors. For example, neonatal administration of the antiestrogen, MER-25, prevented both the masculinizing and defeminizing effects of androgen (or estrogen) in male rats.[74,75] It similarly prevented defeminization caused by the synthetic estrogen, RU-2858, in female rats.[76] In contrast, Gottlieb et al.[77] reported that MER-25 had no effect on androgen-induced defeminization in female hamsters. Also note that the interpretation of these results should be tempered with the fact that antiestrogens may be partial estrogen agonists, and may have defeminizing effects even when given alone (e.g., nafoxidine[78,79] and CI-628[73]).

The aromatization inhibitor, ATD, has been used in several studies of behavioral sexual differentiation. When given neonatally to male rats, it typically prevents defeminization, but not masculinization, normally caused by circulating androgens.[73,80-82] Consistent with the proposed mechanisms of action of aromatization inhibitors vs. antiestrogens,[3] CI-628, but not ATD, prevented the defeminization caused by neonatally administered estrogen in female rats[68] and hamsters.[83] Prenatally administered ATD can prevent defeminization in male rats, and even enhance subsequent lordosis responding in females.[84] However, even prenatally administered ATD may not prevent masculinization in male rats.[85] This has led to a more general hypothesis stating that defeminization in rats and hamsters occurs primarily (although not exclusively) postnatally, via estrogenic metabolites of androgen, whereas masculinization is more determined by prenatal direct androgenic effects.[80] Conflicting with this hypothesis are several reports of perinatal treatment with an aromatization inhibitor blocking subsequent ejaculatory behavior in rats and hamsters.[83,86,87] Related to this, note that these antagonists typically do not prevent androgenic effects on penile development and other nonneural androgenic target tissues.[86] In summary, most research using steroid antagonists has supported the importance of aromatization for at least some aspects of sexual differentiation.

IMPLICATIONS FOR THE FUTURE

Initial optimism that steroid hormone antagonists, particularly antiestrogens, could be a critical tool in determining the mechanisms of action of steroid hormones has faded somewhat. Research on estrogen effects on responses such as uterine growth began with a relatively simple model of antiestrogenic action: direct competitive inhibition of the antiestrogen for estrogen receptor sites. This has given way to more elaborate models that now include receptor depletion, disruption of nuclear processing, and separate antiestrogen receptors, among others. Similarly, it is now recognized that these antagonists may be antagonistic only for some tissues while they may actually be agonistic for others. Further, one antiestrogen may be antagonistic in a paradigm where another is agonistic. The very word "antiestrogen" becomes of questionable validity at this point. As is often the case, while the last few years have greatly increased knowledge of the specific effects of antiestrogens, a more general explanation of their action seems even further away.

The neurobehavioral data are no less clouded. Antiestrogens may be antagonistic for lordosis while agonistic for food intake. Actions of antiestrogens on neural tissue, in terms of inhibition of uptake, receptor depletion, or prevention of induction of progestin receptors, often do not correlate well with the accompanying behavioral inhibition. Antagonistic effects on male behavior are even less consistent.

Beyond the straightforward fact that these antagonists can inhibit certain steroid-stimulated responses and also can compete with steroid hormones for binding and retention at the cellular level, very little else can be offered as a comprehensive explanation of steroid antagonist action. This lack of understanding has reduced the effective use of antagonists in answering the broader questions of how the steroid hormones themselves act. Of perhaps even greater concern is the recent finding, mentioned above, that the classic two-step mechanisms of steroid uptake may be in error: cytoplasmic steroid receptors may be an artifact of isolation procedures.

There seems to have been a decline in behavioral research using steroid hormone antagonists in recent years. The problems cited above are undoubtedly a cause of this decline. A novel and comprehensive explanation of steroid antagonist effects would surely spur a resurgence of interest in their use.

REFERENCES

1. BEACH, F. 1981. Historical origins on modern research on hormones and behavior. Horm. Behav. **15:** 325-376.
2. CALLANTINE, M. R., L. E. CLEMENS & Y. SHIH. 1968. Displacement of 17β-estradiol from uterine receptor sites by an estrogen antagonist. Proc. Soc. Exp. Biol. Med. **128:** 383-386.
3. LIEBERBURG, I., G. WALLACH & B. S. MCEWEN. 1977. The effects of an inhibitor of aromatization (1,4,6-androstatriene-3,17-dione) and an anti-estrogen (CI-628) on in vivo formed testosterone metabolites recovered from neonatal rat brain tissues and purified nuclei. Implications for sexual differentiation of the rat brain. Brain Res. **128:** 176-181.
4. CLARK, C. R. & N. W. NOWELL. 1979. The effect of the antiestrogen CI-628 on androgen-induced aggressive behavior in castrated male mice. Horm. Behav. **12:** 205-210.
5. LUTTGE, W. G. 1979. Anti-estrogen inhibition of testosterone-stimulated aggression in mice. Experientia **32:** 273.
6. ROY, E. J. & G. N. WADE. 1976. Estrogenic effects of an anti-estrogen, MER-25, on eating and body weight in rats. J. Comp. Physiol. Psychol. **90:** 156-166.
7. MEYERSON, B. J. & L. LINDSTROM. 1968. Effect of an oestrogen antagonist ethamoxytriphetol (MER-25) on oestrous behaviour in rats. Acta Endocrinol. **59:** 41-48.
8. ARAI, Y. & R. A. GORSKI. 1968. Effect of anti-estrogen on steroid induced sexual receptivity in ovariectomized rats. Physiol. Behav. **3:** 351-353.
9. KOMISARUK, B. R. & C. BEYER. 1972. Differential antagonism by MER-25, of behavioral and morphological effects of estradiol benzoate in rats. Horm. Behav. **3:** 63-70.
10. POWERS, J. B. 1975. Anti-estrogen suppression of the lordosis response in female rats. Horm. Behav. **6:** 379-392.
11. ROSS, J. W., D. C. PAUP, M. BRANT-ZAWADSKI, J. R. MARSHALL & R. A. GORSKI. 1973. Effects of cis- and trans-clomiphene in the induction of sexual behavior. Endocrinology **93:** 681-685.
12. SODERSTEN, P. 1974. Effects of an estrogen antagonist, MER-25 on mounting behavior and lordosis behavior in the female rat. Horm. Behav. **5:** 111-122.
13. WADE, G. N. & J. D. BLAUSTEIN. 1978. Effects of anti-estrogen on neural estradiol binding and on behaviors in female rats. Endocrinology **102:** 245-251.
14. RUH, T. S. & M. F. RUH. 1974. The effect of anti-estrogens on the nuclear binding of the estrogen receptor. Steroids **24:** 209-224.
15. FEDER, H. H., I. T. LANDAU & W. A. WALKER. 1979. Anatomical and biochemical substrates of the actions of estrogens and anti-estrogens on brain tissues that regulate female sex behavior in rodents. *In* Endocrine Control of Sexual Behavior. C. Beyer, Ed.: 317-340. Raven Press. New York, N.Y.
16. CLARK, J. H., J. N. ANDERSON & E. J. PECK. 1973. Estrogen receptor-anti-estrogen complex: atypical binding by uterine nuclei and effects on uterine growth. Steroids **22:** 707-718.
17. CLARK, J. N., E. J. PECK, JR. & J. N. ANDERSON. 1974. Oestrogen receptors and antagonism of steroid hormone action. Nature **251:** 446-448.

18. SUTHERLAND, R. L. & L. C. MURPHY. 1982. Mechanisms of oestrogen antagonism by nonsteroidal antiestrogens. Mol. Cell. Endocrinol. **25:** 5-23.
19. CIDLOWSKI, J. A. & T. G. MULDOON. 1976. Dissimilar effects of anti-estrogens upon estrogen receptors in responsive tissues of male and female rats. Biol. Reprod. **15:** 381-389.
20. SUTHERLAND, R., J. MESTER & E-E. BAULIEU. 1977. Tamoxifen is a potent 'pure' antiestrogen in chick oviduct. Nature **267:** 434-435.
21. DIX, C. J. & V. C. JORDAN. 1980. Modulation of rat uterine steroid hormone receptors by estrogen and antiestrogen. Endocrinology **107:** 2011-2020.
22. HORWITZ, K. B. & W. L. MCGUIRE. 1978. Nuclear mechanisms of estrogen action. J. Biol. Chem. **253:** 8185-8191.
23. BAUDENDISTEL, L. J. & T. S. RUH. 1976. Anti-estrogen action: differential nuclear retention and extractability of the estrogen receptor. Steroids **28:** 223-237.
24. RUH, T. S. & L. J. BAUDENDISTEL. 1978. Anti-estrogen modulation of the salt-resistant nuclear estrogen receptor. Endocrinology **102:** 1838-1846.
25. MARKAVERICH, B. M., S. UPCHURCH, S. A. MCCORMACK, S. R. GLASSER & J. H. CLARK. 1981. Differential stimulation of uterine cells by nafoxidine and clomiphene: relationship between nuclear estrogen receptors and type II estrogen binding sites and cellular growth. Biol. Reprod. **24:** 171-181.
26. SUTHERLAND, R. L., L. C. MURPHY, M. S. FOO, M. D. GREEN, A. M. WHYBOURNE & Z. S. KROZOWSKI. 1980. High affinity anti-oestrogen binding site distinct from the oestrogen receptor. Nature **288:** 273-275.
27. SUDO, K., F. J. MONSMA, JR. & B. S. KATZENELLENBOGEN. 1983. Anti-estrogen-binding sites distinct from the estrogen receptor: subcellular localization, ligand specificity, and distribution in tissues of the rat. Endocrinology **112:** 425-434.
28. BLACK, L. J. & R. L. GOODE. 1981. Evidence for biological action of the anti-estrogens LY117018 and tamoxifen by different mechanisms. Endocrinology **109:** 987-989.
29. KATZENELLENBOGEN, B. S. & E. R. FERGUSON. 1975. Anti-estrogen action in the uterus: ineffectiveness of nuclear bound estradiol after antiestrogen. Endocrinology **97:** 1-12.
30. KING, W. J. & G. L. GREENE. 1984. Monoclonal antibodies localize oestrogen receptor in the nuclei of target cells. Nature **307:** 745-747.
31. WELSHONS, W. V., M. E. LIEBERMAN & J. GORSKI. 1984. Nuclear localization of unoccupied oestrogen receptors. Nature **307:** 747-749.
32. LUTTGE, W. G. 1975. Effects of anti-estrogens on testosterone stimulated male sexual behavior and peripheral target tissues in the castrate male rat. Physiol. Behav. **14:** 839-846.
33. WHALEN, R. E. & B. B. GORZALKA. 1973. Effects of an estrogen antagonist on behavior and on estrogen retention in neural and peripheral target tissues. Physiol. Behav. **10:** 35-40.
34. CHAZAL, G., M. FAUDON, F. GOGAN & W. ROTSZTEJN. 1975. Effects of two estradiol antagonists upon estradiol uptake in the rat brain and peripheral tissues. Brain Res. **89:** 245-254.
35. LANDAU, I. T. 1977. Relationships between the effects of the anti-estrogen, CI-628, on sexual behavior, uterine growth, and cell nuclear estrogen retention after estradiol-17β-benzoate administration in the ovariectomized rat. Brain Res. **133:** 119-138.
36. ROY, E. J. & G. N. WADE. 1977. Binding of ^{3}H-estradiol by brain cell nuclei and female rat sexual behavior: Inhibition by anti-estrogens. Brain Res. **126:** 73-88.
37. WHALEN, R. E., J. V. MARTIN & K. L. OLSEN. 1975. Effect of oestrogen antagonist on hypothalamic oestrogen receptors. Nature **258:** 742-743.
38. ETGEN, A. M. 1979. Anti-estrogens: effects of tamoxifen, nafoxidine, and CI-628 on sexual behavior, cytoplasmic receptors, and nuclear binding of estrogen. Horm. Behav. **13:** 97-112.
39. WALKER, W. A. & H. H. FEDER. 1977. Inhibitory and facilitatory effects of various anti-estrogens on the induction of female sexual behavior by estradiol benzoate in guinea pigs. Brain Res. **134:** 455-465.
40. ROY, E. J. 1978. Anti-estrogens and nuclear estrogen receptors in the brain. Cur. Stud. Hypothalamic Function **1:** 204-213.

41. ROY, E. J. & B. S. MCEWEN. 1979. Oestrogen receptors in cell nuclei of the hypothalamus-preoptic area-amygdala following an injection of oestradiol or the antioestrogen CI-628. J. Endocrinol. **83:** 285-293.
42. FEDER, H. H. & L. P. MORIN. 1974. Supression of lordosis in guinea pigs by ethamoxytriphetol (MER-25) given at long intervals (36-46 Hr) after estradiol benzoate treatment. Horm. Behav. **5:** 63-71.
43. BLAUSTEIN, J. D., S. D. DUDLEY, J. M. GRAY, E. J. ROY & G. N. WADE. 1979. Long-term retention of estrogen by brain cell nuclei and female rat sexual behavior. Brain Res. **173:** 355-359.
44. PARSONS, B., B. S. MCEWEN & D. W. PFAFF. 1982. A discontinuous schedule of estrogen treatment is sufficient to activate progesterone-facilitated feminine sexual behavior and to increase cytosol receptors for progestin in the hypothalamus of the rat. Endocrinology **110:** 613-619.
45. FEDER, H. H. & B. L. MARRONE. 1977. Progesterone: its role in the central nervous system as a facilitator and inhibitor of sexual behavior and gonadotropin release. Ann. N.Y. Acad. **286:** 331-354.
46. BLAUSTEIN, J. D. & H. H. FEDER. 1979. Cytoplasmic progestin-receptors in guinea pig brain: characteristics and relationship to the induction of sexual behavior. Brain Res. **169:** 481-497.
47. MACLUSKY, N. J. & B. S. MCEWEN. 1980. Progestin receptors in rat brain: distribution and properties of cytoplasmic progestin-binding sites. Endocrinology **106:** 192-202.
48. ROY, E. J., N. J. MACLUSKY & B. S. MCEWEN. 1979. Antiestrogen inhibits the induction of progestin receptors by estradiol in the hypothalamus-preoptic area and pituitary. Endocrinology **104:** 1333-1336.
49. LANDAU, I. T. 1981. Inhibition of lordosis in rats by the antiestrogen CI-628 in the absence of progesterone. J. Comp. Physiol. Psychol. **95:** 270-277.
50. WILCOX, J. N. & H. H. FEDER. 1983. Long-term priming with a low dosage of estradiol benzoate or an anti-estrogen (enclomiphene) increases nuclear progestin receptor levels in brain. Brain Res. **266:** 243-251.
51. WHALEN, R. E., C. BATTIE & W. G. LUTTGE. 1972. Anti-estrogen inhibition of androgen induced sexual receptivity in rats. Behav. Biol. **7:** 311-320.
52. SODERSTEN, P. & K. LARRSON. 1974. Lordosis behavior in castrated male rats treated with estradiol benzoate or testosterone propionate in combination with an estrogen antagonist, MER-25, and in intact male rats. Horm. Behav. **5:** 13-18.
53. BROWN, T. J. & J. D. BLAUSTEIN. 1984. Inhibition of sexual behavior in female guinea pigs by a progestin receptor antagonist. Brain Res. **301:** 343-349.
54. YAHR, P. & S. A. GERLING. 1978. Aromatization and androgen stimulation of sexual behavior in male and female rats. Horm. Behav. **10:** 128-142.
55. BLOCH, G. J. & J. M. DAVIDSON. 1971. Behavioral and somatic responses to the antiandrogen cyproterone. Horm. Behav. **2:** 11-25.
56. SODERSTEN, P., G. GRAY, D. A. DAMASSA, E. R. SMITH & J. M. DAVIDSON. 1975. Effects of a non-steroidal anti-androgen on sexual behavior and pituitary-gonadal function in the male rat. Endocrinology **97:** 1468-1475.
57. CHRISTENSEN, L. W. & L. G. CLEMENS. 1975. Blockade of T-induced mounting behavior in the male rat with intracranial application of the aromatization inhibitor androst-1,4,6-triene-3,17-dione. Endocrinology **97:** 1545-1551.
58. LANDAU, I. T. 1980. Facilitation of male sexual behavior in adult male rats by the aromatization inhibitor, 1,4,6-androstatriene-3,17-dione (ATD). Physiol. Behav. **25:** 173-177.
59. MORALI, G., K. LARSSON & C. BEYER. 1977. Inhibition of testosterone-induced sexual behavior in the castrated male rate by aromatase blockers. Horm. Behav. **9:** 203-213.
60. SIMON, N. G., R. GANDELMAN & S. M. HOWARD. 1981. MER-25 does not inhibit the activation of aggression by testosterone in adult Rockland-Swiss mice. Psychoneuroendocrinology **2:** 131-137.
61. BAUM, M. J. & J. T. M. VREEBURG. 1976. Differential effects of the anti-estrogen MER-25 and the three 5α-reduced androgens on mounting and lordosis behavior in the rat. Horm. Behav. **7:** 87-104.

62. BEYER, C., G. MORALI, F. NAFTOLIN, K. LARRSON & G. PEREZ-PALACIOS. 1976. Effect of some anti-estrogens and aromatase inhibitors on androgen induced sexual behavior in castrated male rats. Horm. Behav. **7:** 353-363.
63. FERGUSON, E. R. & B. S. KATZENELLENBOGEN. 1977. A comparative study of antiestrogen action: temporal patterns of antagonism of estrogen stimulated uterine growth and effects on estrogen receptor levels. Endocrinology **100:** 1242-1251.
64. GLADUE, B. A. & L. G. CLEMENS. 1980. Flutamide inhibits testosterone-induced masculine sexual behavior in male and female rats. Endocrinology **106:** 1917-1922.
65. LUTTGE, W. G. 1976. Intracerebral implantation of the anti-estrogen CN-69,725-27: effects on female sexual behavior in rats. Pharmacol. Biochem. Behav. **4:** 685-688.
66. BAUM, M. J. 1979. Differentiation of coital behavior in mammals: A comparative analysis. Neurosci. Biobehav. Rev. **3:** 265-284.
67. GLADUE, B.A. & L. G. CLEMENS. 1978. Androgenic influences on feminine sexual behavior in male and female rats: defeminization blocked by prenatal antiandrogen treatment. Endocrinology **103:** 1702-1709.
68. MCEWEN, B. S., I. LIEBERBURG, C. CHAPTAL, P. G. DAVIS, L. C. KREY, N. J. MACLUSKY & E. J. ROY. 1979. Attenuating the defeminization of the neonatal rat brain: Mechanisms of action of cyproterone acetate, 1,4,6-androstatriene-3,17-dione and a synthetic progestin, R5020. Horm. Behav. **13:** 269-281.
69. WARD, I. 1972. Female sexual behavior in male rats treated prenatally with an antiandrogen. Physiol. Behav. **8:** 53-56.
70. BAUM, M. J., J. A. CANICK, M. A. ERSKINE, C. A. GALLAGHER & J. H. SHIM. 1983. Normal differentiation of masculine sexual behavior in male ferrets despite neonatal inhibition of brain aromatase or 5-alpha-reductase activity. Neuroendocrinology **36:** 277-284.
71. OLSEN, K. L. 1983. Effects of 17 beta-hydroxy-17 alpha-methyl-estra-4,9,11-triene-3-one (R-1881): evidence for direct involvement of androgens in the defeminization of behaviour in rats. J. Endocrinol. **98:** 431-438.
72. MCEWEN, B. S., L. PLAPINGER, C. CHAPTEL, J. GERLACH & G. WALLACH. 1975. Role of fetonatal estrogen binding protein in associations of estrogen with neonatal brain cell nuclear receptors. Brain Res. **96:** 400-406.
73. MCEWEN, B. S., I. LIEBERBURG, C. CHAPTAL & L. C. KREY. 1977. Aromatization: Important for sexual differentiation of the neonatal rat brain. Horm. Behav. **9:** 249-263.
74. BOOTH, J. E. 1977. Sexual behavior of male rats injected with the anti-oestrogen MER-25 during infancy. Physiol. Behav. **19:** 35-39.
75. SODERSTEN, P. 1978. Effects of anti-oestrogen treatment of neonatal male rats on lordosis behaviour and mounting behaviour in the adult. J. Endocrinol. **76:** 241-249.
76. DOUGHTY, C., J. E. BOOTH, P. G. MCDONALD & R. F. PARROTT. 1975. Inhibition, by the anti-oestrogen MER-25, of defeminization induced by the synthetic oestrogen RU-2858. J. Endocrinol. **67:** 459-460.
77. GOTTLIEB, H., A. A. GERALL & A. THIEL. 1974. Receptivity in female hamsters following neonatal testosterone, testosterone propionate, and MER-25. Physiol. Behav. **12:** 61-68.
78. ETGEN, A. M. 1981. Differential effects of two estrogen antagonists on the development of masculine and feminine sexual behavior in hamsters. Horm. Behav. **15:** 299-311.
79. ETGEN, A. M. & R. E. WHALEN. 1979. Masculinization and defeminization induced in female hamsters by neonatal treatment with estradiol, RU-2858, and nafoxidine. Horm. Behav. **12:** 211-217.
80. DAVIS, P. G., C. V. CHAPTAL & B. S. MCEWEN. 1979. Independence of masculine and feminine sexual behavior in rats. Horm. Behav. **12:** 12-19.
81. FADEM, B. H. & R. J. BARFIELD. 1981. Neonatal hormonal influences on the development of proceptive and receptive feminine sexual behavior in rats. Horm. Behav. **15:** 282-288.
82. VREEBURG, J. T. M., P. D. M. VAN DER VAART & P. VAN DER SCHOOT. 1977. Prevention of central defeminization but not masculinization in male rats by inhibition neonatally of oestrogen biosynthesis. J. Endocrinol. **74:** 375-382.
83. RUPPERT, P. H. & L. G. CLEMENS. 1981. The role of aromatization in the development of sexual behavior of the female hamster *(Mesocricetus auratus)*. Horm. Behav. **15:** 68-76.

84. CLEMENS, L. G. & B. A. GLADUE. 1978. Feminine sexual behavior in rats enhanced by prenatal inhibition of androgen aromatization. Horm. Behav. **11:** 190-201.
85. WHALEN R. E. & K. L. OLSEN. 1981. Role of aromatization in sexual differentiation: Effects of prenatal ATD treatment and neonatal castration. Horm. Behav. **15:** 107-122.
86. BOOTH, J. E. 1978. Effects of aromatization inhibitor androst-4-ene-3,6,17-trione on sexual differentiation induced by testosterone in the neonatally castrated rat. J. Endocrinol. **79:** 69-76.
87. GLADUE, B. A. & L. G. CLEMENS. 1980. Masculinization diminished by disruption of prenatal estrogen biosynthesis in male rats. Physiol. Behav. **25:** 589-593.

Sex Steroids and Energy Balance: Sites and Mechanisms of Action[a]

GEORGE N. WADE

Division of Neuroscience and Behavior
Department of Psychology
University of Massachusetts
Amherst, Massachusetts 01003

INTRODUCTION

In addition to their well-known effects on reproductive behaviors, gonadal steroids also have some rather striking actions on the regulation of energy metabolism in a wide range of vertebrate species. It has been suggested that these hormone actions on both regulatory behaviors and metabolic processes represent adaptive mechanisms which allow animals to anticipate some of the predictable changes in metabolic demands which they experience during their lifetimes (e.g., lactation or changes of season).[1,2]

Energy balance is the difference between the intake and expenditure of metabolic fuels. Body energy stores (particularly triglycerides in white adipose tissue depots) fluctuate in a predictable fashion when there is an imbalance between energy intake and expenditure. Although energy intake is simply the amount of food (metabolic fuel) which is eaten, digested, and absorbed, energy may be expended in a wide variety of ways. For the sake of convenience, energy expenditure is often divided into obligatory and facultative components.[3] Obligatory energy expenditure (obligatory thermogenesis) includes both the basal metabolic rate (the cost of maintaining minimum body functions) and the thermic effect of feeding (the cost of assimilating and storing ingested nutrients). Facultative energy expenditure includes behaviors such as voluntary exercise and thermoregulatory behaviors (e.g., nest-building) in addition to controllable metabolic heat production as in thermoregulatory or diet-induced thermogenesis.[3,4]

It is clear that gonadal steroids influence both energy intake and expenditure and, thus, body weight and energy stores. I will first describe some of the effects of sex hormones on the behavioral and physiological determinants of energy balance in laboratory rodents such as rats and hamsters. Then I will discuss a typical strategy for studying steroid actions on behaviors and the usefulness of this approach in studying energy balance. Finally, I will suggest some potentially promising directions for future research. The discussion will reflect the fact that much more is known about the actions of estrogens than about either progestins or androgens.

---

[a]Original research and preparation of the manuscript supported by National Institutes of Health Research Grants NS 10873, AM 20785, and AM 32976 and by National Institute of Mental Health Research Scientist Development Award MH 00321.

HORMONE EFFECTS ON ENERGY BALANCE

Estradiol and Progesterone

Estrogens, including estradiol, induce a negative energy balance and reduce body energy stores, whereas progesterone has the opposite effect and increases body weight in estrogen-primed animals. Although all body components are affected by fluctuations in circulating levels of estradiol and progesterone, the changes that are seen in white adipose tissue mass are energetically most significant.[2,5]

Food intake plays a role in these steroid-induced shifts in energy balance. In gonadectomized or intact animals, estradiol decreases food intake, and progesterone, in the presence of estradiol, increases it. (Note that progesterone has no effect on energy balance in the absence of estradiol.) For some time it was widely assumed that these changes in food intake were of primary importance in causing the shifts in body weight and energy storage. However, a number of findings now indicate that factors other than food intake also play a significant role.

Pair-feeding experiments, in which animals have their hormone levels manipulated while they are fed the same amount as controls, reveal that estradiol and progesterone can influence body weight and composition in the absence of changes in food intake.[6,7] Similar results are often seen during naturally occurring changes in hormone levels. For example, pregnant hamsters lose substantial amounts of energy (body fat) without changing their food intake.[8] Thus, there are significant hormone-induced shifts in energy balance without changes in food intake, indicating that estradiol and progesterone must also affect energy expenditure.

It has been known for decades that ovarian hormones modulate behaviors that influence energy expenditure in rodents.[9] Estradiol stimulates voluntary exercise and reduces thermoregulatory nest-building, whereas progesterone has just the opposite effects (in the presence of estradiol). In addition to these behavioral actions, ovarian steroids have perhaps even more significant effects on physiological/metabolic thermoregulatory mechanisms. For example, estradiol both increases heat production (oxygen consumption) and activates physiological heat-loss mechanisms.[10-12]

What we see, then, is that ovarian steroids induce a coordinated pattern of behavioral and metabolic changes which result in an imbalance in energy intake and expenditure (TABLE 1). Thus, estradiol both reduces food intake and concurrently stimulates a variety of processes that increase energy expenditure. This imbalance causes a substantial reduction in body energy stores (largely adipose tissue). Note that the hormone-induced changes in intake and expenditure reinforce one another. There seem to be no short-term counterregulatory changes which would tend to offset the energetic imbalance (e.g., as there are during involuntary undereating or during forced exercise). However, with prolonged hormone treatment there are some poorly understood compensatory changes which stabilize body weight.

Testosterone

The effects of androgens on body weight and energy balance vary with a number of factors including species, hormone levels, and the androgen being used. Treatment of castrated male rats or hamsters with physiological doses of testosterone increases lean body mass (primarily muscle) and reduces carcass lipid content.[13,14] In testosterone-treated rats the anabolic effects obscure the decreases in body fat content, and there is a net gain in body weight. In hamsters the magnitude of these effects is reversed, and there is a net loss of weight during testosterone treatment.

Testosterone, like estradiol and progesterone, affects body weight via both behavioral and metabolic changes. Treatment with physiological doses of testosterone increases food intake (especially protein consumption) in castrated male rats, but not hamsters.[13,14] At the same time, testosterone increases energy expenditure, including voluntary exercise. Relatively little is known about androgen action on facultative thermogenesis or thermoregulatory behaviors.

STRATEGIES FOR STUDYING HORMONE ACTION

A typical strategy for studying molecular actions of steroid hormones on behaviors might include the following steps: (a) The potential site(s) of hormone action must be identified first. This usually involves identifying brain regions with high densities of hormone receptors using autoradiographic or scintillation counting methods. (b) Then intracranial hormone implants can be used to determine at which of these loci the hormone may act to influence the behavior that is under investigation. (c) Complementary ablation experiments can determine whether lesions of these structures dampen the behavioral responses to hormone treatments. (d) Often a series of correlative studies is done in which receptor levels or hormone-receptor interactions are manipulated (e.g., using hormone antagonists), and behavioral effects are described. This sort of work allows inferences to be made about the role of hormone receptors in the behavioral actions of the hormones. (e) Once these steps are completed, it is then possible to use pharmacological, histochemical, and biochemical techniques to study post-receptor mechanisms of hormone action.

This sort of strategy has been quite successful in studying hormone effects on copulatory behaviors in rodents, as the other papers in this volume clearly demonstrate. It is likely that this approach will be of some value in understanding some of the effects of gonadal steroids on energy balance. However, as we will see, there are a number of factors which will limit its usefulness. The principal problem is that energy balance (and even individual components such as food intake) is determined by multiple, redundant mechanisms and that gonadal steroids affect nearly all of them. It is likely that changes in any one of these processes would be sufficient to alter energy balance (or food intake). However, in our experience none of them seems to be essential for the weight gains or losses to occur.

SITES AND MECHANISMS OF ACTION

Perhaps the most straightforward case is that of the estradiol-induced increases in voluntary exercise in female rats. Estradiol implants in the vicinity of the medial preoptic area, a region with a high density of estrogen receptors, stimulate running-wheel activity in ovariectomized rats.[15] Lesions in this area block the effects of systemic estradiol treatments on activity without causing a general debilitation.[16] Concurrent treatment with estrogen antagonists also blocks estradiol-induced activity in neurologically intact rats.[17] These findings are certainly consistent with the possibility that estradiol acts via estrogen receptors in the medial preoptic area to stimulate voluntary exercise.

Treatment with estradiol induces a variety of neurochemical changes in this part of the brain, but no one has attempted to relate these molecular effects to changes in locomotor activity. This, of course, points to an especially significant problem in behavioral endocrinology—that of linking specific neurochemical changes to specific behavioral events. Each hormone-sensitive brain region probably mediates several behavioral, endocrine, and/or metabolic rseponses to the hormone, and one must be extremely cautious in attempting to causally relate any one behavioral response to a particular (series of) neurochemical event(s). The problem of specificity arises no matter what end points or neural target tissues are under study. Although one could attempt to lessen this limitation by working with increasingly smaller pieces of brain, there is no a priori assurance that this approach will do anything other than retain most of the same problems in a preparation which is merely more difficult to study.

This is an example of a simple case. Hormone actions on food intake are a great deal more complicated. Early work indicated that estradiol implants in the vicinity of the ventromedial hypothalamus (VMH), another region with a high density of estrogen receptors, decreased food intake in ovariectomized rats.[15] Later work demonstrated a close temporal correlation between changes in food intake and subcellular distribution of hypothalamic estrogen receptors after treatment with an estrogen mimetic (FIGURE 1).[28] These findings are consistent with a VMH, receptor-mediated action of estradiol on food intake and the possibility that some of the neurochemical effects of estradiol in the VMH could be related to changes in intake.

However, the VMH is not the sole site of estrogen action on eating or body weight, because lesions of the VMH do not block the effects of estradiol on either measure.[19] One possibility is that estradiol could act at additional neural loci to affect eating and

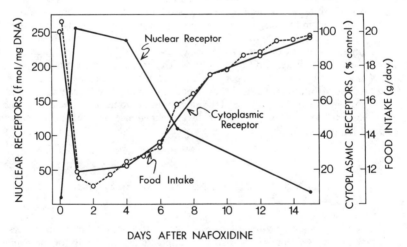

FIGURE 1. Time course of changes in food intake and subcellular distribution of hypothalamic estrogen receptors in ovariectomized rats given a single injection of 4 mg nafoxidine. (Data from Wade and Blaustein[28] and Wade and E. J. Roy, unpublished.)

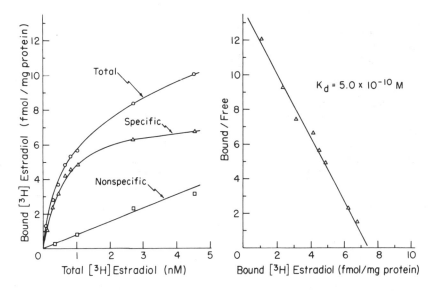

FIGURE 2. Saturation analysis and Scatchard plot of estradiol binding in rat parametrial white adipose tissue. (Data from Wade and Gray.[26])

body weight.[20] Even if this is the case, it appears as though estradiol can act on nonneural peripheral tissues to influence energy intake and storage. Work from at least two laboratories now indicates that estrogen mimetics can cause decreases in eating and body weight when given in doses which are too low to affect hypothalamic estrogen receptors, suggesting that peripheral actions alone are sufficient (although perhaps not necessary)[21] to influence energy balance.[22,23]

As we have pointed out before,[2,5] there are a number of metabolically significant peripheral tissues that are also target tissues for estradiol. They include the pituitary gland, liver, pancreas, kidney, skeletal muscle, and adipose tissues. Our own work has focused on white adipose tissues because of their responsiveness to ovarian hormones and their significance as energy storage depots. Briefly, this work demonstrates that adipose tissues are bona fide target tissues with high-affinity, limited-capacity receptors for both estrogens and progestins (FIGURE 2).[23-26] Several lines of research suggest that estrogens decrease white adipose tissue fat stores by decreasing uptake of circulating lipids (via changes in lipoprotein lipase activity) and by increasing hydrolysis and release of intracellular triglycerides (via changes in hormone-sensitive lipase activity) (TABLE 1).[2,5] We have suggested that estradiol acts via these "gatekeeping" enzymes to cause a redistribution of metabolic fuels away from white adipose tissue storage depots and into tissues that expend energy (e.g., brown adipose tissue, skeletal muscle, uterus) (FIGURE 3).[2,5] To the extent that comparisons have been made, progesterone seems to have the opposite effects.

The point of all this is that estradiol and progesterone have profound effects on the disposition and utilization of circulating and intracellular metabolic fuels. An enormous body of literature indicates that food intake is sensitive to the supply of utilizable metabolic fuels (although neither the exact signal(s) nor the detector(s) for the signal(s) has been specified). Thus, in addition to any direct effects in the brain, estradiol probably also affects food intake by altering peripheral metabolic processes.

TABLE 1. Effects of Estradiol on Energy Disposition in Rats[a]

ENERGY INTAKE:
 ↓ Food Intake

ENERGY STORAGE:
 ↓ White Adipose Tissue Lipid Stores
  ↓ Lipoprotein Lipase Activity and Fatty Acid Uptake
  ↑ Hormone-Sensitive Lipase Activity and Fatty Acid Release

ENERGY EXPENDITURE:
 BEHAVIORAL: METABOLIC:
  ↑ Voluntary Exercise   ↑ Resting Heat Production
  ↓ Thermoregulatory Nest-building   ↑ Heat Loss

[a]Note that estradiol acts independently via multiple mechanisms to influence energy balance. It concurrently decreases intake, increases expenditure, and reduces storage of lipids.

This extensive redundancy whereby estradiol affects food intake and energy metabolism via multiple sites and mechanisms of action is a characteristic of neuroendocrine systems regulating energy balance in general. Redundant mechanisms may have evolved, because the consequences of a failure to regulate one's energy balance for a few days are certainly more serious than, for example, a failure to mate, especially in small, polyestrous homeotherms.

Another major difficulty in studying mechanisms of hormone action on food intake has been the absence of effective hormone antagonists. None of the synthetic antiestrogens which antagonize the effects of estradiol on copulatory behavior has any effect on estradiol-induced decreases in food intake. In fact, all of the antiestrogens that

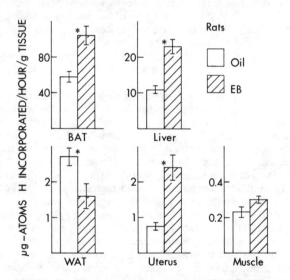

FIGURE 3. Incorporation of tritium (from [^{3}H]$_2$O) into newly synthesized fatty acids in ovariectomized rats treated with estradiol benzoate (EB) or sesame oil vehicle. BAT-interscapular brown adipose tissue; WAT-parametrial white adipose tissue; Muscle-gastrocnemius muscle. (Data from Edens and Wade.[31])

have been examined (MER-25, CI-628, nafoxidine, tamoxifen, en-clomiphene, LY117018) are devoid of antiestrogenic activity and, in fact, are full estrogen agonists for food intake and body weight regulation.[22,27,28] Thus, those of us working on hormone and energy balance have lacked a tool which has been extremely valuable in studying receptor and post-receptor events in other neuroendocrine responses to estrogens.

Less is known about the sites and mechanisms of progestin and androgen actions on energy balance, but there are a number of points worth mentioning. Although the sites of progesterone action have not been identified, nearly all of the central and peripheral tissues that contain estrogen receptors also have progestin receptors.[18,24,25] Note that in all these tissues estradiol priming is necessary to induce high levels of progestin receptors (FIGURE 4) which may explain why progesterone has minimal effects on energy balance in the absence of estrogens.[18,24] In the regulation of energy balance progesterone tends to counteract the actions of estradiol. It is not clear whether progesterone is simply acting to undo the molecular events induced by estradiol or

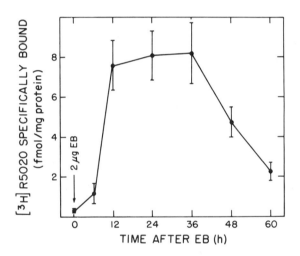

FIGURE 4. Concentration of progestin receptor, indicated by *in vitro* binding of [^{3}H]R5020, in parametrial white adipose tissue of ovariectomized rats after a single injection of 2 µg estradiol benzoate (EB). (Data from Gray and Wade.[24])

whether progesterone acts via independent pathways and mechanisms. (Of course, these possibilities are not mutually exclusive.)

Several lines of evidence indicate that the adiposity-reducing actions of testosterone are probably mediated by aromatized (estrogenic) metabolites of the hormone: (a) Aromatizable (e.g., testosterone), but not nonaromatizable (e.g., dihydrotestosterone, DHT), androgens stimulate exercise, reduce carcass lipid stores, and (in high doses) reduce food intake in castrated male rats. (b) In male rats, many tissues, including hypothalamus and adipose tissues, contain high concentrations of estrogen receptors. (c) Several of the effects of testosterone in adipose tissue are mimicked by estradiol and blocked or attenuated by concurrent treatment with an aromatase inhibitor.[13,25,27,29,30]

On the other hand, the anabolic and appetite-stimulating actions of testosterone are probably not mediated by either aromatized or 5α-reduced (e.g., DHT) metabolites. Neither of these effects is mimicked by estradiol, and testosterone is substantially more effective than DHT in stimulating weight gain and food intake.[13,29] It is likely that these anabolic effects of (unmetabolized) testosterone are due to direct actions in peripheral tissues such as kidney or skeletal muscle.[5,13,29]

FUTURE DIRECTIONS

Given that estrogens, progestins, and androgens seem to affect nearly every aspect of energy metabolism, there is certainly no shortage of potential directions for future research. Some examples with estrogens follow.

Estrogens and Food Intake

Estradiol decreases food intake in part by direct actions in the brain. One approach to this problem would be to correlate post-receptor neurochemical events with estradiol-induced changes, much as those who are studying reproductive behaviors are doing. However, for this approach to be of any value, two confounding factors must be overcome: (a) We must find a way to dissociate the neurochemical responses related to feeding from those related to other neuroendocrine actions of estradiol. (b) In addition, we must be able to study central actions of the hormone in the absence of peripheral actions which may also alter food intake.

It should be possible to dissociate central and peripheral hormone actions by using intracerebral hormone implants which minimize leakage to the periphery.[21] We could apply estradiol to central sites and look for neurochemical events such as changes in neurotransmitters or their receptors. In order to discriminate between central effects on food intake and other responses it might be possible to make use of the fact that the various antiestrogens are full estrogen agonists for food intake. Thus, estradiol-induced neurochemical events related to feeding should be mimicked, not antagonized, by central application of antiestrogens. Of course, finding an antiestrogen that actually antagonizes the actions of estradiol on food intake would be enormously helpful.

To understand how peripheral hormone actions affect food intake we need to know: (a) What are the metabolic effects of estradiol, and how are they induced? (b) What are the metabolic signals that control food intake? To answer the first question we must continue to determine at which sites estradiol acts to alter metabolic fuel disposition. (It is important to keep in mind the fact that hormones could also act in the brain to influence peripheral fuel metabolism.[5]) *In vivo* and *in vitro* studies can then begin to delineate the molecular mechanisms by which these changes are accomplished (e.g., via "gatekeeper" enzymes). The second question, What are the metabolic signals for feeding and satiety? has been the subject of a great deal of speculation and research but remains unresolved. (Once again, redundancy may be a major problem.)

Estrogens and Energy Expenditure

Although estradiol-induced activity has an effect on energy balance, by far the most significant means of energy expenditure is obligatory and facultative heat production.[3,4] Recall that estradiol stimulates resting thermogenesis (oxygen consumption) and heat loss in rats,[10–12] but little is known about the site(s) of this increase in energy expenditure. One intriguing possibility is that brown adipose tissue could be a significant contributor to estradiol-induced thermogenesis. Within the last several years, brown adipose tissue has been recognized as an important site for nonshivering thermogenesis; it may also play a significant role in the dissipation of excess ingested energy and in regulation of body weight.[3,4] Preliminary data indicate that estradiol treatment affects brown adipose tissue lipid metabolism in ovariectomized rats.[10,31,32] However, recent work from this laboratory (J. E. Schneider, J. F. McElroy, and G. N. Wade, unpublished data) indicates that ovarian steroids do not affect brown adipose tissue thermogenesis in rats or hamsters. Thus, gonadal effects on metabolic rate seem to be mediated by tissues other than brown adipose tissue.

CONCLUSIONS

In summary, gonadal steroids induce a series of coordinated shifts in the intake, tissue distribution, and expenditure of metabolic fuels. As with reproductive behaviors, some of these effects are due to direct hormone actions in the brain. Therefore, the same sorts of approaches should be useful in studying the molecular events mediating sex steroid effects on reproductive behaviors and on energy metabolism. Of course, a problem for all of us is determining which molecular events are responsible for which behavioral and metabolic responses.

Hormone effects on reproductive behaviors and energy balance differ in that there is probably a greater redundancy in the sites and mechanisms of steroid action on energy balance, perhaps because of the serious and fairly immediate consequences of a failure to regulate energy metabolism. Thus, hormones can act in the brain to influence both behaviors and peripheral metabolic processes; at the same time they can act on nonneural peripheral tissues to affect the distribution and utilization of metabolic fuels which may then feed back to alter regulatory behaviors. To study these phenomena, concepts and techniques from regulatory biology, nutrition, and metabolism will also be necessary.

It is possible that hybrid research such as this could be of value to fields which do not normally overlap. It could contribute to our understanding of the mechanisms regulating energy balance in mammals, particularly hormonal influences. In addition, studies of steroid actions on a variety of functions, such as reproductive behaviors and behavioral regulation of energy balance, could reveal some general principles of hormone-behavior interactions.

ACKNOWLEDGMENTS

I am grateful to Jeff Blaustein and Harvey Feder for helpful comments and discussions during the preparation of the manuscript.

REFERENCES

1. WADE, G. N. 1976. Sex hormones, regulatory behaviors, and body weight. *In* Advances in the Study of Behavior, Vol. 6. J. S. Rosenblatt, R. A. Hinde, E. Shaw & C. G. Beer, Eds.: 201-279. Academic Press. New York, N.Y.
2. WADE, G. N., J. M. GRAY & T. J. BARTNESS. 1985. Gonadal influences on adiposity. Int. J. Obesity. **9:** Supplement 1, 83-92.
3. HIMMS-HAGEN, J. 1981. Nonshivering thermogenesis, brown adipose tissue and obesity. *In* Nutritional Factors: Modulating Effects on Metabolic Processes. R. F. Beers & E. G. Bassett, Eds.: 85-99. Raven Press. New York, N.Y.
4. GIRARDIER, L. & M. J. STOCK. 1983. Mammalian Thermogenesis. Chapman & Hall. London, England.
5. WADE, G. N. & J. M. GRAY. 1979. Gonadal effects on food intake and adiposity: a metabolic hypothesis. Physiol. Behav. **22:** 583-593.
6. HERVEY, E. & G. R. HERVEY. 1968. Energy storage in female rats treated with progesterone in the absence of increased food intake. J. Physiol. **200:** 118P-119P.
7. ROY, E. J. & G. N. WADE. 1977. Role of food intake in estradiol-induced body weight changes in female rats. Horm. Behav. **8:** 265-274.
8. WADE, G. N., G. JENNINGS & P. TRAYHURN. 1986. Energy balance and brown adipose tissue thermogenesis during pregnancy in Syrian hamsters. Am. J. Physiol. **250:** In Press.
9. KINDER, E. F. 1927. A study of the nest-building activity of the albino rat. J. Exp. Zool. **47:** 117-161.
10. BARTNESS, T. J. & G. N. WADE. 1984. Effects of interscapular brown adipose tissue denervation on body weight and energy metabolism in ovariectomized and estradiol-treated rats. Behav. Neurosci. **98:** 674-685.
11. LAUDENSLAGER, M. L., H. J. CARLISLE & S. E. CALVANO. 1982. Increased heat loss in ovariectomized hypothyroid rats treated with estradiol. Am. J. Physiol. **243:** R70-R76.
12. LAUDENSLAGER, M. L., C. W. WILKINSON, H. J. CARLISLE & H. T. HAMMEL. 1980. Energy balance in ovariectomized rats with and without estrogen replacement. Am. J. Physiol. **238:** R400-R405.
13. SIEGEL, L. I., A. A. NUNEZ & G. N. WADE. 1981. Effects of androgens on dietary self-selection and carcass composition in male rats. J. Comp. Physiol. Psychol. **95:** 529-539.
14. SLUSSER, W. N. & G. N. WADE. 1981. Testicular effects on food intake, body weight, and body composition in male hamsters. Physiol. Behav. **27:** 637-640.
15. WADE, G. N. & I. ZUCKER. 1970. Modulation of food intake and locomotor activity in female rats by diencephalic hormone implants. J. Comp. Physiol. Psychol. **72:** 328-336.
16. KING, J. M. 1979. Effects of lesions of the amygdala, preoptic area, and hypothalamus on estradiol-induced activity in the female rat. J. Comp. Physiol. Psychol. **93:** 360-367.
17. ROY, E. J. & G. N. WADE. 1975. Role of estrogens in androgen-induced spontaneous activity in male rats. J. Comp. Physiol. Psychol. **89:** 573-579.
18. MACLUSKY, N. J. & B. S. MCEWEN. 1978. Oestrogen modulates progestin receptor concentrations in some brain regions but not in others. Nature **274:** 276-278.
19. BEATTY, W. W., D. A. O'BRIANT & T. R. VILBERG. 1975. Effects of ovariectomy and estradiol injections on food intake and body weight in rats with ventromedial hypothalamic lesions. Pharmacol. Biochem. Behav. **3:** 539-544.
20. DONOHOE, T. P. & R. STEVENS. 1981. Modulation of food intake by amygdaloid estradiol benzoate implants in female rats. Physiol. Behav. **27:** 105-114.
21. NUNEZ, A. A., J. M. GRAY & G. N. WADE. 1980. Food intake and adipose tissue lipoprotein lipase activity after intrahypothalamic estradiol benzoate implants in rats. Physiol. Behav. **25:** 595-598.
22. BOWMAN, S. P., A. LEAKE, M. MILLER & I. D. MORRIS. 1981. Agonist and antagonist activity of en-clomiphene upon oestrogen-mediated events in the uterus, pituitary gland and brain of the rat. J. Endocrinol. **88:** 367-374.
23. GRAY, J. M., S. D. DUDLEY & G. N. WADE. 1981. *In vivo* cell nuclear binding of 17β-[^{3}H]estradiol in rat adipose tissues. Am. J. Physiol. **240:** E43-E46.
24. GRAY, J. M. & G. N. WADE. 1979. Cytoplasmic progestin binding in rat adipose tissues. Endocrinology **104:** 1377-1382.

25. GRAY, J. M. & G. N. WADE. 1980. Cytoplasmic estrogen, but not progestin, receptors in male rat adipose tissues. Am. J. Physiol. **239:** E237-E241.
26. WADE, G. N. & J. M. GRAY. 1978. Cytoplasmic 17β-[^{3}H]estradiol binding in rat adipose tissues. Endocrinology **103:** 1695-1701.
27. ROY, E. J. & G. N. WADE. 1976. Estrogenic effects of an antiestrogen, MER-25, on eating and body weight in rats. J. Comp. Physiol. Psychol. **90:** 156-166.
28. WADE, G. N. & J. D. BLAUSTEIN. 1978. Effects of an antiestrogen on neural estradiol binding and on behaviors in female rats. Endocrinology **102:** 245-251.
29. GENTRY, R. T. & G. N. WADE. 1976. Androgenic control of food intake and body weight in male rats. J. Comp. Physiol. Psychol. **90:** 18-25.
30. GRAY, J. M., A. A. NUNEZ, L. I. SIEGEL & G. N. WADE. 1979. Effect of testosterone on body weight and adipose tissues: role of aromatization. Physiol. Behav. **23:** 465-469.
31. EDENS, N. K. & G. N. WADE. 1983. Effects of estradiol on tissue distribution of newly-synthesized fatty acids in rats and hamsters. Physiol. Behav. **31:** 703-709.
32. KEMNITZ, J. W., Z. GLICK & G. A. BRAY. 1983. Ovarian hormones influence brown adipose tissue. Pharmacol. Biochem. Behav. **18:** 563-566.

Steroid Receptors and Hormone Action in the Brain[a]

JEFFREY D. BLAUSTEIN

Division of Neuroscience and Behavior
Department of Psychology
University of Massachusetts
Amherst, Massachusetts 01003

The discovery in 1969 of binding proteins (receptors) for estradiol in the brain[1,2] opened the door for studies of the mechanisms of hormone action on behavior. Since that time, a great deal of work has been done on the cellular mechanisms by which the sex steroid hormones regulate reproductive behavior. Much of this work has been aimed at resolving if and how intracellular receptors for estradiol and progesterone are involved in mediating the effects of these hormones on behavior.

The notion that receptors for estradiol and progesterone might mediate the effects of these two hormones on sexual behavior is derived from work on the cellular mechanism of action of estradiol and progesterone in the chick oviduct[3] and rodent uterus[4,5] (FIGURE 1). In these tissues, the actions of estradiol and progesterone require interaction with their respective hormone receptors. Evidence suggests that steroid hormones pass freely through the cell membrane. According to the traditional model of steroid action, in target cells there are receptors in the cytoplasm that bind the hormone, thereby forming a hormone-receptor complex. This binding causes a conformational change in the receptor resulting in its translocation to the cell nucleus. Once it is present in the nucleus, the hormone-receptor complex may bind to acceptor sites on the chromatin, causing changes in gene expression. This may then lead to a change in protein synthesis and consequently to altered cellular function.

Two recent experiments have suggested another model of steroid action. These experiments provided evidence that the apparent cytoplasmic localization of estrogen receptors in target tissues in the absence of estradiol may be an artifact of homogenization of the tissue.[6,7] The authors of both papers have suggested that all estrogen receptors (and perhaps, other steroid hormone receptors) normally may be localized in cell nuclei, but they may be either tightly or loosely associated with nuclear components depending on whether receptor-ligand binding has occurred.

A third possibility has been raised that the intracellular receptors are present in equilibrium between the cytoplasm and the nucleus of the cell.[8] According to this scheme, steroid binding with consequent binding to chromatin acceptor sites causes an increase in the concentration of receptors in the cell nucleus as the receptors move from the cytoplasm to reestablish the initial equilibrium.

These three models differ as to the subcellular compartment in which they claim the unoccupied receptors are located. However, each of the models is in agreement

---

[a] The research that was performed in the author's laboratory was supported by Grant BNS 13050 from the National Science Foundation and Grant NS 19327 from the National Institutes of Health.

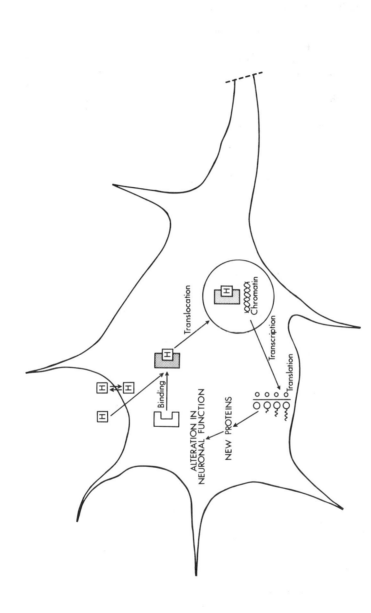

FIGURE 1. Model of the hypothesized neuronal mechanism of action of steroid hormones based on the model originally developed for hormone action in the chick oviduct and rodent uterus. Steroid hormones may freely diffuse through the cell membrane. If there are intracellular receptors present in the cell, the steroid hormone may bind, causing a conformational change. The hormone–receptor complex is then translocated into the cell nucleus where it may bind to acceptor sites on the chromatin. This may cause changes in gene expression, leading to the synthesis of new proteins, and ultimately to a change in the function of the cell.

with the idea that the unoccupied receptors either are not present in the nucleus or at least are not tightly associated with it. In all three models, presence of the hormone causes a greater or tighter association of receptors with nuclear components, presumably the chromatin. Because there are a number of unanswered questions regarding these alternate models of steroid mechanism of action, it may be premature to depart from the traditional model and its associated nomenclature. However, if one of the new models were to prove accurate, then the nomenclature used throughout this paper would require some modification; those receptors that are called cytosol receptors in this paper would be those receptors that are loosely associated with the cell nuclei. Similarly, those receptors that are called nuclear receptors would be those receptors that are tightly associated with the cell nuclei.

Estrogen Receptors and Behavior

When neural estrogen receptors were first characterized, they were found to have high affinity and be similar physicochemically to estrogen receptors that had been documented and characterized in peripheral reproductive tissues.[9] It was soon learned through [^{3}H]estradiol autoradiography[10] and *in vitro* binding studies[11] that the receptors are present in high concentration in those areas that had been implicated in estradiol's effects on sexual behavior in rats and guinea pigs (e.g., mediobasal hypothalamus[12]). As in peripheral target tissues, estradiol treatment causes the depletion of cytosol receptors and the accumulation of cell nuclear estrogen receptors in these neural areas.[13]

There have been several attempts to correlate the presence of estrogen receptors and of cell nuclear binding in the hypothalamus with estradiol's effects on sexual behavior. The possible role of cell nuclear binding of the estradiol-receptor complexes in sexual behavior has been studied primarily through the use of estrogen antagonists. In general, treatments that block the nuclear accumulation of estradiol-estrogen receptor complexes are effective in blocking estradiol's actions on sexual behavior.[14]

Most of the cell nuclear bound [^{3}H]estradiol that results from a [^{3}H]estradiol injection is lost rapidly from cell nuclei such that, by the time of a progesterone injection 24 hr later, there is little or no estradiol (or presumably nuclear estrogen receptors) remaining in cell nuclei.[15] This suggests that estradiol has a triggering effect, and estrogen receptor complexes need not be present in hypothalamic cell nuclei at the time that sexual behavior is expressed. However, the results of studies, that demonstrated that estrogen antagonists administered just prior to the progesterone injection inhibited sexual behavior, were at odds with this finding.[16] In an attempt to reconcile these disparate findings, we injected female rats with [^{3}H]estradiol and measured cell nuclear binding of [^{3}H]estradiol 24 hr later.[17] Other rats were injected with unlabeled estradiol followed 19 hr later by progesterone; these animals were tested for sexual behavior 24 hr after the estradiol. We found low, but detectable, levels of [^{3}H]estradiol in hypothalamic cell nuclei 24 hr after injection, at the time that animals display sexual receptivity. Furthermore, injection of either of two estrogen antagonists just prior to the progesterone injection both inhibited estradiol's priming action on sexual behavior and displaced the cell nuclear bound [^{3}H]estradiol in the hypothalamus. Similar conclusions were reached in experiments with guinea pigs.[18] These experiments suggest that estradiol (and presumably nuclear estrogen receptors) must be present in hypothalamic cell nuclei at least at the start of the period of sexual behavior.

In another attempt to link the cell nuclear binding of estradiol and sexual behavior, it was found that experimentally induced diabetes mellitus results in a decrease in the level of sexual behavior after estradiol and progesterone treatment.[19] Diabetic rats also show reduced cell nuclear binding of [^{3}H]estradiol[19] and reduced levels of hypothalamic nuclear estrogen receptors[20] after estradiol injection. Because all of the deficits can be reversed by treatment with insulin, Wade and co-workers have suggested that diabetes may inhibit sexual behavior by interfering with nuclear estrogen receptor accumulation.

Therefore, evidence from three types of studies is in agreement with the notion that estrogen receptors are involved in estradiol's effects on sexual behavior in rodents. Specifically, (1) estrogen receptors are present in neuroanatomical areas that have been demonstrated to be sites of action for estradiol's effect on sexual behavior; (2) estrogen receptor antagonists that inhibit estradiol's interaction with its receptor block estradiol's effect on sexual behavior, even if administered very close to the time of the facilitatory progesterone injection; and (3) diabetes inhibits sexual behavior, perhaps by interfering with cell nuclear binding of hypothalamic estrogen receptors.

PROGESTIN RECEPTORS AND SEXUAL BEHAVIOR

After a sufficient period of estrogen priming, progesterone facilitates the display of rodent sexual behavior with a short latency (within 1 hr in rats[12] and guinea pigs[21] with some conditions). Guinea pigs remain sexually receptive for 8-10 hr after which heat terminates, and they become refractory to further stimulation by progesterone[16] (desensitization; TABLE 1). A model of progesterone's mechanism of action must be able to account for all of these effects of progesterone on sexual behavior.

Once the presence of neural progestin receptors in the brain was established with biochemical and autoradiographic techniques,[22-26] the hypothesis was tested that estradiol induces responsiveness to progesterone by increasing the level of cytosol progestin receptors. It was learned that in ovariectomized rats and guinea pigs, treatment with estradiol causes an increase in the concentration of cytosol progestin receptors in some brain regions.[23,27] Estradiol induces progestin receptors in the mediobasal hypothalamus and preoptic area, and to a lesser extent, the midbrain, all areas that have been implicated as possible sites of action for progesterone's effects on sexual behavior.[12,27] Furthermore, the time course for estradiol's induction of cytosol progestin

TABLE 1 Typical Procedures Used in the Study of Cellular Mechanisms of Progesterone Action on Female Sexual Behavior

	Facilitation of Sexual Behavior by Progesterone		
0 hr		40 hr	Lordosis
Estradiol		Oil	No
Estradiol		Progesterone	Yes

	Progesterone Desensitization of Progesterone-Facilitated Sexual Behavior			
0 hr	40 hr	Lordosis	64 hr	Lordosis
Estradiol	Oil	No	Progesterone	Yes
Estradiol	Progesterone	Yes	Progesterone	No

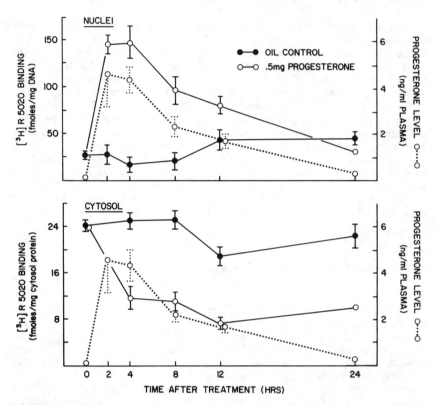

FIGURE 2. Time course of depletion of cytosol progestin receptors (mean ± SEM) and accumulation of nuclear progestin receptors in pooled hypothalamus-preoptic area-septum and plasma levels of progesterone in ovariectomized guinea pigs injected with 1.6 μg estradiol benzoate, followed 40 hr later by oil or 0.5 mg progesterone. Animals were killed at various times after oil or progesterone injection, their blood plasma was assayed for progesterone concentration, and the hypothalamus—preoptic area—septum was assayed for either cytosol or nuclear progestin receptors. Similarly treated guinea pigs display lordosis from approximately 4 hr to 12-14 hr after progesterone injection. (Reprinted with permission from Blaustein and Feder.[32])

receptors closely parallels that for estradiol's induction of responsiveness to progesterone, suggesting that the two events might be causally linked.[27,28]

We have postulated that progesterone's desensitization effect on sexual behavior may be due to down-regulation of cytosol progestin receptors.[29-32] A day after progesterone injection, hypothalamic cytosol progestin receptors are present at a level much lower than at the time of injection[30-33] (FIGURE 2, bottom panel, 24 hr). Furthermore, a supplemental injection of estradiol at the time of the first progesterone injection offsets progesterone's desensitization effect and restores the level of cytosol progestin receptors a day later, apparently by increasing the synthesis of cytosol progestin receptors[29] (FIGURE 3B, 0 hr).

The hypothesis that accumulation of elevated levels of nuclear progestin receptors is critically involved in progesterone's effects on sexual behavior has been tested in a

TABLE 2 RU 486 Inhibition of Progesterone-Facilitated Sexual Behavior[a]

Group	0 hr	39 hr	40 hr	n	Percentage Responding[b]	Heat Duration[c] (hr)	Heat Duration[d] (hr)	Latency to Lordosis[d] (hr)	Maximum Lordosis Duration[d] (sec)
I	EB	V	$P_{0.1}$	10	90	7.30 (±0.90)	8.11 (±0.42)	4.44 (±0.63)	12.33 (±1.08)
II	EB	$RU_{0.5}$	$P_{0.1}$	9	100	8.00 (±0.99)	8.00 (±0.99)	4.56 (±0.34)	14.89 (±1.66)
III	EB	RU_5	$P_{0.1}$	9	11	0.56 (±0.56)	5.00 (—)	7.00 (—)	10.00 (—)
IV	EB	V	V	10	0	—	—	—	—

[a]EB = estradiol benzoate, V = vehicle, P = progesterone, RU = RU 486. Hourly tests for lordosis began just prior to the injection at 39 hr and continued until all animals were nonresponsive. Values reported are means (± SEM). Subscripts are doses in milligrams.
[b]I vs. III: $p = 0.002$, Fisher's Exact Probability Test.
[c]I vs. IV: $p = 0.002$, Mann–Whitney U Test.
[d]Includes responding animals only.

variety of experiments. Progesterone injection in estrogen-primed rats and guinea pigs results in accumulation of progestin receptors in cell nuclei.[32,34] Injection of a progesterone antagonist, RU 486, which decreases the concentration of available cytosol progestin receptors, blocks progesterone's facilitatory effect on sexual behavior in guinea pigs[35] (TABLE 2), an effect that can be overcome by increasing the dose of progesterone used (TABLE 3). In progesterone-injected guinea pigs that are refractory to a progesterone injection, a second progesterone injection results in less accumulation of nuclear progestin receptors that in guinea pigs that had not received the intervening progesterone injection.[32] As discussed above, in progesterone-treated guinea pigs an injection of a supplemental dose of estradiol partially blocks progesterone's down-regulation of its cytosol receptors. Consistent with a role for nuclear progestin receptors in progesterone's effects on sexual behavior, progesterone injection in these animals

TABLE 3 A Large Dose of Progesterone Overcomes RU 486-Induced Inhibition of Progesterone-Facilitated Sexual Behavior[a]

Group	0 hr	39 hr	40 hr	n	Percentage Responding[b]	Heat Duration[c] (hr)	Heat Duration[d] (hr)	Latency to Lordosis[d] (hr)	Maximum Lordosis Duration[d,e] (sec)
I	EB	RU_5	$P_{0.1}$	10	40	2.00 (±1.19)	5.00 (±2.38)	5.75 (±1.75)	5.75 (±1.44)
II	EB	RU_5	P_5	9	100	9.33 (±0.77)	9.33 (±0.77)	3.67 (±0.50)	10.67 (±1.12)

[a]Values reported are means (± SEM). Abbreviations are the same as in TABLE 2.
[b]$p = 0.02$, Fisher's Exact Probability Test.
[c]$p < 0.02$, Mann–Whitney U Test.
[d]Includes responding animals only.
[e]$p < 0.02$, Student's t Test.

results in both elevated levels of nuclear progestin receptors (FIGURE 3A, 4 hr) and facilitation of lordosis.[29]

Progesterone-injected guinea pigs are not insensitive to subsequent progesterone treatment; they are only hyposensitive. That is to say, although progesterone-injected guinea pigs do not subsequently respond to a second, moderate dose of progesterone (0.5 mg), they respond to a pharmacological dose[30] (20 mg). Furthermore, doses of progesterone that result in the expression of sexual receptivity, also cause increased accumulation of nuclear progestin receptors. This finding, together with the fact that progesterone does not decrease sensitivity to other substances that can facilitate lordosis in rats (but that do not interact with progestin receptors such as methysergide or luteinizing hormone-releasing hormone,[36,37] suggests that the desensitization effect of progesterone is a specific induction of a hyposensitivity to progesterone.

We have also suggested that the duration of the period of sexual receptivity may be determined by the duration of the retention of progestin receptors by cell nuclei.[29,32,38] The retention of elevated levels of nuclear progestin receptors after a progesterone

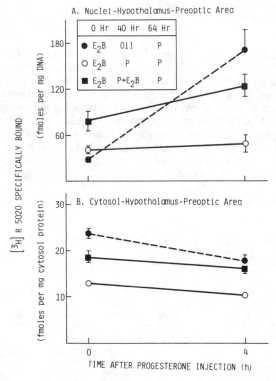

FIGURE 3. Mean (± SEM) progestin receptor concentration in cell nuclei or cytosol from pooled hypothalamus-preoptic area of guinea pigs injected with 2 μg estradiol benzoate followed 40 hr later by either oil, 0.5 mg progesterone, or 0.5 mg progesterone and 10 μg estradiol benzoate. At 64 hr, animals were killed (time = 0 hr) or injected with 0.5 mg progesterone and killed 4 hr later, and progestin receptors were measured. (Reprinted with permission from Blaustein.[29])

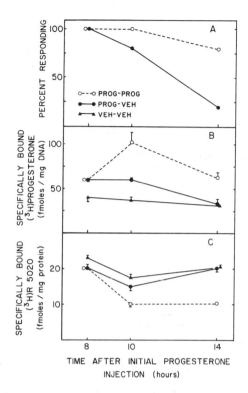

FIGURE 4. (A) Percentage of guinea pigs responding during tests for lordosis. Animals were injected with 50 μg progesterone or oil 48 hr after receiving 4 μg estradiol benzoate. Eight hours later, animals received a supplemental injection of 500 μg progesterone or oil (VEH). (B) Mediobasal hypothalamus-preoptic area nuclear progestin receptor levels (± SEM) in guinea pigs treated as in (A). Guinea pigs received 50 μg progesterone or oil 48 hr after receiving 4 μg estradiol benzoate. Eight hours later, brains from some animals were assayed for progestin receptors. Other animals received the supplemental injection of 500 μg progesterone or oil and were killed 10 or 14 hr after the initial progesterone injection. (C) Mediobasal hypothalamus-preoptic area cytosol receptor levels in animals described in (B). (Reprinted with permission from Brown and Blaustein.[38])

injection correlates temporally with heat duration[29] (FIGURE 2). Heat duration can be extended either by injecting a supplemental dose of estradiol at the time of the progesterone injection,[32] by providing a supplemental dose of progesterone while the animals are still sexually receptive[39] (FIGURE 4), or by implanting and leaving *in situ* a Silastic capsule containing progesterone in guinea pigs that already have a capsule containing estradiol (T. J. Brown and J. D. Blaustein, unpublished observations). In all of these cases, the retention of elevated levels of nuclear progestin receptors is prolonged.

At this time, it is not clear what cellular mechanisms are involved in the decrease in the concentration of progestin receptors from cell nuclei. Experiments are in progress to determine if there is a dependence of the level of nuclear progestin receptors on the level of both cytosol progestin receptors and tissue levels of progesterone. We also hope to determine if the loss nuclear progestin receptors is due to degradation (perhaps by proteolysis) or inactivation (conversion to a form that does not bind to the hormone, perhaps by dephosphorylation[40] of the receptors. In light of the relationship between the level of nuclear progestin receptors and heat duration, the study of the regulation of the level of nuclear progestin receptors should provide important information on the regulation of heat duration by gonadal steroid hormones.

To summarize, there seems to be a link between the level of cytosol progestin receptors in the hypothalamus and behavioral sensitivity to progesterone. The evidence suggests that the cellular site of action of progesterone for sexual behavior is the cell nucleus where, as in other tissues, it may cause changes in gene expression, ultimately leading to changes in neuronal function. Finally, the evidence suggests that the duration of heat is determined, at least in part, by the duration of elevated levels of progestin receptors in cell nuclei.

ALTERNATIVE MECHANISMS

The experiments discussed so far support the hypothesis that intracellular steroid receptors are involved in mediating the effects of sex steroid hormones on behavior. However, it is possible that steroid hormones also have direct effects on membrane receptors. Binding proteins that may be membrane-bound receptors have been documented for estradiol in some target tissues,[41] and uncharacterized binding proteins have been documented for estrogens and progestins in membranes from neural tissues.[42] The question of the involvement of these receptors in behavior unfortunately has not been addressed.

Steroid hormones have some direct effects at the level of the cell membrane in the brain. For example, rapid electrophysiological effects have been observed after steroid hormones were applied locally to the septum-preoptic area, suggesting an action at the cell membrane level.[43,44] Second, under some circumstances, estradiol can decrease the concentration of serotonin binding sites when incubated, *in vitro*, with a crude membrane fraction from hypothalamus.[45] At present, it is not known if this membrane effect of estradiol is related to estradiol's effects on sexual behavior.

Recent experiments have implicated cyclic AMP in the mechanism of progesterone action. A variety of potentiators of cyclic AMP's action can substitute for progesterone in the facilitation of sexual behavior in female rats.[46] Inhibitors of cyclic AMP's action can block progesterone's effect. Although these experiments are consistent with the possibility that progesterone acts through a cyclic nucleotide (and perhaps a membrane receptor mechanism), they are not strong evidence for it. Rather, these studies demonstrate that at any one of the many stages of hormone action on behavior, cyclic nucleotides may be involved. This is not surprising, considering that steroid hormones influence a variety of neurotransmitter systems.[47] As discussed in other papers in this volume,[39] some hormonal effects on behavior may involve modulation of neurotransmitter systems. Because some of these neurotransmitter systems and neuromodulators (including luteinizing hormone-releasing hormone) operate through a cyclic nucleotide mechanism,[48] it would be expected that influencing the cyclic nucleotides might facilitate sexual behavior.

WHY RECEPTORS?

The presence of multiple steroid receptor systems may allow for multiple levels or fine-tuning of control of steroid hormone sensitivity. To start at the simplest possible case, if an animal with no specific detectors for steroid hormones secreted hormone (estradiol and progesterone), then all cells of the body would either respond or not respond to those hormones. This would be true because there would be no way of differentiating target cells from nontarget cells. However, the presence of receptors in all cells would serve to make all cells targets for the hormones, while the presence of receptors in *some* cells would make a subset of all of the cells targets for the hormones. If only a subset of estradiol receptor-containing cells responded to estradiol with an increase in the concentrations of progestin receptors (E-P cells), then we would have three types of cells—estrogen target cells, progesterone target cells, and those estrogen target cells that are also capable of responding to progesterone under appropriate conditions (after treatment with estradiol). This addition would give the system increased control; some cells would only respond to progesterone under conditions of recent treatment with estradiol. In addition, some cells would always respond to estradiol, some cells would always respond to progesterone, and some cells would not respond to either hormone under any condition.

Superimposed upon this system may be the regulation of levels of receptors for estradiol and progesterone by various neurotransmitter systems. For example, an E-P neuron that is innervated by a noradrenergic neuron might only respond to progesterone in the presence of appropriate noradrenergic stimulation. Other neurotransmitters might have similar relationships with either estradiol, progesterone, or E-P-sensitive cells. A steroid-sensitive neuron that is regulated by one or more hormones, and is also regulated by a variety of neurotransmitters, should be able to modulate sensitivity to a given hormone quickly to meet the ever-changing demands of the environment.

Although there is little direct evidence at the present, it is possible that all of these events occur in the brain. It is known that some neurons contain estrogen receptors, some neurons contain progestin receptors, and in some neurons the concentration of progestin receptors increases after estradiol treatment. However, it has not been demonstrated that the two types of receptor are present in the same cell. The fact that they are present together in certain nonneural homogeneous cell lines[49] suggests that they could coexist in the brain as well. Similarly, although we know that neurotransmitters can influence the concentration of steroid hormone receptors, it has not yet been shown that this results in the predicted change in the sensitivity to the particular hormone. Although the notion that these interactions provide a way for the environment to exert precise control over steroid hormone sensitivity is attractive, at present it is only speculation.

FUTURE DIRECTIONS

Many correlations have been made between hormone receptors and behavior, suggesting that receptors are involved in mediating the effects of steroid hormones on sexual behavior. Although the temporal correlations are tighter for progesterone than for estradiol, it seems that binding of steroids to receptors and subsequent translocation of receptors to cell nuclei are critically involved in mediating the effects of both hormones on behavior. It is unfortunate that, unlike the case for some other nonbehavioral hormone-sensitive systems (e.g., ovalbumin synthesis in the chick oviduct), some of the critical experiments cannot be performed. That is, even if a primary genomic event that is dependent on steroid hormone receptors were found, it could not (at this time) be determined that it is related to a behavioral response. However, if proteins are found that are uniquely involved with hormonal effects on behavior, it might be possible to study their regulation and make inferences about the regulation of behavior. Nevertheless, if we are interested in hormonal mechanisms of behavior, at present, we seem to be limited to the correlational types of experiments that are being done.

The availability of monoclonal antibodies for steroid hormone receptors[6] may provide a new technique to measure steroid hormone receptors directly, rather than based on their ability to bind a radioactively labeled steroid. Immunochemical techniques might be used to address problems such as the fate of the progestin receptors once they are no longer present in the cell nucleus in a form that binds progestins. Also, this type of technique could be useful for determining what happens to steroid receptors that have been decreased in concentration by neurotransmitter antagonists. In each of these cases, determining if the receptors are modified, losing their ability to bind their respective hormone, or are degraded could give us insight into the cellular mechanisms involved in this regulation.

There are also some very interesting questions that need to be answered that may tell us more about the integrative capacities of the nervous system. One of the critical questions is how hormones and neurotransmitters interact. There are numerous examples of ways in which hormones may act through neurotransmitter systems to exert their effects. These include the dependence of hormonally induced behavioral changes on a particular neurotransmitter,[39] modulation of enzymes involved in neurotransmitter metabolism (and consequently, neurotransmitter levels) by steroid hormones,[47] and effects of steroids on neurotransmitter receptor concentrations.[47] In addition, there are now a number of examples of modulation of steroid hormone receptors (and presumably of sensitivity to the hormone) by neurotransmitters.[50,51] Precise techniques that can determine the relationship between steroid-sensitive systems and neurotransmitter systems are essential.

More and more emphasis is being placed on the study of hormone mechanisms in smaller subdivisions of the brain.[47] Recently, tissue microdissection techniques have been developed with which steroid hormone receptors can be measured in very small neuroanatomical regions, such as the ventromedial nucleus of the hypothalamus (VMN).[52,53] Although this type of procedure could be used to map some of the hormone-neurotransmitter relationships, it is not without its problems. Although a defined nucleus such as the VMN is more homogeneous than the whole hypothalamus, it is still composed of many cell types. These studies are therefore plagued by many of the same problems as studies using such grosser dissection. In fact, they may have other problems as well. The first problem is that tissues from a large number of animals must be pooled for a single data point for steroid receptor assays.[52,53] Second, because dissection must be precise, there is a greater potential for dissection error, which may manifest itself in larger variation from sample to sample. This increased error could lead to a failure to detect small differences that might be seen with a technique with more repeatability.

A more powerful approach that has been used by several research groups to map relationships between steroid-sensitive systems and neurotransmitter systems is steroid autoradiography in conjunction with other histological techniques, such as dye-tracing.

Using a combination of [³H]estradiol autoradiography with fluorescent dye retrograde tracing, Morrell and Pfaff[54] have shown that some estradiol-concentrating neurons in the ventrolateral aspect of the VMN project to the dorsal midbrain. These neurons could provide the steroid-sensitive neuroanatomical connections that have been postulated to be involved in the hormonal regulation of sexual behavior between the VMN and the midbrain central gray.[55] Similarly, by combining [³H]estradiol autoradiography with immunohistochemistry, Sar[56] has identified neurons which both concentrate [³H]estradiol and contain the rate-limiting enzyme for catecholamine synthesis, tyrosine hydroxylase. This finding provides evidence that estradiol may act directly on catecholaminergic neurons. Using a combined [³H]estradiol autoradiography-fluorescence microscopy technique, Heritage et al.[57] have identified catecholaminergic nerve terminals in proximity to [³H]estradiol-concentrating cells in some regions. Some of these cells could be the type in which noradrenergic inhibitors decrease the concentration of cytosol estrogen receptors (J. D. Blaustein, unpublished observations) and of estradiol-induced progestin receptors.[58]

It may be possible to map the projections of specific neurons that, for example, have noradrenergic or dopaminergic inputs and also contain estrogen or progestin receptors. By combining steroid receptor autoradiography or a monoclonal antibody immunocytochemical technique[6] with a neurotransmitter receptor mapping technique,[47] we may be able to determine precisely which neurotransmitter receptors are present on estrogen- and progesterone-sensitive neurons. Furthermore, if one or more specific proteins are found that are specifically involved in sexual behavior, then neurons that contain them could be localized by an immunocytochemical technique. By dye-tracing procedures, their projections could be mapped. It might then be possible to selectively destroy some of these projections and determine their role in sexual behavior. Further investigation of these types of neuroanatomical interactions can be expected to give us clearer insights into the hormone-neurotransmitter interactions involved in sexual behavior.

ACKNOWLEDGMENTS

I thank George Wade and Ted Brown for their helpful comments on the manuscript. I am also grateful to Jay Rosenblatt and the other faculty, students, and postdoctoral fellows of the Institute of Animal Behavior of Rutgers University for providing an environment in which I was able to begin work on the cellular mechanisms of hormonal regulation of sexual behavior in guinea pigs.

REFERENCES

1. EISENFELD, A. J. 1969. Hypothalamic estradiol binding macromolecules. Nature **224**: 1202-1203.
2. KAHWANAGO, I., W. L. HEINRICHS & W. L. HERRMANN. 1969. Isolation of oestradiol receptors from bovine hypothalamus and anterior pituitary gland. Nature **223**: 313-314.
3. O'MALLEY, B. W. & A. R. MEANS. 1974. Female steroid hormones and target cell nuclei. Science **183**: 610-624.
4. GORSKI, J. & F. GANNON. 1976. Current models of steroid hormone action: a critique. Annu. Rev. Physiol. **38**: 425-450.

5. JENSEN, E. V. & E. R. DESOMBRE. 1973. Estrogen-receptor interaction. Science 182: 126-134.
6. KING, W. J. & G. L. GREENE. 1984. Monoclonal antibodies localize oestrogen receptor in the nuclei of target cells. Nature 307: 745-747.
7. WELSHONS, W. V., M. E. LIEBERMAN & J. GORSKI. 1984. Nuclear localization of unoccupied oestrogen receptors. Nature 307: 747-749.
8. SHERIDAN, P., J. M. BUCHANAN, V. C. ANSELMO & P. M. MARTIN. 1979. Equilibrium: the intracellular distribution of steroid receptors. Nature 282: 579-582.
9. FEDER, H. H., I. T. LANDAU & W. A. WALKER. 1979. Anatomical and biochemical substrates of the actions of estrogens and antiestrogens on brain tissues that regulate female sexual behavior of rodents. *In* Endocrine Control of Sexual Behavior. C. Beyer, Ed.: 371-340. Raven Press. New York, N.Y.
10. PFAFF, D. W. & M. KEINER. 1973. Atlas of estradiol-concentrating cells in the central nervous system of the female rat. J. Comp. Neurol. 151: 121-158.
11. MCEWEN, B. S., P. G. DAVIS, B. PARSONS & D. W. PFAFF. 1979. The brain as a target for steroid hormone action. Annu. Rev. Neurosci. 2: 65-112.
12. BARFIELD, R. J., B. S. RUBIN, J. H. GLASER & P. G. DAVIS. 1983. Site of action of ovarian hormones in the regulation of oestrous responsiveness in rats. *In* Hormones and Behaviour in Higher Vertebrates. J. Balthazart, E. Prove & R. Gilles, Eds.: 2-17.
13. ROY, E. J. & B. S. MCEWEN. 1977. An exchange assay for estrogen receptors in cell nuclei of the adult rat brain. Steroids 30: 657-669.
14. ROY, E. J. & G. N. WADE. 1977. Binding of [^{3}H]estradiol by brain cell nuclei and female rat sexual behavior: inhibition by antiestrogens. Brain Res. 126: 73-87.
15. MCEWEN, B. S., D. W. PFAFF, C. CHAPTAL & V. N. LUINE. 1975. Brain cell nuclear retention of [^{3}H]estradiol in doses able to promote lordosis: temporal and regional aspects. Brain Res. 86: 155-161.
16. FEDER, H. H. & L. P. MORIN. 1974. Suppression of lordosis in guinea pigs by ethamoxytriphetol (MER-25) given at long intervals (34 - 46 h) after estradiol benzoate treatment. Horm. Behav. 5: 63-71.
17. BLAUSTEIN, J. D., S. D. DUDLEY, J. M. GRAY, E. J. ROY & G. N. WADE. 1979. Long-term retention of estradiol by brain cell nuclei and female rat sexual behavior. Brain Res. 173: 355-359.
18. WALKER, W. A. & H. H. FEDER. 1979. Long term effects of estrogen action are crucial for the display of lordosis in female guinea pigs: antagonism by antiestrogens and correlations with *in vitro* cytoplasmic binding activity. Endocrinology 104: 89-96.
19. GENTRY, R. T., G. N. WADE & J. D. BLAUSTEIN. 1977. Binding of [^{3}H]estradiol by brain cell nuclei and female rat sexual behavior: inhibition by experimental diabetes. Brain Res. 135: 135-146.
20. SIEGEL, L. I. & G. N. WADE. 1979. Insulin withdrawal impairs sexual receptivity and retention of brain cell nuclear estrogen receptors in diabetic rats. Neuroendocrinology. 29: 200-206.
21. MORIN, L. P. & H. H. FEDER. 1974. Hypothalamic progesterone implants and facilitation of lordosis behavior in estrogen-primed ovariectomized guinea pigs. Brain Res. 70: 81-93.
22. BLAUSTEIN, J. D. & G. N. WADE. 1978. Progestin binding by brain and pituitary cell nuclei and female rat sexual behavior. Brain Res. 140: 360-367.
23. MACLUSKY, N. J. & B. S. MCEWEN. 1978. Oestrogen modulates progestin receptor concentrations in some rat brain regions, but not in others. Nature 274: 276-277.
24. MOGUILEWSKY, M. & J. P. RAYNAUD. 1977. Progestin binding sites in the rat hypothalamus, pituitary and uterus. Steroids 30: 99-109.
25. SAR, M. & W. E. STUMPF. 1973. Neurons of the hypothalamus concentrate [^{3}H]progesterone or its metabolites. Science 182: 1266-1268.
26. WAREMBOURG, M. 1978. Radioautographic study of the brain and pituitary after [^{3}H]progesterone injection into estrogen-primed ovariectomized guinea pigs. Neurosci. Lett. 7: 1-5.
27. BLAUSTEIN, J. D. & H. H. FEDER. 1979. Cytoplasmic progestin receptors in guinea pig brain: characteristics and relationship to the induction of sexual behavior. Brain Res. 169: 481-497.

28. PARSONS, B., N. J. MACLUSKY, L. KREY, D. W. PFAFF & B. S. MCEWEN. 1980. The temporal relationship between estrogen-inducible progestin receptors in the female rat brain and the time course of estrogen activation of mating behavior. Endocrinology **107**: 774-779.
29. BLAUSTEIN, J. D. 1982. Alteration of sensitivity to progesterone facilitation of lordosis in guinea pigs by modulation of hypothalamic progestin receptors. Brain Res. **243**: 287-300.
30. BLAUSTEIN, J. D. 1982. Progesterone in high doses may overcome progesterone's desensitization effect on lordosis by translocation of hypothalamic progestin receptors. Horm. Behav. **16**: 175-190.
31. BLAUSTEIN, J. D. & H. H. FEDER. 1979. Cytoplasmic progestin receptors in female guinea pig brain and their relationship to refractoriness in expression of female sexual behavior. Brain Res. **177**: 489-498.
32. BLAUSTEIN, J. D. & H. H. FEDER. 1980. Nuclear progestin receptors in guinea pig brain measured by an *in vitro* exchange assay after hormonal treatments that affect lordosis. Endocrinology **106**: 1061-1069.
33. MOGUILEWSKY, M. & J. P. RAYNAUD. 1977. The relevance of hypothalamic and hypophyseal progestin receptor regulation in the induction and inhibition of sexual behavior in the female rat. Endocrinology **105**: 516-522.
34. RAINBOW, T. C., M. Y. MCGINNIS, L. C. KREY & B. S. MCEWEN. 1982. Nuclear progestin receptors in rat brain and pituitary. Neuroendocrinology. **34**: 426-431.
35. BROWN, T. J. & J. D. BLAUSTEIN. 1984. Inhibition of sexual behavior in female guinea pigs by a progestin receptor antagonist. Brain Res. **301**: 343-349.
36. GILCHRIST, S. & J. D. BLAUSTEIN. 1984. The desensitization effect of progesterone on female rat sexual behavior is not due to interference with estrogen priming. Physiol. Behav. **32**: 879-882.
37. RODRIGUEZ-SIERRA, J. F. & G. A. DAVIS. 1978. Progesterone does not inhibit lordosis through interference with estrogen priming. Life Sci. **22**: 373-378.
38. BROWN, T. J. & J. D. BLAUSTEIN. 1984. Supplemental progesterone delays heat termination and the loss of progestin receptors from hypothalamic cell nuclei in female guinea pigs. Neuroendocrinology **39**: 384-391.
39. CROWLEY, W. R. Reproductive neuroendocrine regulation in the female rat by central catecholamine-neuropeptide interactions: a local control hypothesis. Ann. N.Y. Acad. Sci., this volume.
40. GRODY, W. W., W. T. SCHRADER & B. W. O'MALLEY. 1982. Activation, transformation, and subunit structure of steroid hormone receptors. Endocr. Rev. **3**: 141-163.
41. SZEGO, C. M. & R. J. PIETRAS. 1981. Membrane recognition and effector sites in steroid hormone action. *In* Biochemical Actions of Hormones. G. Litwack, Ed. **III**: 307-463. Academic Press. New York, N.Y.
42. TOWLE, A. C. & P. Y. SZE. 1983. Steroid binding to synaptic plasma membrane: differential binding of glucocorticoids and gonadal steroids. J. Steroid Biochem. **18**: 135-143.
43. KELLEY, M. J., R. L. MOSS, C. A. DUDLEY & C. P. FAWCETT. 1977. The specificity of the response of preoptic-septal area neurons to estrogen: 17α-estradiol versus 17β-estradiol and the response of extrahypothalamic neurons. Exp. Brain Res. **30**: 43-52.
44. POULAIN, P. & B. CARETTE. 1981. Pressure injections of drugs on single neurons *in vivo*: technical considerations and application to the study of estradiol's effects. Brain Bull. **7**: 33-40.
45. BIEGON, A. & B. S. MCEWEN. 1982. Modulation by estradiol of serotonin$_1$ receptors in brain. J. Neurosci. **2**: 199-205.
46. BEYER, C., E. CANCHOLA & K. LARSSON. 1981. Facilitation of lordosis behavior in the ovariectomized estrogen primed rat by dibutyryl cAMP. Physiol. Behav. **26**: 249-251.
47. MCEWEN, B. S., A. BIEGON, C. T. FISCHETTE, V. N. LUINE, B. PARSONS & T. C. RAINBOW. 1984. Toward a neurochemical basis of steroid hormone action. *In* Frontiers in Neuroendocrinology. L. Martini & W. F. Ganong, Eds. Vol. **8**: 153-176. Raven Press. New York, N.Y.
48. BEYER, C., P. GOMORA, E. CANCHOLA & Y. SANDOVAL. 1982. Pharmacological evidence that LH-RH action on lordosis behavior is mediated through a rise in cAMP. Horm. Behav. **16**: 107-112.

49. HAUG, H. 1979. Progesterone suppression of estrogen-stimulated prolactin secretion and estrogen receptor levels in rat pituitary cells. Endocrinology **104**: 429-437.
50. CARDINALI, D. P., M. I. VACAS, M. N. RITTA & P. V. GLEIMAN. 1983. Neurotransmitter-controlled steroid hormone receptors in the central nervous system. Neurochem. Int. **5**: 185-192.
51. NOCK, B. & H. H. FEDER. 1981. Neurotransmitter modulation of steroid action in target cells that mediate reproduction and reproductive behavior. Neurosci. Biobehav. Rev. **5**: 437-447.
52. PARSONS, B., T. C. RAINBOW, N. J. MACLUSKY & B. S. MCEWEN. 1982. Progestin receptor levels in rat hypothalamic and limbic nuclei. J. Neurosci. **2**: 1446-1452.
53. RAINBOW, T. C., B. PARSONS, N. J. MACLUSKY & B. S. MCEWEN. 1982. Estradiol receptor levels in rat hypothalamic and limbic nuclei. J. Neurosci. **2**: 1439-1445.
54. MORRELL, J. I. & D. W. PFAFF. 1982. Characterization of estrogen-concentrating hypothalamic neurons by their axonal projections. Science **217**: 1273-1275.
55. HARLAN, R. E., B. D. SHIVERS & D. W. PFAFF. 1983. Midbrain microinfusions of prolactin increase the estrogen-dependent behavior, lordosis. Science **219**: 1451-1453.
56. SAR, M. 1984. Estradiol is concentrated in tyrosine hydroxylase-containing neurons of the hypothalamus. Science **233**: 938-940.
57. HERITAGE, A. S., L. D. GRANT & W. E. STUMPF. 1977. [^{3}H]-Estradiol in catecholamine neurons of rat brain stem: combined localization by autoradiography and formaldehyde-induced fluorescence. J. Comp. Neurol. **176**: 607-630.
58. NOCK, B., J. D. BLAUSTEIN & H. H. FEDER. 1981. Changes in noradrenergic transmission alter the concentration of cytoplasmic progestin receptors in hypothalamus. Brain Res. **207**: 371-396.
59. FEDER, H. H. & B. L. MARRONE. 1977. Progesterone: its role in the central nervous system as a facilitator and inhibitor of sexual behavior and gonadotropin release. Ann. N.Y. Acad. Sci. **286**: 331-354.
60. KATO, J. & T. ONOUCHI. 1977. Specific progesterone receptors in the hypothalamus and anterior hypophysis of the rat. Endocrinology **101**: 920-928.
61. SEIKI, K. & M. HATTORI. 1973. *In vivo* uptake of progesterone by the hypothalamus and pituitary of the female ovariectomized rat and its relationship to cytoplasmic progesterone-binding protein. Endocrinol. Japonica **20**: 111-119.

Noradrenergic Regulation of Progestin Receptors: New Findings, New Questions[a]

BRUCE NOCK[b]

*The Rockefeller University
New York, New York 10021*

INTRODUCTION

The steroid hormones that are secreted by the cortex of the adrenal gland and by the gonads are important components of the internal signaling system that coordinates behavior with physiology, and behavior and physiology with the external environment. The list of processes that are influenced by steroids is long. Perhaps the best-known effects are on processes related to reproduction but steroids also affect such things as body weight, brain structure, mood, and affective state. Basic processes such as learning, memory, and sensory function are also sensitive to steroids (see References 1-5).

Responses to steroid hormones, however, are not "hard-wired" and many factors can influence the form and magnitude of responses to steroids. For example, time of day and season of the year can have profound effects on the sensitivity of reproductive processes to steroids.[6-14] Shifts in responsiveness to steroids also occur during early development, puberty, and old age.[15-19]

We[20] recently suggested that neurotransmitters might mediate some changes in responsiveness to steroids. This idea grew from findings that indicate that changes in the activity of neurons that innervate some steroid target tissues affect processes related to hormone action in postsynaptic cells. This phenomenon has been most extensively studied in the pineal gland by Cardinali and associates. Interruption of the neural input into the pineal gland, for example, has been shown to decrease the number of steroid receptors in pinealocytes, and to decrease steroid induction of RNA and protein synthesis.[21-25] However, neurotransmitters appear to influence the sensitivity of other tissues to steroids as well (see Reference 20). In the pages that follow, the evidence for noradrenergic regulation of progestin action in guinea pig brain and its importance for the hormone-dependent lordosis response are discussed in light of recent findings.

[a] Supported by National Institutes of Health Grant NS06966.
[b] Correspondence address: Department of Psychiatry, Washington University, School of Medicine, St. Louis, Missouri 63110.

BACKGROUND: NORADRENERGIC REGULATION OF LORDOSIS

Our interest in whether noradrenergic transmission affects steroid action in brain grew from work with the lordosis response of female guinea pigs. This reflex-like behavior is normally displayed during the 7-10 hr prior to ovulation and functions to facilitate copulation by males. Lordosis is strictly dependent on the synergistic action of ovarian estrogen and progestin. Ovariectomized guinea pigs usually do not display lordosis unless they are treated with estrogen (e.g., estradiol) followed 24-60 hr later by progestin (e.g. progesterone).[26-28] This behavioral response to steroid treatment is as reliable as endocrine and peripheral organ responses to steroids and has been a useful model for studying ovarian steroid action in brain.

Noradrenergic function was implicated in the regulation of guinea pig lordosis behavior when it was found that females treated with estradiol and progesterone do not display lordosis when norepinephrine is depleted in brain by inhibiting norepinephrine synthesis (using the dopamine-β-hydroxylase inhibitor U-14,624) or when noradrenergic receptors are blocked by injection of phenoxybenzamine.[29,30] In other experiments, it was found that treatment with the α-noradrenergic receptor agonist clonidine potentiates lordosis responding in estradiol- and progesterone-primed animals and that this drug can actually be substituted for progesterone to induce lordosis in ovariectomized females that are primed with estradiol alone.[29,31] Subsequent experiments established that these drug effects are attributable to actions on a specific α-noradrenergic receptor subtype, α_1-receptors, in brain.[32] Stimulation of α_1-receptors potentiates lordosis responding, while a decrease in α_1-receptor activation interferes with lordosis responding.

Prior to the discovery that noradrenergic transmission facilitates lordosis, experiments by Morin and Feder[33,34] established that the hypothalamus was the primary site where estradiol and progesterone act to induce lordosis in guinea pigs. Small amounts of crystalline hormone implanted into the hypothalamus induced lordosis responding but implants into other brain areas did not. Studies of the intracellular mechanism of action of estradiol and progesterone in the hypothalamus indicated that these hormones probably induce lordosis through a genomic action.[35-37] This fact is important because the hypothalamus does not contain noradrenergic cell bodies. Therefore, estradiol and progesterone apparently do not act in noradrenergic cell bodies to induce lordosis. However, the hypothalamus is richly innervated by noradrenergic fibers.[38] In fact, some steroid-concentrating cells in the hypothalamus appear to be surrounded by noradrenergic terminals.[39,40] This anatomical relationship suggested the possibility that noradrenergic neurons innervating the hypothalamus might affect lordosis by altering the sensitivity of postsynaptic cells to estradiol and/or progesterone. This idea was supported by evidence indicating transsynaptic regulation of steroid action[21,23,24] and other biochemical events[41-43] in peripheral organs by neurotransmitters.

NORADRENERGIC REGULATION OF UNOCCUPIED PROGESTIN RECEPTORS

The first evidence supporting the idea of noradrenergic regulation of steroid action in guinea pig brain was provided by experiments in which cytosol progestin receptors were assayed after drug treatments that inhibit or potentiate lordosis responding. In those experiments, guinea pigs were injected with estradiol 34 hr prior to drug injection in order to approximate as closely as possible experiments with lordosis. Under these conditions, blockade of α_1-receptors or inhibition of norepinephrine synthesis decreased (by about 30%) the concentration of cytosol progestin receptors, but only in the hypothalamus. Drug treatment had no effect on progestin receptors in other brain areas, including preoptic area, midbrain, and cerebral cortex. Progestin receptor concentration was significantly decreased in the hypothalamus by 4 hr and remained depressed for at least 12 hr after α_1-receptor blockade. Direct stimulation of α_1-receptors by injection of clonidine reversed the effects of norepinephrine synthesis inhibition and restored cytosol progestin receptor concentration in the hypothalamus to control levels. Because changes in receptor number usually result in altered sensitivity to hormone, it was concluded that noradrenergic transmission could influence the sensitivity of postsynaptic cells in the hypothalamus to progestins.[20,32,44] However, recent experiments with nuclear progestin receptors indicate that it might not be the total number of cellular progestin receptors that noradrenergic transmission modulates.

Nuclear progestin receptors were assayed in the hypothalamus of estradiol-primed guinea pigs at 5 hr after administration of drugs that interfere with noradrenergic transmission. The nuclear receptors measured in this experiment were presumably unoccupied since the animals were not treated with progesterone and radioimmunoassay indicated that the drugs did not release adrenal progesterone. Interestingly, however, the concentration of nuclear progestin receptors was found to be higher in drug-treated animals than in vehicle-treated animals.[45] Apparently, noradrenergic function in some way influences where hypothalamic progestin receptors partition upon cell disruption. Interference with noradrenergic transmission decreases the number of receptors in cytosol and increases the number of receptors found in nuclear preparations. There are two ways this effect of noradrenergic transmission might be described, depending on whether unoccupied receptors normally reside in the cytoplasm or the nucleus of target cells.

In the past, it was considered almost dogma that most, if not all, unoccupied steroid receptors reside in the cytoplasm of steroid target cells. According to this model, the receptors measured in cytosol are cytoplasmic receptors. When interpreted within this framework, the finding that interference with noradrenergic transmission decreases the number of progestin receptors in cytosol and increases the number of receptors in nuclear preparations suggests that noradrenergic transmission affects the distribution of unoccupied receptors within hypothalamic cells. Depressed noradrenergic function favors a shift in the distribution of the receptors out of the cytoplasm and into the nuclear compartment.

Recently, however, the idea that unoccupied receptors reside in the cytoplasm has been challenged. On the basis of new evidence, it has been argued that virtually all unoccupied steroid receptors are in the cell nucleus. In the absence of steroid, according to this model, the loose association of the receptors with nuclear elements causes the receptors to partition into the cytosol when the cell is disrupted.[46,47] When interpreted within this framework, the finding that interference with noradrenergic transmission decreases the number of receptors in cytosol and increases the number in nuclear preparations suggests that noradrenergic transmission affects the strength of the association of unoccupied receptors with nuclear elements. A reduction in noradrenergic transmission favors a stronger association between the unoccupied receptor and nuclear elements; therefore, fewer receptors are extracted into cytosol and more receptors are seen in nuclear preparations.

At present, both interpretations of the drug effects on unoccupied progestin receptors are viable and until additional information is available concerning the location of unoccupied receptors within target cells, it will be difficult to choose between these alternatives. Also, at present we can say little concerning differences in the functional consequences of these alternatives. However, regardless of whether noradrenergic transmission affects the distribution of progestin receptors or their strength of association with nuclear elements it remains clear that some aspect of progestin receptor dynamics in hypothalamic cells is sensitive to noradrenergic function.

NEW QUESTIONS, NEW TOOLS, NEW PERSPECTIVES

An important question is whether noradrenergic-induced changes in hypothalamic progestin receptors affect the sensitivity of hypothalamic cells to progestin. That is, does the noradrenergic-induced change in progestin receptors have physiological or behavioral consequences? We[20,32] previously pointed out the parallel of drug effects on lordosis and on the concentration of cytosol progestin receptors in the hypothalamus—the principal site where progesterone acts to induce lordosis. Female guinea pigs injected with drugs that depress noradrenergic function have fewer hypothalamic cytosol progestin receptors after treatment with estradiol than control females and do not show lordosis after sequential treatment with estradiol followed by progesterone. Activation of α_1-receptors by injection of clonidine reverses the effects of norepinephrine synthesis inhibition on hypothalamic cytosol progestin receptors and on lordosis. However, because the change in the number of cytosol receptors now appears to be at least partially attributable to a change in receptor distribution or in the association of the receptor with nuclear elements rather than to a change in the total number of cellular receptors, is it still reasonable to suppose that noradrenergic function might affect hypothalamic sensitivity to progestins? I believe it is.

There are many ways a change in receptor distribution or association with nuclear elements might alter sensitivity to progestin. For example, it is conceivable that this kind of change might alter the access of progestin to the receptor or the binding of progestin to the receptor. It is also possible that the change in receptor distribution reflects a noradrenergic-induced alteration in nuclear elements with which the receptors normally interact. That is, the observed changes in the distribution of unoccupied progestin receptors might be secondary to noradrenergic-induced alterations in, for example, nuclear acceptor sites for the receptors. If so, noradrenergic function might influence the interaction of the progestin-receptor complex with nuclear elements. These are important questions for future research but they are questions that probably cannot be answered by conventional binding assays which are suited to questions concerning receptor number. A recent experiment with nuclear progestin receptors illustrates this point.

Nuclear progestin receptors were assayed in the hypothalamus of estradiol-primed guinea pigs that were injected with progesterone and with drugs that interfere with noradrenergic transmission. Under these conditions, the concentration of nuclear receptors was similar in drug- and vehicle-treated animals.[45] However, measurement of nuclear receptors gives no clue as to whether the receptors are occupied or unoccupied; exchange assay procedures used to measure nuclear receptors measure both. There is no reason to assume the same number of *occupied* receptors in drug- and vehicle-treated animals. There is also no reason to assume that the receptors interact with chromatin in the same way in drug- and vehicle-treated animals but, again, this might not be reflected by measurement of nuclear receptors. Thus, it appears that one outcome of the finding that noradrenergic transmission affects progestin receptor distribution rather than receptor number is that it raises new questions that probably require new techniques to answer. Quantitative autoradiography might prove useful since this would provide information concerning the amount of progestin concentrated in hypothalamic cells rather than just the number of receptors.

With regard to lordosis behavior, we have exercised great caution in the past in extrapolating a cause and effect relationship from the correlation of drug effects on progestin receptors and on lordosis. At present, there is no reason to abandon this caution. However, there is also no reason to abandon the idea that noradrenergic transmission might regulate lordosis through effects on progestin action in hypothalamic cells. Noradrenergic transmission does alter some process related to progestin receptors in hypothalamic cells. Until more is known concerning the nature of the noradrenergic-progestin receptor interaction in the hypothalamus, it would be premature to draw firm conclusions.

It would also be a mistake to overemphasize the importance of the question of whether or not noradrenergic function affects lordosis through effects on progestin action. Of much more significance is the basic idea that neurotransmitters can influence steroid action in brain target neurons. Although relatively new, the concept is rapidly gaining support. For example, A. Clark and E. Roy (personal communication) recently found that blockade of α_1-receptors decreases the concentration of nuclear estrogen receptors in guinea pig hypothalamus and preoptic area. Also, Shani et al.[48,49] and Woolley and associates[50-52] have shown neurotransmitter effects on estrogen uptake and estrogen receptors in rat brain.

There is a great deal that is not known concerning neurotransmitter effects on steroid action. At present, for example, we know almost nothing concerning the mechanism(s) by which neurotransmitters influence steroid action. We[20,32] previously speculated on the possibility that neurotransmitters might exert some effects on steroid action through phosphorylation of nuclear proteins or even through phosphorylation of steroid receptors themselves, but other mechanisms are also conceivable (see Reference 20). We also know little about the functional consequences of neurotransmitter-induced changes in sensitivity to steroids although there are numerous known phenomena that might be relatable to the operation of such a mechanism. For example, neurotransmitter-induced changes in sensitivity to steroids might underlie circadian and seasonal changes in behavioral and endocrine responsiveness to steroids and shifts in sensitivity to steroids that occur during early development, puberty, and old age.[7-13,15-19] From the finding that neurotransmission can influence the action of steroids in postsynaptic cells, it is not a large conceptual jump to envisioning this as an important mechanism by which environmental, behavioral, and emotional events can rapidly and selectively influence steroid-dependent processes.

ACKNOWLEDGMENTS

I thank Drs. Harvey Feder, Bruce McEwen, and Lynn O'Connor for helpful discussion of the manuscript.

REFERENCES

1. ADLER, N. T. (Ed.). 1981. Neuroendocrinology of Reproduction. Plenum Press. New York, N.Y.

2. ARNOLD, A. P. & R. A. GORSKI. 1984. Gonadal steroid induction of structural sex differences in the central nervous system. Annu. Rev. Neurosci. **7:** 413-442.
3. LESHNER, A. I. 1978. An Introduction to Behavioral Endocrinology. Oxford University Press. New York, N.Y.
4. SILVER, R. & H. H. FEDER (Eds.). 1979. Hormones and Reproductive Behavior. W. H. Freeman and Co. San Francisco, Calif.
5. YOUNG, W. C. (Ed.). 1961. Sex and Internal Secretions. 3rd edit. Williams & Wilkins. Baltimore, Md.
6. BEACH, F. A. & G. LEVINSON. 1949. Diurnal variations in the mating behavior of male rats. Proc. Soc. Exp. Biol. Med. **72:** 78-80.
7. HANSEN, S., P. SODERSTEN & B. SREBRO. 1978. A daily rhythm in the behavioral sensitivity of the female rat to oestradiol. J. Endocrinol. **77:** 381-388.
8. HANSEN, S., P. SODERTEN, B. ENROTH, B. SREBRO & K. HOLE. 1979. A sexually dimorphic rhythm in oestradiol-activated lordosis behavior in the rat. J. Endocrinol. **83:** 267-274.
9. HARLAN, R. E., B. D. SHIVERS, R. L. MOSS, J. E. SHRYNE & R. A. GORSKI. 1980. Sexual performance as a function of time of day in male and female rats. Biol. Reprod. **23:** 64-71.
10. HINDE, R. A. & E. STEEL. 1978. The influence of daylength and male vocalizations on the estrogen-dependent behavior of female canaries and budgerigars, with discussion of data from other species. *In* Advances in the Study of Behavior, Vol. 8. J. S. Rosenblatt, R. A. Hinde, C. Beer & M.-C. Busnel, Eds. Academic Press. New York, N.Y.
11. MORIN, L. P., K. M. FITZGERALD, B. RUSAK & I. ZUCKER. 1977. Circadian organization and neural mediation of hamster reproductive rhythms. Psychoneuroendocrinology **1:** 265-279.
12. MORIN, L. P. & I. ZUCKER. 1978. Photoperiodic regulation of copulatory behavior in the male hamster. J. Endocrinol. **77:** 249-258.
13. ROBERTS, J. S. 1973. Functional integrity of the oxytocin-releasing reflex in goats: dependence on estrogen. Endocrinology **53:** 1309-1314.
14. SPELSBERG, T. C., P. A. BOYD & F. HALBERG. 1978. Circannual rhythms in progesterone receptor levels and function. *In* Steroid Hormone Receptor Systems. W. W. Leavitt & J. H. Clark, Eds. Plenum Press. New York, N.Y.
15. DAVIDSON, J. M. 1974. Hypothalamic-pituitary regulation of puberty, evidence from animal experimentation. *In* The Control of the Onset of Puberty. M. M. Grumbach, G. D. Grave & F. E. Mayer, Eds. John Wiley. New York, N.Y.
16. GOLDMAN, B. D. 1981. Puberty. *In* Neuroendocrinology of Reproduction. N. T. Adler, Ed. Plenum Press. New York, N.Y.
17. MACLUSKY, N. J., I. LIEBERBURG & B. S. MCEWEN. 1979. Development of steroid receptor systems in the rodent brain. *In* Ontogeny of Receptors and Reproductive Hormone Action. T. H. Hamilton, J. H. Clark & W. A. Sadler, Eds. Raven Press. New York, N.Y.
18. ROTH, G. S. 1979. Hormone action during aging: alterations and mechanisms. Mech. Ageing Dev. **9:** 497-514.
19. ROTH, G. S. 1979. Hormone receptor changes during adulthood and senescence: significance for aging research. Fed. Proc. **38:** 1910-1914.
20. NOCK, B. & H. H. FEDER. 1981. Neurotransmitter modulation of steroid action in target cells that mediate reproduction and reproductive behavior. Neurosci. Biobehav. Rev. **5:** 437-447.
21. CARDINALI, D. P. 1979. Models in neuroendocrinology. Neurohumoral pathways to the pineal gland. Trends Neurosci. **2:** 250-253.
22. CARDINALI, D. P., C. A. NAGLE & J. M. ROSNER. 1975. Control of estrogen and androgen receptors in the rat pineal gland by catecholamine transmitters. Life Sci. **16:** 93-106.
23. CARDINALI, D. P., E. GOMEZ & J. M. ROSNER. 1976. Changes in [^{3}H]leucine incorporation into pineal proteins following estradiol or testosterone administration: involvement of the sympathetic superior cervical ganglion. Endocrinology **94:** 849-858.
24. CARDINALI, D. P., C. A. NAGLE & J. M. ROSNER. 1976. Pineal-gonad relationships. Nature of the feedback mechanism at the level of the pineal gland. *In* Neuroendocrine Regulation of Fertility. T. C. Anand Kumar, Ed. S. Karger. Basel, Switzerland.

25. NAGLE, C. A., D. P. CARDINALI & J. M. ROSNER. 1975. Testosterone effects on protein synthesis in the rat pineal gland. Life Sci. **16:** 81-91.
26. FEDER, H. H. 1978. Specificity of steroid hormone activation of sexual behavior in rodents. *In* Biological Determinants of Sexual Behavior. J. B. Hutchison, Ed. John Wiley. New York, N.Y.
27. FEDER, H. H., J. D. BLAUSTEIN & B. NOCK. 1979. Oestrogen-progestin regulation of female sexual behavior in guinea pigs. J. Biochem. **11:** 873-877.
28. YOUNG, W. C. 1969. Psychobiology of sexual behavior in the guinea pig. *In* Advances in the Study of Behavior, Vol. 2. D. S. Lehrman, R. A. Hinde & E. Shaw, Eds. Academic Press. New York, N.Y.
29. CROWLEY, W. R., H. H. FEDER & L. P. MORIN. 1976. Role of monoamines in sexual behavior of the female guinea pig. Pharmacol. Biochem. Behav. **4:** 67-71.
30. NOCK, B. & H. H. FEDER. 1979. Noradrenergic transmission and female sexual behavior of guinea pigs. Brain Res. **166:** 369-380.
31. CROWLEY, W. R., B. NOCK & H. H. FEDER. 1978. Facilitation of lordosis behavior of clonidine in female guinea pigs. Pharmacol. Biochem. Behav. **8:** 207-209.
32. NOCK, B. & H. H. FEDER. 1986. α_1-Noradrenergic regulation of hypothalamic progestin receptors and guinea pig lordosis behavior. Brain Res. **310:** 77-85.
33. MORIN, L. P. & H. H. FEDER. 1974. Hypothalamic progesterone implants and facilitation of lordosis behavior in estrogen-primed ovariectomized guinea pigs. Brain Res. **70:** 81-93.
34. MORIN, L. P. & H. H. FEDER. 1974. Intracranial estradiol benzoate implants and lordosis behavior of ovariectomized guinea pigs. Brain Res. **70:** 95-102.
35. BLAUSTEIN, J. D. & H. H. FEDER. 1979. Cytoplasmic progestin receptors in guinea pig brain: characteristics and relationship to the induction of sexual behavior. Brain Res. **169:** 481-497.
36. BLAUSTEIN, J. D. & H. H. FEDER. 1980. Nuclear progestin receptors in guinea pig brain measured by an *in vitro* exchange assay after hormonal treatments that affect lordosis. Endocrinology **106:** 1061-1069.
37. MCEWEN, B. S. 1979. Steroid hormone interactions with the brain: cellular and molecular aspects. Rev. Neurosci. **4:** 1-30.
38. LINDVALL, O. & A. BJORKLUND. 1978. Organization of catecholamine neurons in the rat central nervous system. *In* Handbook of Pharmacology, Vol. 9, Chemical Pathways in the Brain. L. L. Iversen, S. D. Iversen & S. H. Snyder, Eds. Plenum Press. New York, N.Y.
39. GRANT, L. D. & W. E. STUMPF. 1975. Hormone uptake sites in relation to CNS biogenic amine systems. *In* Anatomical Neuroendocrinology. W. E. Stumpf & L. D. Grant, Eds. S. Karger. Basel, Switzerland.
40. HERITAGE, A. S., W. E. STUMPF, M. SAR & L. D. GRANT. 1980. Brainstem catecholamine neurons are target sites for sex steroid hormones. Science **207:** 1377-1379.
41. COSTA, E., A. GUIDOTTI & I. HANBAUER. 1974. Do cyclic nucleotides promote the trans-synaptic induction of tyrosine hydroxylase? Life Sci. **14:** 1169-1188.
42. GREENGARD, P. 1976. Possible role of cyclic nucleotide and phosphorylated membrane proteins in postsynaptic actions of neurotransmitters. Nature (London) **260:** 101-108.
43. ROMERO, J. A., M. ZATZ & J. AXELROD. 1975. Beta-adrenergic stimulation of pineal *N*-acetyltransferase: adenosine 3':5'-cyclic monophosphate stimulates both RNA and protein synthesis. Proc. Natl. Acad. Sci. (Wash.) **72:** 2107-2111.
44. NOCK, B., J. D. BLAUSTEIN & H. H. FEDER. 1981. Changes in noradrenergic transmission alter the concentration of cytoplasmic progestin receptors in hypothalamus. Brain Res. **207:** 371-396.
45. BLAUSTEIN, J. D. 1985. Noradrenergic inhibitors cause accumulation of nuclear progestin receptors in guinea pig hypothalamus. Brain Res. **325:** 89-98.
46. KING, W. J. & G. L. GREENE. 1984. Monoclonal antibodies localized oestrogen receptors in the nuclei of target cells. Nature **307:** 745-747.
47. WELSHONS, W. V., M. E. LIEBERMAN & J. GORSKI. 1984. Nuclear localization of unoccupied oestrogen receptors. Nature **307:** 747-749.
48. SHANI, J., Y. GIVANT, F. G. SULMAN, U. EYLATH & B. ECKSTEIN. 1971. Competition of phenothiazines with oestradiol for oestradiol receptors in rat brain. Neuroendocrinology **8:** 307-316.

49. SHANI, J., Z. ROTH, Y. GIVANT, G. GOLDHABER & F. G. SULMAN. 1976. Competition between [^{3}H]estradiol and prolactin-releasing phenothiazines for estradiol receptors in vitro. Isr. J. Med. Sci. **12:** 1338-1339.
50. GIETZEN, D. W., W. G. HOPE & D. E. WOOLLEY. 1983. Dopaminergic agonists increase [^{3}H]estradiol binding in hypothalamus of female rats, but not of males. Life Sci. **33:** 2221-2228.
51. THOMPSON, M. A., D. E. WOOLLEY, D. W. GIETZEN & S. CONWAY. 1983. Catecholamine synthesis inhibitors acutely modulate [^{3}H]estradiol binding by specific brain areas and pituitary in ovariectomized rats. Endocrinology **113:** 855-865.
52. WOOLLEY, D. E., W. G. HOPE, D. W. GIETZEN, M. T. THOMPSON & S. B. CONWAY. 1982. Bromocriptine increases ^{3}H-estradiol uptake in brain and pituitary of female, but not of male, gonadectomized adrenalectomized rats. Proc. West. Pharmacol. Soc. **25:** 437-441.

Reproductive Neuroendocrine Regulation in the Female Rat by Central Catecholamine-Neuropeptide Interactions: A Local Control Hypothesis[a]

WILLIAM R. CROWLEY

Department of Pharmacology
University of Tennessee Center
for the Health Sciences
Memphis, Tennessee 38163

INTRODUCTION AND OBJECTIVES

The importance of the ovarian hormones, estradiol and progesterone, for the regulation of reproductive behavior and for the control over the secretion of luteinizing hormone (LH), follicle-stimulating hormone, and prolactin from the anterior pituitary gland is unquestioned. Equally clear is the need for understanding the neurochemical mechanisms underlying the neuroendocrine and neurobehavioral actions of the ovarian hormones. In female mammals, of which the rat is the most extensively investigated, the ovarian hormones integrate the appearance of reproductive behavior with ovulation.[1,2] Mating behavior in the female rat is characterized by the lordosis reflex (receptive posture), as well as an active soliciting component (proceptive behavior), and is induced by the sequential and synergistic actions of estradiol followed by progesterone. During the estrous cycle, both ovarian hormones also exert inhibitory effects on the secretion of LH, in addition to stimulatory effects on gonadotropin and prolactin secretion. Detailed descriptions of these behavioral and neuroendocrine phenomena are provided elsewhere.[3,4]

That these steroid hormones act in the brain and in the pituitary gland is clear, and there has been a substantial research effort devoted to identifying the neural substrates for steroid hormone action, with regard to both neuroanatomical localization and cellular and subcellular mechanisms (cf. References 3-8 for reviews). The objectives of this paper are (1) to review recent approaches toward identifying neurons that may be the primary targets of the ovarian hormones in regulating and integrating mating behavior and the preovulatory LH surge in the female rat, and (2) to present models that attempt to explain the mechanisms by which ovarian hormones affect these neural systems.

---

[a] Research from the author's laboratory was supported by National Institutes of Health (NIH) Grant HD-13703 and NIH Research Career Development Award HD-00366.

NEUROCHEMICAL ACTIONS OF OVARIAN STEROIDS: CENTRAL CATECHOLAMINES AND NEUROPEPTIDES AS TARGETS FOR FEEDBACK REGULATION OF LH SECRETION

The stimulation of the preovulatory LH surge by the ovarian hormones occurs via effects in the medial preoptic area and medial basal hypothalamus, probably to enhance the secretion of LH-releasing hormone (LHRH),[3,5,6,9] and also through sensitization of LHRH receptor mechanisms in the pituitary gland.[5,10] The identification of the chemical (i.e., neurotransmitter) nature of those hypothalamic-preoptic neurons that are affected directly by the ovarian hormones remains to be achieved. It now appears, however, that LHRH cell bodies do not accumulate estrogen,[11] although direct steroid effects on the LHRH nerve terminals to affect release are possible.[12-14] It seems probable that ovarian hormones act through other neuronal systems to alter the neurosecretion of LHRH. In this regard, most evidence points to the central catecholamines, norepinephrine (NE) and epinephrine (E), as important mediators of LH release induced by estradiol and progesterone.[3,15]

Initial pharmacological studies showed that a variety of catecholamine antagonists block the preovulatory LH surge and the LH surge induced in ovariectomized rats by sequential treatment with estradiol followed by progesterone.[16,17] The pharmacological profile of this effect suggested NE, rather than dopamine (DA), as the critical catecholamine, but did not eliminate a possible role for the less-abundant catecholamine, E. Recent studies show that selective depletion of E with synthesis inhibitors also disrupts LH release on proestrus and after estrogen-progesterone treatment.[18,19] These data therefore suggest that ovarian hormones activate central noradrenergic and/or adrenergic systems. This would be consistent with the generally stimulatory influence of these catecholamines on LH release.[3,15]

NE and E probably stimulate LH release by directly affecting LHRH neurons. Anatomical studies show catecholaminergic (presumably NE or E) nerve terminals in the vicinity of LHRH cell bodies in the preoptic area and also near LHRH nerve terminals in the median eminence.[20] LH release is acutely disrupted after the catecholamine innervation to these areas is destroyed by central administration of the neurotoxin, 6-hydroxydopamine.[21-23] NE and E neurons may control various aspects of LHRH synthesis and degradation as well as release. Both catecholamines stimulate the release of LHRH from isolated medial basal hypothalamus and median eminence fragments, *in vitro*.[24,25] Prior to the LH surge on proestrus or after progesterone administration to estrogen-primed ovariectomized rats, LHRH levels in the medial eminence rise significantly,[26-28] a phenomenon that may reflect altered synthesis or degradation of the neuropeptide. This accumulation can be totally prevented by depletion of both NE and E[26,27,29] and partially prevented by decreases of only E.[29]

Corroborating the neuropharmacological studies are those in which NE and E turnover rates have been measured on proestrus or after estrogen-progesterone treatment[29-35] and correlated with changes in LH secretion. Such turnover measurements provide estimates of the state of activity within catecholaminergic systems.[36] In general, NE and E activities in areas such as the medial preoptic nucleus and the arcuate-median eminence region are increased during the LH surge.[29-35] Both systems also are activated during the earlier period of LHRH accumulation in the median eminence.[29] Considered together, these studies support the concept that ovarian hormones activate NE and E neurons that innervate the preoptic area and medial basal hypothalamus to regulate LHRH neurosecretion.

An important question that remains to be answered concerns the means by which the ovarian hormones alter noradrenergic and adrenergic activity. Recent studies have demonstrated that a substantial number of NE and E cell bodies in the brainstem accumulate estradiol,[37] so it is conceivable that the steroids affect catecholamine transmission by direct actions at the cell body. Possible mechanisms could include membrane effects to alter firing rates as well as direction of the synthesis of molecules important for catecholamine synthesis and release.

Alternative modes of steroid-catecholamine interaction should also be considered. For example, estradiol and progesterone are concentrated most avidly by neurons in the preoptic-medial hypothalamic region,[38-41] which receives substantial NE and E innervation.[42,43] Because these areas are the sites for stimulation of LH release by the ovarian hormones, the steroids appear to be acting primarily in regions containing the catecholamine nerve terminals, rather than in areas with catecholamine cell bodies. It is unknown whether estradiol or progesterone is taken up into nerve terminals; some reported effects on catecholamine release and reuptake[44,45] could involve this action, but the specific hypothalamic uptake observed in autoradiograms is in the nuclei of nerve cells.[38-40] Thus, while ovarian hormones clearly affect catecholaminergic neurotransmission, the aforementioned observations suggest that estradiol or progesterone might not act directly on or within the catecholamine neuron.

An alternative possibility, therefore, is that the ovarian hormones may affect intrahypothalamic, noncatecholamine neurons that in turn regulate the release of NE and E from the ascending systems. Such "local control" over catecholamine release has been investigated extensively in the autonomic nervous system[46,47] and to some extent in the brain.[48] Two systems have been proposed to exert such presynaptic regulation over catecholamine release in the brain, the endogenous opioid neuropeptides (β-endorphin, met- and leu-enkephalin, dynorphins A and B, α-neoendorphin) and the inhibitory amino acid neurotransmitter, γ-aminobutyric acid (GABA).

There is increasing evidence that these systems may be involved in steroid hormone feedback mechanisms.[3] For example, endogenous opioid peptidergic systems extensively innervate the preoptic area and medial basal hypothalamus (reviewed in Reference 49) and therefore are in a position to influence neuroendocrine events. A large body of evidence shows that opiate alkaloids, such as morphine, and the opioid peptides, such as β-endorphin, consistently decrease LH secretion (reviewed in References 3 and 50). Conversely, opiate receptor blockers, such as naloxone and naltrexone, increase LH release when given to ovariectomized, steroid-pretreated rats.[51] These latter effects are due to stimulation of LHRH release.[52-54]

Strong evidence now indicates that opioids exert such effects via modulation of the catecholaminergic controls over LHRH. As for progesterone's stimulation of LH, the increase of LH seen after treatment with opiate antagonists can be blocked by inhibition of NE and E synthesis.[51,55,56] These findings suggest (1) that opiate antagonists increase LH by stimulating NE and E release and (2) that this occurs through the removal of a tonic inhibitory influence of an endogenous opioid system. Direct tests of these hypotheses have shown that naloxone increases NE and E release from hypothalamic fragments *in vitro,* concomitant with increased LHRH release,[54] and that naloxone also increases hypothalamic NE and E turnover during the elevation of LH *in vivo.*[57] Conversely, morphine depresses catecholamine turnover in these regions.[57,58]

From this work, it seems clear that endogenous opioids influence LHRH and LH secretion via an interaction with central NE and E neurons. FIGURE 1 depicts the anatomical and functional relationships that may exist. In this scheme, it is proposed that LHRH cell bodies in the preoptic area and LHRH nerve terminals in the median eminence are innervated by NE and E neurons, and that the release of these cate-

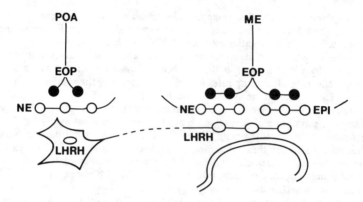

FIGURE 1. Model for catecholamine-LHRH-opioid interactions in control over LH release in the female rat. Abbreviations: POA, preoptic area; ME, median eminence; EOP, endogenous opioid peptide; NE, norepinephrine; EPI, epinephrine; LHRH, luteinizing hormone-releasing hormone. This is based on a similar model proposed in Reference 3. See text for details.

cholamines stimulates LHRH secretion into the portal circulation. The activity of the catecholamine systems is under tonic inhibition by an endogenous opioid peptide(s) via opiate receptors located on the NE- and E-containing nerve terminals. The model proposes further that steroids may act at the level of the endogenous opioid neuron to remove its suppressive influence on catecholamine release, and thus to allow LHRH to be secreted. In this scheme, therefore, the steroids act functionally as opiate antagonists. Indeed, it is intriguing to note the neuroendocrine, behavioral, and neurochemical similarities in the pharmacological actions of progesterone and opiate receptor blockers, such as naloxone (TABLE 1).

A similar role has been proposed for the amino acid, GABA. As for opioids, GABAergic systems densely innervate the hypothalamus,[59,60] and under some conditions, GABA or GABA agonists inhibit LH release.[61,62] Concomitant with GABA-induced depression of LH in ovariectomized animals and GABA agonist-induced blockade of the estrogen/progesterone-induced LH surge is a marked decrease of NE and E turnover.[61,63] Despite these similarities in GABA and opioid actions, preliminary observations suggest that the two systems act independently and in parallel.[63] The

TABLE 1. Neuroendocrine, Behavioral, and Neurochemical Actions of Progesterone and Opiate Antagonists[a]

Effect	Progesterone	Opiate Antagonists
1. Increase LH release in estrogen-primed rats	Yes	Yes
2. Induce lordosis behavior in estrogen-primed rats	Yes	Yes
3. Increase NE turnover in preoptic area, medial basal hypothalamus	Yes	Yes
4. Increase E turnover in preoptic area, medial basal hypothalamus	Yes	Yes
5. Increase LHRH release, in vitro	Yes	Yes

[a]See text for references.

inhibitory actions of GABA on LH and on NE and E turnover appear to be exerted at a presynaptically located GABA B receptor, which differs from the classic postsynaptic GABA A receptor in that the major consequence of occupation of the GABA B receptor is inhibition of central neurotransmitter release.[64,65]

Summary and Conclusions

The local control processes hypothesized here and elsewhere[3] probably cannot be applied to all facets of LH secretion, but the model does identify types of neuronal systems whose activity may be affected by the ovarian hormones. The model is also consistent with the observations that gonadal steroids are concentrated in brain regions containing NE and E nerve terminals and with the fact that ovarian hormonal effects on catecholamine turnover are usually confined to only a few discrete innervation sites.[33-35] Such circumscribed changes in neurotransmitter release could be the consequence of steroid actions directly on the "regulator" neurons (i.e., endogenous opioids or GABA) that gate activity of the ascending systems.

Morphologic evidence consistent with this scheme has been provided by the recent observations that histochemically identified β-endorphin, dynorphin, and GABA neurons in the hypothalamus are labeled by [³H]estradiol,[66,67] and that the levels and turnover of GABA, and the concentrations of β-endorphin and met-enkephalin, are altered by treatment with estrogen.[68-70] It is also interesting to note that opioids and GABA have been proposed to mediate the inhibitory feedback effects of testosterone on LH in males,[71,72] and the inhibitory effect of estrogen on LH in females,[68,72] respectively. An important task for future investigations will be to unravel the mechanisms, whether through classical genomic or through novel nongenomic actions, underlying the effects of estradiol and progesterone on these targeted neuronal systems.

NEUROCHEMICAL ACTIONS OF OVARIAN STEROIDS: DO SIMILAR MECHANISMS MEDIATE HORMONAL EFFECTS ON FEMININE MATING BEHAVIOR?

The neural areas involved in the induction of sexual receptivity by estradiol and progesterone differ from those controlling gonadotropin and prolactin (PRL) release. For example, recent research has focused on the ventromedial nucleus of the hypothalamus as a specific locus for the actions of ovarian hormones on lordosis behavior.[4,73] This does not exclude the possibility, however, that similar neurochemical mechanisms underlie ovarian hormonal regulation of sexual behavior and LH release. Indeed, neuropharmacological studies suggest that catecholamines, LHRH, and opioid peptides all play important roles in the control of lordosis behavior that seem to parallel their actions on LH release. This is interesting in view of the fact that the appearance of mating behavior is closely linked to the preovulatory LH surge during the estrous cycle. This section reviews some of the evidence that these neurotransmitter systems participate in the neural control of lordosis behavior, as a prelude to assessing whether they are targets for the ovarian hormones in the regulation of sexual receptivity.

Catecholamines

Microinfusion of NE, E, or agonists at the β-adrenergic receptor to the medial preoptic area or medial basal hypothalamus induces lordosis behavior in nonreceptive, estrogen-pretreated rats.[74] Agonists at the α-receptor have the opposite effect to depress lordosis behavior in receptive rats.[74] This suggests a stimulatory noradrenergic (or adrenergic) influence on lordosis that is mediated through β-adrenergic receptors, perhaps coupled with inhibitory α-adrenergic mechanisms in the preoptic area and hypothalamus. Other investigators[75] have proposed that noradrenergic neurons convey important somatosensory information from the genital region, based on their observation that interruption of the ascending NE tracts has the same inhibitory effect on lordosis as does applying local anesthesia to the perivaginal area. Consistent with this idea is the observation that genital tract stimulation activates both ascending and descending NE systems.[76]

A number of investigators have demonstrated the importance of the mesencephalon in control of lordosis behavior.[4,77–79] The central gray region in particular appears to be an important site for the convergence of ascending and descending influences on the expression of lordosis behavior.[4] It is interesting to note that a dense projection of noradrenergic axons, emanating from the ascending ventral NE bundle, innervates the central gray,[80] but it is unknown at present whether these fibers carry lordosis-relevant information. Hansen and Ross[81] have also found that 6-hydroxydopamine-induced degeneration of the bulbo-spinal NE tracts decreases the performance of lordosis in steroid-treated rats. Noradrenergic neurons, therefore, may relay information to the preoptic area and medial hypothalamus, to the midbrain, and to the spinal cord for integration of neuroendocrine control mechanisms with the sensorimotor aspects of mating behavior.

Luteinizing Hormone-Releasing Hormone

Systemic administration of LHRH to estrogen-primed female rats increases lordosis behavior, and this effect is not mediated secondarily by LH or progesterone secretion.[82–84] This action has been localized to the medial preoptic area, medial basal hypothalamus,[85] and mesencephalic central gray.[86,87] In addition, application of LHRH antiserum to the central gray disrupts ongoing lordosis behavior,[86–88] suggesting that unimpeded LHRH action in this area is essential for the appearance of female mating behavior. Immunocytochemical studies reveal the presence of LHRH-positive nerve terminals in these three regions (reviewed in References 3, 89, and 90; see also References 91-93). Thus, in addition to its role as a hypophysiotropic hormone, LHRH may serve as a peptidergic transmitter or modulator in the neural circuitry controlling lordosis behavior. The behaviorally relevant LHRH receptor appears from pharmacological analysis to differ from that on pituitary gonadotrope membranes.[87,88,94]

Opioids

A large number of studies have shown inhibition of masculine copulatory performance in rats following administration of opiate agonists and opioid peptides (e.g.,

References 95, and 96), but the effect of opiates on feminine mating behavior has received only scant attention. However, the long-acting opiate receptor antagonist, naltrexone, increases lordosis behavior in ovariectomized, estrogen-primed rats after systemic administration.[97] Similarly, the potent opiate receptor antagonist, naloxone, which is ineffective systemically,[98] also elicits lordosis behavior after administration directly to the mesencephalic central gray. Conversely, administration of β-endorphin specifically to the central gray abolishes lordosis behavior in receptive rats.[88,99] Lordosis behavior is also enhanced by infusing antiserum to β-endorphin, but not antisera to dynorphin or met-enkephalin into the central gray, implicating β-endorphin as the critical opioid neuropeptide.[88]

Opioids may play a similar role in the female hamster. Noble and co-workers have shown that while morphine did not alter lordosis responses, it did diminish the lateral displacement response, which is considered a measure of somatosensory influences.[100] Conversely, naloxone did not increase lordosis behavior in female hamsters, but did have a weak effect to increase lateral displacement.[100] It should be noted that naloxone is relatively short acting and a longer duration of opiate receptor blockade may be required to affect mating behavior in this species, as well as in the rat.

Catecholamine-Neuropeptide Interactions

The evidence reviewed above shows that NE (and possibly E), LHRH and the opioid neuropeptides participate in the neural control over lordosis behavior. As summarized in TABLE 2, the effects of these systems on sexual behavior run parallel to their effects on LH secretion, raising the tantalizing prospect that they may mediate both the behavioral (lordosis) and neuroendocrine (LH surge) actions of estradiol and progesterone in the female rat.

Evidence that these three systems interact in the control of LH secretion has already been discussed in this paper. There have been few attempts to examine whether such interrelationships also regulate reproductive behavior. Notable exceptions are the recent reports by Sirinathsinghji and co-workers,[88,99] in which the effect of naloxone or β-endorphin antiserum in the central gray to induce lordosis behavior was prevented by pretreatment with LHRH antiserum or an LHRH antagonist. Conversely, the inhibition of lordosis by intra-central gray β-endorphin was reversed by injections of LHRH to the same site. These findings suggest that, as for LH, the inhibitory opioid

TABLE 2. Summary of Effects of Norepinephrine, Epinephrine, LHRH, and Opioid Peptides and Antagonists on Reproductive Neuroendocrine Processes

Neurotransmitter	Effect on LH Release	Effect on Lordosis
1. Norepinephrine	↑↓[a]	↑↓[a]
2. Epinephrine	↑	↑
3. LHRH	↑	↑
4. Opioid neuropeptides	↓	↓
5. Opiate antagonists	↑	↑

[a]Differential effects may be mediated by different receptor subtypes and are influenced by ovarian hormonal milieu. See text for references.

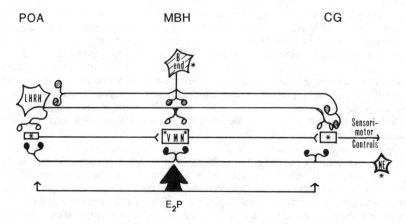

FIGURE 2. Model for interactions of ovarian hormones, catecholamines, and neuropeptides in control over lordosis behavior in the female rat. Abbreviations: POA, preoptic area; MBH, medial basal hypothalamus; CG, central gray; B-end, β-endorphin; VMN, ventromedial nucleus; NE, norepinephrine; LHRH, luteinizing hormone-releasing hormone; E_2, estradiol; P, progesterone. * denotes identified site of ovarian hormone uptake. See text for details.

tone over lordosis behavior is mediated by suppression of LHRH release at central synapses, in this case, the central gray.

FIGURE 2 presents in schematic form a working hypothesis on the functional interactions among catecholamine, LH-RH, and opioid neurons that may form part of the neural system controlling the lordosis reflex in response to ovarian hormones. The principal target for estradiol (E_2) and progesterone (P) in inducing lordosis behavior appears to be the ventromedial nucleus (VMN),[4,74] but additional sites for uptake and binding of ovarian hormones that may be relevent to behavior are also found in the preoptic area (POA) and mesencephalic central gray (CG) (denoted by arrows[38,39]). Neurons that at present are unidentified neurochemically (shown as open boxes) may be among those affected directly by the steroids, and such cells may integrate and transmit information to lower brainstem and spinal sensorimotor centers, as proposed by Pfaff.[4]

FIGURE 2 further identifies some components of the POA, MBH (medial basal hypothalamus), and CG that influence these centers. For example, LHRH-containing neurosecretory cells may innervate the POA, VMN, and CG, and when released from these nerve terminals, the peptide may function as a neurotransmitter to promote lordosis behavior. NE neurons ascending from the medulla, perhaps carrying important sensory cues, also innervate these three areas.[42,43] As for LH secretion, the excitatory noradrenergic influence on lordosis may be mediated by enhanced LHRH release in these areas, but this has not been directly tested. Additional components of the model are β-endorphin cells, which are found in the arcuate nucleus and which project to the POA, VMN, and CG.[49] As noted above, β-endorphin may act to decrease LHRH release in the CG and one may conjecture that a similar effect also occurs in the POA and MBH. Again drawing upon the work with LH secretion, opioid suppression of LHRH (and lordosis) may be accomplished by tonic inhibition of the release of NE or E in these three areas. This also has not yet been directly tested.

The two models (FIGURES 1 and 2) presented in this review are similar in proposing that ovarian hormones alter the activity of local regulators over catecholamine and

LHRH release. Such mechanisms are consistent with results in the neuroendocrine literature, but some of the key experiments on behavior have not yet been performed. That similar mechanisms appear to participate in control of both LH secretion and sexual receptivity may be significant when viewed from a temporal perspective. The neurochemical mechanisms that are activated by ovarian hormones on the afternoon of proestrus, which may involve local control over ascending catecholamine release, result in the increased synthesis and release of LHRH. The peptide is discharged into the portal circulation to trigger LH release and perhaps concomitantly or in response to the later major release of progesterone, from nerve terminals present in other regions of the brain to facilitate lordosis behavior. This could provide a basic framework for the temporal linkage of receptivity and ovulation. Obviously, a fruitful topic for further research will be to examine in closer detail the mechanisms that may participate in this coordinated regulation of reproductive neuroendocrine events.

REFERENCES

1. FEDER, H. H. 1981. Estrous cyclicity in mammals. *In* Neuroendocrinology of Reproduction. Physiology and Behavior. N. T. Adler, Ed.: 279-348. Plenum Press. New York, N.Y.
2. GORSKI, R. A. 1979. The neuroendocrinology of reproduction: an overview. Biol. Reprod. **20:** 111-127.
3. KALRA, S. P. & P. S. KALRA. 1983. Neural regulation of luteinizing hormone secretion in the rat. Endocr. Rev. **4:** 311-351.
4. PFAFF, D. W. 1983. Impact of estrogen on hypothalamic nerve cells: ultrastructural, chemical and electrical effects. Recent Prog. Horm. Res. **39:** 127-179.
5. GOODMAN, R. L. & E. KNOBIL. 1981. The sites of action of ovarian steroids in the regulation of LH secretion. Neuroendocrinology **32:** 57-63.
6. FINK, G. 1979. Feedback actions of target hormones on hypothalamus and pituitary with special reference to gonadal steroids. Annu. Rev. Physiol. **41:** 571-585.
7. MCEWEN, B. S., A. BIEGON, P. G. DAVIS, L. C. KREY, V. N. LUINE, M. MCGINNIS, C. M. PADEN, B. PARSONS & T. C. RAINBOW. 1982. Steroid hormones: hormonal signals which alter brain cell properties and functions. Recent Prog. Horm. Res. **38:** 41-92.
8. MCEWEN, B. S. & B. PARSONS. 1982. Gonadal steroid action on the brain: neurochemistry and neuropharmacology. Annu. Rev. Pharmacol. Toxicol. **22:** 555-598.
9. RAMIREZ, V. D., K. KIM & D. DLUZEN. 1985. Progesterone action on the LHRH and the nigrostriatal dopamine neuronal systems: *In vitro* and *in vivo* studies. Recent Prog. Horm. Res. **41:** 421-465.
10. SAVOY-MOORE, R. T., N. B. SCHWARTZ, J. A. DUNEON & J. C. MARSHALL. 1980. Pituitary gonadotropin releasing hormone receptors during the rat estrous cycle. Science **209:** 942-944.
11. SHIVERS, B. D., R. E. HARLAN, J. I. MORRELL & D. W. PFAFF. 1983. LHRH—absence of oestradiol concentration in cell nuclei of LH-RH immunoreactive neurons. Nature **304:** 345-347.
12. RASMUSSEN, D. D. & S. S. C. YEN. 1983. Progesterone and 20-α-hydroxyprogesterone stimulate the *in vitro* release of GNRH by the isolated mediobasal hypothalamus. Life Sci. **32:** 1523-1530.
13. BENNETT, G. W., J. A. EDWARDSON, D. HOLLAND, S. L. JEFFCOATE & M. WHITE. 1975. Release of immunoreactive luteinising hormone-releasing hormone and thyrotropin-releasing hormone from hypothalamic synaptosomes. Nature **257:** 323-325.
14. DROUVA, S. V., E. LAPLANTE, J. P. GAUTRON & C. KORDON. 1984. Effects of 17 β-estradiol on LH-RH release from rat mediobasal hypothalamic slices. Neuroendocrinology **38:** 152-157.

15. RAMIREZ, V. D., H. H. FEDER & C. H. SAWYER. 1984. The role of brain catecholamines in the regulation of LH secretion. A critical inquiry. *In* Frontiers in Neuroendocrinology, Vol. 8. L. Martini and W. F. Ganona, Eds. Raven Press. New York, N.Y.
16. KALRA, S. P. & S. M. MCCANN. 1974. Effects of drugs modifying catecholamine synthesis on plasma LH and ovulation in the rat. Neuroendocrinology **15:** 79-91.
17. KALRA, S. P. S. P. KALRA, L. KRULICH, C. P. FAWCETT & S. M. MCCANN. 1972. Involvement of norepinephrine in transmission of the stimulatory influence of progesterone on gonadotropin release. Endocrinology **90:** 1168-1176.
18. CROWLEY, W. R., L. C. TERRY & M. D. JOHNSON. 1982. Evidence for the involvement of central epinephrine systems in the regulation of luteinizing hormone, prolactin, and growth hormone release in female rats. Endocrinology **110:** 1102-1107.
19. COEN, C. W. & M. C. COOMBS. 1983. Effects of manipulating catecholamines on the incidence of the preovulatory surge of luteinizing hormone and ovulation in the rat: evidence for a necessary involvement of hypothalamic adrenaline in the normal or 'midnight' surge. Neuroscience **10:** 187-206.
20. JENNES, L., W. C. BECKMAN, W. E. STUMPF & R. GRZANNA. 1982. Anatomical relationships of serotoninergic and noradrenergic projections with the GNRH system in septum and hypothalamus. Exp. Brain Res. **46:** 331-338.
21. MARTINOVIC, J. V. & S. M. MCCANN. 1977. Effect of lesions in the ventral nonadrenergic tract produced by microinjection of 6-hydroxydopamine on gonadotropin release in the rat. Endocrinology **100:** 1206-1213.
22. SIMPKINS, J. W., J. P. ADVIS, C. A. HODSON & J. MEITES. 1979. Blockade of steroid-induced luteinizing hormone release by selective depletion of anterior hypothalamic norepinephrine activity. Endocrinology **104:** 506-509.
23. HANCKE, J. L. & W. WUTTKE. 1979. Effects of chemical lesion of the ventral noradrenergic bundle or of the medial preoptic area on preovulatory LH release in rats. Exp. Brain Res. **35:** 127-134.
24. NEGRO-VILAR, A., S. R. OJEDA & S. M. MCCANN. 1979. Catecholaminergic modulation of luteinizing hormone releasing hormone release by median eminence terminals, *in vitro*. Endocrinology **104:** 1749-1757.
25. NEGRO-VILAR, A. 1982. The median eminence as a model to study presynaptic regulation of neural peptide release. Peptides **3:** 305-310.
26. SIMPKINS, J. W., P. S. KALRA & S. P. KALRA. 1980. Temporal alterations in luteinizing hormone-releasing hormone concentrations in several discrete brain regions. Effects of estrogen-progesterone and norepinephrine synthesis inhibition. Endocrinology **107:** 573-577.
27. ADVIS, J. P., J. E. KRAUSS & J. F. MCKELVY. 1983. Evidence that endopeptidase-catalyzed luteinizing hormone releasing hormone cleavage contributes to the regulation of median eminence LHRH levels during positive steroid feedback. Endocrinology **112:** 1147-1149.
28. KALRA, P. S. & S. P. KALRA. 1977. Temporal changes in the hypothalamic and serum luteinizing hormone-releasing hormone (LH-RH) levels and the circulating ovarian steroids during the rat oestrous cycle. Acta Endocrinol. **85:** 449-455.
29. ADLER, B. A., M. D. JOHNSON, C. O. LYNCH & W. R. CROWLEY. 1983. Evidence that norepinephrine and epinephrine systems mediate the stimulatory effects of ovarian hormones on luteinizing hormone and luteinizing hormone-releasing hormone. Endocrinology **113:** 1431-1438.
30. LÖFSTRÖM, A. 1977. Catecholamine turnover alterations in discrete areas of the median eminence of the 4 and 5-day cyclic rat. Brain Res. **120:** 113-131.
31. RANCE, M., P. M. WISE, M. K. SELMANOFF & C. A. BARRACLOUGH. 1981. Catecholamine turnover rates in discrete hypothalamic areas and associated changes in median eminence luteinizing hormone-releasing hormone and serum gonadotropins on proestrus and diestrous day 1. Endocrinology **108:** 1795-1802.
32. COOMBS, M. C. & C. W. COEN. 1983. Adrenaline turnover rates in the medial preoptic area and mediobasal hypothalamus in relation to the release of luteinizing hormone in female rats. Neuroscience **10:** 207-210.
33. HONMA, K. & W. WUTTKE. 1980. Norepinephrine and dopamine turnover rates in the

medial preoptic area and the mediobasal hypothalamus of the rat brain after various endocrinological manipulations. Endocrinology 106: 1848-1853.
34. WISE, P. M., M. RANCE & C. A. BARRACLOUGH. 1981. Effects of estradiol and progesterone on catecholamine turnover rates in discrete hypothalamic regions in ovariectomized rats. Endocrinology 105: 2186-2193.
35. CROWLEY, W. R. 1982. Effects of ovarian hormones on norepinephrine and dopamine turnover in individual hypothalamic and extrahypothalamic nuclei. Neuroendocrinology 34: 381-386.
36. WEINER, N. I. 1974. A critical assessment of the methods for the determination of monoamine synthesis turnover rates in vivo. Adv. Biochem. Psychopharmacol. 12: 143-152.
37. SAR, M. & W. E. STUMPF. 1981. Central noradrenergic neurones concentrate ^{3}H-oestradiol. Nature 289: 500-502.
38. STUMPF, W. E., M. SAR & D. A. KEEFER. 1975. Atlas of estrogen target cells in rat brain. *In* Anatomical Neuroendocrinology. W. E. Stumpf & L. D. Grant, Eds.: 104-119. Karger. Basel, Switzerland.
39. PFAFF, D. W. & M. KEINER. 1974. Atlas of estradiol concentrating cells in the central nervous system of the female rat. J. Comp. Neurol. 151: 121-158.
40. WAREMBOURG, M. 1978. Uptake of ^{3}H-labelled synthetic progestin by rat brain and pituitary. A radioautography study. Neurosci. Lett. 9: 329-332.
41. PARSONS, B., T. C. RAINBOW, N. J. MACLUSKEY & B. S. MCEWEN. 1982. Progestin receptor levels in rat hypothalamic and limbic nuclei. J. Neurosci. 2: 1446-1452.
42. JACOBOWITZ, D. M. & M. PALKOVITS. 1974. Topographic atlas of catecholamine and acetylcholinesterase-containing neurons in the brain. I. Forebrain (telencephalon, diencephalon). J. Comp. Neurol. 157: 13-28.
43. MOORE, R. Y. & F. E. BLOOM. 1979. Central catecholamine neuron systems: anatomy and physiology of the norepinephrine and epinephrine systems. Annu. Rev. Neurosci. 2: 113-168.
44. PAUL, S. M., J. AXELROD, J. M. SAAVEDRA & P. SKOLNICK. 1979. Estrogen-induced efflux of endogenous catecholamines from the hypothalamus in vitro. Brain Res. 178: 499-505.
45. DAVIS, B. F., C. F. DAVIS & A. E. HALARIS. 1977. Variations in the uptake of ^{3}H-norepinephrine during the rat estrous cycle. Life Sci. 21: 1387-1394.
46. LANGER, S. Z. 1981. Presynaptic regulation of the release of catecholamines. Pharmacol. Rev. 32: 337-362.
47. WESTFALL, T. O. 1977. Local regulation of adrenergic neurotransmission. Physiol. Rev. 57: 659-726.
48. TAUBE, H. D., K. STARKE & E. BOROWSKI. 1977. Presynaptic receptor systems on the noradrenergic neurons at rat brain. Naunyn-Schmiedeberg's Arch. Pharmacol. 299: 123-141.
49. AKIL, H., S. J. WATSON, E. YOUNG, M. E. LEWIS, H. KHACHATURIAN & J. M. WALKER. 1984. Endogenous opioids: biology and function. Annu. Rev. Neurosci. 7: 232-255.
50. VAN VUGT, D. A. & J. MEITES. 1980. Influence of endogenous opiates on anterior pituitary function. Fed. Proc. 39: 2533-2538.
51. KALRA, S. P. & J. W. SIMPKINS. 1981. Evidence for noradrenergic mediation of opioid effects on luteinizing hormone secretion. Endocrinology 109: 776-782.
52. BLANK, M. S. & D. L. ROBERTS. 1982. Antagonist of gonadotropin releasing hormone blocks naloxone-induced elevation in serum luteinizing hormone. Neuroendocrinology 35: 309-312.
53. WILKES, M. M. & S. S. C. YEN. 1981. Augmentation by naloxone of efflux of LRF from superfused medial basal hypothalamus. Life Sci. 28: 2355-2359.
54. LEADEM, C. A., W. R. CROWLEY, J. W. SIMPKINS & S. P. KALRA. 1985. Effects of naloxone on catecholamine and LHRH release from the perifused hypothalamus of the steroid-primed rat. Neuroendocrinology. 40: 497-500.
55. VAN VUGT, D. A., C. F. AYLSWORTH, P. W. SYLVESTER, F. C. LEUNG & J. MEITES. 1981. Evidence for hypothalamic noradrenergic involvement in naloxone-induced stimulation of luteinizing hormone release. Neuroendocrinology 33: 261-264.

56. KALRA, S. P. & W. R. CROWLEY. 1982. Epinephrine synthesis inhibitors block naloxone-induced LH release. Endocrinology **111:** 1403-1405.
57. ADLER, B. A. & W. R. CROWLEY. 1984. Modulation of luteinizing hormone release and catecholamine activity by opiate in the female rat. Neuroendocrinology **38:** 248-253.
58. BARRACLOUGH, C. A. & A. AKABORI. 1984. Interaction of the hypothalamic noradrenergic and opiate systems on LH and prolactin secretion. Society for Neuroscience Abstracts, Vol. 10, No. 273.1, p. 928.
59. VINCENT, S. R., T. HOKFELT & J. T. WU. 1982. GABA neuron systems in hypothalamus and the pituitary gland. Neuroendocrinology **34:** 117-125.
60. TAPPAZ, M. L., M. WASSEF, W. H. OERTEL, L. PAUT & D. F. PUJOL. 1983. Light and election microscopic immunocytochemistry of glutamic acid decarboxylase (GAD) in the basal hypothalamus: morphological evidence for neuroendocrine γ-aminobutyrate (GABA). Neuroscience **9:** 271-287.
61. LAMBERTS, R., E. VIJAYAN, M. GRAF, T. MANSKY & W. WUTTKE. 1983. Involvement of preoptic-anterior hypothalamic GABA neurons in the regulation of pituitary LH and prolactin release. Exp. Brain Res. **52:** 356-362.
62. DONOSO, A. O. & A. M. BANZAN. 1984. Effects of increase of brain GABA levels on the hypothalamic-pituitary-luteinizing hormone axis in rats. Acta Endocrinol. **106:** 298-304.
63. ADLER, B. A. & W. R. CROWLEY. 1986. Evidence for gamma-aminobutyric acid (GABA) regulation of luteinizing hormone secretion and hypothalamic catecholamine activity in the female rat. Endocrinology **118:** 91-97.
64. BOWERY, N. G., G. W. PRICE, A. L. HUDSON, D. R. HILL, G. D. WILKIN & M. J. TURNBULL. 1984. GABA receptor multiplicity. Neuropharmacology **23:** 219-231.
65. BOWERY, M. G., D. R. HILL, A. L. HUDSON, A. DOBLE, D. N. J. SHAW & M. TURNBULL. 1980. (−) Baclofen decreases neurotransmitter release in the mammalian CNS by an action at a novel GABA receptor. Nature **283:** 92-94.
66. MORRELL, J. I., J. MCGINTY & D. W. PFAFF. 1985. A subset of β-endorphin- or dynorphin-containing neurons in the medial basal hypothalamus accumulates estradiol. Neuroendocrinology **41:** 417-426.
67. SAR, M., W. E. STUMPF & M. L. TAPPAZ. 1983. Localization of ^{3}H-estradiol in preoptic GABAergic neurons. Fed. Proc. **42:** 495.
68. MANSKY, T., P. MESTRES-VENTURA & W. WUTTKE. 1982. Involvement of GABA in the feedback action of estradiol on gonadotropin and prolactin release: hypothalamic GABA and catecholamine turnover rates. Brain Res. **231:** 353-364.
69. DUPONT, A., N. BARDEN, L. CUSAN, Y. MERAND, F. LABRIE & H. VAUDRY. 1980. β-Endorphin and met-enkephalin: their distribution, modulation by estrogen and haloperidol, and role in neuroendocrine control. Fed. Proc. **39:** 2544-2550.
70. WARDLAW, S. L., L. THORON & A. G. FRANTZ. 1982. Effects of sex steroids on brain β-endorphin. Brain Res. **245:** 327-331.
71. CICERO, T. J., B. A. SCHAINKER & E. R. MEYER. 1979. Endogenous opioids participate in the regulation of the hypothalamic-pituitary-luteinizing hormone axis and testosterone's negative feedback control of luteinizing hormone. Endocrinology **104:** 1286-1291.
72. VAN VUGT, D. A., P. W. SYLVESTER, C. F. AYLSWORTH & J. MEITES. 1982. Counteraction of gonadal steroid inhibition of luteinizing hormone release by naloxone. Neuroendocrinology **34:** 274-278.
73. RUBIN, B. S. & R. J. BARFIELD. 1983. Introduction of estrous behavior in ovariectomized rats by sequential replacement of estrogen and progesterone to the ventromedial hypothalamus. Neuroendocrinology **37:** 218-224.
74. FOREMAN, M. M. & R. L. MOSS. 1978. Role of hypothalamic alpha and beta adrenergic receptors in the control of lordotic behavior in the ovariectomized-estrogen-primed rat. Pharmacol. Biochem. Behav. **9:** 235-241.
75. HANSEN, S., E. J. STANFIELD & B. J. EVERITT. 1980. The role of ventral bundle noradrenergic neurons in sensory components of sexual behavior and coitus-induced pseudopregnancy. Nature **286:** 152-154.
76. CROWLEY, W. R., J. F. RODRIGUEZ-SIERRA & B. R. KOMISARUK. 1977. Monoaminergic mediation of the antinociceptive effect of vaginal stimulation in rats. Brain Res. **137:** 67-84.

77. CARRER, H. F. 1978. Mesencephalic participation in the control of sexual behavior in the female rat. J. Comp. Physiol. Psychol. **92:** 877-887.
78. SAKUMA, Y. & D. W. PFAFF. 1979. Facilitation of female reproductive behavior from mesencephalic central gray in the rat. Am. J. Physiol. **237:** R278-R284.
79. SAKUMA, Y. & D. W. PFAFF. 1979. Mesencephalic mechanisms for integration of female reproductive behavior in the rat. Am. J. Physiol. **237:** R285-R290.
80. ROIZEN, M. F. & D. M. JACOBOWITZ. 1976. Studies on the origin of innervation of the noradrenergic area bordering on the nucleus raphe dorsalis. Brain Res. **101:** 561-568.
81. HANSEN, S. & S. B. ROSS. 1983. Role of descending monoaminergic neurons in the control of sexual behavior: effects of intrathecal infusions of 6-hydroxydopamine and 5,7-dihydroxytryptamine. Brain Res. **268:** 285-290.
82. PFAFF, D. W. 1973. Luteinizing hormone-releasing factor potentiates lordosis behavior in hypophysectomized ovariectomized female rats. Science **182:** 1148-1149.
83. MOSS, R. L. & S. M. MCCANN. 1983. Induction of mating behavior in rats by luteinizing hormone-releasing factor. Science **181:** 177-179.
84. MOSS, R. L. & S. M. MCCANN. 1979. Action of luteinizing hormone-releasing factor (LRF) in the initiation of lordosis behavior in the estrone-primed ovariectomized female rat. Neuroendocrinology **17:** 309-318.
85. MOSS, R. L. & M. M. FOREMAN. 1976. Potentiation of lordosis behavior by intrahypothalamic infusion of synthetic luteinizing hormone-releasing hormone. Neuroendocrinology **20:** 176-181.
86. SAKUMA, T. & D. W. PFAFF. 1981. LH-RH in the mesencephalic central gray can potentiate lordosis reflex of female rats. Nature **283:** 566-567.
87. SAKUMA, Y. & D. W. PFAFF. 1983. Modulation of the lordosis reflex of female rats by LHRH, its antiserum and analogs in the mesencephalic central gray. Neuroendocrinology **36:** 218-224.
88. SIRINATHSINGHJI, D. J. S. 1984. Modulation of lordosis behavior of female rats by naloxone, β-endorphin and its antiserum in the mesencephalic central gray: possible mediation via Gn RH. Neuroendocrinology **39:** 222-230.
89. STERNBERGER, L. A. & G. E. HOFFMAN. 1978. Immunocytology of luteinizing hormone-releasing hormone. Neuroendocrinology **25:** 111-128.
90. SILVERMAN, A. J., L. C. KREY & E. A. ZIMMERMAN. 1979. A comparative study of the luteinizing hormone releasing hormone (LHRH) neuronal networks in mammals. Biol. Reprod. **20:** 98-110.
91. KAWANO, J. & S. DAIKOKU. 1981. Immunohistochemical demonstration of LHRH neurons and their pathways in the rat hypothalamus. Neuroendocrinology **32:** 179-186.
92. SHIVERS, B. D., R. E. HARLAN, J. I. MORRELL & D. W. PFAFF. 1983. Immunocytochemical localization of luteinizing hormone-releasing hormone in male and female rat brains. Neuroendocrinology **36:** 1-12.
93. SAMSON, W. K., S. M. MCCANN, L. CHERL, C. A. DUDLEY & R. L. MOSS. 1980. Intra and extrahypothalamic luteinizing hormone-releasing hormone (LHRH) distribution in the rat with special reference to mesencephalic sites which contain both LHRH and single neurons responsive to LHRH. Neuroendocrinology **31:** 66-72.
94. KASTIN, A. J., D. H. COY, A. V. SCHALLY & J. E. ZADINA. 1980. Dissociation of effects of LH/RH analogues on pituitary regulation and reproductive behavior. Pharmacol. Biochem. Behav. **13:** 913-914.
95. MEYERSON, B. J. & L. TERENIUS. 1977. β-Endorphin and male sexual behavior. Eur. J. Pharmacol. **42:** 191-192.
96. PELLEGRINI QUARONDOTTI, B., M. G. CORDA, E. PAGLIETTI, G. BIGGIO & G. L. GESSA. 1978. Inhibition of copulatory behavior in male rats by D-ala^2-met enkephalinamide. Life Sci. **23:** 673-678.
97. ALLEN, D. L., K. J. RENNER & V. N. LUINE. 1985. Naltrexone facilitation of sexual receptivity in the rat. Horm. Behav. **19:** 98-103.
98. WIESNER, J. B. & R. L. MOSS. 1982. β-Endorphin suppression of mating behavior and plasma luteinizing hormone in the female rat. Society for Neuroscience Abstracts, Vol. 8, No. 267.4, p. 930.

99. SIRINATHSINGHJI, D. J. S., P. E. WHITTINGTON, A. AUDSLEY & H. M. FRASER. 1983. β-Endorphin regulates lordosis in female rats by modulating LHRH release. Nature **301:** 62-64.
100. OSTROWSKI, N. L., J. M. STAPLETON, R. G. NOBLE & L. D. REID. 1979. Morphine and naloxone effects on sexual behavior of the female golden hamster. Pharmacol. Biochem. Behav. **11:** 673-681.

Hormone Effects on Serotonin-Dependent Behaviors

LYNN H. O'CONNOR[a]

The Rockefeller University
New York, New York 10021

CHRISTINE T. FISCHETTE

Hoffmann-La Roche, Inc.
Nutley, New Jersey 07110

It is well known that gonadal hormones are involved in controlling reproductive behaviors and neuroendocrine events. In addition they also play a more subtle, but significant, role as modulators of a great many systems, such as an organism's responsiveness to environmental changes, mood state, clinical response to drugs, and neurological and neuropsychiatric disorders. An important question is, what are the mechanisms by which hormones regulate neuronal activity? The main focus of this paper is to examine the interaction between gonadal hormones and a specific neurotransmitter system in the brain, the serotonin (5HT) system.

The range of techniques used to study hormone effects on the 5HT system is extremely broad and includes behavioral, neurochemical, neuroanatomical, and neurophysiological techniques. In this paper we describe the interactions of gonadal hormones and 5HT as indicated by studies of two different behaviors. These behaviors, myoclonus in the guinea pig and the serotonin behavioral syndrome in the rat, are behaviors that are not normally displayed by healthy animals. Instead, they represent visible outputs of a hyperactive 5HT system. However, through the study of these behaviors, important clues have been gained about the nature of 5HT-steroid interactions in naturally occurring behaviors such as those involved in reproduction in rodents. These behavioral models can also be used to investigate hormonal mechanisms in the brain, and to investigate 5HT neurotransmission directly. In addition, some of the features of these behaviors strongly parallel situations seen clinically, and are therefore all the more interesting and valuable.

This paper is organized as follows: First, we describe the data obtained from studies of hormone effects on each of these behaviors. Second, we examine possible mechanisms by which these interactions produce alterations in behavior. Finally, we examine some implications of this research for rodent reproductive behavior and for neurologic and neuropsychiatric disorders.

---

[a] Address correspondence to: Dr. Lynn H. O'Connor, Dept. of Psychiatry, Washington Univ. School of Medicine, 4940 Audubon Ave., St. Louis, MO 63110.

MYOCLONUS

In guinea pigs, myoclonus is an abnormal behavior induced by drugs that cause intense central 5HT receptor activation.[1-4] The 5HT precursor 5-hydroxytryptophan (5HTP) has been used extensiely to induce myoclonus in guinea pigs. Typically, about 20 min after injection of 5HTP, guinea pigs show arrhythmic jerking of the head and neck. By 1 hr after 5HTP administration, these jerks are rhythmic (one to two per second) and involve the whole body.

Interest in the effects of ovarian hormones on myoclonus was stimulated by the finding that hormones and 5HT interact in the regulation of female reproductive behavior. In the early 1960s pharmacologic studies by Meyerson[5,6] suggested that 5HT neurotransmission exerts an inhibitory influence on female reproductive behavior. A wealth of studies have strongly reinforced Meyerson's hypothesis (for recent comprehensive reviews see References 7-9). One difficulty in examining the interaction of 5HT and hormones using female reproductive behavior as an end point is that female reproductive behavior is an elaborate behavior influenced by many neurotransmitter systems. Studies of hormone effects on myoclonus have been carried out in an attempt to examine hormone effects on a behavior that is more directly related to changes in 5HT transmission.

Interestingly, neither prolonged treatments nor high doses of steroids are required in order to influence myoclonus. The same relatively low doses of estradiol (E_2) and rogesterone (P) used to elicit sexual receptivity in ovariectomized female guinea pigs also modulate the induction of myoclonus by 5HTP. E_2 has similar effects on myoclonus in males and females. In both sexes, pretreatment with a single injection of E_2 46 hr prior to 5HTP injection enhances the myoclonic effects of this drug.[10,11] In contrast, P has opposite effects on myoclonus in males and females. In gonadectomized E_2-primed females, treatment with P suppresses myoclonus, whereas in E_2-primed males, P enhances myoclonus.[10,11]

In addition to sex differences in the effects of P on 5HTP-induced myoclonus, there are also sex differences in sensitivity to 5HTP. In the absence of steroids, 5HTP-induced myoclonus is greater in gonadectomized males than females. Castration does not alter myoclonus in males, suggesting that myoclonus is not sensitive to endogenous androgens[11] (see below for a comparison with androgen effects on the serotonin behavioral syndrome in rats).

Because myoclonic responding is an index of central 5HT activity, it can give us insight into hormone effects on 5HT neurotransmission. For example, because E_2 enhances myoclonus induced by 5HTP, it is possible that E_2 increases 5HT activity in some brain areas. Also, these findings raise a number of hypotheses about P action on 5HT systems. Because P suppressed myoclonus in E_2-primed females, it is likely that P reverses the effects of E_2 on 5HT activity. In contrast, P enhanced myoclonic responding by E_2-primed males. Therefore, it is possible that P has opposite effects on 5HT transmission in males and females.

Available evidence suggests that in females, the effects of P on myoclonus are probably mediated by intracellular actions of P.[10] This hypothesis is based on the temporal correlation of effects of P on myoclonus with behavioral effects of P that are presumed to be mediated by intracellular progestin receptors. Because the time course of sexual receptivity is correlated with the latency to increased cytosol progestin receptors and with the presence of progestin in cell nuclei, it has been suggested that intracellular progestin receptors are involved in P effects on sexual receptivity.[12,13] The inhibitory effects of P on myoclonus correlate closely with the time course of P effects on receptivity. Normally, E_2-primed female guinea pigs become sexually receptive approximately 4-5 hr after P injection. This period of receptivity lasts about 5-6 hr. P inhibits myoclonus when 5HTP is given during the period of receptivity, but not when 5HTP is given just prior to or just after this period.[11] This temporal correlation suggests the progestin receptor mechanisms proposed to mediate receptivity are also involved in the inhibitory effects of P on myoclonus in females. The involvement of intracellular E_2-inducible P receptors is also suggested by the finding that treatment with P alone does not decrease 5HTP-induced myoclonus in females. The current concept of P action following binding to intracellular progestin receptors is that these complexes bind to intranuclear sites on the genome and alter gene expression in a quantitative and qualitative fashion (see Reference 14 for a recent review). Thus, effects of steroids on 5HTP-induced myoclonus that appear to be mediated by intracellular receptors also involve genomic actions of steroids.

As noted above, P has opposite effects on myoclonus in males and females. Because males responded to P, it is unlikely that sex differences in P effects on myoclonus involve decreased sensitivity to P by males. Rather, it seems possible that certain genomic effects of P might be different in males and females. Sex differences in P action might involve differences in effects of P on gene transcription (RNA synthesis) in males and females or perhaps some other intrinsic sex difference (anatomical, biochemical, etc.).

SEROTONIN BEHAVIORAL SYNDROME

In rats, the serotonin behavioral syndrome is a constellation of characteristic behavioral responses which are produced upon elevation of 5HT content or activation of 5HT receptors.[15] This syndrome is behaviorally very different from that induced by increased 5HT activity in guinea pigs. The serotonin behavioral syndrome is a group of stereotyped responses that include resting tremor, reciprocal forepaw treading, lateral head movements, splayed hind legs and toes, Straub tail (rigid, erect, or s-shaped tail), hypersensitivity to delicate stroking of the back (observed in the pargyline/L-tryptophan-induced syndrome, not observed with the direct acting agonist 5-methoxy-*N,N*-dimethyltryptamine), hyperactivity, hyperreactivity, salivation, and rigidity. For quantification purposes the syndrome is rated in terms of numbers of animals displaying the syndrome (frequency of response—at least four of the first six symptoms must be present) or on severity of symptoms (intensity score—0 = absent, 1 = mild, 2 = definitely present, 3 = severe).

The effects of hormones upon this behavior were investigated as an outgrowth of steroid effects on 5HT-1 receptor binding experiments and represent an attempt to correlate binding studies with some physiological response (see below). Using a standard dose regimen of 50 mg/kg of the monoamine oxidase; (MAO) inhibitor pargyline and 50 mg/kg of the 5HT precursor L-tryptophan, the effects of various hormonal states were investigated in both male and female rats. These studies were particularly appropriate since there is a sex difference in the frequency of animals displaying this syndrome. Although both sexes will exhibit the syndrome, females are much more sensitive than males across a range of doses of L-tryptophan (20-80 mg/kg) in pargyline-pretreated rats.[16] The results of a series of experiments[17,18] indicate that the hormonal environment can determine the response of an organism to a particular drug paradigm. It was found, for example, that the sex difference in frequency of animals exhibiting the syndrome was eliminated by castration of the male in both the adult and prepubertal state, while ovariectomy had no effect. Both gonadectomized groups resembled the intact female group in displaying the syndrome. The sex difference is restored by administration of dihydrotestosterone (DHT), the nonaromatizable androgen, to male rats in a time (dose)-dependent manner. DHT treatment of females, however, is much less effective (this may reflect sex differences in DHT metabolism itself, or a sex difference in the central androgen response). Because castration and DHT replacement affected this behavior, the testicular feminized mutant (tfm/y) was chosen to investigate sex steroid-dependent receptor mechanisms. As is well known, this pseudohermaphrodite is a genetic male deficient in androgen receptors and is phenotypically female despite normal-to-high levels of plasma androgens. In this model both estrogen and estrogen receptor systems appear to function normally.[19] Tfm/y rates respond similarly to their littermate King-Holtzman females indicating that androgen receptors are necessary for the expression of the subsensitivity of the male rat to the drug challenge. Although DHT treatment was effective in reducing the incidence or the intensity of rats displaying the syndrome, treatment with estradiol benzoate was not. No differences were found between gonadectomized rats given 2×10 μg E_2 and their respective oil-treated controls. Neither was the sex difference in response affected by neonatal hormone manipulations. Males and females treated neonatally with androstatriendione (ATD) and E_2, respectively, procedures which "feminize" males and "defeminize" females with respect to lordosis behavior and gonadotropin secretion (lordosis quotients were determined in all rats), did not differ from their respective same sex control-treated littermates when the animals were tested as adults. Therefore, like the lordosis response, the serotonin behavioral syndrome is affected by "activational" effects of hormones, but in contrast to lordosis, it has little of the "organizational" influences of hormones when manipulations are performed at this "critical period" of development (postnatal days 0-5). Of course, hormones may affect the expression of this behavior at another "critical period" of development.

A completely different picture is seen in 5HTP-induced myoclonus in the guinea pig. The incidence of myoclonus is greater in gonadectomized males than in gonadectomized females. Both sexes are affected by estrogen and progesterone (although differentially). As stated above, the sex difference in the serotonin behavioral syndrome is found in intact rats, is eliminated by gonadectomy, and is influenced by androgens. These differences between hormone effects on myoclonus and the serotonin behavioral syndrome might reflect species differences in hormone effects on 5HT transmission in certain brain nuclei or perhaps involvement of different 5HT receptor subtypes in myoclonus and the serotonin behavioral syndrome. A number of investigators have correlated 5 HT-mediated behaviors with 5HT receptor subtypes, but this issue has not been completely resolved.[20-23]

MECHANISMS INVOLVED IN STEROID ACTION ON 5HT-DEPENDENT BEHAVIORS

It is currently thought that the serotonin behavioral syndrome and myoclonus are mediated by brainstem and spinal areas.[24,25] Steroids could influence these areas directly or they could modulate forebrain structures which in turn modulate the downstream sensory and motor components of these behaviors. There are many ways that steroids could influence neurotransmission. Some of the parameters include precursor availability (tryptophan or 5HTP uptake), 5HT synthesis, release, reuptake, catabolism, turnover, postsynaptic receptor sites, the cascade of postsynaptic receptor activating events, and drug metabolism. Data obtained with regard to the serotonin behavioral syndrome indicate that a number of these possibilities are not responsible for the sex difference observed in intact rats. A sex difference in postsynaptic (5HT-1?) receptors appears to be at least partially responsible for the observed effects since females are more sensitive than males to the direct acting receptor agonist 5-methoxy-N,N,-dimethyltryptamine at the threshold dose for the response.[26] Other receptor agonists are under study.

Both estrogen and androgens affect 5HT-1 receptor binding in rat forebrain using either the Palkovits punch technique or grossly dissected brain regions.[27] Experiments are under way to localize these effects to discrete nuclei using quantitative autoradiography or both 5HT-1 and 5HT-2 receptors. Where hormones affect these subtypes will give us clues as to the loci that modulate or control these behaviors.

IMPLICATIONS FOR BEHAVIOR

The two 5HT-mediated responses are good *in vivo* models which allow us to study 5HT neurotransmission and hormone effects upon the brain in general. The study of abnormal behaviors that are mediated by a hyperactive 5HT system can also tell us about hormone effects on normal behaviors. For example, there are both similarities and differences between hormone effects on myoclonus and on female reproductive behavior. P modulates the effects of 5HTP on both receptivity and on myoclonus. However, it enhances the effects of 5HTP on receptivity,[28] and inhibits the effects of 5HTP on myoclonus in E_2-primed females.[10] Differences in pathways or receptor subtypes mediating these behaviors may account for these findings. Despite these differences, there appears to be a basic similarity in P effects on 5HT systems underlying these two behaviors. On the basis of effects of sequential treatment with E_2 and P on 5HTP-induced myoclonus in females, it seems likely that in certain pathways P reverses the increase in 5HT activity caused by E_2. Similarly, it has been hypothesized that one way by which P facilitates sexual receptivity in E_2-primed females is by decreasing activity in 5HT pathways that are inhibitory to receptivity.[29] Information gained through the use of these behaviors also suggests future directions for research on hormonal mechanisms underlying behavior. For example, sequential treatment with E_2 and P does not reliably elicit feminine behaviors in male guinea pigs. Interestingly, in contrast to P effects in female guinea pigs, P enhanced E_2 action on myoclonus in males. This finding raises the possibility that sequential treatment with E_2 and P does not facilitate feminine receptivity behaviors in male guinea pigs because P enhances rather than suppresses particular 5HT activities in male guinea pigs.

These areas of study are not without clinical significance. 5HT-mediated behaviors can be useful models for investigating the role of hormones in the etiology and treatment of neurological disorders and neuropsychiatric disorders. One illustration of this point is the example of hormone effects in human myoclonus. In humans, postanoxic intention myoclonus and progressive myoclonus epilepsy are incapacitating involuntary movement disorders that appear to be related, in part, to imbalances in the activity of brain 5HT systems. The decreased concentration of 5HIAA in cerebrospinal fluids in these patients and the dramatic clinical response to treatment with 5HTP support this hypothesis (for recent reviews see References 30 and 31). Case reports indicate that in some women there are marked changes in the severity of this disorder over the course of the menstrual cycle and that these changes appear to be correlated with blood levels of E_2 and not P.[30] Interestingly, E_2 treatment (after ovariectomy) has been reported to cause a dramatic lessening of intention myoclonus.[30] Perhaps E_2 treatment alleviates intention myoclonus in humans by enhancing 5HT neurotransmission. Thus, despite the apparently opposite role of 5HT transmission in human intention myoclonus and guinea pig myoclonus, the enhancement of 5HT activity by E_2 in guinea pig myoclonus may be a useful model for studying the beneficial effects of E_2 treatment in some patients with intention myoclonus.

The serotonergic system, as well as other systems, is implicated in the pathogenesis of depression. Females are more susceptible to this illness than males.[32] Patients who fail to respond to conventional treatments of depression sometimes respond beneficially to gonadal hormone treatments.[33] Antidepressant drugs can affect both the behavioral syndrome[34] and the 5HT-mediated head shake response[35-37] in rats. Many antidepressants down-regulate 5HT-2 receptors as well as β-adrenergic receptors (these systems appear to be linked). In addition, the ability of certain antidepressant drugs to down-regulate 5HT-2 receptors depends on sex and hormonal state.[38,39] Therefore, the sex of the organism and the presence or absence of hormones may affect clinical responses to antidepressant drugs. How steroids, neurotransmitter receptors, and antidepressant drugs interact in the brain may well prove to be a highly significant and clinically useful area of research. The two 5HT-mediated behaviors may help to elucidate these mechanisms.

REFERENCES

1. KLAWANS, H. L., C. GOETZ & W. J. WEINER. 1973. 5-Hydroxytryptophan-induced myoclonus in guinea pigs and the possible role of serotonin in infantile myoclonus. Neurology **23:** 1234-1240.
2. KLAWANS, H., D. D'AMICO, P. A. NAUSIEDA & W. J. WEINER. 1977. The specificity of neuroleptic- and methysergide-induced behavioral hypersensitivity. Psychopharmacology **55:** 49-52.
3. LUSCOMBE, G., P. JENNER & C. D. MARSDEN. 1981. Pharmacological analysis of the myoclonus induced by 5-hydroxytryptophan in the guinea pig suggests the presence of multiple 5-hydroxytryptamine receptors in the brain. Neuropharmacology **20:** 819-831.
4. VOLKMAN, P. H., S. A. LORENS, G. H. KINDEL & J. Z. GINOS. 1978. L-5-Hydroxytryptophan-induced myoclonus in guinea pigs: a model for the study of central serotonin-dopamine interactions. Neuropharmacology **17:** 947-955.
5. MEYERSON, B. J. 1964. Central nervous monoamines and hormone induced estrus behaviour in the spayed rat. Acta Physiol. Scand. Suppl. **241:** 5-29.
6. MEYERSON, B. J. 1966. The effect of imipramine and related antidepressive drugs on estrus behaviour in ovariectomised rats activated by progesterone, reserpine or tetrabenazine in combination with estrogen. Acta Physiol. Scand. **67:** 411-422.
7. CARTER, S. S. & J. M. DAVIS. 1977. Biogenic amines, reproductive hormones and female sexual behavior: a review. Biobehav. Rev. **1:** 213-224.
8. CROWLEY, W. R. & F. P. ZEMLAN. 1981. The neurochemical control of mating behavior. *In* Neuroendocrinology of Reproduction Physiology and Behavior. N. T. Adler, Ed.: 451-484. Plenum Press. New York, N.Y.
9. MCEWEN, B. S. & B. PARSONS. 1982. Gonadal steroid action on the brain: neurochemistry and neuropharmacology. Annu. Rev. Pharmacol. Toxicol. **22:** 555-598.
10. O'CONNOR, L. H. & H. H. FEDER. 1984. Estradiol and progesterone influence a serotonin mediated behavioral syndrome (myoclonus) in female guinea pigs: comparisons with steroid effects on reproductive behavior. Brain Res. **293:** 119-125.

11. O'CONNOR, L. H. & H. H. FEDER. 1985. Estradiol and progesterone influence L-5-hydroxytryptophan-induced myoclonus in male guinea pigs: sex differences in serotonin-steroid interactions. Brain Res. **330:** 121-125.
12. BLAUSTEIN, J. D. & H. H. FEDER. 1979. Cytoplasmic progestin-receptors in guinea pig brain: characteristics and relationship to the induction of sexual behavior. Brain Res. **169:** 481-497.
13. BLAUSTEIN, J. D. & H. H. FEDER. 1980. Nuclear progestin receptors in guinea pig brain measured by an *in vitro* exchange assay after hormonal treatments that affect lordosis. Endocrinology **106:** 1064-1069.
14. SPELSBERG, T. C., B. B. LITTLEFIELD, R. SEELKE, G. MARTIN DANI, H. TOYODA, P. BOYD-LEINEN, C. THRALL & O. LIAN KON. 1983. Role of specific chromosomal proteins and DNA sequences in the nuclear binding sites for steroid receptors. Recent Prog. Horm. Res. **39:** 463-513.
15. JACOBS, B. L. 1976. An animal model for studying central serotonergic synapses. Life Sci. **19:** 777-786.
16. BIEGON, A., M. SEGAL & D. SAMUEL. 1979. Sex differences in behavioral and thermal responses to pargyline and tryptophan. Psychopharmacology **61:** 77-80.
17. FISCHETTE, C. A., A. BIEGON & B. MCEWEN. 1982. Hormonal manipulation of the serotonin behavioral syndrome. Soc. Neurosci. **8:** 101.
18. FISCHETTE, C. A., A. BIEGON & B. MCEWEN. 1985. Sex steroid modulation of the serotonin behavioral syndrome. Life Sci. **35:** 1197-1206.
19. KREY, L. C., I. LIEBERBURG, N. J. MACLUSKY, P. G. DAVIS & R. ROBBINS. 1982. Testosterone increases cell nuclear estrogen receptor levels in the brain of Stanley Gumbreck pseudohermaphrodite male rat: implications for testosterone modulation of neuroendocrine activity. Endocrinology **110:** 2168-2176.
20. PEROUTKA, S. J., R. M. LEBOVITZ & S. H. SNYDER. 1981. Two distinct central serotonin receptors with different physiological functions. Science **212:** 827-829.
21. GREEN, A., K. O'SHAUGNHNESSY, M. HAMMOND, M. SCHACTER & D. G. GRAHME-SMITH. 1983. Inhibition of 5-hydroxytryptamine mediated behavior by the putative 5HT-2 antagonist pirenperone. Neuropharmacology **22:** 573-578.
22. LUCKI, I., M. NOBLER & A. FRAZER. 1984. Differential actions of serotonin antagonists on two behavioral models of serotonin receptor activation in the rat. J. Pharmacol. Exp. Ther. **228:** 133-139.
23. LUSCOMBE, G., P. JENNER & D. MARSDEN. 1984. 5-Hydroxytryptamine (5-HT)-dependent myoclonus in guinea pigs is induced through brainstem 5-HT-1 receptors. Neurosci. Lett. **44:** 241-246.
24. JACOBS, B. L. & H. KLEMFUSS. 1975. Brain stem and spinal cord mediation of a serotonergic behavioral syndrome. Brain Res. **100:** 450-457.
25. CHADWICK, D., M. HALLETT, P. JENNER & D. MARSDEN. 1978. 5-Hydroxytryptophan-induced myoclonus in guinea pigs. J. Neurological Sci. **35:** 157-165.
26. FISCHETTE, C. F., K. RENNER & B. MCEWEN. 1983. Mechanisms involved in the sex difference observed in the serotonin behavioral syndrome. Soc. Neurosci. **9:** 125.
27. FISCHETTE, C. T., A. BIEGON & B. MCEWEN. 1983. Sex differences in serotonin receptor binding in rat brain. Science **222:** 333-335.
28. SIETNIEKS, A. & B. J. MEYERSON. 1982. Enhancement by progesterone of 5-hydroxytryptophan inhibition of the copulatory response in the female rat. Neuroendocrinology **35:** 321-326.
29. KOW, L.-M., C. MALSBURY & D. W. PFAFF. 1974. Effects of progesterone on female reproductive behavior in rats: possible modes of action and role in behavioral sex differences. *In* Reproductive Behavior. W. Montagna & W. Sadler, eds.: 179-210. Plenum Press. New York, N.Y.
30. FAHN, S. 1979. Posthypoxic action myoclonus: review of the literature and report of two new cases with response to valproate and estrogen. Adv. Neurol. **26:** 49-82.
31. VAN WOERT, M. H. & D. ROSENBAUM. 1979. L-5-Hydroxytryptophan therapy in myoclonus. Adv. Neurol. **26:** 107-115.
32. WEISSMAN, M. M. & G. L. KLERMAN. 1977. Sex differences and the epidemiology of depression. Arch. Gen. Psychiatry **34:** 98-111.
33. VOGEL, W., E. L. KLAIBER & D. H. BROVERMAN. 1978. Roles of the gonadal steroid

hormones in psychiatric depression in men and women. Prog. Neuro-Psychopharmacol. **2:** 487-503.
34. LUCKI, I. & A. FRAZER. 1982. Prevention of the serotonin syndrome in rats by repeated administration of monoamine oxidase inhibitors but not tricyclic antidepressants. Psychopharmacology **77:** 205-211.
35. FRIEDMAN, E. & A. DALLOB. 1979. Enhanced serotonin receptor activity after chronic treatment with imipramine or amitriptyline. Commun. Psychopharmacol. **3:** 89-92.
36. ANTELMAN, S. M., L. A. DEGIOVANNI, D. KOCAN, J. M. PEREL & L. A. CHIODO. 1983. Amitriptyline sensitization of a 5HT-mediated behavior depends on the passage of time and not repeated treatment. Life Sci. **33:** 1727-1730.
37. GREEN, A. R., D. J. HEAL, P. JOHNSON, B. E. LAURENCE & V. L. NIEMGAONKAR. 1983. Antidepressant treatments: effects in rodents on dose response curves of 5-hydroxytryptamine and dopamine mediated behaviors and 5HT-2 receptor number in frontal cortex. Br. J. Pharmacol. **80:** 377-385.
38. KENDALL, D. A., G. M. STANCEL & S. J. ENNA. 1981. Imipramine: effects of ovarian steroids on modifications in serotonin receptor binding. Science **211:** 1183-1185.
39. KENDALL, D. A., G. M. STANCEL & S. J. ENNA. 1982. The influence of sex hormones on antidepressant induced alterations in neurotransmitter receptor binding. J. Neurosci. **2:** 354-360.

Hormonal Aspects of the Morphological Differentiation of Neuronal Clonal Cell Lines

CHRISTIAN P. REBOULLEAU

Institute of Animal Behavior
Rutgers-The State University of New Jersey
Newark, New Jersey 07102

MODEL SYSTEMS

The biochemical processes that control cell differentiation during brain development can be studied at several levels. The whole-animal model is primarily used to investigate the biochemical changes resulting from interactive processes that are mediated by various tissues and organs. In this case, overall, final changes are measured and it is often difficult to identify the various mechanisms that cause certain cells to differentiate the way they do. In an effort to understand the changes mediated by certain functionally important biomolecules in whole animals, one can try to replicate these changes *in vitro* using tissue explants maintained in culture medium. Although this technique allows one to investigate the role of a specific molecule suspected of playing an important role in the process of development, its drawback is that tissue culture explants are generally heterogeneous in nature and therefore the cell type on which a particular chemical may act is open to question. Similarly, while biochemical changes can easily be measured in these systems, morphological change cannot be measured readily. To remedy this problem, primary cultures of dissociated tissues are often used. These systems allow the monitoring of cell morphology, but they too are often heterogeneous and the number of cells that one can use and their life span in culture are often limited. A different level of investigation, making use of theoretically homogeneous clonal cell lines, is available. Since such cell lines permit the collection of large numbers of cells that can multiply indefinitely, from a practical point of view, this model is more convenient to study minute macromolecular changes that take place during morphological differentiation. All these approaches are useful and complementary. The following discussion, however, will be limited to work done on clonal cell lines and will focus on the biochemical changes that arise as morphological differentiation proceeds in cell lines of neuronal origin.

Clonal cell lines of neuronal origin have been obtained in several ways: (1) by cloning single cells isolated from spontaneous tumors such as neuroblastoma C 1300 of neural crest origin[1,2] or pheochromocytoma PC 12, a sympathetic ganglion-like cell line[3]; (2) by cloning single cells from isolated normal embryonic tissue[4,5]; (3) through viral transformation and cloning of cells isolated from embryonic brain tissue[6]; and (4) by cloning single cells isolated from chemically induced tumors such as rat central nervous system neuroblastomas B35, B50, B65, B103, and B104.[7] It must be remembered at this point, that none of these techniques afford normal cells. Viral transformation involves incorporation of the viral genome into the genetic material of the cell, imposing artificial, foreign constraints upon the genetic expression of the cell. Similarly, the changes in genetic expression that have taken place in cells originated from tumors, spontaneous or chemically induced, cannot be assessed with certainty. Even cells isolated from normal embryonic tissues undergo a certain amount of transformation in order to reach their "immortal" status; it is also unlikely that the genotypic configuration of such cells corresponds to any real stage of normal development. The only thing that is known for sure is that these cells exhibit seemingly undifferentiated phenotypes and that under certain circumstances, they undergo certain changes that lead to a new phenotype more characteristic of mature neurons. The biochemical changes involved during this "development" are what one studies under the assumption that such changes are similar to those that take place during the normal course of development.

THE INDUCTION OF MORPHOLOGICAL DIFFERENTIATION

Most of the neuronal clonal cell lines isolated to date grow in culture as a monolayer and present a fibroblast-like morphology. Under the influence of an appropriate stimulus, the morphological differentiation of such cell types is characterized by the outgrowth of neurites. This is not to say that one has obtained a terminally differentiated cell with all the characteristics of a mature normal neuron, since in many cases, upon withdrawal of the stimulus, the cells regress to their original state. The morphological change that occurs is more akin to a step forward toward terminal differentiation, such a step involving a finite number of biochemical changes that are visually characterized by a morphological change that gives the cells a more neuron-like appearance. In many cases, the outgrowth of neurites is not accompanied by dendritic formations, only long axons are present.

The morphological differentiation of neuroblastoma can be easily induced by agents that cause the accumulation of cyclic-AMP in the cell. Dibutyryl-adenosine-3′,5′-cyclic-phosphate[8,9] and phosphodiesterase inhibitors are such agents.[10] Similarly, agents that stimulate adenylate cyclase such as prostaglandins[11] also induce differentiation. The use of serumless medium to promote neurite outgrowth has also been shown to affect not only cyclic-AMP levels but also cell adhesiveness and cell division.[12] All these treatments eventually affect DNA synthesis. Conversely, the use of DNA synthesis-blocking agents such as cytosine arabinoside, fluoro-deoxyuridine,[13] and 6-thioguanine[14] also promotes neurite extension.

Certain differentiation-promoting agents exert their primary effects on the cell membrane. Such agents tend to be more specific in terms of the identity of the clones that are affected; for example, dimethyl sulfoxide (DMSO) has a positive effect on clones N1E-115, N18, and NS-20 but has no effect on N1A-103.[15] Similarly, central nervous system (CNS) clone B50 is insensitive to DMSO (Reboulleau, unpublished observation). High levels of extracellular calcium also promote neurite extension in neuroblastoma B50.[16] This effect may not be mediated by increases in the level of intracellular calcium since the calcium ionophore A 23187 was ineffective in promoting neurite extension in this clone. Others, however, have observed neurite extension in neuroblastoma N2A upon treatment with A 23187.[17] Recent evidence also suggests that Ca^{2+} is involved in the nerve growth factor-stimulated differentiation of PC 12 cells.[18,19] It has been suggested that calcium channels located in the growth cone of neurites may play a regulatory role in their extension.[20] Some studies have indicated that a close relationship may exist between calcium and other ions during differentiation. The ionophore valinomycin has been reported to cause neurite extension[21] and cytoplasmic calcium, while not itself promoting neurite extension in PC 12 pheochromocytoma cells, has been postulated to be involved in the sodium-dependent neurite outgrowth observed in these cells.[22] However, the inducer of choice for PC 12 is nerve growth factor,[3,23] while in neuroblastoma C 1300, according to circumstances, this same agent may have either differentiating or proliferating properties.[24]

EFFECTS OF STEROID HORMONES ON CELL DIFFERENTIATION

Some clones of neuroblastoma C 1300 are sensitive to differentiating conditions that are known to prevail during the late stages of neural crest tissue development. For example, hydrocortisone administered together with L-thyroxine and growth hormone causes morphological differentiation.[25] Hydrocortisone administered alone does not promote differentiation but the potent glucocorticoid dexamethasone does.[26] Upon dexamethasone treatment, a large percentage of the cells retain their processes for at least 48 hr after removal of the inducer. Morphological differentiation induced by dexamethasone is accompanied by an increase in tyrosine hydroxylase activity and dopamine content.[27] Other steroid hormones have not been shown to possess these properties in clonal cell lines, but there is evidence that estrogen induces neurite outgrowth stimulation in preoptic area and hypothalamus brain explants maintained in tissue culture.[28–30] Such behavior is confined to cells that have developed estrogen receptors.[31] It is likely that sensitivity of clonal cell lines to estrogens has not been observed to date for lack of such receptors in these cells. It would therefore be of great interest to isolate cell lines that possess estrogen receptors. Examples of steroid hormone effects in non-neuronal cell lines have been documented. Both estradiol and testosterone facilitate the expression of the globin gene in dimethyl sulfoxide-induced murine erythroleukemia cells[32] while hydrocortisone has an inhibitory effect on this process.[32,33] While it has not been shown that this effect is receptor mediated, it is likely that steroid hormones play a role in the modulation of certain genes that are linked to the expression of a fully differentiated genotype. This is an agreement with the proposition that in hypothalamic neurons, the final cell division must be completed before the onset of the critical period of hormone sensitivity for sexual differentiation.[34,35] It is therefore likely that in order to show sensitivity to sex hormones, neurons must undergo their final cell division and probably must possess steroid hormone receptors, a situation that may require some manipulations to prevail in an appropriate clonal cell line.

CHANGES IN GENE EXPRESSION ACCOMPANYING THE DIFFERENTIATION OF CLONAL NERVE CELLS

Early studies have revealed that the RNA content of the developing nervous system increases during the process of differentiation.[36] Under certain conditions, the same phenomenon is observed in neuroblastoma.[37-39] The process of neurite formation, however, does take place in the presence of actinomycin D, implying that the synthesis of RNA may not be necessary to fiber formation.[11,40,41] More recent studies have shown, however, that under the influence of agents such as dibutyryl-cyclic-AMP, increases in the activity of tyrosine hydroxylase and choline acetyltransferase as well as changes in chromosomal nonhistone proteins can be demonstrated,[11,42-46] and an increase in cyclic-AMP receptor protein has been shown to occur.[47] On the other hand, the activity of the enzyme dopamine hydroxylase seems to be independent of cyclic-AMP production,[48] but multiple forms of the enzyme have been detected in C 1300 clones.[49] A two-dimensional gel electrophoresis study of proteins extracted from PC 12 cells revealed no changes in distribution between nondifferentiated and nerve growth factor-induced cells.[50] On that basis it has been argued that dividing-cell precursors are not qualitatively distinct from cells with extensive neurites,[51] and it has been suggested that "undifferentiated neuronal cells" already show neuron-like characteristics as demonstrated by electrical excitability and neuron-specific cell surface proteins. This suggests that such cells are already differentiated to a significant extent, neurite extension being a final modification process.[52] The evidence in certain clones of neuroblastoma is rather different. Prashad et al.[53] have shown, using two-dimensional gel electrophoresis, that certain proteins exhibit changes in their ionic charges upon cell differentiation. The proteins involved differ depending on whether the clones are adrenergic, cholinergic, or neither. The changes observed in these ionic charges may be attributable to post-synthetic modification processes such as phosphorylation, a phenomenon to be expected when the differentiation paradigm involves raising the concentration of cyclic-AMP in the cell. Other types of modification may also be involved; for example, protein carboxymethylation has been shown to increase during differentiation.[54] An important question remaining is whether such protein modifications are related to the process of morphological differentiation or merely reflect other mechanisms elicited by the inducer but separate from the differentiation process; for example, it has been shown that in PC 12 cells, acetylcholine esterase activity increases when the cells are differentiated with nerve grow factor, but it has also been shown that the activity of the enzyme still increases when a series of PC 12 variants that do not extend neurites are treated with the differentiating agent.[55] This indicates that the activity of acetylcholine esterase and morphological differentiation can be dissociated from each other. Such findings underscore the danger of overinterpreting correlative studies. When one considers the changes in protein activities that take place with morphological differentiation it is difficult to assess whether new genes are being expressed or old ones simply modulated. Recent studies have demonstrated differences in distribution of mRNA between proliferating cells and cells treated to differentiate.[56,57] Specifically, the mRNAs for actin and tubulin have been shown to undergo a biphasic regulation dependent on dibutyryl-cyclic-AMP treatment.[58] It has also been shown that concomitant with morphological differentiation, certain poly-A$^+$-mRNAs associated with proliferation disappear while new ones appear.[59] Furthermore, due to the use of stage-specific cDNA probes to detect poly-A$^+$-mRNA and in view of the fact that there exists in the brain a rather large quantity of poly-A$^-$-mRNA, the complexity of mRNA detected and the number of different mRNAs may be underestimated by this method when compared to results obtained by saturation analysis.[60,61] Nevertheless, there exists one possible candidate as a differentiated stage-specific protein, a "tubulin association factor." This is a large 170-kilodalton protein that assists in the formation of neurotubules.[62]

CELL MEMBRANE CHANGES ASSOCIATED WITH DIFFERENTIATION

Since the process of neurite extension is closely related to the adhesive properties of the cell, either in the context of cell-cell interactions or cell-culture substratum interactions, it is not surprising that the cell surface presents clear differences in the distribution of its components when one compares proliferating cells and axon-forming cells. Studies aimed at measuring cell adhesion of the PC 12 pheochromocytoma cell line which forms cholinergic synapses with the L6 muscle cell line show no differences in rate of adhesion to L6 cells or a control such as Ruppin sarcoma RR 1022 cells.[63] However, in the presence of nerve growth factor, PC 12 displays an accelerated rate of adhesion when co-cultured with smooth muscle, heart muscle, or fibroblast cells as compared to L6 cells.[64] Central nervous system B103 cells have been shown to adhere preferentially to chick retina, tectum, telencephalon, or rat cerebral cortex rather than to chick or rat liver cells, chick embryo fibroblasts, or Chinese hamster ovary cells.[65,66] These types of interactions are mediated by a variety of glycoprotein molecules. N-CAM (neutral cell adhesion molecule) mediates cellular adhesion in the absence of Ca^{2+},[67] while a Ca^{2+}-dependent glycoprotein that also mediates cellular adhesion has been identified.[68-70] The ontogeny of such molecules during cell differentiation has not been studied in detail. Some cell surface glycopeptides sensitive to trypsin have been identified in differentiated cells but found to be absent in proliferating cells.[71] Furthermore, the membrane glycopeptide pattern of clones capable of forming axons differs from that of clones unable to do so.[72] Some of these differences seem to be attributable to differences in the complexity of oligosaccharides of the glycosylated proteins in those two cell types.[73-75] A 200-kilodalton surface glycoprotein that has been identified primarily in differentiated cells[76-78] has been linked to cell adhesiveness, neurite extension, and the expression of a morphologically differentiated phenotype; these changes can be inhibited by the presence of the glycosylation inhibitor tunicamycin.[79]

The distribution of specific gluco- and mannopyranoside residues on the cell membrane has been investigated with concanavalin A. These studies show in differentiated cells a lack of receptor lateral diffusion indicative of a greater membrane rigidity.[80,81]

Glycolipid components of membranes such as certain gangliosides are affected in their distribution pattern during induction of morphological differentiation. Upon cyclic-AMP treatment, increased concentration of gangliosides in various neuroblastoma cell lines has been reported.[82] It has been suggested that the ganglioside G_{D1a} may play a vital role in the process of neurite extension,[83] while the ganglioside G_{M1} has been shown to promote neurite extension in a murine neuroblastoma cell line.[84]

A large number of membrane-bound proteins and enzymes are affected by membrane changes taking place during morphological differentiation. For example, Na^+/K^+-ATPase, acetylcholine esterase, and choline acetyltransferase activities are all found to increase during differentiation. Although this is sometimes attributed to topographical changes in the distribution of membrane proteins,[85] it must be kept in mind that such changes may be due to a specific action of the inducing agent used and may be unrelated to the process of morphological differentiation as demonstrated elsewhere.[55] Each case should therefore be treated independently of all others and the meaning of specific correlations thoroughly assessed.

Neuronal cell lines have been shown to possess neurotransmitter receptor proteins of various types,[86–93] but little is known of the ontogeny of these receptors. In the nonneuronal murine erythroleukemia cell line, the concentration of β-adrenergic receptors has been shown to increase approximately fourfold during butyrate-, DMSO-, or hexamethylene-bis-acetamide-induced differentiation[94]; the reason for these apparent changes in receptor concentration should be investigated.

CONCLUSIONS

The evidence presented points to the usefulness of clonal cell lines to study specific molecular interactions and identify mechanisms regulating these interactions. Nevertheless, for reasons cited earlier, it must be borne in mind that such cell lines must be used with caution when one attempts to relate phenomena observed in the cell lines to events taking place during normal development. The usefulness of clonal neuronal cell lines probably lies in developing models that must ultimately be tested *in vivo* or on isolated homogenous cell populations from brain, surrounded by appropriate stimuli. The complete spectrum of developmental changes free from the constraints imposed by transforming agents could then be studied. Until such systems become experimentally feasible, a reasonable course of action lies in designing experiments aimed at understanding the molecular mechanisms influencing the process of neurite extension (a key component of neuronal differentiation since it involves arrest of cell division). Special attention will have to focus on attempting to reach a developmental stage that is not reversible upon the removal of the differentiating agent, so that these neuronal cell lines can be made much more useful than they are at present. Such a goal has not yet been attained conclusively.

REFERENCES

1. AUGUSTI-TOCCO, G. & G. SATO. 1969. Proc. Natl. Acad. Sci. USA **64:** 311-315.
2. SCHUBERT, D., S. HUMPHREYS, C. BARONI & M. COHN. 1969. Proc. Natl. Acad. Sci. USA **64:** 316-323.
3. GREENE, L. & A. S. TISCHLER. 1976. Proc. Natl. Acad. Sci. USA **73:** 2424-2428.
4. BULLOCH, K., W. B. STALLCUP & M. COHN. 1977. Brain Res. **135:** 25-36.
5. BULLOCH, K., W. B. STALLCUP & M. COHN. 1978. Life Sci. **22:** 495-504.
6. DE VITRY, F., M. CAMIER, P. CZERNICHOW, P. BENDA, P. COHEN & A. TIXIER-VIDAL. 1974. Proc. Natl. Acad. Sci. USA **71:** 3575-3579.
7. SCHUBERT, D., S. HEINEMANN, W. CARLISLE, H. TARIKAS, B. KIMES, J. PATRICK, J. H. STEINBACH, W. CULP & B. L. BRANDT. 1974. Nature **249:** 224-227.
8. PRASAD, K. N. & A. W. HSIE. 1971. Nature New Biol. **233:** 141-142.
9. FURMANSKI, P., O. J. SILVERMAN & M. LUBIN. 1971. Nature **233:** 413-415.
10. PRASAD, K. N. & J. R. SHEPPARD. 1972. Exp. Cell Res. **73:** 436-440.
11. PRASAD, K. N. 1972. Nature New Biol. **236:** 49-52.
12. SHEPPARD, J. R. & K. N. PRASAD. 1973. Life Sci. **12:** 431-439.

13. KLEBE, R. J. & F. H. RUDDLE. 1969. J. Cell Biol. **43:** 69a.
14. PRASAD, K. N. 1973. Int. J. Cancer **12:** 631-635.
15. KIMHI, Y., C. PALFREY, I. SPECTOR, Y. BARAK & U. Z. LITTAUER. 1976. Proc. Natl. Acad. Sci. USA **73:** 462-466.
16. REBOULLEAU, C. P. 1986. J. Neurochem. **46:** 920-930.
17. GUREVICH, V. S., N. G. PATAPOVA, U. I. OGURECHNIKOV & T. V. IGNASHEVA. 1982. Byull. Eksp. Biol. Med. **94:** 118-119.
18. TRAYNOR, A. E. & D. SCHUBERT. 1984. Dev. Brain Res. **14:** 197-204.
19. TRAYNOR, A. E. 1984. Dev. Brain Res. **14:** 205-210.
20. ANGLISTER, L., I. C. FARBER, A. SHAHAR & A. GRINVALD. 1982. Dev. Biol. **94:** 351-365.
21. KOIKE, T. 1978. Biochim. Biophys. Acta **509:** 429-439.
22. KOIKE, T. 1983. Biochim. Biophys. Acta **763:** 258-264.
23. TISCHLER, A. S. & L. A. GREENE. 1978. Lab. Invest. **39:** 77-89.
24. REVOLETTA, R. P. & R. H. BUTLER. 1980. J. Cell. Physiol. **104:** 27-34.
25. DE VELLIS, J., D. INGLISH, R. COLE & J. MOLSON. 1971. *In* Influence of Hormones on the Nervous System. D. H. Ford, Ed.: 25. International Society of Psychoneuroendocrinology. Karger. Basel, Switzerland.
26. SANDQUIST, D., T. H. WILLIAMS, S. K. SAHU & S. KATAOKA. 1978. Exp. Cell Res. **113:** 375-381.
27. SANDQUIST, D., A. C. BLACK, S. K. SAHU, L. WILLIAMS & T. H. WILLIAMS. 1979. Exp. Cell Res. **123:** 417-421.
28. TORAN-ALLERAND, C. D. 1976. Brain Res. **106:** 407-412.
29. TORAN-ALLERAND, C. D. 1978. Am. Zool. **18:** 553-565.
30. TORAN-ALLERAND, C. D. 1980. Brain Res. **189:** 413-427.
31. TORAN-ALLERAND, C. D., J. L. GERLACH & B. S. MCEWEN. 1980. Brain Res. **184:** 517-522.
32. LO, S-C., R. AFT, J. ROSS & G. C. MUELLER. 1978. Cell **15:** 447-453.
33. SCHER, W., D. TSUEI, S. SASSA, P. PRICE, N. GUBELMAN & C. FRIEND. 1978. Proc. Natl. Acad. Sci. USA **75:** 3851-3855.
34. IFFT, J. D. 1972. J. Comp. Neurol. **144:** 193-204.
35. ALTMAN, J. & S. BEYER. 1979. J. Comp. Neurol. **182:** 945-1016.
36. JUDES, C., M. SENSENBRENNER, M. JACOB & P. MANDEL. 1973. Brain Res. **51:** 241-251.
37. MILLER, C. A. & E. M. LEVINE. 1972. Science **177:** 799-802.
38. AUGUSTI-TOCCO, G., E. PARISI, F. ZUCCO, L. CASOLA & M. ROMANO. 1973. *In* Tissue Culture of the Nervous System. G. Sato, Ed.: 87-106. Plenum. New York, N.Y.
39. PRASAD, K. N., K. GILMER & S. KUMAR. 1973. Proc. Soc. Exp. Biol. Med. **143:** 1168-1171.
40. SCHUBERT, D. & F. JACOB. 1970. Proc. Natl. Acad. Sci. USA **67:** 247-254.
41. SCHUBERT, D., S. HUMPHREYS, F. DE VITRY & F. JACOB. 1971. Dev. Biol. **25:** 514-546.
42. WAYMIRE, J. C., N. WEINER & K. N. PRASAD. 1972. Proc. Natl. Acad. Sci. USA **69:** 2241-2245.
43. SIMAN-TOV. R. & L. SACHS. 1972. Eur. J. Biochem. **30:** 123-129.
44. RICHELSON, E. 1973. Nature New Biol. **242:** 175-177.
45. LLOYD, T. & X. O. BREAKEFIELD. 1974. Nature **252:** 719-720.
46. ZORNETZER, M. S. & G. S. STEIN. 1975. Proc. Natl. Acad. Sci. USA **72:** 3119-3123.
47. LOHMANN, S. M., G. SCHWOCH, G. REISER, R. POST & D. WALTER. 1983. EMBO J. **2:** 153-160.
48. HAMPRECHT, B., J. TRABER & F. LAMPRECHT. 1974. FEBS Lett. **42:** 221-226.
49. FRAEYMAN, N. H., E. J. VAN DE VELDE, F. H. DE SMET & A. F. DE SCHAEPDRYVER. 1982. J. Neurochem. **39:** 1179-1184.
50. GARRELS, J. & D. SCHUBERT. 1979. J. Biol. Chem. **254:** 7978-7985.
51. PATRICK, J., S. HEINEMANN & D. SCHUBERT. 1978. Annu. Rev. Neurosci. **1:** 417-443.
52. AKESON, R. 1979. Curr. Top. Dev. Biol. **13:** 215-236.
53. PRASHAD, N., B. WISCHMEYER, C. EVETTS, F. BASIN & R. ROSENBERG. 1977. Cell Differ. **6:** 147-157.
54. KLOOG, Y., J. AXELROD J. & I. SPECTOR. 1983. J. Neurochem. **40:** 522-529.
55. GREENE, L. A. & A. RUCKENSTEIN. 1981. J. Biol. Chem. **256:** 6363-6367.
56. CROIZAT, B., F. BERTHELOT, A. FELSANI & F. GROS. 1977. Eur. J. Biochem. **74:** 405-412.
57. FELSANI, A., F. BERTHELOT, F. GROS & B. CROIZAT. 1978. Eur. J. Biochem. **92:** 569-577.

58. GINZBURG, I., S. RYBAK, Y. KIMHI & U. Z. LITTAUER. 1983. Proc. Natl. Acad. Sci. USA **80:** 4243-4247.
59. GROUSE, L. D., B. K. SCHRIER, C. H. LETENDRE, M. Y. ZUBAIRI & P. G. NELSON. 1980. J. Biol. Chem. **255:** 3871-3877.
60. KAPLAN, B. B. & C. E. FINCH. 1978. *In* Molecular Approaches to Neurobiology. I. R. Brown, Ed.: 71-98. Academic Press. New York, N.Y.
61. SCHRIER, B. K., M. Y. ZUBAIRI, C. H. LETENDRE & L. D. GROUSE. 1978. Differentiation *12:* 23-30.
62. SEEDS, N. W. & R. B. MACCIONI. 1978. J. Cell Biol. **76:** 547-555.
63. SCHUBERT, D., S. HEINEMANN & Y. KODOKORO. 1977. Proc. Natl. Acad. Sci. USA **74:** 2579-2583.
64. SCHUBERT, D. & C. WHITLOCK. 1977. Proc. Natl. Acad. Sci. USA **74:** 4055-4058.
65. GLASER, L., R. MERRELL, D. GOTTLIEB, D. LITTMAN, M. W. PULLIAM & R. A. BRADSHAW. 1976. *In* Surface Membrane Receptors. R. A. Bradshaw, W. A. Frasier, R. C. Merrell, D. I. Gottlieb & R. A. Hogue-Angeletti, Eds: 109-131. Plenum. New York, N.Y.
66. SANTALA, B., D. I. GOTTLIEB, D. LITTMAN & L. GLASER. 1977. J. Biol. Chem. **252:** 7625-7634.
67. HOFFMAN, S. & G. M. EDELMAN. 1983. Proc. Natl. Acad. Sci. USA **80:** 5762-5766.
68. TAKEICHI, M., H. S. OZAKI, K. TOKUNAGA & T. S. OKADA. 1979. Dev. Biol. **70:** 195-205.
69. URUSHIHARA, H., H. S. OZAKI & M. TAKEICHI. 1979. Dev. Biol. **70:** 206-216.
70. GRUNWALD, G. B., R. L. GELLER & J. LILIEN. 1980. J. Cell Biol. **85:** 766-776.
71. BROWN, J. C. 1971. Exp. Cell Res. **69:** 440-442.
72. GLICK, M. C., Y. KIMHI & U. Z. LITTAUER. 1973. Proc. Natl. Acad. Sci. USA **70:** 1682-1687.
73. TRUDING, R., M. L. SCHELANSKI, M. P. DANIELS & P. MORELL. 1974. J. Biol. Chem. **249:** 3973-3982.
74. TRUDING, R., M. L. SCHELANSKI & P. MORELL. 1975. J. Biol. Chem. **250:** 9348-9354.
75. GARVICAN, J. H. & G. L. BROWN. 1977. Eur. J. Biochem. **76:** 251-261.
76. AKESON, R. & W-C. HSU. 1978. Exp. Cell Res. **115:** 367-377.
77. LITTAUER, U. Z., M. Y. GIOVANNI & M. C. GLICK. 1979. Biochem. Biophys. Res. Commun. **88:** 933-939.
78. LITTAUER, U. Z., M. Y. GIOVANNI & M. C. GLICK. 1980. J. Biol. Chem. **255:** 5448-5453.
79. RICHTER-LANDSBERG, C. & D. DUSKIN. 1983. Exp. Cell Res. **149:** 335-345.
80. ROSENBERG, S. B. & F. C. CHARALAMPOUS. 1977. Arch. Biochem. Biophys. **181:** 117-127.
81. DENIS-DONINI, S., M. ESTENOZ & G. AUGUSTI-TOCCO. 1978. Cell Differ. **7:** 193-201.
82. REBEL, G., J. ROBERT & P. MANDEL. 1980. *In* Structure and Function of Gangliosides. L. Svennerholm, P. Mandel, H. Dreyfus & P. F. Urban, Eds. Adv. Exp. Med. Biol. **125:** 159-166. Plenum. New York, N.Y.
83. SVENNERHOLM, L. 1983. Neurochem. Int. **5:** 549-552.
84. FACCI, L., A. LEON, G. TOFFANO, S. SONNINO, R. GHIDONI & G. TETTAMANTI. 1984. J. Neurochem. **42:** 299-305.
85. CHARALAMPOUS, F. S. 1977. Arch. Biochem. Biophys. **183:** 399-407.
86. MATSUZAWA, H. & M. NIRENBERG. 1975. Proc. Natl. Acad. Sci. USA **72:** 3472-3476.
87. SHARMA, S. K., W. A. KLEE & M. NIRENBERG. 1977. Proc. Natl. Acad. Sci. USA **74:** 3365-3369.
88. NATHANSON, N. M., W. L. KLEIN & M. NIRENBERG. 1978. Proc. Natl. Acad. Sci. USA **75:** 1788-1791.
89. STRANGE, P. G., N. J. BIRDSALL & A. S. V. BURGEN. 1978. Biochem. J. **172:** 495-501.
90. SABOL, S. L. & M. NIRENBERG. 1979. J. Biol. Chem. **254:** 1913-1920.
91. KENIMER, J. G. & M. NIRENBERG. 1981. Mol. Pharmacol. **20:** 585-591.
92. KAHN, D. J., J. C. MITRIUS & D. C. U'PRICHARD. 1982. Mol. Pharmacol. **21:** 17-26.
93. REBOULLEAU, C. P. Submitted for publication.
94. SCHMITT, H., M. GUYAUX, R. POCHET & R. KRAM. 1980. Proc. Natl. Acad. Sci. USA **77:** 4065-4068.

Analysis of Steroid Action on Gene Expression in the Brain[a]

JOSIAH N. WILCOX[b]

Center for Reproductive Sciences
Columbia University
College of Physicians and Surgeons
New York, New York 10032

INTRODUCTION

The effect of hormones on animal behavior depends, at least in part, on the modification of gene expression in specific target cells of the brain. The steroid hormones, estradiol and progesterone, affect gene expression by binding to specific receptor proteins upon entering a target cell. In the nucleus the steroid-receptor complex binds to specific acceptor sites on the chromatin and exerts physiological/behavioral actions by altering gene expression. This occurs by modification of specific gene transcription resulting in synthesis of mRNAs which are then translated into new proteins in the cytoplasm of the cell. Steroid-dependent proteins synthesized as a result of this process are thought to be the ultimate effectors of the physiological/behavioral response. This hypothesis has been supported by studies showing increased protein synthesis in the brain after steroid administration. In many cases the behavioral effects of steroids are blocked by inhibitors of protein synthesis.

Previously, researchers have analyzed peptide or enzyme levels as a way to follow steroid action on gene expression in the brain. Sometimes, indirect assays of peptide levels are used and these depend on incorporation of radioactive precursors into newly synthesized proteins. Alternatively, peptide levels may be measured directly by radioimmunoassay if a specific antibody is available. Relating peptide levels to alteration of gene expression by steroids is complicated by the possibility of transport or release of neuropeptides from the site of synthesis. Reduced levels of a neuropeptide at its site of synthesis may not reflect decreased synthesis but may rather reflect increased synthesis coupled with active transport of the neuropeptide to other regions of the brain.

With the advent of recombinant DNA technology it is possible to overcome this problem by measuring the immediate product of altered gene expression directly, the mRNA. While mRNA levels closely parallel the synthesis of neuropeptides there is no evidence that mRNAs are transported in the brain and so are not subject to the

---

[a]This research was the result of work conducted while the author was a postdoctoral fellow in the laboratory of Dr. James L. Roberts at Columbia University. The author was supported as a training fellow on the Center for Reproductive Sciences training grant HD07093.

[b]Current address: Department of Molecular Biology, Genentech, Inc., 460 Point San Bruno Blvd., South San Francisco, Ca. 94080.

same problems of interpretation as peptide levels. Thus, alteration of the level of mRNA for a given neuropeptide after steroid treatment strongly suggests that the expression of that neuropeptide gene has been affected.

In the following brief review I want to demonstrate how recombinant DNA technology has revolutionized neurobiology and how this research can be applied to the study of steroid hormone action in the brain. Since the primary focus of my current research is the regulation of pro-opiomelanocortin gene expression in the brain, much of the following discussion shall deal with the expression of this gene.

cDNA PROBES

Specific populations of mRNAs can be detected and quantitated using specific complementary DNA probes *(cDNA probes)*. cDNA probes are strings of nucleotides that are complementary to the unique sequence of nucleotides of a specific mRNA species. When added to a mixture of mRNAs these probes will hybridize to their corresponding mRNA by specific base pairing. This means that each nucleotide base of the mRNA binds to its complementary base on the cDNA probe. Since the nucleotide strands are long, in the hundreds or thousands of bases, and the sequence of nucleotide bases of the mRNA is unique and complementary to the cDNA, the cDNA probe binds to the mRNA very much like a zipper. The hybridization of a cDNA probe to its corresponding mRNA is very specific and the hybrids formed are quite stable. Using radioactive cDNA probes it is possible to estimate the amount of a specific mRNA in a mixture of nucleic acids by the binding of the cDNA probe to its corresponding mRNA.

DOT BLOTS

One technique that has been developed to measure very small amounts of a specific mRNA species is that of *filter hybridizations* sometimes referred to as *dot blots*. This technique is based on the hybridization of radioactive cDNA probes to mRNA immobilized on nitrocellulose filters.[1] The tissue to be analyzed is homogenized and a nucleic acid fraction purified by extraction in phenol followed by ethanol precipitation. This nucleic acid preparation contains a mixture of the DNA and RNA obtained from each dissection. The nucleic acid mixture is spotted onto a nitrocellulose filter which is baked to bind the nucleic acids to the filter. mRNAs bound to nitrocellulose filters in this manner are quite stable and yet are available to hybridize to cDNA probes. The nitrocellulose filters are then incubated with a ^{32}P-labeled cDNA probe specific for the mRNA to be assayed under conditions favoring cDNA:mRNA hybridization. At the end of the hybridization step the unbound probe is washed off in salt solutions and the amount of radioactivity on each spot determined by X-ray film autoradiography. The darker the spot on the film the more radioactivity is present which represents the specifically hybridized cDNA probe and therefore the amount of mRNA. The darkness of each spot is quantitated by densitometry and compared to a standard dilution curve which then yields information about relative amounts of specific mRNAs present in the original nucleic acid mixture. This technique is extremely sensitive and in some cases can measure mRNA levels in less than 1 mg of tissue.

REGULATION OF PRO-OPIOMELANOCORTIN IN THE BRAIN

Pro-opiomelanocortin (POMC) is a large precursor protein from which β-LPH, ACTH, MSH, and β-endorphin are derived.[2] Immunocytochemical staining using ACTH or endorphin antibodies indicates that this peptide is present in cell bodies in the periarcuate region of the hypothalamus extending from the retrochiasmatic region to the premammillary nucleus.[3] It was initially thought that the POMC peptides in the brain were derived by transport and uptake of these proteins from the pituitary, a tissue rich in ACTH and β-endorphin. However, it has been shown that the hypothalamus contains POMC mRNA[4-7] and releases β-endorphin, one of the POMC peptides, into the portal blood.[8] Thus the presence of a population of POMC-producing cells in the brain is now accepted.

There is considerable evidence that POMC synthesized in the brain is related to reproductive functioning. β-Endorphin administration has a profound influence on pituitary secretions resulting in decreases in plasma luteinizing hormone, growth hormone, and follicle-stimulating hormone and an increase in plasma prolactin levels (reviewed in Reference 9). It is hypothesized that β-endorphin modifies gonadotropin-releasing hormone (GnRH) release by interacting with GnRH neurons in the median eminence or the preoptic area thus altering gonadotropin secretion. Estrogen administration has been shown to decrease immunoassayable endorphin levels in the basal hypothalamus,[10] but whether this represents a direct effect on endorphin synthesis or release is unclear. β-Endorphin levels decrease in the arcuate nucleus and increase in the median eminence and superchiasmatic nucleus on the afternoon of proestrus in rats[11] suggesting that ovarian hormones may alter the release and transport of this peptide from cell bodies. Thus, the question of direct hormonal action on the synthesis of POMC peptides cannot be answered by assaying for the peptide but rather must be approached by measuring the first product of altered gene expression, the mRNA.

In Dr. James Roberts' laboratory at Columbia University I have used dot blots to measure POMC mRNA levels in the rat brain to determine whether estrogen reduces endorphin levels in the hypothalamus by altering POMC gene expression. Ovariectomized rats were given Silastic implants of estrogen or oil and sacrificed 1 or 3 days later. Total nucleic acids from the arcuate-median eminence of each animal were isolated and spotted on nitrocellulose filters as previously described. The tissue samples originally weighed less than 5 mg and yet as many as five assays could be performed on each sample attesting to the sensitivity of this procedure.

One day of estrogen treatment had no effect on arcuate-median eminence POMC mRNA levels. However, 3 days of estrogen treatment was found to decrease POMC mRNA content 50% compared to oil controls. That estrogen reduced POMC mRNA levels explains the reduction in POMC peptides seen previously and suggests that estrogen acts to decrease POMC gene expression in the brain. β-Endorphin has been shown to be inhibitory to the display of female sexual behavior in rats[12] so it is of interest that estrogen treatment that would facilitate lordosis also inhibits POMC biosynthesis.

IN SITU cDNA:mRNA HYBRIDIZATION

A major drawback to the use of dot blots to assess mRNA levels in the brain is that it can only assess overall tissue content of a specific mRNA. If the tissue dissection contains a heterogeneous population of cells expressing the gene of interest (each population being regulated differently), such information would not be apparent if total mRNA is assayed since the observed level of the specific mRNA would be an average of all cell types assayed together. One way around this problem would be to make smaller and smaller dissections in which the regulation of the gene of interest is more homogeneous. This is often not possible even in microdissected brain areas, since brain tissue is so complex in nature. For example, less than 10% of the POMC cells in the hypothalamus contain estrogen receptors.[13] We know that estrogen causes reductions in POMC mRNA levels in the rat hypothalamus, but is estrogen having a slight effect in all POMC cells equally or is it having a large effect in only that population of POMC cells that contain estrogen receptors? These kinds of questions can only be answered by assays of POMC mRNA levels in individual cells using a technique that I have been developing in Dr. Roberts' laboratory called *in situ* cDNA:mRNA hybridization.

In situ cDNA:mRNA hybridization is a technique that has been developed for the visualization of cDNA:mRNA hybrids at the level of the individual cell. It is conceptually very similar to the dot blot technique except that the cDNA probe is allowed to hybridize to mRNA fixed, *in situ*, within frozen tissue sections. The unbound probe is washed off after hybridization and the tissue section is dipped in liquid photographic emulsion and exposed in the dark for various periods of time. Upon development and counterstaining the density and distribution of silver grains over the tissue section yield information regarding the presence or absence of a specific mRNA in a particular cell. The technique was originally developed to detect viral DNA in tissue sections[14] and was refined in Dr. Roberts' laboratory for the detection of POMC mRNA in histological sections of the rat pituitary.[15]

I have been able to use the *in situ* cDNA:mRNA hybridization technique to localize POMC mRNA containing cells in the brain. Hybridization of a ^{3}H-labeled POMC cDNA probe to a section of rat brain results in cDNA:mRNA hybrid formation, represented by clusters of silver grains, in cells in the periarcuate region of the hypothalamus (FIGURE 1). Closer examination of the cells showing positive hybridization to the cDNA probe shows discrete localization of the silver grains over the cytoplasm of positive cells with little or no label over others (FIGURE 2). Serial sections processed for ACTH immunocytochemistry and POMC *in situ* hybridization indicate that cDNA:mRNA hybrids have formed in cells containing POMC peptides (FIGURE 3). The co-localization of both POMC peptides and mRNA in the same cells by this technique is very good evidence that the POMC peptides are synthesized in the hypothalamus and did not arrive there by transport from other regions of the brain or pituitary.

While the *in situ* technique can be very useful in answering anatomical questions, its greatest value to neuroendocrinology lies in its ability to resolve cDNA:mRNA hybrids at the level of the individual cell. It therefore possesses potential as a semi-quantitative assay for mRNA levels in individual neurons in the brain. However, to use this technique to study regulation of gene expression in the brain it will be necessary to first divide the hybridizing cells into functional subgroups that can be located in experimental and control tissues. Secondary techniques must be employed to define related cell types if there are no clear anatomical distinctions. With this problem in mind, studies are under way designed to make the *in situ* technique compatible with

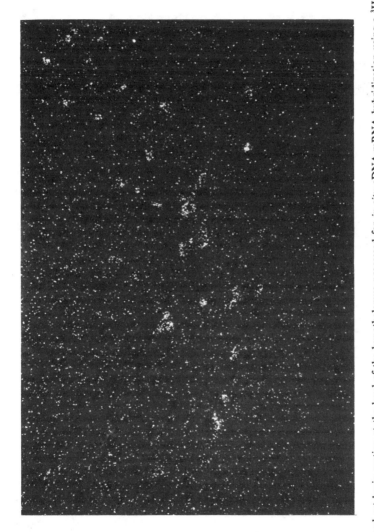

FIGURE 1. Coronal rat brain section at the level of the hypothalamus processed for *in situ* cDNA:mRNA hybridization using a ^{3}H-labeled POMC cDNA probe. Illumination is by a combination of bright field and polarized light epiluminescence so that the silver grains appear white. POMC mRNA positive cells are seen in the periarcuate region lateral to the third ventricle. Exposure time, 10 days. (Original magnification, × 100; reduced to 75% of original size.)

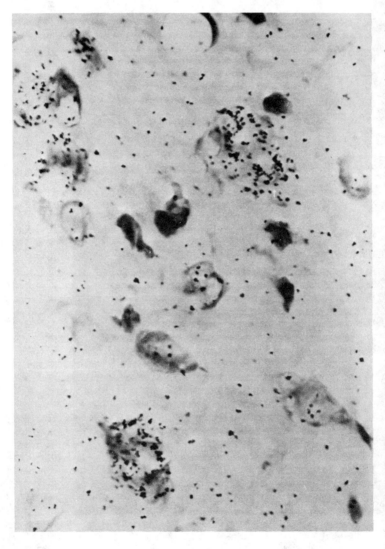

FIGURE 2. Bright field photomicrograph showing the cellular distribution of silver grains in the rat arcuate nucleus after *in situ* cDNA:mRNA hybridization to a ^{3}H-labeled POMC cDNA probe. Exposure time, 14 days. (Original magnification, × 500; reduced to 80% of original size.)

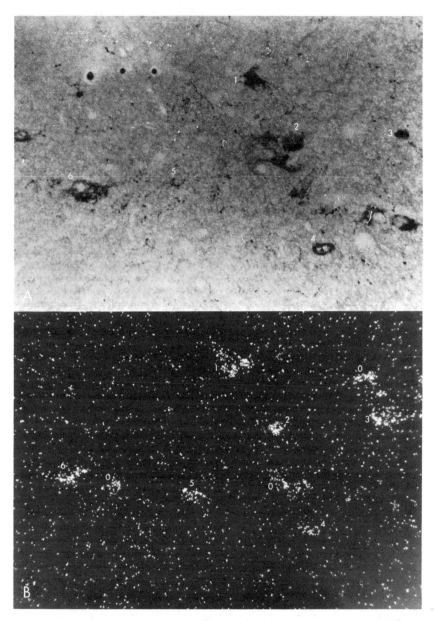

FIGURE 3. Ten-micrometer serial sections from rat hypothalamus showing that POMC mRNA is co-localized in cells containing POMC peptides. Section A was processed for immunohistochemistry using an ACTH antibody (one of the POMC peptides). Section B was processed for *in situ* hybridization using a ^{3}H-labeled POMC cDNA probe and was photographed using polarized light epiluminescence so that only the silver grains are illuminated an appear white. The numbered cells were found to be common to the two sections and contain both POMC mRNA and POMC peptides; cells labeled 0 on the *in situ* section were found to be POMC peptide positive on the next section serial to the *in situ* section. (Original magnification, × 250; reduced to 67% of original size.)

immunocytochemistry on the same tissue section. We have determined that *in situ* hybridization is compatible with retrograde fluorescent tracing techniques used to map axonal projections in the brain. Thus it is now possible to begin to study regulation of a specific mRNA in subsets of cells that project to any given region of the brain.

REFERENCES

1. THOMAS, P. S. 1977. Hybridization of denatured RNA transferred or dotted to nitrocellulose paper. Methods Enzymol. **100:** 255-266.
2. EIPPER, B. A. & R. E. MAINS. 1980. Structure and biosynthesis of proadrenocorticotropin/endorphin and related peptides. Endocr. Rev. **1:** 1-27.
3. JOSEPH, S. A. 1980. Immunoreactive adrenocorticotropin in rat brain: a neuroanatomical study using antiserum generated against synthetic ACTH 1-39. Am. J. Anat. **158:** 533-548.
4. LIOTTA, A. S., D. GILDERSLEEVE, M. J. BROWNSTEIN & D. T. KRIEGER. 1979. Biosynthesis in vitro of immunoreactive 31,000-dalton corticotropin/beta endorphin-like material by bovine hypothalamus. Proc. Natl. Acad. Sci. USA **76:** 1448-1452.
5. HERBERT, E., N. BIRNBERG, J. C. LISSITSKY, O. CIVELLI & M. UHLER. 1981. Proopiomelanocortin: a model for the regulation of expression of neuropeptides in pituitary and brain. Neurosci. Newslett. **12:** 16-27.
6. CIVELLI, O., N. BIRNBERG & E. HERBERT. 1982. Detection and quantitation of proopiomelanocortin mRNA in pituitary and brain tissues from different species. J. Biol. Chem. **257:** 6783-6787.
7. GEE, C. E., C. L. C. CHEN, J. L. ROBERTS, R. THOMPSON & S. J. WATSON. 1983. Identification of propiomelanocortin neurones in rat hypothalamus by in situ cDNA-mRNA hybridization. Nature **306:** 374-376.
8. WARDLAW, S. L., W. B. WEHRENBERG, M. FERIN, P. W. CARMEL & A. G. FRANTZ. 1980. High levels of beta-endorphin in hypophyseal portal blood. Endocrinology **106:** 1323-1326.
9. MEITES, J., J. F. BRUNI, D. A. VAN VUGT & A. F. SMITH. 1979. Relation of endogenous opiod peptides and morphine to neuroendocrine functions. Life Sci. **24:** 1325-1336.
10. WARDLAW, S. L., L. THORON & A. G. FRANTZ. 1982. Effects of sex steroids on brain beta-endorphin. Brain Res. **245:** 327-331.
11. BARDEN, N., Y. MERAND, D. ROULEAU, M. GARON & A. DUPONT. 1981. Changes in the beta-endorphin content of discrete hypothalamic nuclei during the estrous cycle of the rat. Brain Res. **204:** 441-445.
12. SIRINATHSINGHJI, D. J. S. 1984. Modulation of lordosis behavior of female rats by naloxone, beta-endorphin and its antiserum in the mesencephalic central gray: possible mediation via GnRH. Neuroendocrinology **39:** 222-230.
13. MORRELL, J. I., J. MCGINTY & D. W. PFAFF. Some steroid hormone concentrating cells in the medial basal hypothalamus (MBH) and anterior pituitary contain beta endorphin or dynorphin. Abstract (#27.1) presented at the 13th Annual Meeting of the Society for Neuroscience, 1983.
14. BRAHIC, M. & A. T. HAASE. 1978. Detection of viral sequences of low reiteration frequency by in situ hybridization. Proc. Natl. Acad. Sci. USA **75:** 6125-6129.
15. GEE, C. E. & J. L. ROBERTS. 1983. In situ hybridization histochemistry: a technique for the study of gene expression in single cells. DNA **2:** 157-163.

Index of Contributors

Adler, Norman T., 21–32

Balthazart, Jacques, 308–324
Barfield, Ronald J., 33–43
Beer, Colin G., xiii–xiv
Beyer, Carlos, 268–269, 270–281
Blaustein, Jeffrey D., 400–414
Buntin, John D., 252–267
Burger, Joanna, 170–186

Cheng, Mei-Fang, ix, xi, 1–3, 4–12
Crews, David, 187–198
Crowley, William R., 423–436

Diakow, Carol, xiii–xiv, 325–330

Erikson, Carl J., 13–20

Feder, Harvey H., ix, xi, 352–353
Fischette, Christine T., 437–444
Fleming, Alison, 234–251

Gibson, Marie J., 53–63
González-Mariscal, Gabriela, 270–281

Harding, Cheryl F., 371–378
Herrenkohl, Lorraine Roth, 120–128

Inman, Barbee L., 44–52

Johns, Margaret A., 148–157

Komisaruk, Barry R., ix, xi, 1–3, 64–75, 268–269

Landau, I. Theodore, 379–388
Lazar, J. Wayne, xiii–xiv
Lenington, Sarah, 141–147
Lott, Dale F., 129–134

Mayer, Anne D., 216–225
Michel, George F., 158–169
Moore, Celia L., 108–119
Morin, L. P., 331–351

Nock, Bruce, 415–422
Numan, Michael, 226–233

O'Connor, Lynn H., 437–444

Reboulleau, Christian P., 445–452
Rodriquez-Sierra, Jorge F., 293–307
Rosenblatt, Jay S., 199–201

Schumacher, Michael, 308–324
Siegel, Harold I., ix, xi, 202–215
Steinman, Judith L., 64–75
Stern, Judith M., 95–107
Storey, Anne, 135–140

Terkel, Joseph, 76–94
Thomas, David A., 33–43
Thornton, Janice E., 362–370
Toner, James Patrick, Jr., 21–32

Wade, George N., 389–399
Whalen, Richard E., 354–361
Williams, Christina L., 282–292
Wilcox, Joseiah N., 453–460

Subject Index

Androgen metabolism, role in activation of male behavior, 371–377
Androgens, effects on energy balance, 389–397
Antiestrogens, 379–382
Avian breeding cycle, role of female vocalizations in the, 44–52

Behavior
  avian parental, prolactin and, 252–264
  copulatory, effects on sperm, 21–30
  estrogens and, 282–290
  hormonal control of, 354–358
  incubation, avian, prolactin and, 252–264
  individual behavioral response mediates endocrine changes, 4–11
  male, role of androgen metabolism in activation of, 371–377
  maternal, *see* Maternal behavior
  mating, neuroendocrine mechanisms regulating, 188–192
  nest-cooing, in female ring doves, 5–9
  regulator of hormonal mechanisms and reproduction, 1–198
  reproductive
    approaches to study of, 325–329
    as a phenotypic correlate of genotype, 141–147
    biochemical bases of hormonal action on, 352–460
    prenatal stress and, 120–128
  reproductive social, intraspecific variation in, 129–134
  serotonin-dependent, hormone effects on, 437–442
  sexual
    effect of estrogen and progesterone on, 400–411
    heterotypical, 362–368
    hormones and, 268–351
  sexual differentiation of, 108–119
  steroid hormone antagonists and, 379–384
Behavioral control of reproduction, genital-stimulation and, 64–75
Biological clocks, 331–348
Birds
  colonial, role of perception in habitat selection in, 170–186
  female vocalizations, role of, 44–52
  prolactin and parental behavior in, 252–264
  See also Doves, ring
Brain, the
  analysis of steroid action on gene expression in, 453–460

estrogens and development of, 282–290
regulation of pro-opiomelanocortin in, 455
steroid receptors and hormone action in, 400–411
Bruce effect, 135–140

Catecholamine-neuropeptide interactions, 423–431
Cell lines, neuronal clonal, differentiation of, 445–450
Clocks, biological, 331–348
Clonal cell lines, neuronal, differentiation of, 445–450
Contact stimulation, maternal, psychobiology and, 95–107
Copulatory behavior, effects on sperm, 21–30

Doves, ring
  behavioral response mediates endocrine changes in, 1–11
  events between male courtship and female's follicular growth, 5
  experiential influences on parental care, 158–169
  female, social induction of ovarian response in, 13–18

Endocrine changes mediated by behavioral response, 4–11
Endocrinology, comparative behavioral, 187–198
Energy balance, sex steroids and, 389–397
Estradiol
  effect on behavior, 400–411
  effect on energy balance, 389–397
Estrogen
  and development of brain and behavior, 282–290
  effect on behavior, 400–411
  effect on energy balance, 389–397
  organizational effects of, 293–306

Feedback, proprioceptive, 7–9
Fertility, effects of copulatory behavior on, in rats, 21–30
Follicular growth, female ring dove, male courtship and, 5

Gene expression in the brain, steroid action on, 453–460
Genital stimulation as trigger for neuroendocrine and behavioral control of reproduction, 64–75

463

Genital tract
  behavioral responses to stimulation of the, 66–73
  neuroendocrine responses to stimulation of the, 64–66
Gonadal hormones, see Hormones, gonadal
Gonadal steroids, energy balance and, 389–397

Habitat selection, role of perception in, 170–186
Heterotypical sexual behavior, 363–368
Homeokinetics, 331
Homografts, preoptic, analysis of, 53–63
Hormonal factors, 129–134
Hormones
  control of behavior, 354–358
  effects on serotonin-dependent behaviors, 437–442
  gonadal
    dimorphic sexual behavior and, 362–363
    lordosis in males and, 365
    mounting by females and, 364–365
    maternal behavior in rats, 234–248
    parental behavior and, 199–267
    regulators of reproductive behavior, 199–351
    sex steroids and energy balance, 389–397
    sexual behavior and, 268
    steroid receptors and hormone action in the brain, 400–411
Hypothalamic cyclic AMP, lordosis in rodents and, 270–278
Hypothalamus-pituitary-ovarian axis, 5, 9

Imprinting, ovarian activity and, 15
Incubator behavior, avian, prolactin and, 252–264

Lactational diestrus not a pseudopregnancy, 86–87
Lordosis, 365
  antagonism of estrogen effect on, 379–380
  hypothalamic cyclic AMP in facilitation of, 270–278
  noradrenergic regulation of, 416

Maternal behavior, 199–201
  hormonal basis of, in rats, 202–215
  neural circuitry of, 228–231
  neural control of, 243–278
  rat, psychobiology of, 234–248
  role of medial preoptic area in regulation of, in rats, 226–233
Maternal responsiveness during the prepartum period in laboratory rats, 216–225
Mating behavior, neuroendocrine mechanisms regulating, 188–191

Metabolic factors, 129–134
Metabolism, androgen, role in activation of male behavior, 371–377
Morphological differentiation of neuronal clonal cell lines, 445–450
Mounting by females, 364
Myoclonus, 438–439

Nest-cooing behavior, in female ring doves, 5–9
Nest defense during prepartum period in laboratory rats, 216–225
Neurochemical actions of ovarian steroids, 424–431
Neuroendocrine control of reproduction, genital stimulation and, 64–75
Neuroendocrine regulation, reproductive, 423–431
  role of preoptic area in, 53–63
Neuroendocrinology of induced pseudopregnancy, 76–94
Neuron clonal cell lines, differentiation of, 445–450
Neuropeptide-catecholamine interactions, 423–431
Noradrenergic regulation of progestin receptors, 415–419
Nutritional effects, 132

Ovarian activity
  imprinting and, 15
  male inhibition of, 16–17
Ovarian response, social induction of, in female ring dove, 13–18
Oxytocin, maternal behavior and, 211–212

Parental behavior
  avian, prolactin and, 252–264
  hormones and, 199–267
Parental care, ring dove, experiential influences on, 158–169
Perception, role of, in habitat selection, 170–186
Physiological time, concept of, 331–348
Pregnancy disruptions, advantages of, 135–139
Prenatal stress, as disruptive factor in reproductive psychophysiology, 120–128
Preoptic area
  medial, role of, in regulation of maternal behavior in rats, 226–233
  role in neuroendocrine regulation of reproduction, 53–63
Preoptic homografts, functional, analysis of, 53–63
Prepartum period, maternal responsiveness and nest defense during, in laboratory rats, 216–225

SUBJECT INDEX

Primer effect, 15
Progesterone
  effect on behavior, 400–411
  effect on energy balance, 389–397
Progestin
  noradrenergic regulation of progestin receptors, 415–419
Prolactin, parental behavior and, 252–264
Prolactin surges
  copulatory behavior and, 77–78
  mnemonic control over, 82–86
  photoperiodicity of, 81–82
Pro-opiomelanocortin, regulation of, in the brain, 455
Proprioceptive feedback, effects of, 7–9
Pseudopregnancy
  adrenergic and cholinergic mediation of, 78–79
  electrophysiological mediation of, 79–80
  induced, neuroendocrinology of, 76–94
  lactational diestrus is not, 86–87
  noncoitally induced, 87–90
Psychobiology
  maternal, contact stimulation and, 95–107
  of rat maternal behavior, 234–248
Puberty, organizational effects of estrogen at, 293–306

Quail, sexual differentiation in, 308–321

Rats
  contact stimulation and maternal psychobiology in, 95–107
  effects of copulatory behavior on sperm in, 21–30
  hormonal basis of maternal behavior in, 202–215
  laboratory, maternal responsiveness and nest defense during prepartum period in, 216–225
  mating calls, 33–40
  psychobiology of maternal behavior in, 234–248
  role of medial preoptic area in regulation of maternal behavior in, 226–233
  role of ultrasonic vocalizations in regulation of reproduction in, 33–42
Reproductive behavior, see Behavior, reproductive
Reproductive neuroendocrine regulation, 423–431
Rhythms, biological, 331–348
Ring doves, see Doves, ring
Rodents, lordosis in, hypothalamic cyclic AMP and, 270–278

Serotonin behavioral syndrome, 439–440
Sex steroids, energy balance and, 389–397
Sexual behavior, see Behavior, sexual
Sexual differentiation, 308–321
  of behavior, 108–119
  steroid hormone antagonists and, 382–383
Social behavior, reproductive, intraspecific variation in, 129–134
Social induction of ovarian response in female ring dove, 13–18
Social interaction, behavioral response mediates endocrine changes induced by, 4–11
Sperm, effects of copulatory behavior on, 21–30
Steroid action on gene expression in the brain, 453–460
Steroid hormone antagonists, behavior and, 379–384
Steroid hormones, effects on cell differentiation, 447
Steroids, sex, and energy balance, 389–397
Stimulation, contact, and maternal psychobiology, 95–107
Stress, prenatal, reproductive behavior and, 120–128

Testosterone, effect on energy balance, 389–397
Time, physiological, concept of, 331–348

Ultrasonic vocalizations, role in regulation of reproduction in rats, 33–42

Vocalizations
  female, their role in avian breeding cycle, 44–52
  ultrasonic, role in regulation of reproduction in rats, 33–42
Vomeronasal organ, role of, 148–157

Women, contact stimulation and maternal psychobiology in, 95–107